Functional Nanoparticles for Bioanalysis, Nanomedicine, and Bioelectronic Devices Volume 2

ACS SYMPOSIUM SERIES **1113**

Functional Nanoparticles for Bioanalysis, Nanomedicine, and Bioelectronic Devices Volume 2

Maria Hepel, Editor
State University of New York at Potsdam
Potsdam, New York

Chuan-Jian Zhong, Editor
State University of New York at Binghamton
Binghamton, New York

Sponsored by the
ACS Division of Colloid and Surface Chemistry

American Chemical Society, Washington, DC

Distributed in print by Oxford University Press, Inc.

Library of Congress Cataloging-in-Publication Data

Functional nanoparticles for bioanalysis, nanomedicine, and bioelectronic devices
/ Maria Hepel, editor, State University of New York at Potsdam, Potsdam, New York,
Chuan-Jian Zhong, editor, State University of New York at Binghamton, Binghamton,
New York ; sponsored by the ACS Division of Colloid and Surface Chemistry.
    pages cm. -- (ACS symposium series ; 1113)
 Includes bibliographical references and index.
 ISBN 978-0-8412-2828-3 (alk. paper)
 1. Nanoparticles--Congresses. 2. Biotechnology--Congresses. I. Hepel, Maria, editor of
compilation. II. Zhong, Chuan-Jian, editor of compilation. III. American Chemical Society.
Division of Colloid and Surface Chemistry, sponsoring body.
 TP248.25.N35F86 2012
 660.6--dc23

                      2012041200

The paper used in this publication meets the minimum requirements of American National
Standard for Information Sciences—Permanence of Paper for Printed Library Materials,
ANSI Z39.48n1984.

Copyright © 2012 American Chemical Society

Distributed in print by Oxford University Press, Inc.

All Rights Reserved. Reprographic copying beyond that permitted by Sections 107 or 108
of the U.S. Copyright Act is allowed for internal use only, provided that a per-chapter fee of
$40.25 plus $0.75 per page is paid to the Copyright Clearance Center, Inc., 222 Rosewood
Drive, Danvers, MA 01923, USA. Republication or reproduction for sale of pages in this
book is permitted only under license from ACS. Direct these and other permission requests
to ACS Copyright Office, Publications Division, 1155 16th Street, N.W., Washington, DC
20036.

The citation of trade names and/or names of manufacturers in this publication is not to be
construed as an endorsement or as approval by ACS of the commercial products or services
referenced herein; nor should the mere reference herein to any drawing, specification,
chemical process, or other data be regarded as a license or as a conveyance of any right
or permission to the holder, reader, or any other person or corporation, to manufacture,
reproduce, use, or sell any patented invention or copyrighted work that may in any way be
related thereto. Registered names, trademarks, etc., used in this publication, even without
specific indication thereof, are not to be considered unprotected by law.

PRINTED IN THE UNITED STATES OF AMERICA

Foreword

The ACS Symposium Series was first published in 1974 to provide a mechanism for publishing symposia quickly in book form. The purpose of the series is to publish timely, comprehensive books developed from the ACS sponsored symposia based on current scientific research. Occasionally, books are developed from symposia sponsored by other organizations when the topic is of keen interest to the chemistry audience.

Before agreeing to publish a book, the proposed table of contents is reviewed for appropriate and comprehensive coverage and for interest to the audience. Some papers may be excluded to better focus the book; others may be added to provide comprehensiveness. When appropriate, overview or introductory chapters are added. Drafts of chapters are peer-reviewed prior to final acceptance or rejection, and manuscripts are prepared in camera-ready format.

As a rule, only original research papers and original review papers are included in the volumes. Verbatim reproductions of previous published papers are not accepted.

ACS Books Department

Contents

Indexes

Preface

Current applications of nanotechnology span from new materials for energy production or conversion (including photovoltaics, fuel cells, etc.), to a variety of consumer products and novel pharmacological formulations, and expand now extensively to biomedical research, medical diagnostics and therapy. In this book, a comprehensive overview of the progress achieved in the development of functional nanoparticles for novel bioanalytical techniques and assays, enhancement of medical diagnostic imaging, and a variety of cancer treatment therapies is presented, and future trends of nanotechnology applications in these areas are evaluated. Particular emphasis is placed on the functionalization of metal, semiconductor and insulator nanoparticles for targeting cancer cells, delivering drugs and chemotherapeutic agents, preventing or reducing body inflammation response, averting biofouling and cytotoxicity, and enabling cell membrane crossing. Progress in the development of nanoparticles enhancing the diagnostic imaging is discussed in detail and this includes nanoparticles for Magnetic Resonance Imaging (MRI), Computed Tomography (CT) scan, megasonic imaging, and novel photoacoustic and Raman imaging. The image enhancing nanoparticles enable precise in-surgery viewing of tumors aiding surgical tumor removal in such complex cases as the brain tumor neuroblastoma, requiring ultra-precise incisions to remove cancer protrusions. Among novel imaging techniques, the Raman imaging provides distinctive features, such as chemical identification and multiplexing capabilities. It is noteworthy that Raman imaging is only possible owing to the extraordinary signal amplification enabled by the interaction of laser light with functionalized gold or silver nanoparticles administered to the targeted tissue. On the other hand, the interaction of nanoparticles with electromagnetic fields is also a key element in radiotherapy, photodynamic therapy, and hyperthermal cancer destruction. In these treatments, the nanoparticles specially optimized for efficient absorbance of electromagnetic irradiation are administered to a body, either locally to a tumor tissue or systemically into the blood stream when targeting had been developed so that the nanoparticles can recognize the tumor and accumulate in it. The tuning of nanoparticles to achieve absorbance of electromagnetic radiation at a specific wavelength in UV, visible, or NIR spectral region, requires an in-depth knowledge of optical and electronic properties of nanoparticles. Several Chapters are devoted to the synthesis and analysis of particles' size and shape dependencies of optoelectronic properties. In the case of gold and silver nanoparticles, the main absorption of electromagnetic energy in the visible region is due to the excitation of a collective oscillation of surface free electrons which is called the surface plasmon resonance. The optical property depends on the nanoparticle size, shape, surface, interparticle spacing, and the medium. Very strong effect on the absorption frequency has been observed with different aspect ratios of the nanorods. Interesting properties of fluorescent semiconductor core-shell nanoparticles, their broad across-the-spectrum size-dependent absorbance tunability, biocompatible functionalization, and biomedical applications are the focus of several Chapters.

The book covers comprehensively the two new trends of the recent development: (i) the theranostic multimodality (where theranostics = therapy + diagnostics), which combines two important nanocarrier functionalities: therapeutic drug delivery and enhanced diagnostics (e.g. by adding imaging-enhancement markers), and (ii) the race for the development of nanoparticle-enhanced biosensors which can serve as a useful, inexpensive, and easy-to-operate bioelectronic device in patients point-of-care.

Since the field of functional nanoparticles is rapidly expanding, with new discoveries reported every year, we hope that a wide audience will find this book very attractive as a source of valuable information concerning the synthesis of nanoparticles, their functionalization, and construction of biosensors and bioelectronic devices. We truly hope that researchers, students, medical doctors, and other professional workers will find the two volumes of this large collection of intriguing ideas, new data, and future developments, as a useful source to build on in further studies.

Maria Hepel

Chuan-Jian Zhong

Chapter 1

Persistent Luminescence Nanoparticles for Diagnostics and Imaging

Thomas Maldiney, Daniel Scherman, and Cyrille Richard*

Unité de Pharmacologie Chimique et Génétique et d'Imagerie, CNRS UMR 8151, Paris, F-75270 cedex France; Inserm, U 1022, Paris, F-75270 cedex France; Université Paris Descartes, Sorbonne Paris Cité, Faculté des Sciences Pharmaceutiques et Biologiques, Paris, F-75270 cedex France; ENSCP, Paris, F-75231 cedex France
*E-mail: cyrille.richard@parisdescartes.fr

Persistent luminescence is the property of some materials that continue to emit light for minutes, sometimes hours and days, after the end of the excitation. Such materials are mainly used in traffic signs, emergency signage, watches and clocks, luminous paints, as well as textile printing. Our group has recently suggested the use of this optical property for *in vivo* bioimaging applications in living animals. However, to fulfill such complicated task, the persistent luminescence material should possess at least two criteria : an emission located in the tissue transparency window, and sizes comprised in the nanometer scale. Indeed, contrary to most of the conventional optical probes such as fluorescent molecules and quantum dots, that require continuous excitation to fluoresce, persistent luminescence nanoparticles possess the ability to store the excitation energy and to emit light for a long period of time. This property is of particular interest for *in vivo* bioimaging applications since it allows complete avoidance of the autofluorescence signal coming from endogeneous chromophors also excited when using fluorescent probes. The synthesis of persistent luminescence nanoparticles with appropriate optical properties, the chemical modification of their surface, as well as exemples of *in vitro* and *in vivo* bioimaging applications are reported in this chapter.

© 2012 American Chemical Society

Introduction

Persistent luminescence relates to a particular optical phenomenon, in which the excitation light is stored by the material within a few minutes to be slowly released upon thermal activation to emitting centers, thereby producing a light emission which can last for hours (*1*). The excitation irradiation may be visible or ultraviolet (UV) light, electron beam, plasma beam, X-rays, or gamma radiations. Persistent luminescence has been, and still is, unfortunately in a misleading manner, called phosphorescence because of the long emission time. Phosphorescence may be an appropriate term to be used in the context of light emission from organic compounds involving triplet-to-singlet transitions. The long decay time of persistent luminescence, however, is due to the storage of the excitation energy by traps, and to its subsequent release by thermal activation. On the contrary, fluorescence is a simple process in which the fluorophore absorbs energy from incident light to reach an excited state. While returning back to the ground state, the absorbed energy is released as radiation. Since some energy is lost in this process, the emission energy is lower than the incident energy, or in other words, the wavelength of the emitted light is longer than the one of the incident light. The phenomenon of persistent luminescence has been known for over a thousand years. Descriptions have been found of ancient Chinese paintings that remained visible during the night, by mixing the colors with a special kind of pearl shell (*1*). The first scientifically described observation of persistent luminescence dates back to 1602, when V. Casciarolo observed strong luminescence from a mineral barite, $BaSO_4$, later known as the Bologna stone. Natural impurities in the stone were responsible for the long duration of the afterglow. Until the end of the 20th century, very little research was done on the phenomenon of persistent luminescence. For many decades, zinc sulfide (ZnS) doped with copper, and later codoped with cobalt, has been the most famous and widely used persistent phosphor (*2, 3*). ZnS was used in many commercial products including watch dials, luminous paints and glow-in-the-dark toys. However, the brightness and lifetime that could be achieved with this material were rather low for practical purposes. To tackle this problem, traces of radioactive elements such as promethium or tritium were often introduced in the powders to stimulate the brightness and lifetime of the light emission (*4*). But even then, a commercial glow-in-the-dark object had to contain a large amount of luminescent material to yield an acceptable afterglow. Materials with persistent luminescence properties usually consist in an inorganic host matrix which is activated by introducing rare earth ions as dopants in the host lattice (*5*). In 1996, Matsuzawa *et al.* published an article that sent a shockwave through the field of persistent luminescence, relatively unpopular at that time (*6*). By codoping the green-emitting phosphor $SrAl_2O_4:Eu^{2+}$ with the rare earth cation dysprosium (Dy^{3+}), they were able to create a material that emit bright light for hours after ending the excitation (*7*). They found an afterglow with both a far higher initial intensity and a much longer lifetime, compared to ZnS:Cu,Co. Their discovery marked the beginning of a renewed search for different and better persistent luminescence materials. Initially, this research was concentrated on alkaline earth aluminates, and it took a few years before other types of compounds came into view. In 2001, Lin *et al.* reported a bright and long lasting afterglow in $Sr_2MgSi_2O_7$:Eu^{2+}, Dy^{3+} (*8*). Shortly afterwards, other doped disilicates were found to exhibit an equally long afterglow. Today, almost fifteen years after the discovery of $SrAl_2O_4$:Eu^{2+}, Dy^{3+}, the research for new persistent luminescence compounds has become increasingly popular (*9*). Even if the optical properties of these materials are extremely interesting, until the last five years, no article reported the use of persistent luminescence for general bioimaging purposes. This book chapter is dedicated to the recent discovery of persistent luminescence nanoparticles (PLNP), able to emit light in the tissue transparency window, between 600 and 900 nm, and to their use as photonic nanoprobes for optical diagnosis and bioimaging applications.

In Vivo Optical Imaging

Optical imaging techniques have widely served biologists trying to understand the structural and functional processes in cells, tissues, and living organisms (*10*). In recent years, the scientific community demonstrated a growing interest toward optical *in vivo* imaging systems because of their high potential for repeated imaging without exposure to harmful ionizing radiations such as X-rays. Current commercial probes, from organic dyes to inorganic quantum dots (QDs), still suffer from important limitations (*11*). Organic dyes display a poor photostability, and broad absorption/emission spectra (*12*). Broad emission spectra limit multiplexing, i.e. the simultaneous use of different dyes to monitor concomitant biological processes. On the contrary, semiconductor-based quantum dots possess high photostability, tunable emission spectra, high quantum yields, and narrow emission bandwidths. For these reasons, they have already been widely used in biological applications (*13*). However, most of these materials are downconverting as they emit visible fluorescence when excited by UV or short-wavelength visible light. In addition, the use of UV light to monitor living processes in cells and tissues is responsible for some potential drawbacks as long-term irradiation of living cells and organisms may not only cause DNA damage and cell death but also induce significant autofluorescence from endogeneous fluorescent molecules which reduces the signal-to-background ratio. Furthermore, *in vivo* fluorescence imaging is limited by the short penetration depth of the excitation light. In general, NIR radiation is harmless to cells and minimizes autofluorescence from biological tissues, thereby increasing the signal-to-noise ratio significantly (*14*). Lately, NIR fluorescent dyes as well as NIR QDs (*15*) were shown to enable deep tissue penetration and low autofluorescence as compared to visible fluorophores. However, they still carry critical drawbacks with regard to signal detection and processing. First, NIR detectors require appropriate filters when the excitation and emission wavelengths are too close to each other. Then, NIR dyes are very likely to photobleach which impairs stable and reproducible measurements. Finally, QDs are said to be cytotoxic (*16*). The search for alternative bioimaging probes, emitting in the tissue transparency window (*17*) is thus of particular interest. This chapter will report how persistent luminescence can be exploited to create a new generation of photonic nanoprobe intended for diagnosis and multiple optical imaging applications both *in vitro* and *in vivo*.

In Vivo Optical Imaging with Persistent Luminescence Nanoparticles

As mentioned in the introduction, the main characteristics of persistent luminescence materials, compared to the other optical nanoprobes based on fluorescence phenomenon, resides in their ability to store energy allowing their excitation before the injection to small animals, and to be followed under the appropriate photon-counting system (cooled GaAs intensified charge-coupled device camera, Biospace Lab, Paris, France) without the need for constant illumination. The principle of *in vivo* imaging with PLNP is summarized in Figure 1 (*18*).

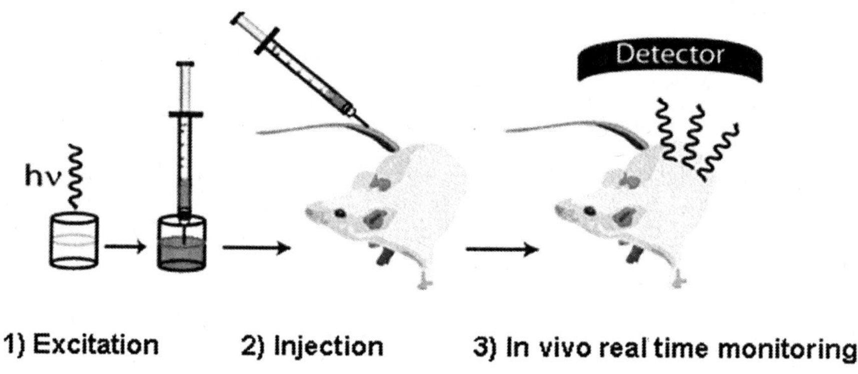

Figure 1. Principle of using PLNP for in vivo imaging.

Silicates as Persistent Luminescence Materials for Bioimaging

Most of persistent luminescence materials reported so far display green or blue emission. However, for *in vivo* imaging applications, the material should emit light in the tissue transparency window to avoid complete signal absorption by fatty tissues, water and hemoglobin (*16*). In 2003, Wang *et al.* reported the synthesis of $MgSiO_3$ doped with Eu^{2+}, Dy^{3+} and Mn^{2+} showing persistent luminescence properties at 660 nm (*19*). This composition initiated the development of near-infrared PLNP from our group, with formula $Ca_{0.2}Mg_{0.9}Zn_{0.9}Si_2O_6$, doped with Eu^{2+}, Mn^{2+}, and Dy^{3+}, and intended for optical imaging in living animals.

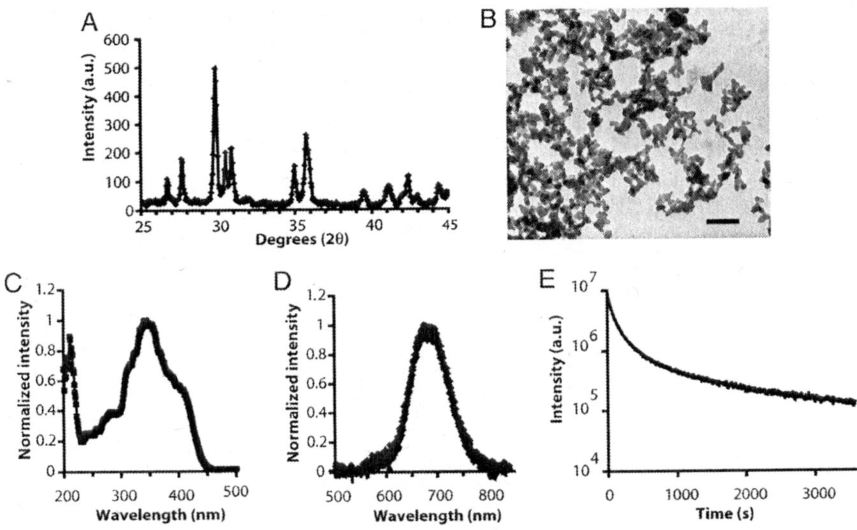

Figure 2. Characteristics of Eu^{2+}, Dy^{3+}, Mn^{2+} doped $Ca_{0.2}Mg_{0.9}Zn_{0.9}Si_2O_6$. (A) PLNP showed a clinoenstatite-like structure. (B) Transmission electronic microscopy image. (scale bar: 200 nm.) (C) Excitation spectrum. (D) Long afterglow emission spectrum. (E)Persistent luminescence decline curve after a 5 minutes UV exposure of PLNP. Reproduced with permission from ref. (30). Copyright (2012) National Academy of Sciences, U.S.A.

Sol-Gel Synthesis of Persistent Luminescence Nanoparticles for in Vivo Imaging

The solid-state process used by Wang *et al.* (*19*) to synthesize persistent luminescence materials leads to the formation of micrometric particles which are very likely to cause embolism after systemic injection, thus impeding any bioimaging applications. Thereby, PLNP are synthesized by a Sol-Gel approach in order to decrease the particles size (*20–22*) and obtain final cores in the nanometric range, compatible with *in vivo* applications. Raw materials are magnesium nitrate [$Mg(NO_3)_3,6H_2O$], zinc chloride ($ZnCl_2$), calcium chloride ($CaCl_2,2H_2O$), europium chloride ($EuCl_3$, $6H_2O$), dysprosium nitrate ($Dy(NO_3)_3,5H_2O$), manganese chloride ($MnCl_2,4H_2O$), and tetraethoxysilane (TEOS). The salts are dissolved in deionized water acidified at pH 2 by addition of concentrated nitric acid. TEOS was then added rapidly, and the solution was stirred vigorously at room temperature for 1h. The solution was heated at 70°C for 2h until the sol-gel transition occurred. The gel was then dried at 110°C for 20h, and sintered at 1100°C for 10h under reducing atmosphere (90% Ar, 10% H_2) (*23*). Nanometer-sized particles of $Ca_{0.2}Mg_{0.9}Zn_{0.9}Si_2O_6$ (Figure 2A) were obtained by basic wet grinding of the solid for 15 minutes with a mortar and pestle in a minimum volume of 5 mM NaOH solution. Hydroxylation was then performed overnight by dispersing the ground powder in 50 mL of the same

NaOH solution to get hydroxyl-PLNP. Nanoparticles with a diameter of 180 nm were selected from the whole polydisperse colloidal suspension by centrifugation on a SANYO MSE Mistral 1000 at 4500 rpm for 5 minutes (centrifugation time was lengthened to 30 min in order to obtain 120 nm PLNP). They were located in the supernatant (assessed by Dynamic Light Scattering). The supernatants were gathered and concentrated to a final 5 mg/mL suspension. Nanoparticles with a diameter of 80 nm were selected from the 120 nm concentrated suspension by centrifugation on an Eppendorf MiniSpin Plus at 8000 rpm for 5 minutes (Figure 2B). Following the same approach, the centrifugation step was repeated 4 times and the resulting suspension concentrated to a final amount of 5 mg/mL (*24*). Optical properties of $Ca_{0.2}Mg_{0.9}Zn_{0.9}Si_2O_6$: Eu^{2+}, Mn^{2+}, Dy^{3+} are characterized by an excitation spectrum with a maximum peaking at 350 nm (Figure 2C) responsible for a near-inrafred persistent luminescence centered on 685 nm (Figure 4D and E).

The luminescence instensity of $Ca_{0.2}Mg_{0.9}Zn_{0.9}Si_2O_6$: Eu^{2+}, Dy^{3+}, Mn^{2+} as a function of time and particle diameter (80, 120 and 180 nm) was also investigated. Figure 3 shows the influence of diameter on the emission behavior of hydroxyl-PLNP during the first thirteen minutes after the excitation. Luminescence intensity decreases with the diameter of the particle: 180 nm crystals were approximately 5 times brighter than 80 nm ones (*24*).

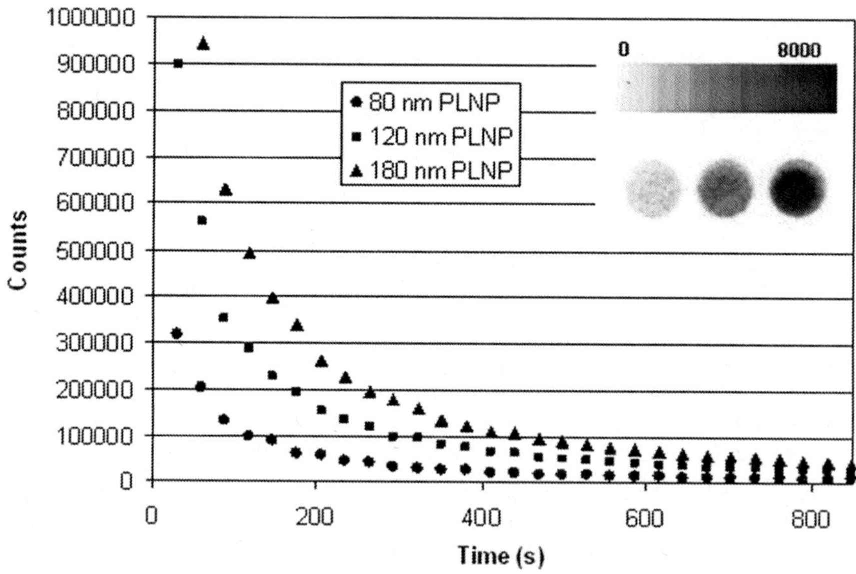

Figure 3. Luminescence decline curve of hydroxyl-PLNP in water. The insert represents the corresponding image obtained after a 15 minutes acquisition (from left to right: 80 nm, 120 nm, and 180 nm hydroxyl-PLNP; luminescence intensity is expressed in grey unit: 1 unit = 2800 photons per s.cm².steradians). Reproduced with permission from ref. (24). Copyright (2012) American Chemical Society.

Mechanism of Persistent Luminescence in Diopside-Doped Materials

With the aim of understanding the persistent luminescence mechanism in silicate host and improving the optical properties of PLNP, Lecointre *et al.* have recently studied the influence of doping conditions on the persistent luminescence properties of diopside binary compounds, with formula $CaMgSi_2O_6$, displaying a structure very similar to $Ca_{0.2}Zn_{0.9}Mg_{0.9}Si_2O_6$ (*25*). Wavelength-resolved thermally stimulated luminescence (TSL) measurements from 30 to 650 K after X-ray irradiation were performed on $CaMgSi_2O_6$: Mn^{2+}, $CaMgSi_2O_6$: Mn^{2+},Eu^{2+} and $CaMgSi_2O_6$: Mn^{2+},Dy^{3+}. The wavelength-resolved TSL spectrum of $CaMgSi_2O_6$: Mn^{2+} is presented in Figure 4A. Two strong peaks are observed at low temperature, corresponding to shallow traps, and two weaker peaks appear at higher temperature. Emissions corresponding to all these TSL peaks are mostly located at 680 nm, which means that they originate almost exclusively from Mn^{2+} in Mg site. The similarity of the TSL emission and the persistent luminescence spectrum suggests that the same mechanism should occur. The luminescent center Mn^{2+} may act as a hole trap in this compound following the equation $Mn^{2+} + h^+ \rightarrow Mn^{3+}$. In particular, Mn^{2+} in Mg site appears as a much better hole trap than Mn^{2+} in Ca site. In this case, the TSL peaks are related to electron traps which might be intrinsic defects in the lattice. The authors have tested this hypothesis by doping the material with Eu^{2+} and Dy^{3+}, which can act as hole and electron traps, respectively. The TSL spectrum of $CaMgSi_2O_6$: Mn^{2+},Eu^{2+} is displayed in Figure 4B. At low temperature, peaks position appears at the same temperature meaning that the same electron traps are involved but the emission spectra changed. Indeed, the emission of Mn^{2+} in Mg site at 680 nm is still observed, with lower intensity, but it appears the emission of Mn^{2+} in Ca site at 580 nm and the emission of Eu^{2+} at 450 nm with higher intensity. In this compound, Eu^{2+} also plays the role of hole trap, following $Eu^{2+} + h^+ \rightarrow Eu^{3+}$, and seems to be the favored recombination center in low temperature TSL. After recombination of the electron with Eu^{3+}, ($Eu^{3+} + e^- \rightarrow Eu^{2+*}$) the excited Eu^{2+} transfers its energy to the two different Mn^{2+}, which explains why both Mn^{2+} emission bands are observed in this case. The TSL spectrum of $CaMgSi_2O_6$: Mn^{2+},Dy^{3+} is shown in Figure 4C. First, only the emission of Mn^{2+} in Mg site is observed and no luminescence of Dy^{3+} appears, meaning that Dy^{3+} does not play the role of recombination center. Moreover, unlike Mn^{2+},Eu^{2+} doped compound, the addition of Dy^{3+} strongly modifies the TSL peaks position. Peaks at low temperature are much weaker whereas a new broad peak with high intensity appears around 480 K. This means that Dy^{3+} plays the role of a deeper electron trap, following $Dy^{3+} + e^- \rightarrow Dy^{2+}$, than intrinsic electron traps in the mechanism. Finally persistent luminescence of Mn^{2+} in Mg site, corresponding to the emission at 680 nm, was measured for an hour. The Mn^{2+},Dy^{3+} doped compound shows a much stronger persistent luminescence than the two others (Figure 4D). $CaMgSi_2O_6$: Mn^{2+},Dy^{3+} appears to be a good persistent luminescence phosphor for *in vivo* small animal imaging whereas $CaMgSi_2O_6$:Mn^{2+} and $CaMgSi_2O_6$:Mn^{2+},Eu^{2+} do not. The effect of Dy^{3+} which creates new high temperature peak at 480 K explains the better persistent luminescence observed. Indeed the low temperature tail of the glow peak at 480

K is located in the region of interest for persistent luminescence, between 300 and 420 K. In the Mn^{2+},Eu^{2+} doped compound, the quenching of the red persistent luminescence is due to the competition between Eu^{2+} and Mn^{2+} ions to trap a hole.

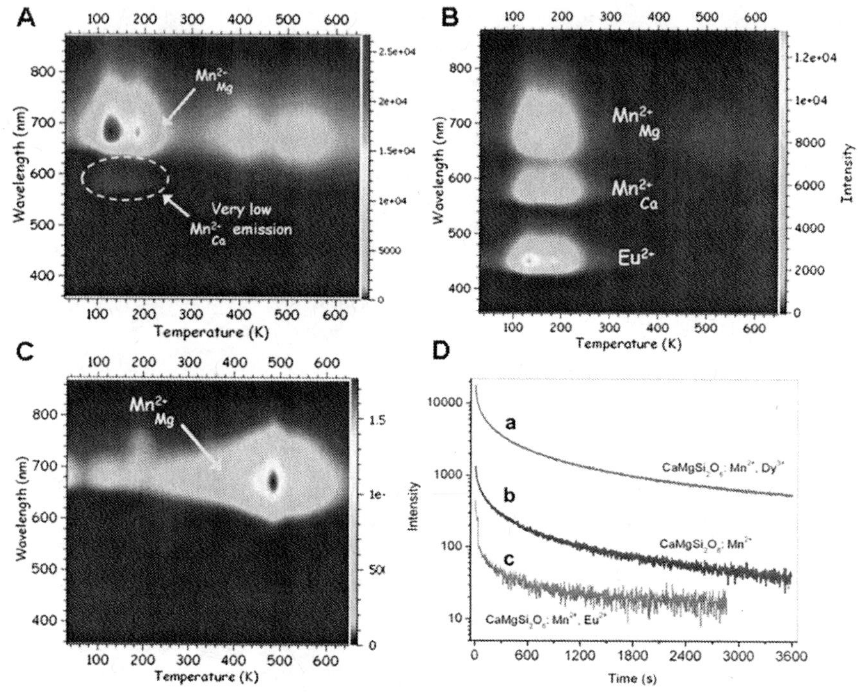

Figure 4. (A) Wavelength-resolved TSL spectrum of CaMgSi$_2$O$_6$: Mn^{2+}. (B) Wavelength-resolved TSL spectrum of CaMgSi$_2$O$_6$: Mn^{2+},Eu $^{2+}$. (C) Wavelength-resolved TSL spectrum of CaMgSi$_2$O$_6$: Mn^{2+},Dy^{3+}. (D) Persistent luminescence spectra at 680 nm: (a) CaMgSi$_2$O$_6$: Mn^{2+},Dy^{3+}, (b) CaMgSi$_2$O$_6$: Mn^{2+}, (c) CaMgSi$_2$O$_6$: Mn^{2+},Eu $^{2+}$. Reproduced with permission from ref. (25). Copyright (2012) Elsevier.

In Vivo Imaging with $Ca_{0.2}$ $Zn_{0.9}Mg_{0.9}Si_2O_6$ Doped Nanoparticles

We first demonstrated that the light level produced by $Ca_{0.2}Zn_{0.9}Mg_{0.9}Si_2O_6$ doped with Eu^{2+}, Mn^{2+}, and Dy^{3+} was sufficient to yield a localizable signal under a few millimeters of tissues, such as for subcutaneous or intramuscular injection (Figure 5A). For this purpose, the excitation of the PLNP suspension before injection was accomplished by direct exposure to a 6-W UV lamp for 5 min at a distance of 2 cm. The lowest detectable dose was then evaluated by injecting suspensions (20 µl) at different concentrations (100, 10, and 1 µg/ml) corresponding approximately to 2 x 10^{10} (2 µg), 2 x 10^9 (200 ng), and 2 x

10^8 (20 ng) PLNP in three different localizations on the mouse back. The two highest doses (2 μg and 200 ng) were easily monitored by using a 2 minutes acquisition time (Figure 5B). The lowest administrated dose (20 ng) also produced a detectable signal with a satisfactory signal-to-noise ratio superior to 5.

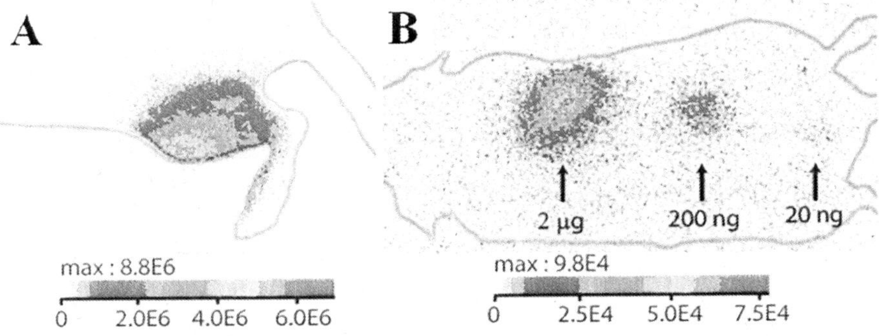

Figure 5. (A) Image of an intramuscular injection (200 μg) corresponding to a 90-s acquisition. (B) Image of three subcutaneous injections of PLNP (2 μg, 200 ng, 20 ng). The different localizations are labeled with arrows, and the corresponding PLNP amounts are indicated. The acquisition was performed during the 2 min after injection. The luminous intensity is expressed in photons per s.cm².steradians (sr). Reproduced with permission from ref. (30). Copyright (2012) National Academy of Sciences, U.S.A.

Surface Modification of Persistent Luminescence Nanoparticles for in Vivo Imaging

To ensure and broaden *in vivo* applications, optical nanoprobes require surface mofication via chemical reaction. We used classical coating procedures developed for silicate materials (*26*). After thermal treatment, partial erosion of PLNP with aqueous sodium hydroxide led to surface free hydroxyl groups, which were used for covalent linkage of different functional groups (Scheme 1). The free hydroxyl groups gave to PLNP a natural negative surface charge at neutral pH, which can be assessed by zeta potential measurements (-34.3 mV). The hydroxyl-PLNP were then reacted with 3-aminopropyltriethoxysilane (APTES), which was coupled to the surface. This reaction provided positively charged PLNP (referred to as amino-PLNP) resulting from the presence of free amino groups (NH_2) at the surface. The success of the grafting procedure was confirmed by zeta potential measurements (+35.8 mV at pH 7) and a positive test with trinitrobenzene sulfonate (TNBS). The APTES in excess was removed by several sedimentation washing procedures. The surface charge of the amino-PLNP was reversed by reaction with diglycolic anhydride, which reacted with amines to give free carboxyl groups. These nanoparticles, referred to as carboxyl-PLNP, displayed negative surface charge at neutral pH (-37.3 mV). We also conducted peptide coupling of amines with MeO-PEG$_{5000}$-COOH [carboxyl-methoxy-polyethyleneglycol (F.W.): 5,000 g/mol]. This reaction led to neutral particles (+5.1 mV), partly originating from the charge-shielding effect of PEG. The unreacted reagents were always removed by three or more centrifugation-washing steps. We have thus obtained three types of nanoparticles bearing different surface charges. The next step was to determined whether the persistent luminescence could provide real-time in vivo biodistribution imaging in mice after systemic tail vein injection of 1 mg of PLNP (corresponding to 10^{13} nanoparticles).

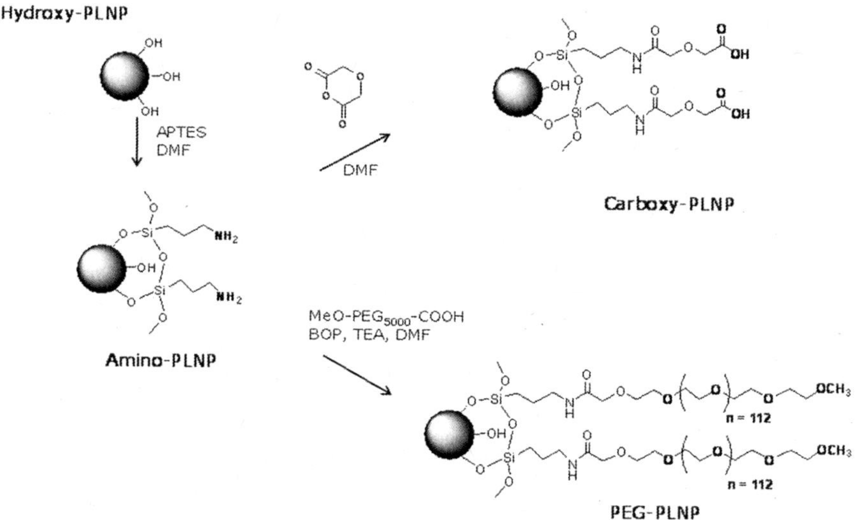

Scheme 1. PLNP surface modification with different molecules

Biodistribution images obtained 30 minutes after systemic injection of 180 nm amino-PLNP, carboxy-PLNP and PEG-PLNP are reported in Figure 6. An important lung retention was observed for amino-PLNP displaying positive surface charge (Figure 6a). Two reasons can explain this biodistribution pattern (*27*). The first one is a nonspecific electrostatic interaction of amino-PLNP with the negative charges displayed by plasmatic membranes of capillary endothelial cells, such as sulfated proteoglycans and glycosylaminoglycans. Another explanation could be provided by aggregation of amino-PLNP with negatively charged blood components, leading to the trapping of agglomerates in the narrow lung capillaries. Negative carboxyl-PLNP were rapidly cleared from the blood by liver uptake (Figure 6b). This biodistribution pattern presumably resulted from a rapid opsonisation and uptake of PLNP by phagocytic cells of the reticuloendothelial system (RES). Such behaviour after systemic injection is generally observed for nanoparticles or liposomes (*28, 29*). Neutral PEG-PLNP were able to circulate longer, as assessed by a diffuse signal throughout the mouse body that lasted during all of the acquisition time (Figure 6c). However, at a longer acquisition time (60 minutes) the distribution clearly showed liver accumulation (*30*).

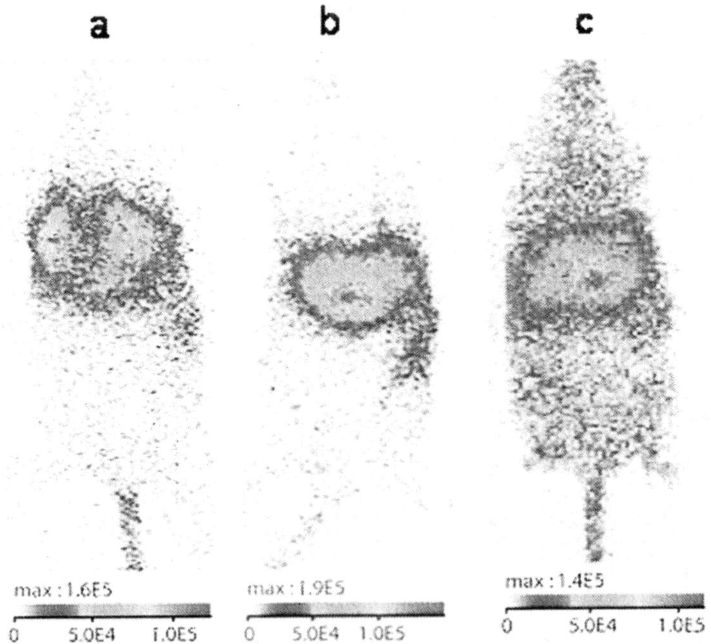

Figure 6. Images of biodistribution PLNP 30 minutes after tail vein injection. (a) amino-PLNP. (b) carboxy-PLNP. (c) PEG-PLNP. Reproduced with permission from ref. (30). Copyright (2012) National Academy of Sciences, U.S.A.

Influence of Nanoparticle Diameter and PEG Chain Length on the Biodistribution in Healthy Mice

The biodistribution of PLNP was studied as a function of nanoparticle diameter (80, 120 and 180 nm) and PEG chain length (5, 10 and 20 kDa). Surface functionalization was achieved following the same chemical routes as the one described earlier. Relative long-term biodistribution in healthy mice was achieved six hours after systemic injection to validate a potential use of this probe for relative long-term *in vivo* applications. Delayed fluorescence from europium ions trapped in the core of the nanoparticles allowed an *ex vivo* quantitative analysis through tissue homogenates after animal sacrifice. As shown in Figure 7, similarly to several classes of nanoparticles, and regardless of surface coverage, long term PLNP biodistribution occurs predominantly within liver and spleen (*31*). Combined uptake in the kidneys and lungs remains below 5 % of the injected dose. The PEG chain length seems to have little influence on the global RES accumulation, i.e. combined uptake in liver and spleen, and is only responsible for a slight change in the liver/spleen ratio, only noticeable for 120 nm PLNP. On the contrary, the nanoparticle core diameter appears to be critical. As shown in Figure 7, decreasing the particle core diameter from 180 nm to 80 nm causes a much lower combined RES uptake of PEG-PLNP from 100 % down to around 35 % of the injected dose (*24*). Any cytotoxicity was observed (MTT assays) using such nanoparticles (data not shown).

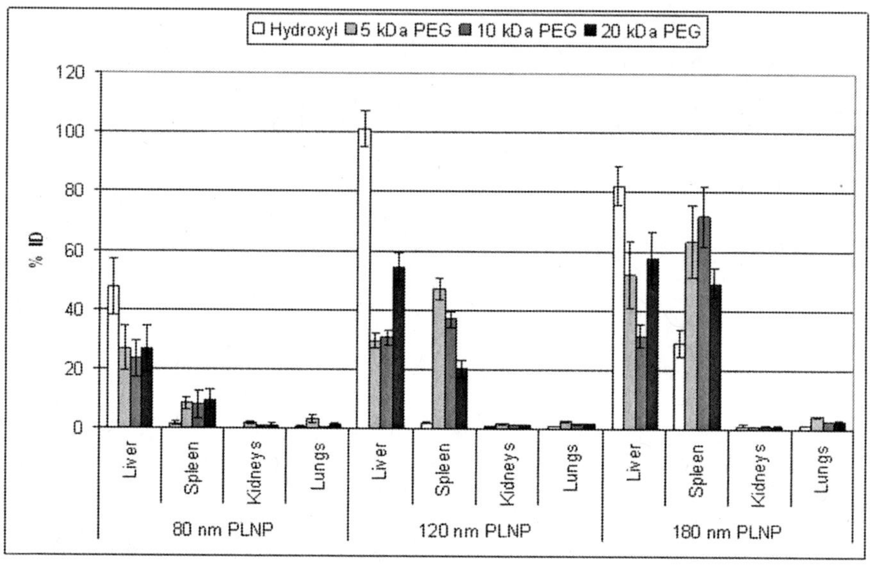

Figure 7. PLNP tissue distribution 6 hours after systemic injection to healthy mice (n=6). Error bars correspond to standard deviation. Reproduced with permission from ref. (24). Copyright (2012) American Chemical Society.

Surface Modification of Persistent Luminescence Nanoparticles with Small Molecules for in Vitro Targeting

Along with the need to evade RES uptake in order to induce a much better distribution of PLNP within the animal, comes the fundamental question associated with the will to aim at a given region of interest *in vivo*: how to conduct the probe to the designated malignancy? This difficult task is generally achieved by taking advantage of the specific and privileged interaction between a ligand, generally small molecules or antibodies, and its pointed target, the receptor (*32*). In order to tackle this task, we described two different examples intended for the specific targeting of PLNP against cancer cells *in vitro*.

The first example relies on the recent discovery of a small molecule, referred to as Rak-2, reported for its high affinity toward PC-3 cells (*33*). We have shown that functionalization of PLNP with Rak-2 molecule through a small 250 daltons PEG spacer (Scheme 2) allows preferential binding of PLNP on PC-3 cells (Figure 8) compared to the control PLNP on which N-propyl amide was grafted instead of Rak-2 (*34*).

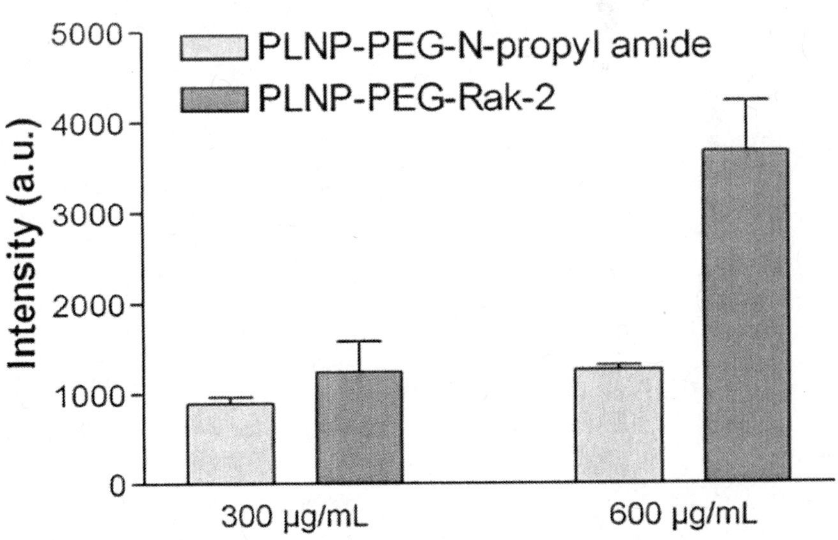

Scheme 2. Functionalization of PLNP with Rak-2

Figure 8. In vitro binding (30 minutes) of 180 nm Rak-2-PLNP on PC-3 cells. Reproduced with permission from ref. (34). Copyright (2012) Elsevier.

Then, taking advantage of the avidin-biotin technology, the second approach exploits biotinylated PLNP to target malignant cells transduced with a fusion protein called lodavin, composed of both avidin and the endocytotic low-density lipoprotein (LDL) receptor. Indeed, Lesch et al. recently produced a lentiviral vector, carrying the gene coding for lodavin fusion protein, that could induce long-term expression of avidin on the surface of transduced glioma cells, referred to as BT4C + (*35*). We evaluated the ability of biotin-PEG-functionalized PLNP (*36*) (Scheme 3) to preferentially target transduced BT4C cells, displaying avidin on the outer membrane, compared to wild type BT4C, referred to as BT4C -. Additional control PLNP was obtained after biotin coverage with neutravidin.

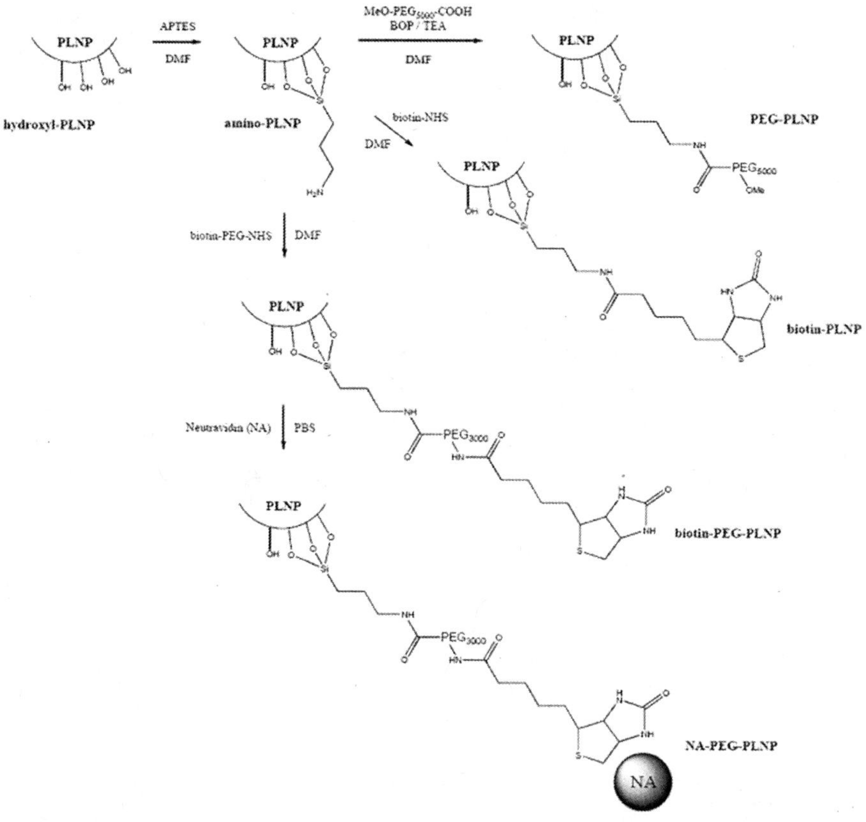

Scheme 3. Functionalization of PLNP with biotin and neutravidin

Results showed that biotinylated PLNP were more retained on glioma cells expressing the avidin fusion protein than on control cells (Figure 9). The significant decrease in PLNP binding on BT4C + cells after biotin saturation with neutravidin (NA-PEG-PLNP) ensured that the binding of biotin-PEG-PLNP on BT4C + cells, expressing the fusion protein, occurred through a specific interaction between biotin, grafted on the surface of the particle, and the avidin moiety of lodavin (*37*).

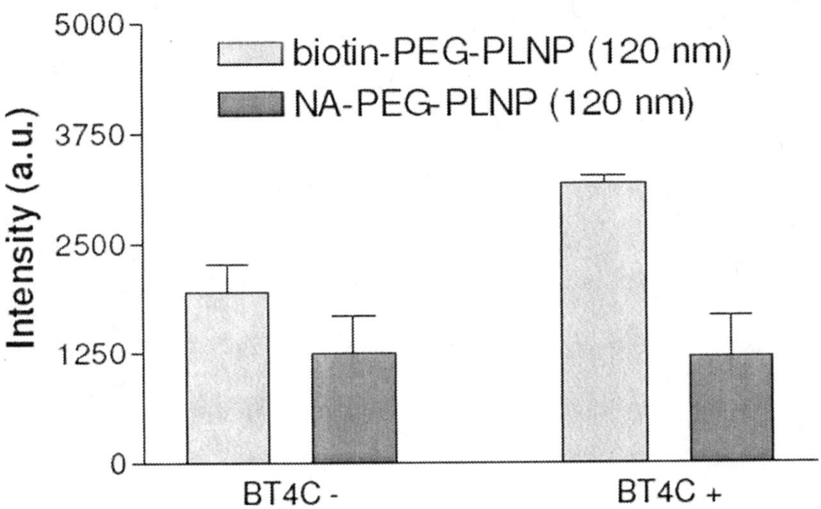

Figure 9. In vitro binding of PLNP (150 µg/mL) on BT4C cells following a 15 minutes incubation period. Reproduced with permission from ref. (37). Copyright (2012) American Chemical Society.

In Vitro Diagnostic Based on Persistent Luminescence Nanoparticles and Fluorescence Resonance Energy Transfer

Fluorescence resonance energy transfer (FRET) is an electrodynamic phenomenon that can be explained using classical physics. FRET occurs between a donor molecule in the excited state and an acceptor molecule in the ground state. The donor molecule typically emits at shorter wavelengths that overlap with the absorption spectrum of the acceptor. Energy transfer occurs without the appearance of a photon and is the result of long range dipole–dipole interactions between the donor and acceptor. The distance dependency of FRET allows measurement of the distances between donors and acceptors. FRET has been widely used in all applications of fluorescence for the past 50 years including medical diagnostics, DNA analysis, and optical imaging. At present, steady-state measurements are often used to measure binding interactions. Distances are usually obtained from time-resolved measurements (*38*).

In 2011, Yan *et al.* reported the first example of fluorescence resonance energy transfer (FRET) inhibition assay for α-fetoprotein (AFP) excreted during cancer cell growth using water-soluble functionalized persistent luminescence nanoparticles (*39*). AFP is the serum biomarker of hepatocellular carcinoma (HCC), which is the sixth most common cancer worldwide in terms of numbers of cases and has almost the lowest survival rates because of its very poor prognosis. Serum levels of AFP often increase under conditions such as periods of rapid liver cancer cell growth, cirrhosis, and chronic active hepatitis as well as carbon tetrachloride intoxication. Therefore, detection of serum levels of AFP can lead to the early diagnosis of HCC (*40*). To probe AFP with high selectivity and sensitivity, Yan *et al.* designed an inhibition assay based on the modulation of FRET between polyethyleneimine (PEI)-capped PLNP (PEI-PLNP) and Ab-AuNPs via inhibition of the interactions between PEI-PLNP and Ab-AuNPs by AFP (Figure 10A). The authors employed Eu^{2+} and Dy^{3+}-doped $Ca_{1.86}Mg_{0.14}ZnSi_2O_7$ nanoparticles as an example of PLNP because their photoluminescence (PL) emission spectrum have maximum overlap with the absorption spectrum of AFP-antibody-gold nanoparticle conjugates (Ab-AuNPs), resulting in maximum FRET efficiency (Figure 10B). The amino group from PEI on the PLNP rendered the nanoparticles water-soluble and also enabled conjugation of the PLNPs with Ab-AuNPs to obtain the FRET inhibition probe (PEI-PLNPs/Ab-AuNPs) for AFP detection (Figure 10C). To obtain better FRET efficiency, AuNPs were used as the quencher because of their high molar adsorption coefficient and easy surface modification. The electrostatic interaction between the positively charged NH_3^+ groups of the PEI-PLNPs (zeta potential = +14.25 mV) and the negatively charged antibodies on the surface of the AuNPs (zeta potential = -28.95 mV) in PBS buffer at pH 7.4 is responsible for the formation of the FRET inhibition probe PEI-PLNPs/Ab-AuNPs (ζ potential = -0.8 mV). The addition of AFP to PEI-PLNPs/Ab-AuNPs in PBS buffer at pH 7.4 led to an obvious recovery of the PL emission of PEI-PLNPs associated with the desorption of Ab-AuNPs from PEI-PLNPs. This effect is the result of the competition of AFP with PEI-PLNPs for Ab-AuNPs due to the strong and specific affinity between AFP and the antibody (Figure 10D).

This PEI-PLNPs/Ab-AuNPs probe was employed to detect AFP in human serum samples (Figure 11). To avoid fluorescent background noise originating from *in situ* excitation, a solution with a proper amount of the FRET inhibition probe was irradiated for 10 min before detection, eliminating the need for further excitation during the photoluminescence analysis. Since the long-lasting afterglow nature of PLNPs allows detection without external illumination, the autofluorescence and scattering light from biological matrices encountered under *in situ* excitation are eliminated. PEI-PLNPs/Ab-AuNPs probe was used to monitor AFP excretion by Bel-7402 cells (Figure 11 a-c), malignant HCC cells known for their ability to excrete large amounts of AFP, L-O2 cells (Figure 11 d-f), healthy hepatic cells giving a positive result for AFP detection, and 3T3 cells as negative control (Figure 11 g-i). The PEI-PLNPs/Ab-AuNPs probe was also employed to stain cells cultured for 22, 46, and 70 h. The images were acquired after a two hours incubation period of the probe with the cells (Figure 11). The PEI-PLNPs were excited with a UV lamp before incubation with Ab-AuNPs. After the incubation, excitation was stopped to avoid the autofluorescence and scattering light from the biological matrices during signal acquisition. The results showed a much greater amount of excreted AFP for Bel-7402 cells than for L-O2 cells (Figure 11 b-c and e-f). The bioconjugates of PEI-PLNPs with Ab-AuNPs provide a sensitive and specific persistent luminescence probe for the detection of AFP in biological fluids and for the optical imaging of AFP excreted during cancer cell growth. The long-lasting afterglow nature of the PEI-PLNPs/Ab-AuNPs probe allows detection and imaging without external illumination, thereby eliminating the autofluorescence and scattering light from biological matrices encountered under *in situ* excitation.

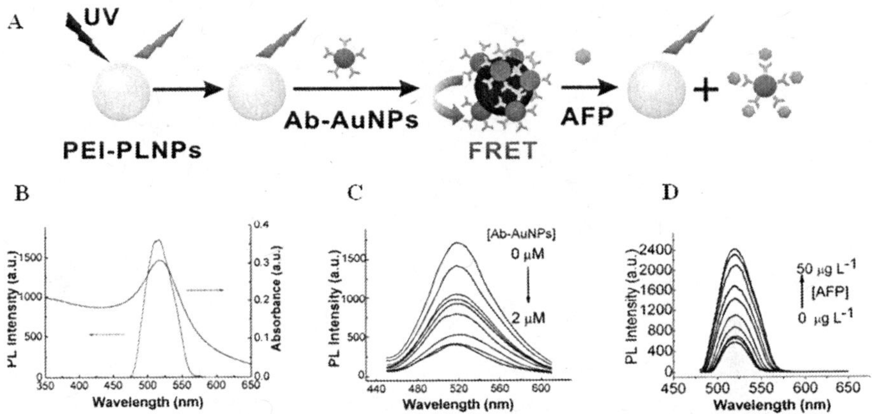

Figure 10. Detection of α-fetoprotein using FRET inhibition assay with PLNP. (A) Schematic illustration of the FRET inhibition assay for AFP based on the photoluminescence quenching of PEI-PLNPs by Ab-AuNPs. (B) Photoluminescence emission spectra of PEI-PLNPs (blue curve) and absorption spectra of Ab-AuNPs (red curve). (C) Quenching effect of Ab-AuNPs on the photoluminescence emission of PEI-PLNPs. (D) Effect of AFP concentration on the photoluminescence emission of PEI-PLNPs/Ab-AuNPs. Reproduced with permission from ref. (39). Copyright (2012) American Chemical Society.

Silicates with Improved Persistent Luminescence Intensity

The main drawback of the first generation of PLNP, with formula $Ca_{0.2}Zn_{0.9}Mg_{0.9}Si_2O_6$:$Eu^{2+}$,$Mn^{2+}$,$Dy^{3+}$, was the inability to gain access to long-term monitoring *in vivo*, mainly due to the decay kinetics after several decades of minutes. The need to work on new materials with improved optical characteristics was investigated in our group and in others. For this purpose, we prepared and evaluated two silicate-based hosts, diopside $CaMgSi_2O_6$ doped with various lanthanides (Ln^{3+}) and nitridosilicate $Ca_2Si_5N_8$ doped with Eu^{2+},Tm^{3+}, as possible alternative to the first generation of PLNP.

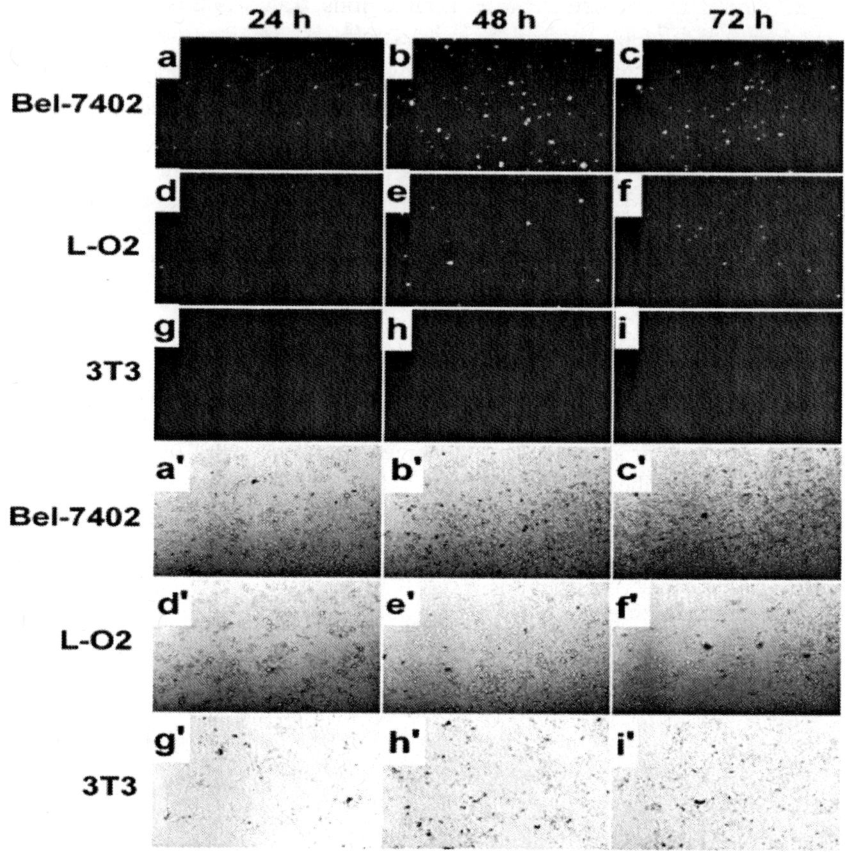

Figure 11. Detection of α-fetoprotein excretion by malignant cells (a-i) Fluorescence images of Bel-7402, L-O2, and 3T3 cells stained with the FRET inhibition probe (PEI-PLNPs/Ab-AuNPs) after the cells had been cultured for 22, 46, and 70 h, respectively and (a'-i') the corresponding bright-field images. Reproduced with permission from ref. (39). Copyright (2012) American Chemical Society.

Trap Depth Optimization to Improve Optical Properties of Diopside-Based Nanophosphors for Medical Imaging

Since electrons trap depth control the optical properties of PLNP (*25, 41*), we synthesized several Mn^{2+} doped diopside nanoparticles, either co-doped with trivalent lanthanide ions $CaMgSi_2O_6:Mn^{2+},Ln^{3+}$ (Ln = Dy, Pr, Ce, Nd), only excitable with X-rays, hereafter referred to as CMSO:Ln, or tri-doped with divalent europium and trivalent lanthanide ions $CaMgSi_2O_6:Mn^{2+},Eu^{2+},Ln^{3+}$, to enable UV excitation, hereafter referred to as CMSO:Eu,Ln (*42*). For this purpose, TSL measurements were used to evaluate both the amount and depth of electron traps. Experiments were performed after 10 minutes of X-rays excitation at 10K with a heating rate of 10 K/min from 30 K to 650 K. Photons were collected via an Acton SpectraPro monochromator coupled with a Princeton CCD camera cooled at -70°C. The normalized intensity of the TSL spectra are presented in Figure 12A. Taking into account the work published in the last years by P. Dorenbos (*43, 44*), we developed with this silicate diopside matrix a similar procedure, i.e. to link decay times to the TSL experiments. The results presented in Figure 13A show the variation of the maximum of the TSL peak when varying the lanthanide dopant, chosen between Ce^{3+}, Pr^{3+}, Nd^{3+}, and Dy^{3+}. The decay profiles are presented in Figure 12B. With this approach it was possible to establish which Ln^{3+} leads to an intense persistent luminescence i.e. where the TSL occurs around or slightly above room temperature.

The situation where TSL peaks are located far below or far above room temperature is not optimal for a strong persistent luminescence. The low temperature peaks only contribute to the short time part of the afterglow, while thermal energy at room temperature is not sufficient to induce a significant electron release from deep traps, which control the intensity and the length of the afterglow. For optimal persistent luminescence properties, TSL peaks should be located in the area of physiological temperature. This is the case for Pr^{3+} doped silicate that gives a TSL peak centered on 353 K, which corresponds to a position just above room temperature and probably explains the intense long-lasting luminescence observed in Figure 12A, as compared to other codopants.

The afterglow dependency, observed with X-rays excitable compounds, remains after additional doping with divalent europium (*42*). We thus identified Pr^{3+} as the optimal electron trap in diopside host, which led to an improved nanomaterial, $CaMgSi_2O_6:Mn^{2+},Eu^{2+},Pr^{3+}$, excitable with UV light, and displaying the most intense afterglow in the near-infrared region. We finally managed to extract nanoparticles with narrow distribution centred on 120 nm from the initial polydisperse powder, and report their application for highly sensitive real-time *in vivo* optical imaging in mice.

To determine which composition was responsible for the best sensitivity for real-time bioimaging, we compared the performances of the two best UV-excitable diopside-based compositions, i.e. CZMSO:Eu,Dy,Mn and CMSO:Eu,Pr,Mn *in vivo*. The Figure 13 shows comparative images obtained from the photon-counting system after 15 minutes acquisitions. Both compositions display the same biodistribution: PLNP almost instantly concentrate within liver. As mentioned above, this trend was already observed with many nanoparticle systems and corresponds to a major uptake of hydroxyl-PLNP by Kupffer cells in liver (*44, 45*). When it comes to signal intensity, the difference between CZMSO:Eu,Dy,Mn and CMSO:Eu,Pr,Mn injected mice appears striking. First, the new composition lightens some of the main pectoral limbs' circulation routes, indistinguishable on the CZMSO-injected mouse (Figure 13) (*46*). Then, optical semi-quantization indicates that CMSO:Eu,Pr returns almost five times more signal through living tissues than the original composition (insert from Figure 13).

We have thus demonstrated the advantage of controlling lanthanide electron trap depth to synthesize PLNP with intense NIR emission for *in vivo* applications and real-time optical imaging in healthy mice. These optimized PLNP are responsible for a significant signal benefit through living tissue and open access to a longer monitoring of the probe.

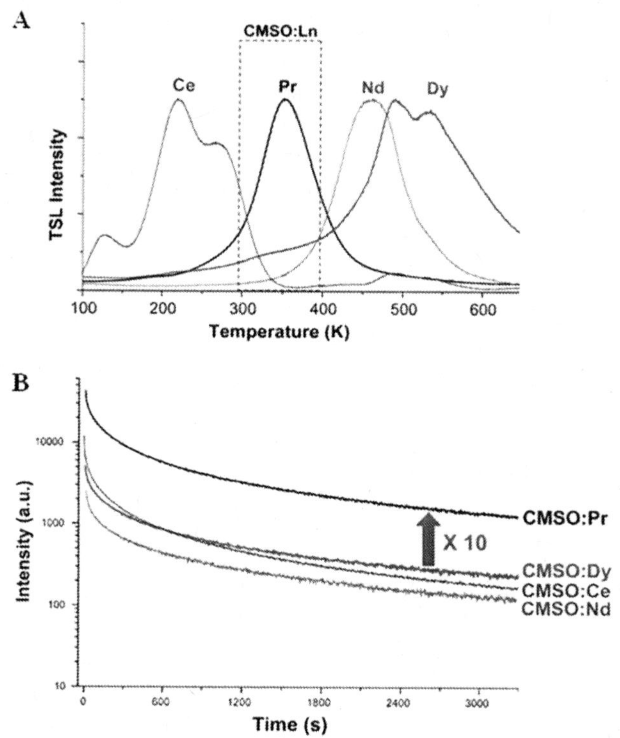

Figure 12. (A) Mn^{2+} TSL intensity at 685 nm of CMSO:Ln (1%, Ln = Ce^{3+}, Nd^{3+}, Dy^{3+}, and Pr^{3+}) materials (normalized curves). The temperature range of interest for room temperature persistent luminescence is delimited by a dotted rectangle.(B) Decay of the Mn^{2+} luminescence intensity at 685 nm of CMSO and rare-earth-codoped CMSO:Ln compounds, recorded after 10 min X-ray irradiation. Reproduced with permission from ref. (42). Copyright (2012) American Chemical Society.

Nitridosilicate-Based Persistent Luminescence Nanoprobes for in Vivo Imaging

Smet *et al.* has recently shown that bulk nitrodosilicate host with formula $Ca_2Si_5N_8:Eu^{2+},Tm^{3+}$ had outstanding long lasting orange-red luminescence peaking at about 610 nm (*47, 48*), in the optical transparency window. However, a solid state synthesis of this material produces a powder with grain sizes in the micrometer range, which limits its direct use in living animals. In order to investigate a possible application of such nitridosilicate materials for *in vivo* bioimaging, a wet grinding of the bulk powder followed by selective sedimentation in alkaline solution was performed to obtain 200 nm hydroxyl-terminated nanoparticles, referred to as CSN-OH. A suspension of this nanomaterial, for a total of 10^{12} nanoparticles dispersed in 200 μL of physiologic serum, was activated for 2 minutes under a UV light (6 W lamp, 254 nm) and injected in the caudal vain of healthy mice. Figure 14A displays the optical images obtained 15 minutes after intravenous injection. The signal coming from CSN-OH is detected with a photon-counting system and returns a clear localization of the probe *in vivo*. (*49*). However, by comparing the luminescence intensities of the different nanoparticles tested for *in vivo* applications, Pr^{3+} doped diopside ($CaMgSi_2O_6{:}Mn^{2+},Eu^{2+},Pr^{3+}$) still displays the best optical characteristics (Figure 14B).

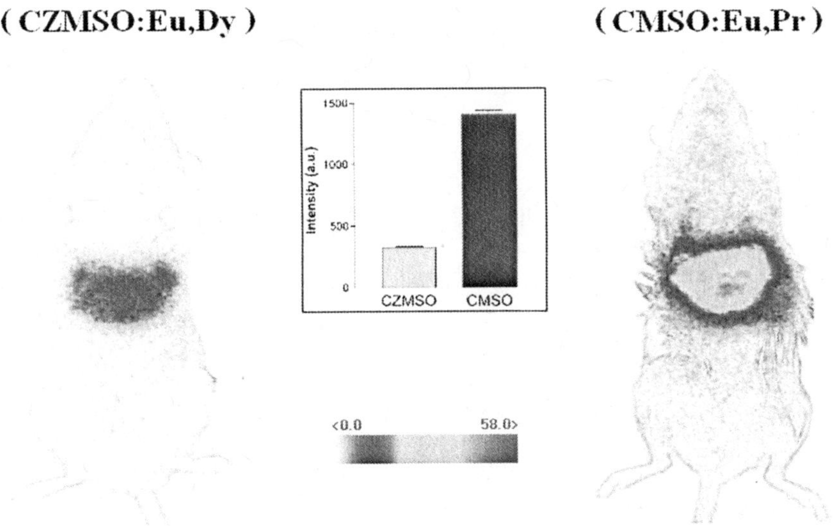

Figure 13. In vivo comparison of CZMSO:Eu,Dy and CMSO:Eu,Pr under the photon counting system. The signal was recorded 15 minutes following systemic injection of the probes (100 μg, excited for 5 minutes under a 6W UV lamp at 254 nm before injection). Luminescence intensity is expressed in false color units (1 unit = 2800 photons per s.cm².steradians). Insert represents the semiquantitative image-based comparison of the signals acquired after 1 minute from the whole mouse body. Reproduced with permission from ref. (42). Copyright (2012) American Chemical Society.

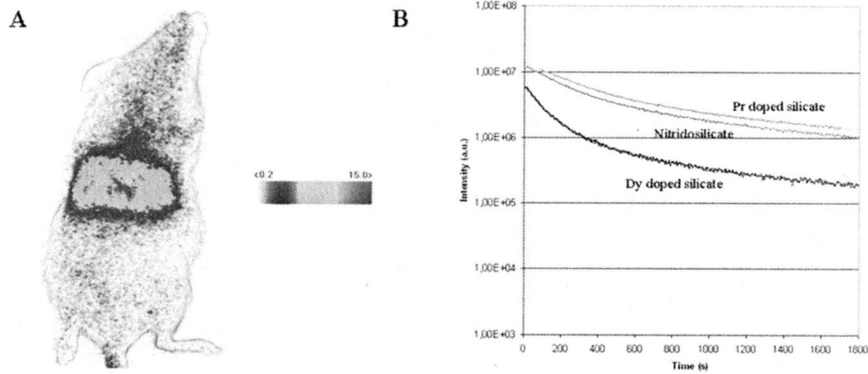

Figure 14. (A) Biodistribution of CSN-OH 15 minutes after tail vein injection. Luminescence intensity is expressed in false color unit (1 unit = 2800 photons per s.cm².steradians). (B) Comparison of luminescence intensity of 10 mg of each type of synthetized nanoparticles. Reproduced with permission from ref. (49). Copyright (2012) The Optical Society.

On the basis of three different examples, we have shown that persistent luminescence nanoparticles could be used as highly sensitive nanoprobe for real-time *in vivo* optical imaging. However, despite serious efforts to improve optical properties of PLNP, and the discovery of a new composition with enhanced persistent luminescence (Figure 14B), we were still unable to perform *in vivo* imaging for more than two hours *in vivo*. Unfortunately, this time threshold constitutes a critical limitation for a large majority of *in vivo* applications in which information need to be monitored for several hours, sometimes several days, after the injection of the probes. In order to overcome this stumbling block, we decided to radically change PLNP base-structure and focus on a different couple of materials/dopants.

Conclusion

We have shown, through this chapter, that persistent luminescence constitutes a powerful technology for diagnosis and imaging applications in living animals. Noteworthy, persistent luminescence nanoparticles emitting in the low tissue absorption window can be easily detected *in vivo* after systemic injection to healthy mice, providing highly sensitive optical detection without any autofluorescence signal from living tissues. Combined efforts to work on the composition and structure of the probe, i.e. nature of dopants and host material, resulted in persistent luminescence nanoparticles with improved optical characteristics. Moreover, surface functionalization of these PLNP was shown to enable visualization of area of interests. This novel nanotechnology should not only introduce and generalize persistent luminescence nanoparticles as one of the most versatile and user-friendly optical nanoprobes, but also be of wide applications to biologists and pharmacologists involved in cancer diagnosis, vascular biology, as well as cell tropism, cell homing, or cell therapy research.

Acknowledgments

The authors wish to thank all their collaborators and colleagues: Q. le Masne de Chermont, J. Seguin, N. Wattier, M. Bessodes (UPCGI, CNRS UMR 8151), B. Viana, A. Lecointre, A. Bessière, D. Gourier, C. Chanéac, C. Rosticher (LCMCP, CNRS UMR 7576), P. Smet (U Gent), and S. Maitrejean (Biospace Lab). This work was supported by grants from the French Région Ile de France, the French National Agency (ANR) in the frame of its program in Nanosciences and Nanotechnologies (NATLURIM project n°ANR-08-NANO-025) and by European community (EC-DG research) through the FP6-Network of Excellence (CLINIGENE: LSHB-CT-2006-018933).

References

1. Newton Harvey, E. *A History of Luminescence: From the Earliest Times until 1900*; American Philosophical Society: Philadelphia, 1957.
2. Hoogenstraaten, W.; Klasens, H. A. *J. Electrochem. Soc.* **1953**, *100*, 366–375.
3. Yen, W. M.; Shionoya, S.; Yamamoto, H. *Phosphor Handbook*, 2nd ed.; CRC Press/Taylor and Francis: Boca Raton, FL, 2007.
4. Wang, D.; Yin, Q.; Li, Y.; Wang, M. *J. Lumin.* **2002**, *97*, 1–6.
5. Kano, T. Luminescence Centres of Rare-Earth Ions. In *Phosphor Handbook*; Shionoya, S., Yen, W. M., Eds.; CRC Press: Boca Raton, FL, 1999; pp 178–200.
6. Matsuzawa, T.; Aoki, Y.; Takeuchi, N.; Murayama, Y. *J. Electrochem. Soc.* **1996**, *143*, 2670–2673.
7. Abbruscato, V. *J. Electrochem. Soc.* **1971**, *118*, 930–933.
8. Lin, Y.; Tang, Z.; Zhang, Z.; Wang, X.; Zhang, J. *J. Mater. Sci. Lett.* **2001**, *20*, 1505–1506.
9. Van den Eeckhout, K.; Smet, P. F.; Poelman, D. *Materials* **2010**, *3*, 2536–2566.
10. (a) Anderson, R. R.; Parrish, J. A. *J. Invest. Dermatol.* **1981**, 77, 13–19. (b) Weissleder, R.; Pittet, M. J. *Nature* **2008**, 452, 580–589.
11. Medintz, I. L.; Uyeda, H. T.; Goldman, E. R.; Mattoussi, H. *Nat Mater* **2005**, *4*, 435–446.
12. Resch-Genger, U.; Grabolle, M.; Cavaliere-Jaricot, S.; Nitschke, R.; Nann, T. *Nat Methods* **2008**, *5*, 763–775.
13. Alivisatos, A. P.; Gu, W.; Larabell, C. *Annu. Rev. Biomed. Eng.* **2005**, *7*, 55–76.
14. Frangioni, J. V. *Curr. Opin. Chem. Biol.* **2003**, *7*, 626–634.
15. Medintz, I. L.; Uyeda, H. T.; Goldman, E. R.; Mattoussi, H. *Nat. Mater.* **2005**, *4*, 435–446.

16. Dubertret, B.; Skourides, P.; Norris, D. J.; Noireaux, V.; Brivanlou, A. H. *Science* **2002**, *298*, 1759–1762.

17. Hale, G. M.; Querry, M. R. *Appl. Opt.* **1973**, *12*, 555–563.

18. Scherman, D.; Bessodes, M.; Chaneac, C.; Gourier, D. L.; Jolivet, J.-P.; le Masne de Chermont, Q.; Maitrejean, S.; Pelle, F. J. PCT Int. Appl. 2007, WO2007048856A1

19. Wang, X. J.; Jia, D.; Yen, W. M. *J Lumin* **2003**, *103*, 34–37.

20. Brinker, C. J.; Scherer, G. W. *Sol-Gel Science: The physics and the chemistry of sol-gel processing*; Academic: London, 1990.

21. Escribano, P.; Julian-Lopez, B.; Planelles-Arago, J.; Cordoncillo, E.; Viana, B.; Sanchez, C. *J. Mater. Chem.* **2008**, *18*, 23–40.

22. Iler, R. K. *The Chemistry of Silica: Solubility, Polymerization, Colloid and Surface Properties, and Biochemistry*; John Wiley and Sons: New York, 1979.

23. le Masne de Chermont, Q.; Richard, C.; Seguin, J.; Chanéac, C.; Bessodes, M.; Scherman, D. *Proc. SPIE* **2009**, *7189*, 71890B/1–71890B/9.

24. Maldiney, T.; Richard, C.; Seguin, J.; Wattier, N.; Bessodes, M.; Scherman, D. *ACS Nano* **2011**, *5*, 854–862.

25. Lecointre, A.; Bessiere, A.; Viana, B.; Gourier, D. *Radiat. Meas.* **2010**, *45*, 497–499.

26. An, Y.; Chen, M.; Xue, Q.; Liu, W. *J. Colloid Interface Sci.* **2007**, *311*, 507–513.

27. Song, Y. K.; Liu, F. *Gene Ther.* **1998**, *5*, 1531–1537.

28. Kamps, J. A.; Morselt, H. W. M.; Swart, P. J.; Meijer, D. K. F.; Scherphof, G. L. *Proc. Natl. Acad. Sci. U.S.A.* **1997**, *94*, 11681–11685.

29. Krieger, M.; Herz, J. *Annu. Rev. Biochem* .**1994**, *63*, 601–637.

30. le Masne de Chermont, Q.; Chanéac, C.; Seguin, J.; Pelle, F.; Maîtrejean, S.; Jolivet, J.-P.; Gourier, D.; Bessodes, M.; Scherman, D. *Proc. Natl. Acad. Sci. U.S.A.* **2007**, *104*, 9266–9271.

31. Minchin, R. F.; Martin, D. J. *Endocrinology* **2010**, *151*, 474–481.

32. Byrne, J. D.; Betancourt, T.; Brannon-Peppas, L. *Adv. Drug Delivery Rev.* **2008**, *60*, 1615–1626.

33. Byk, G.; Partouche, S.; Weiss, A.; Margel, S.; Khandadash, S. *J. Comb Chem* **2010**, *12*, 332–345.

34. Maldiney, T.; Byk, G.; Wattier, N.; Seguin, J.; Khandadash, R.; Bessodes, M.; Richard, C.; Scherman, D. *Int. J. Pharm.* **2012**, *423*, 102–107.

35. Lesch, H. P.; Pikkarainen, J. T.; Kaikkonen, M. U.; Taavitsainen, M.; Samaranayake, H.; Lehtolainen-Dalkilic, P.; Vuorio, T.; Määttä, A. M.; Wirth, T.; Airenne, K. J.; Yla-Herttuala, S. *Hum. Gene Ther.* **2009**, *20*, 871–882.

36. Richard, C.; le Masne de Chermont, Q.; Scherman, D. *Tumori* **2008**, *94*, 264–270.

37. Maldiney, T.; Kaikkonen, M. U.; Seguin, J.; le Masne de Chermont, Q.; Bessodes, M.; Ylä-Herttuala, S.; Scherman, D.; Richard, C. *Bioconjugate Chem* **2012**, *23*, 472–478.

38. Chen, W. *J. Nanosci. Nanotechnol.* **2008**, *8*, 1019–1051.

39. Wu, B.-Y.; Wang, H.-F.; Chen, J.-T.; Yan, X.-P. *J. Am. Chem. Soc.* **2011**, *133*, 686–688.

40. Qin, L.-X.; Tang, Z.-Y. *Cancer Res. Clin. Oncol.* **2004**, *130*, 497.

41. Lecointre, A.; Viana, B.; LeMasne, Q.; Bessière, A.; Chanéac, C.; Gourier, D. *J. Lumin.* **2009**, *129*, 1527–1530.

42. Maldiney, T.; Lecointre, A.; Viana, B.; Bessière, A.; Bessodes, M.; Gourier, D.; Richard, C.; Scherman, D. *J. Am. Chem. Soc.* **2011**, *133*, 11810–11815.

43. Lecointre, A.; Bessière, A.; Bos, A. J. J.; Dorenbos, P.; Viana, B.; Jacquart, S. *J. Phys. Chem. C* **2011**, *115*, 4217–4227.

44. Dorenbos, P.; Krumpel, A. H.; van der Kolk, E.; Boutinaud, P.; Bettinelli, M.; Cavalli, E. *Opt. Mater.* **2010**, *32*, 1681–1685.

45. Kumar, R.; Roy, I.; Ohulchanskky, T. Y.; Vathy, L. A.; Bergey, E. J.; Sajjad, M.; Prasad, P. N. *ACS Nano* **2010**, *4*, 699–704.

46. Maldiney, T.; Lecointre, A.; Viana, B.; Bessière, A.; Gourier, D.; Bessodes, M.; Richard, C.; Scherman, D. *Proc. SPIE* **2012**, *8263*, 826318.

47. Van den Eeckhout, K.; Smet, P. F.; Poelman, D. *J. Lumin.* **2009**, *129*, 1140–1143.

48. Smet, P. F.; Van den Eeckhout, K.; Bos, A. J. J.; Van der Kolk, E.; Dorenbos, P. *J. Lumin.* **2012**, *132*, 682–689.

49. Maldiney, T.; Sraiki, G.; Viana, B.; Gourier, D.; Richard, C.; Scherman, D.; Bessodes, M.; Van den Eeckhout, K.; Poelman, D.; Smet, P. F. *Opt. Mater. Express* **2012**, *2*, 261–268.

Chapter 2

Long-Circulating Polymeric Drug Nanocarriers

Wei Wu[†,‡] and Xiqun Jiang[*,†]

†Laboratory of Mesoscopic Chemistry and Department of Polymer Science & Engineering, College of Chemistry & Chemical Engineering, Nanjing University, Nanjing 210093, P. R. China
‡Nanjing University Yangzhou Institute of Chemistry and Chemical Engineering, Yangzhou, 225000, P. R. China
*E-mail: jiangx@nju.edu.cn

Nano drug delivery systems (NDDS) can transport anticancer agents to tumor sites *via* enhanced permeability and retention (EPR) effect to enhance the therapeutic effect and reduce side effects. To give full play of the EPR effect, long-term circulation of NDDS is needed, since it provides the NDDS with better chance to reach and interact with tumor. The blood circulation time of NDDS is determined by their combination properties, such as size, shape, stiffness, surface shielding *etc.*, all of which are important and needed to be considered in designing long-circulating NDDS.

Thanks to the leaky vasculature and impaired lymphatic drainage of tumor that allow nano drug delivery systems (NDDS) with suitable size to extravasate and accumulate in tumor (so-called enhanced permeability and retention (EPR) effect), well designed NDDS can transport anticancer agents to tumor sites in highly targeting mode to enhance the therapeutic effect and reduce side effects. To take advantage of the EPR effect, the long-term circulation of NDDS is needed, since it provides the NDDS with better chance to reach and interact with tumor. However, the NDDS administered intravenously are often rapidly cleared from blood by the reticuloendothelial system (RES), whose main responsibility is to protect the body from the invasion of extraneous substances. This clearance process is initiated by opsonize, in which the opsonins, such as IgG or complement fragments, adhere to the surface of the NDDS and mediate

© 2012 American Chemical Society

the recognition and phagocytosis of NDDS by RES. Undoubtedly, the natures of NDDS including their size, shape, surface chemistry and chemical composition determine their ability to avoid the adsorption of opsonins and the recognition of RES and their blood circulation time. Therefore, one important way to achieve the prolonged circulation time is to design and modify the NDDS in size, shape, surface chemistry and chemical composition properly.

Here, we overview the achievements in the prolongation of circulation time of polymeric nanoparticles that serve as drug carriers, since polymeric nanoparticle is one sort of the most extensively studied drug carriers and have significant superiority in biocompatibility, biodegradation and tunability (*1, 2*).

PEGylation of Polymeric Nanoparticles

Polyethylene glycol (PEG, Figure 1a) is a kind of hydrophilic, electrically neutral, flexible, biocompatible polymer with low immunogenicity and antigenicity. In 1977, Abuchowski *et al.* first reported their work on increasing the circulation time of bovine liver catalase by covalently combining PEG (*3*). Since then, PEGylation has been a popular strategy to ensure the stealth-shielding and long circulation of nano drug carriers including polymeric nanoparticles in the research field of biomedicine (*4–12*). PEG can interact unstrainedly with water *via* hydrogen bond due to the good fit of their structures, which renders PEG with high hydrophilicity. The hydration and steric exclusion effect of PEG are considered to play important roles in protein repellence (*13*). Depending on the PEG density, the PEG layer on the surface of nanoparticles can exhibit mushroom or brush-like configuration (Figure 1b). At relatively low surface PEG density, mushroom-like configuration is often adopted to maximize PEG coverage. With the increase of the PEG density, the surface gets more crowded and the PEG chains tend to stretch due to the enhanced interchain interactions, resulting in a brush-like configuration (*9, 14, 15*).

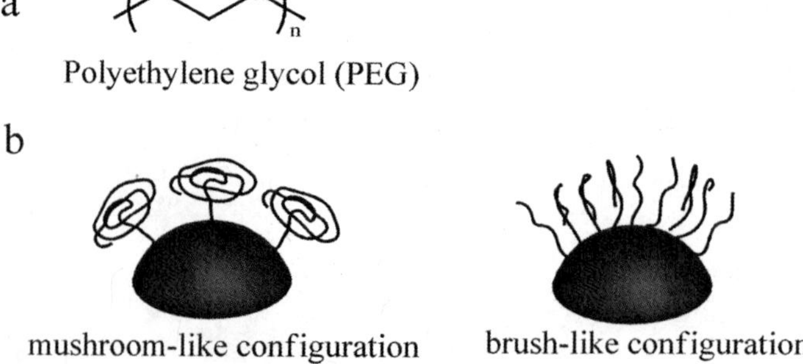

Figure 1. a) Chemical structure of PEG; b) Cartoon of mushroom and brush-like configurations of PEG on nanoparticle surface.

As stated above, PEG layer on the nanoparticle surface can remarkably improve the stealth property of the nanoparticle matrix. The density, thickness and configuration of the PEG chains are crucial determinants of the effect magnitude. Increasing the PEG coverage can increase the protein repellence, however, when the coverage exceeds a certain degree, the stealth property began to decrease due to the decrease of the mobility and flexibility of the PEG chains induced by the crowding condition (*16*). For different systems, the optimal PEG density can be quite different, that is to say, the optimal points are strongly related to the intrinsic features of the nanoparticle matrix, such as size, shape, surface charge and compositions *etc.*. For example, 10 wt% PEG content is reported to be optimal for the dispersibility and stealth property of poly(lactic acid) (PLA) nanoparticles (*17*) or poly(lactide-co-glycolide) (PLGA) nanoparticles (*18*). By contrast, in another published work, the optimal PEG content is calculated to be 5 wt% for the protein resistance of PLA, PLGA and polycaprolatone nanoparticles (*9*).

The PEGylation of the polymeric nanoparticles can be achieved by different methods.

Conjugating Covalently PEG to the Surface of Nanoparticles

The PEGylation of polymeric nanoparticles can be implemented by covalently conjugating PEG to the surface of nanoparticles (*19, 20*). The maximum PEG coverage for this strategy strongly depends on the density of the reactive groups exposed on nanoparticle surface and their reactivity. When the PEG molecule has one group at each terminus, the reactive groups can be remained at the free end of PEG chains after the conjugation and provide the nanoparticles with functional sites for surface conjugation of other functional molecules, such as targeting or therapeutic agents (*20*).

Incorporation of PEG Copolymer into Nanoparticles

Another PEGylation strategy is incorporation of amphiphilic PEG copolymers during nanoparticle formation. This strategy is often started from the synthesis of amphiphilic di- or tri-block copolymer with PEG as a hydrophilic segment followed by assembling the copolymers into nanoparticles. During the assembly of the copolymers, core-shell structure would be formed with the hydrophobic segments as the core of and the PEG moieties as the shell which stabilizes the nanoparticles in aqueous medium and render them desirable stealth property.

Biocompatible and biodegradable hydrophobic polymers, such as PLA, PLGA and poly(ε-caprolactone) *etc.*, are frequently used as the hydrophobic blocks of the amphiphilic copolymers to fabricate polymeric nanoparticles. The synthesis of this kind of copolymers is generally achieved by ring-opening polymerization of the corresponding monomers with PEG as a macroinitiator (*21–23*). The PEGylation efficiency fulfilled by the assembly of PEG-containing amphiphilic copolymers is generally good. For example, it has been reported that the nanoparticles prepared from PEG-PLGA copolymer exhibit much better stealth property than those prepared by directly conjugating PEG to pre-made PLGA nanoparticles (*24*).

Apart from the polyesters mentioned above, the natural polymers, For example gelatin, have also been used to synthesize amphiphilic copolymers with PEG as one block by attaching PEG chains covalently to the backbone of these polymers (*25*).

Although PEG possesses many attractive properties, such as good biocompatibility, low immunogenicity, high hydrophilicity and flexibility, chemical inertness and electrical neutrality, especially high anti-biofouling ability, its intrinsic drawbacks have been progressively perceived by many studies. For example, PEG can induce the complement activation (*26–35*). Many studies showed that a first dose of PEGylated substance injected intravenously would trigger rapid clearance of a subsequent dose of the same substance injected several days after the first injection, which is referred to as the "accelerated blood clearance (ABC) phenomenon". It is generally acknowledged that the ABC of the second dose is mediated by PEG-specific IgM produced in response to the first injection (*27, 33*). The time interval between repeated injections, dose, the species of the loading agents and the physicochemical properties of the PEGylated nanoparticles are the factors influencing the extent of the ABC phenomenon, in other words, the ABC phenomenon can be suppressed by adjusting these factors (*34*). In addition to the complement activation, PEG coating can interfere significantly with the interaction between nanoparticles and cells and the endosomal escape of the nanoparticles, which would adversely affect the intracellular delivery of drug or other bioactive molecules (*36, 37*). It is worth noting that both the drawbacks of PEG coating stated above are closely related to the molecular weight (MW)- and surface density of PEG chains, that is to say, when increasing either the MW or surface density of PEG, the effect of PEG coating on complement activation and cellular uptake of nanoparticles would be accordingly increased (*38–40*).

Poly(*N*-vinylpyrrolidone) (PVP) Stealth Coating of Nanoparticles

Just like PEG, PVP (Figure 2) coating can also effectively prolong the residence of nanoparticles in circulation relative to the naked nanoparticles (*34, 41–44*). Accordingly, PVP has been looked as a promising alternative to PEG in protecting nanoparticles from opsonization and prolonging their circulation time in blood. PVP has good biocompatibility, high hydrophilicity, electrical neutrality, chemical stability and thus is very promising in biomedical applications. PVP is also a cryo/lyoprotectant. This characteristic enables PVP-coated nanoparticles to be readily redispersed in aqueous medium after lyophilization and renders nanoparticles with long-term stability, which is an advantage of PVP over PEG (*42*). However, comparing to the performance of PEG in stealthiness, it has been found that the PVP-coated nanoparticles exhibit shorter circulation half-lives than those of their PEG counterparts, which should be determined by the nature of the corresponding polymer, for example, the larger chain stiffness of PVP than PEG leads to the distinctive protein adsorption pattern between the nanoparticles coated by the two kinds of polymer and eventually the different stealth properties of the nanoparticles (*42*). Despite of the slightly inferior stealthness of PVP with respect to PEG, one unique advantage of PVP is greatly favorable to its application as stealth coating of nanoparticles, namely, the PVP-coated nanoparticles do not trigger ABC phenomenon and have reproducible pharmacokinetics and pharmacodynamics profiles, which has been demonstrated by the studies of Ishihara *et al.* and is greatly significant to prevent unanticipated side effects and preserve pharmacological activity of the encapsulated drug in clinic, though more researches are necessary to corroborate this conclusion (*34*).

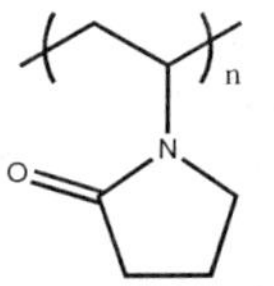

Poly(*N*-vinylpyrrolidone) (PVP)

Figure 2. Chemical structure of PVP.

PVP-coated polymeric nanoparticles are generally prepared by the assembly of the amphiphilic block copolymers containing PVP as a hydrophilic segment in aqueous medium (*34, 41–45*). Upon synthesizing PVP segments by radical polymerization, the introduction of derivable groups at chain ends is very important to their further derivations, which can be achieved by using derivable group-bearing iniators or chain transfer agents in polymerization (*46*). Xanthate-mediated reverse addition fragment transfer (RAFT) polymerization has been demonstrated to be a desirable method to the synthesis of PVP, since by this method, MW can be well controlled and the xanthate end group on synthesized PVP could be readily converted to thiol, hydroxyl and aldehyde groups that are very useful in their functionalizations (*44, 45*).

Polybetaine Stealth Coating of Nanoparticles

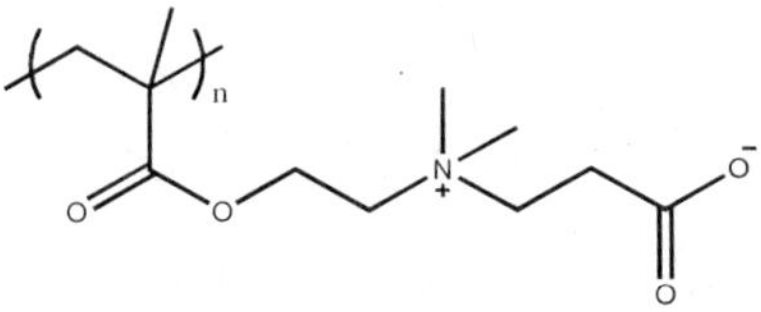

poly(carboxybetaine)

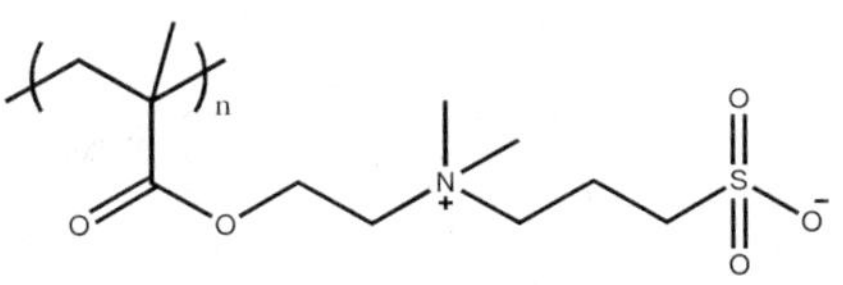

poly(sulfobetaine)

Figure 3. Chemical structures of poly(carboxybetaine) and poly(sulfobetaine).

Polybetaine, such as poly(carboxybetaine) and poly(sulfobetaine) (Figure 3), is a kind of polyzwitterion including anionic and cationic charges on the same repeat unit and can effectively resist opsonization due to their strong hydration *via* electrostatic interactions and zero net electric charge in physiological condition (*47–50*). The polybetaine stealth coating of nanoparticles can be implemented by assemblying the amphiphilic block copolymers containing polybetaine as a hydrophilic segment (*51*) or by surface modifications with polybetaines through surface-initiated radical polymerization (*52, 53*). In addition, the abundant carboxylic acid groups along the chain of poly(carboxybetaine) enable the modified nanoparticles to be functionalized with different bioactive molecules, such as drug or targeting molecules, and thus provide a desirable platform for multifunctional nanomedicines (*47*).

Very recently, it has been found that poly(carboxybetaine) are able to improve the stability of proteins without sacrificing the binding affinity of proteins (*54*). This is an exciting finding and makes a great improvement over the current PEGylation technique, which improves the stability of proteins at the expense of their binding affinity. This finding has significant application potential in the field of protein therapy. It is very important to improve the stability of therapeutic proteins because their low stability in physiological conditions would reduce remarkably their bioavailability.

Size Dependence of the Stealth Property of Nanoparticles

Particle size is one crucial determinant of their *in vivo* behavior including blood circulation time. Small molecules and nanoparticles with diameter less than 10 nm are generally considered to be easily cleared from the body by renal excretion (*55, 56*). On the contrary, too large nanoparticles can not access the tumor through transvascular pores and fenestrations due to the limitation of the pore cutoff sizes (*57*). Therefore, moderate size is a prerequisite for nanoparticles to achieve long blood circulation and good passive tumor targeting. Many studies have been performed to understand how the size affects on the *in vivo* behavior of the nanoparticles that have moderate size between 10-200 nm (*58–60*). Smaller nanoparticles are frequently demonstrated to have better stealth property and thus longer blood circulation than larger ones with similar surface state. As a typical example, Warren C. W. Chan *et al.* studied systematically the pharmacokinetic behavior of the PEGylated gold nanoparticles with different size and revealed that with the same PEG molecular weight, the smaller PEGylated nanoparticles have longer blood circulation half-life (*58*).

Effects of Shape and Stiffness on the Blood Circulation of Nanoparticles

Other natures of nanoparticles, such as shape and stiffness, also play important roles in the *in vivo* behavior. Many studies show that non-spherical shape (*61–64*) and low stiffness (*62, 65*) are favorable to the prolongation of the blood circulation of nano vehicles. However, due to the difficulty in controlling the shape or stiffness of the nano vehicles under the same material, size, surface chemistry and other parameters, more studies are needed to understand the ultimate effect of shape and stiffness on the *in vivo* behavior of the nanovehicles.

Conclusions

The *in vivo* behavior of nanoparticles is determined by their combination properties, such as size, shape, stiffness, surface shielding *etc.*, all of which are needed to be considered in designing long-circulating nanoparticles. Although significant advances have been achieved in particle design and fabrication, more extra effort are required to control more precisely the particle parameters and understand more clearly the interactions between nanoparticles and biological milieu that would direct better the design of the nanoparticles for biomedical applications.

Acknowledgments

This work was supported by National Natural Science Foundation of China (No. 51033002) and the Natural Science Foundation of Jiangsu Province (No. BK2010303).

References

1. Liechty, W. B.; Peppas, N. A. *Eur. J. Pharm. Biopharm.* **2012**, *80*, 241–246.
2. Mehrdad, H.; Mohammad-Ali, S.; Kobra, R. *Macromol. Biosci.* **2012**, *12*, 144–164.
3. Abuchowski, A.; McCoy, J. R.; Palczuk, N. C.; van~Es, T.; Davis, F. F. *J. Biol. Chem.* **1977**, *252*, 3582–3586.
4. Essa, S.; Rabanel, J. M.; Hildgen, P. *Int. J. Pharm.* **2011**, *411*, 178–187.
5. Ebrohimnejad, P.; Dinarvand, R.; Jafari, M. R.; Tabasi, S. A. S.; Atyabi, F. *Int. J. Pharm.* **2011**, *406*, 122–127.
6. Lee, S.-W.; Yun, M.-H.; Jeong, S. W.; In, C.-H.; Kim, J.-Y.; Seo, M.-H.; Pai, C.-M.; Kim, S.-O. *J. Controlled Release* **2011**, *155*, 262–271.
7. Shan, X.; Liu, C.; Yuan, Y.; Xu, F.; Tao, X.; Sheng, Y.; Zhou, H. *Colloids Surf., B* **2009**, *72*, 303–311.
8. Bhattarai, N.; Bhattarai, S. R.; Khil, M. S.; Lee, D. R.; Kim, H. Y. *Eur. Polym. J.* **2003**, *39*, 1603–1608.
9. Gref, R.; Lück, M.; Quellec, P.; Marchand, M.; Dellacherie, E.; Harnisch, S.; Blunk, T.; Müller, R. H. *Colloids Surf., B* **2000**, *18*, 301–313.
10. Allen, T. M.; Hansen, C. B.; de Menezes, D. E. L. *Adv. Drug. Delivery Rev.* **1995**, *16*, 267–284.
11. Allen, T. M.; Hansen, C. *Biochim. Biophys. Acta* **1991**, *1068*, 133–141.
12. Klibanov, A. L.; Maruyama, K.; Torchilin, V. P.; Huang, L. *FEBS Lett.* **1990**, *268*, 235–237.
13. Morra, M. *J. Biomater. Sci., Polym. Ed.* **2000**, *11*, 547–569.

14. Du, H.; Schiavi, S.; Levine, M.; Mishra, J.; Heur, M.; Grabowski, G. A. *Hum. Mol. Genet.* **2001**, *10*, 1639–1648.

15. Tirosh, O.; Barenholz, Y.; Katzhendler, J.; Priev, A. *Biophys. J.* **1998**, *74*, 1371–1379.

16. Storm, G.; Belliot, S.; Daemen, T.; Lasic, D. D. *Adv. Drug Delivery Rev.* **1995**, *17*, 31–48.

17. Sheng, Y.; Yuan, Y.; Liu, C.; Tao, X.; Shan, X.; Xu, F. *J. Mater. Sci.: Mater. Med.* **2009**, *20*, 1881–1891.

18. Beletsi, A.; Panagi, Z.; Avgoustakis, K. *Int. J. Pharm.* **2005**, *298*, 233–241.

19. Gref, R.; Minamitake, Y.; Peracchia, M. *Science* **1994**, *263*, 1600–1603.

20. Suh, J.; Choy, K.-L.; Lai, S. K.; Suk, J. S.; Tang, B. C.; Prabhu, S.; Hanes, J. *Int. J. Nanomed.* **2007**, *2*, 735–741.

21. Hu, Y.; Jiang, X. Q.; Ding, Y.; Zhang, L. Y.; Yang, C. Z.; Zhang, J. F.; Chen, J. N.; Yang, Y. H. *Biomaterials* **2003**, *24*, 2395–2404.

22. Zhang, L. Y.; Hu, Y.; Jang, X. Q.; Yang, C. Z.; Lu, W.; Yang, Y. H. *J. Controlled Release* **2004**, *96*, 135–148.

23. Zhang, L. Y.; Yang, M.; Wang, Q.; Li, Y.; Guo, R.; Jiang, X. Q.; Yang, C. Z.; Liu, B. R. *J. Controlled. Release* **2007**, *119*, 153–162.

24. Betancourt, T.; Byrne, J. D.; Sunaryo, N.; Crowder, S. W.; Kadapakkam, M.; Patel, S.; Casciato, S.; Brannon-Peppas, L. *J. Biomed. Mater. Res., Part A* **2009**, *91*, 263–276.

25. Kaul, G.; Amiji, M. *Pharm. Res.* **2002**, *19*, 1061–1067.

26. Ishida, T.; Atobe, K.; Wang, X.; Kiwada, H. *J. Controlled Release* **2006**, *115*, 251–258.

27. Ishida, T.; Ichihara, M.; Wang, X.; Yamamoto, K.; Kimura, J.; Majima, E.; Kiwada, H. *J. Controlled Release* **2006**, *112*, 15–25.

28. Ishida, T.; Wang, X.; Shimizu, T.; Nawata, K.; Kiwada, H. *J. Controlled Release* **2007**, *122*, 349–355.

29. Ishida, T.; Maeda, R.; Ichihara, M.; Irimura, K.; Kiwada, H. *J. Controlled Release* **2003**, *88*, 35–42.

30. Ishida, T.; Ichihara, M.; Wang, X.; Kiwada, H. *J. Controlled Release* **2006**, *115*, 243–250.

31. Wang, X. Y.; Ishida, T.; Ichihara, M.; Kiwada, H. *J. Controlled Release* **2005**, *104*, 91–102.

32. Wang, X.; Ishida, T.; Kiwada, H. *J. Controlled Release* **2007**, *119*, 236–244.

33. Ishida, T.; Harada, M.; Wang, X. Y.; Ichihara, M.; Irimura, K.; Kiwada, H. *J. Controlled Release* **2005**, *105*, 305–317.

34. Ishihara, T.; Maeda, T.; Sakamoto, H.; Takasaki, N.; Shigyo, M.; Ishida, T.; Kiwada, H.; Mizushima, Y.; Mizushima, T. *Biomacromolecules* **2010**, *11*, 2700–2706.

35. Judge, A.; McClintock, K.; Phelps, J. R.; MacLachlan, I. *Mol. Ther.* **2006**, *13*, 328–337.

36. Du, H.; Chandaroy, P.; Hui, S. W. *Biochim. Biophys. Acta* **1997**, *1326*, 236–248.

37. Romberg, B.; Hennink, W.; Storm, G. *Pharm. Res.* **2008**, *25*, 55–71.

38. Hamad, I.; Hunter, A. C.; Szebeni, J.; Moghimi, S. M. *Mol. Immunol.* **2008**, *46*, 225–232.

39. Liu, Z.; Winters, M.; Holodniy, M.; Dai, H. *Angew. Chem., Int. Ed.* **2007**, *46*, 2023–2027.

40. Zeineldin, R.; Al-Haik, M.; Hudson, L. G. *Nano Lett.* **2009**, *9*, 751–757.

41. Garrec, D. L.; Gori, S.; Luo, L.; Lessard, D.; Smith, D. C.; Yessine, M.-A.; Ranger, M.; Leroux, J.-C. *J. Controlled Release* **2004**, *99*, 83–101.

42. Gaucher, G.; Asahina, K.; Wang, J. H.; Leroux, J.-C. *Biomacromolecules* **2009**, *10*, 408–416.

43. Zhu, Z. S.; Li, Y.; Li, X. L.; Li, R. T.; Jia, Z. J.; Liu, B. R.; Guo, W. H.; Wu, W.; Jiang, X. Q. *J. Controlled Release* **2010**, *142*, 438–446.

44. Zhu, Z. S.; Xie, C.; Liu, Q.; Zhen, X.; Zheng, X. C.; Wu, W.; Li, R. T.; Ding, Y.; Jiang, X. Q.; Liu, B. R. *Biomaterials* **2011**, *32*, 9525–9535.

45. Mishra, A. K.; Patel, V. K.; Vishwakarma, N. K.; Biswas, C. S.; Raula, M.; Misra, A.; Mandal, T. K.; Ray, B. *Macromolecules* **2011**, *44*, 2465–2473.

46. Luo, L. B.; Ranger, M.; Lessard, D. G.; Garrec, L. D.; Gori, S.; Leroux, J.-C.; Rimmer, S.; Smith, D. *Macromolecules* **2004**, *37*, 4008–4013.

47. Jiang, S.; Cao, Z. *Adv. Mater.* **2010**, *22*, 920–932.

48. Shao, Q.; He, Y.; White, A. D.; Jiang, S. *J. Phys. Chem. B* **2010**, *114*, 16625–16631.

49. Chen, S.; Jiang, S.; Mo, Y.; Luo, J.; Tang, J.; Ge, Z. *Colloids Surf., B* **2011**, *85*, 323–329.

50. Zhang, Z.; Vaisocherova, H.; Cheng, G.; Yang, W.; Xue, H.; Jiang, S. *Biomacromolecules* **2008**, *9*, 2686–2692.

51. Cao, Z. Q.; Yu, Q. M.; Xue, H.; Cheng, G.; Jiang, S. Y. *Angew. Chem., Int. Ed.* **2010**, *49*, 3771–3776.

52. Yang, W.; Zhang, L.; Wang, S. L.; White, A. D.; Jiang, S. Y. *Biomaterials* **2009**, *30*, 5617–5621.

53. Zhang, L.; Xue, H.; Gao, C. L.; Carr, L.; Wang, J. N.; Chu, B. C.; Jiang, S. Y. *Biomaterials* **2010**, *31*, 6582–6588.

54. Keefe, A. J.; Jiang, S. Y. *Nature Chem.* **2012**, *4*, 59–63.

55. Rao, J. *ACS Nano* **2008**, *2*, 1984–1986.

56. Kircher, M. F.; Mahmood, U.; King, S. R.; Weissleder, R.; Josephson, L. *Cancer Res.* **2003**, *63*, 8122–8125.

57. Hobbs, S. K.; Monsky, W. L.; Yuan, F.; Roberts, W. G.; Griffith, L.; Torchilin, V. P.; Jain, R. K. *Proc. Natl. Acad. Sci. U.S.A.* **1998**, *95*, 4607–4612.

58. Perrault, S. D.; Walkey, C.; Jennings, T.; Fischer, H. C.; Chan, W. C. W. *Nano Lett.* **2009**, *9*, 1909–1915.

59. Moghimi, S. M.; Hedeman, H.; Muir, I. S.; Illum, L.; Davis, S. S. *Biochim. Biophys. Acta* **1993**, *1157*, 233–240.

60. Fang, C.; Shi, B.; Pei, Y. Y.; Hong, M. H.; Wu, J.; Chen, H. Z. *Eur. J. Pharm. Sci.* **2006**, *27*, 27–36.

61. Muro, S.; Garnacho, C.; Champion, J. A.; Leferovich, J.; Gajewski, C.; Schuchman, E. H.; Mitragotri, S.; Muzykantov, V. R. *Mol. Ther.* **2008**, *16*, 1450–1458.

62. Geng, Y.; Dalhaimer, P.; Cai, S. S.; Tsai, R.; Tewari, M.; Minko, T.; Discher, D. E. *Nat. Nanotechnol.* **2007**, *2*, 249–255.

63. Decuzzi, P.; Godin, B.; Tanaka, T.; Lee, S.-Y.; Chiappini, C.; Liu, X.; Ferrari, M. *J. Controlled Release* **2010**, *141*, 320–327.
64. Devarajan, P. V.; Jindal, A. B.; Patil, R. R.; Mulla, F.; Gaikwad, R. V.; Samad, A. *J. Pharm. Sci.* **2010**, *99*, 2576–2581.
65. Beningo, K. A.; Wang, Y. L. *J. Cell. Sci.* **2002**, *115* (Pt 4), 849–856.

Chapter 3

Targeting Gold Nanoparticles for Cancer Diagnostics and Therapeutics

Andrew J. Coughlin[1] and Jennifer L. West[*,1,2]

[1]Department of Bioengineering, 6100 Main Street, MS-142, Rice University, Houston, Texas 77005
[2]Department of Biomedical Engineering, Room 136 Hudson Hall, Box 90281, Duke University, Durham, North Carolina 27708
[*]E-mail: jlw82@duke.edu

Gold nanoparticles are particularly appealing for cancer medicine because of their plasmonic optical properties that can be exploited for both diagnostic and therapeutic purposes, along with their established biocompatibility and ease of surface modification. Additionally, a variety of targeting moieties has been incorporated on gold nanostructures to achieve high cancer cell specificity and potentially enhanced tumor uptake *in vivo*. Here we describe the synthesis of four gold nanoparticle types: gold colloid, gold–silica nanoshells, gold–gold sulfide nanoparticles, and gold nanorods. Each of these four particles has demonstrated success in both optical imaging and photothermal ablation of cancer. Methods by which each particle can be conjugated to targeting proteins, characterized for their degree of protein conjugation, and then evaluated for their potential to bind cell surface markers overexpressed on tumor cells is also discussed.

Introduction

Although medical advancements have resulted in increased cancer associated five-year survival rates, cancer remains the second leading cause of the death in the United States (*1*). In fact, over 1,500 people die of the disease each day, while 1 in 2 men and 1 in 3 women will be diagnosed with cancer at some point in their lifetime. Globally, nearly 13 million people were diagnosed with cancer and 7.6

© 2012 American Chemical Society

million died from the disease in 2008 alone (*2*). Indeed, improvements in clinical care are needed. Current treatment options such as chemotherapy and radiation are indiscriminate in nature, killing both malignant and healthy cells. Damage to healthy cells results in both acute and chronic side-effects that may effectively limit relevant, therapeutic doses required to eliminate malignancy entirely. Therefore, better cancer detection and treatment are imperative to enhancing patient quality of life and reducing mortality.

Nanotechnology offers a variety of tools that physicians may one day regularly use to combat cancer through both diagnosis and treatment. Nanoparticles are inherently small, providing excellent biological access, as well as an array of unique material properties. For instance, metal nanoparticles exhibit unique optical properties because of their size and freely moving, conduction band electrons. When incident light at wavelengths larger than the physical dimensions of a particle strike its surface, the associated electric field can induce a resonant, collective oscillation of these electrons (also known as surface plasmons). The wavelength at which this phenomenon occurs depends on the particle's makeup and geometry. Although many metal nanoparticles exhibit surface plasmon resonance, gold particles are of particular interest for biological applications because of their established biocompatibility (*3, 4*).

In fact, numerous gold-based nanotechnology platforms have been investigated for cancer medicine including nanorods (*5, 6*), nanocages (*7, 8*), solid nanospheres or colloids (*9, 10*), and gold-dielectric shell-core nanoshells (*11, 12*). Scattering energy associated with gold nanoparticle surface plasmon resonance has been harnessed for imaging cancer cells with modalities such as dark field microscopy (*13, 14*), optical coherence tomography (*11*), and reflectance confocal microscopy (*10*). Additionally, absorption of light by these particles can provide local heating, which has been exploited for photothermal ablation of cancer (*5, 7, 15*). Nanorods, nanocages, and nanoshells are particularly suited for these applications because their geometry and composition can be tailored to scatter or absorb light within the biologically relevant near-infrared (NIR) window (approximately 650 – 900 nm), where human tissue is maximally transmissive (*16*). For instance, nanorods exhibit two plasmon resonance effects, one that occurs along the short axis (transverse band) and another that arises along the long axis (longitudinal band). As nanorod aspect ratios (length divided by width) increase, the longitudinal plasmon band red shifts to the NIR and IR regions of the electromagnetic spectrum (*13, 17*). In the case of hollow gold nanocages, which are fabricated via galvanic replacement of silver nanocubes with gold salt, increasing amounts of gold salt result in decreased wall thicknesses and increased porosity, effectively red shifting extinction profiles (*18*). For example, for silver nanocubes with 30 nm edge lengths, plasmon resonance can be achieved from 400 nm to 1200 nm by increasing gold content. Finally, light interactions with gold nanoshells depend on the particle core-shell ratio, where progressively thinner gold shells over a dielectric core of fixed size results in a red shift of the wavelength at which plasmon resonance occurs (*19*). On the other hand, gold colloids maximally scatter and absorb light within the visible range of 520 – 530 nm, where tissue chromophores exhibit much greater absorption of light. To avoid light-tissue interactions that would occur with visible wavelengths, researchers have relied on second harmonic generation with NIR pulsed laser energy as well as clustering of gold colloids at cell surfaces, which effectively red-shifts the scattering and absorption profiles (*9*).

Delivery of nanoparticles to tumor tissue is an important factor to consider when designing the nanoparticles as well. While passive tumor accumulation of particles via the enhanced permeability and retention (EPR) effect has demonstrated considerable success, recent scientific work has centered on targeting moieties displayed on nanoparticle surfaces to specifically bind overexpressed tumor markers and potentially enhance accumulation *in vivo*, provide molecular information during imaging, or enhance specificity of therapy. Gold nanoparticles are particularly appealing for surface conjugation because of well-established chemistries, especially sulfur-gold interactions. Moieties conjugated to gold nanoparticles have included aptamers (*10*), peptides (*20*), antibodies and their fragments (*14, 21*), and non-antibody ligands, including proteins (*22*) and vitamins (*23*). *In vitro*, these methods have shown high cancer cell specificity, and ongoing work is seeking to evaluate nanoparticle targeting to tumor tissue *in vivo* (*7, 14, 24*).

In this chapter, we describe the synthesis of four different gold-based nanoparticles: gold colloid, gold–silica nanoshells, gold–gold sulfide nanoparticles, and gold nanorods. The gold colloid particles are synthesized via reduction of a gold solution with tetrakis(hydroxymethyl)phosphonium chloride (THPC) and exhibit peak extinction at ~530 nm. On the other hand, gold–silica nanoshells, gold–gold sulfide nanoparticles, and gold nanorods can be designed to maximally scatter and absorb light over a broad range of wavelengths, including the NIR region. Synthesis of the gold–silica nanoshells is a four-step process, including the fabrication of silica cores, core functionalization with amine groups, adsorption of small THPC gold colloid, and formation of the shell with a gold reduction reaction. To synthesize the gold–gold sulfide particles, a one-step self-assembly reaction is performed by combining solutions of sodium thiosulfate ($Na_2S_2O_3$) and gold salt. A final purification step is performed to remove small, gold colloid that also occurs as a result of the reaction. The final group, gold nanorods are fabricated via seed-mediated growth, in which gold salt is reduced onto small, gold seed particles in the presence of surfactant. All four particle types have been employed as diagnostic contrast agents and absorptive agents for photothermal therapy (*9, 12, 13, 24, 25*). In the second part of this chapter, methods for targeting each of these particles to cancer cells and evaluating cancer cell specificity *in vitro* are explained. While we limit our discussion of targeting agents to proteins in this chapter, these methods can be easily adapted to other amine functionalized targeting molecules as required. Additionally, the conjugation methods described have broad applicability to gold surfaces and particulates.

Gold Colloid Synthesis

Gold colloid synthesis, as originally described by Duff *et al.*, can be performed by reduction of chloroauric acid ($HAuCl_4$) with tetrakis(hydroxymethyl)phosphonium chloride (THPC) in water (*26*).   The methods described here produce particles 1 – 2 nm in diameter, but particle size will increase considerably with aging of the suspension. Note that all glassware used for all nanoparticle synthesis discussed in this chapter should be cleaned with aqua regia (75% 12 N HCl and 25% 16 N HNO_3 by volume), followed by thorough rinsing with ultrapure water.

Begin by preparing a 1% $HAuCl_4$ (Alfa Aesar, 99.999%) solution by mixing 1 g of $HAuCl_4$ with 99 ml of ultrapure water. This solution should be stored in an amber glass bottle at room temperature and protected from light. In a glass beaker, mix 400 µl of THPC (TCI America) with 33 ml of ultrapure water. Allow this to stir in the beaker for 2 – 3 min at medium speed on a magnetic stir plate. To another glass beaker with 180 ml of ultrapure water at 18 °C, add 4 ml of the dilute THPC solution as well as 1.2 ml of 1 M sodium hydroxide (NaOH). After stirring this second solution for 5 min at medium speed, increase the stirring rate to highest rate possible, while maintaining a steady vortex. Then, quickly add 6.75 ml of 1% $HAuCl_4$ to ensure even mixing. This addition is best facilitated by first placing the $HAuCl_4$ solution in a separate vessel (such as a plastic conical tube) and then inverting the tube into the $THPC/NaOH/H_2O$ mixture all at once and as quickly as possible.

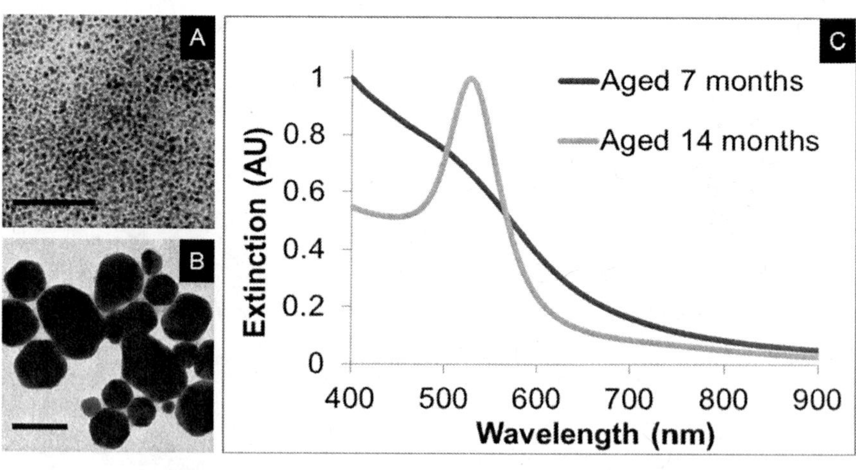

Figure 1. Transmission emission micrographs of THPC gold colloid aged A) 7 months and B) 14 months. C) Associated UV-Vis spectra show a more pronounced peak at 530 nm for the 14 month-old colloid, indicating its larger size. Scale bars = 50 nm.

Allow the final mixture to stir until the color change has stabilized to a light brown color (usually occurs within 1 min). Store the colloid suspension at 4 °C and allow it to age for at least two weeks to achieve ~3 – 5 nm diameter gold particles. At this size, the THPC colloid is appropriate for use in gold–silica nanoshell synthesis in formation of the seed particles as described below. As mentioned, the THPC gold particles will gradually grow larger with aging. One year after synthesis, for example, particles can reach ~40 nm in diameter (Figure 1). If diameters larger than 5 nm are desired sooner after synthesis, the 180 ml volume of ultrapure water and 1% $HAuCl_4$ solution described earlier can be warmed in a water bath or even boiled before mixing.

Gold–Silica Nanoshell Synthesis

Gold–silica nanoshell synthesis is a four-step process that first involves the fabrication of monodisperse silica cores. In the second step, core surfaces are functionalized with positively charged amine groups, which in the third step, serve as adsorption sites for 3 – 5 nm THPC gold colloid. These colloid decorated cores are referred to as nanoshell seed. In the final step, the shell is formed via a gold reduction reaction (*19*).

Amine-Functionalized Silica Core Synthesis

The Stöber method produces monodisperse silica nanoparticles via a basic reduction of tetraethyl orthosilicate (TEOS) in alcohol-based solvents (*27*). The following adaptation yields silica nanospheres 90 – 130 nm in diameter, which is suitable for eventual production of gold nanoshells with NIR extinction. Because size scales with TEOS and ammonium hydroxide (NH_4OH) volumes, more or less of both can be added to achieve other diameters if necessary.

First add 45 ml of 100% ethyl alcohol (EtOH) to 3 glass beakers. Begin stirring at medium speed. To these beakers, add 2.8, 2.9, or 3.0 ml of NH_4OH (Aldrich, 28%). After stirring approximately 1 min, increase the stirring rate. Quickly add 1.5 ml TEOS (Aldrich, 99.999%) to each solution and let stir rapidly for approximately 5 min. Then reduce the stirring rate to medium speed and cover the beakers with parafilm. Allow the reaction to proceed at room temperature for at least 8 h, during which time newly formed silica nanospheres will appear as a white precipitate.

Next, residual NH_4OH must be removed by centrifugation. Split each reaction volume into two 50 ml conical tubes. Centrifuge the silica nanoparticle suspensions for a total of three rounds at 2000g for 30 min each. After each round, carefully decant the supernatant and add back 20 ml of 100% EtOH. Probe sonicate the suspensions at 5 W for 1 min to redisperse the silica nanospheres. The final resuspension should be performed with 10 ml EtOH. If the silica cores do not completely pellet with the suggested speeds and times, additional rounds of centrifugation may be required to achieve transparent or nearly transparent supernatants before decanting.

Now, recombine like silica nanoparticle suspensions and measure the weight percent (wt%) of each according to Equation 1:

$$wt\% = \frac{(dry\ weight\ of\ sample\ volume\ after\ evaporation)}{(wet\ weight\ of\ sample\ volume)} \times 100 \qquad (1)$$

Adjust the silica nanoparticle concentrations to 4 wt% with 100% EtOH. At this point, electron microscopy should be performed to characterize the silica nanoparticle diameters and monodispersity. Only samples with polydispersity values less than 10% should be used for eventual gold nanoshell fabrication. The silica nanoparticle suspensions can be stored at 4 °C until amine functionalization is performed.

To functionalize the silica nanoparticles with amine groups, combine 5 µl of 3-aminopropyl triethoxy silane (APTES, Gelest) for every 1 ml of nanoparticle suspension at 4 wt%. Begin by transferring a desired volume of the silica nanoparticle suspension to a glass beaker and stir at high speed. Add the necessary volume of APTES, cover the beaker with parafilm, and allow the mixture to react overnight at room temperature.

To ensure surface exposure of the amine groups, heat the silica nanoparticle suspensions at ~75 °C for 1 h. Periodically add 100% EtOH during this process to maintain a consistent volume. After an hour, remove the unreacted APTES by further centrifugation (2 rounds at 2000g for 30 min each). Decant the supernatant after each round, add an equivalent volume 100% EtOH, and probe sonicate the samples at 5 W for 1 min to redisperse the pellets. Remeasure the suspension weight percent according to Equation 1 and dilute to 4 wt% with 100% EtOH. Store the amine functionalized silica nanoparticles at 4 °C.

Nanoshell Seed Particle Synthesis

In the third step of gold–silica nanoshell fabrication, gold colloid is adsorbed to the aminated silica nanoparticles to produce what is known as nanoshell seed particles. To one 50 ml conical tube, add 30 ml of THPC colloid, synthesized according to the steps in the section outlined above and aged for at least 2 weeks (3 – 5 nm diameter), followed by 180 µl of aminated silica nanoparticles at 4 wt%. Quickly add 1 ml of 1 M NaCl and vortex at the highest setting for 1 min. Allow this suspension to rock at room temperature overnight. The next day, split the reaction volume into two 50 ml conical tubes and dilute each suspension to 20 ml with ultrapure water. Remove unreacted gold colloid via two rounds of centrifugation at 450g for 15 min. After each round, remove the supernatant, add an equivalent volume of ultrapure water, and probe sonicate at 5 W for 30 sec – 1 min to redisperse the nanoshell seed particles. Next, dilute the nanoshell seed suspensions such that the optical density at 530 nm by UV-Vis spectroscopy equals 0.1. Store the nanoshell seed particle suspension at 4 °C.

Gold Shell Formation

In the final step of gold nanoshell synthesis, gold is reduced on nanoshell seed particle surfaces to form a contiguous shell. First make the gold plating solution by adding 200 ml of ultrapure water to an amber glass bottle followed by 50 mg of potassium carbonate (K_2CO_3) and 4 ml of the 1% $HAuCl_4$ solution, as prepared for gold colloid synthesis. These amounts can be scaled linearly depending on the desired amount of product. Stir the plating solution on a magnetic stir plate until the K_2CO_3 fully dissolves and age the solution 48 – 72 h at room temperature and protected from light prior to use.

Next, determine the optimal ratio of gold plating solution and seed suspension to produce nanoshells with the desired optical properties. Begin by adding 1 ml of the gold plating solution to three 1 ml disposable UV-Vis cuvettes. On the underside of three cuvette caps, add 10 µl of formaldehyde (HCHO, 37%) and place the caps next to the three cuvettes loaded with the gold plating solution. Add 200, 300, or 400 µl of nanoshell seed particle suspension (OD_{530} = 0.1) to each of the three cuvettes. Quickly seal the cuvettes with the HCHO-loaded caps and vortex at high speed for approximately 30 sec. A visible color change should occur (thicker shells will appear more red while thinner shells appear more blue).

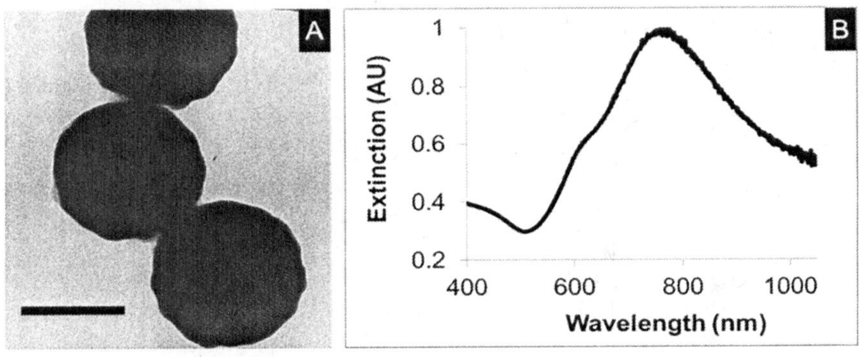

Figure 2. A) Transmission electron micrograph and B) UV-Vis spectra of gold–silica nanoshells with maximum extinction within the NIR window. Scale bar = 100 nm.

Determine the optical properties of each gold nanoshell suspension via UV-Vis spectroscopy. Given the fixed gold plating solution volume and increasing amounts of seed, nanoshells with progressively thinner gold coatings are synthesized. This corresponds to a red shift in the wavelength at which plasmon resonance occurs within the associated UV-Vis spectra. Reattempt the above reaction with varying amounts of seed suspension as needed. Once the appropriate gold plating solution:nanoshell seed volume ratio for desired optical properties is determined, the reaction can be scaled up linearly to produce larger volumes of gold–silica nanoshells. During scale-up reactions, mixing conditions can have profound effects on the gold nanoshell product. Ensure even and thorough mixing by first stirring the gold plating solution and seed particles together 30 sec – 1 min followed by increasing the stirring rate before adding the HCHO. Then mix at high speed until the color change has stabilized (~3 – 5 min).

The residual HCHO should be removed via two rounds of centrifugation as it can alter the gold shells over time and optical properties as a result. Transfer 20 ml of the nanoshell suspension to 50 ml conical tubes and spin for 15 min at 450g. After each round, remove the supernatant and resuspend in 1.8 mM K_2CO_3 in ultrapure water. Probe sonicate the samples at 5 W for 30 sec to redisperse. Store the gold–silica nanoshell suspensions at 4 °C. A sample TEM image and the associated UV-Vis spectra of NIR absorbing gold–silica nanoshells are shown in Figure 2.

Gold–Gold Sulfide Nanoparticle Synthesis

Gold–gold sulfide nanoparticles are formed via a one-step self-assembly reaction upon mixing a sulfur-based precursor (sodium thiosulfate, $Na_2S_2O_3$, or sodium sulfide, Na_2S) and a gold-based precursor ($HAuCl_4$). Depending on the sulfur source used and reagent ratios, particle diameters range 20 – 50 nm with optical extinction attainable within the NIR water window. Early reports first described these particles as nanoshells with a gold shell over a gold-sulfide core, supported by TEM images, electron diffraction data, and optical simulations based on Mie Theory (*28–30*). Yet other investigators maintain that no true core-shell structure arises during synthesis and that NIR extinction is a result of gold particle aggregates (*31–33*). Though this debate remains unsettled, the gold surface layer and NIR absorption of these nanoparticles still afford surface functionalization and use in photothermal ablation (*24, 25*). Below, we describe methods to achieve NIR resonant gold–gold sulfide particles ~20 – 25 nm in diameter, using $Na_2S_2O_3$ as the sulfur-based precursor.

First prepare the precursor solutions of 2 mM $HAuCl_4$ and 1 mM $Na_2S_2O_3$. To make 400 ml solutions, add 26.9 ml 1% $HAuCl_4$, as prepared above for THPC gold colloid, to 373.1 ml of ultrapure water for the gold-based precursor and 99.27 mg $Na_2S_2O_3$ to 400 ml of ultrapure water for the sulfur-based precursor. Store each solution in amber bottles at room temperature and protected from light. The 1 mM $Na_2S_2O_3$ solution must be aged at least 48 h at room temperature before use.

As performed with the gold–silica nanoshells, small scale reactions can be conducted first to determine optimal reagent ratios for achieving gold–gold sulfide nanoparticles with the desired optical properties. To synthesize particles with NIR absorption, begin by mixing 1 mM $Na_2S_2O_3$ with 2 mM $HAuCl_4$ at volumetric ratios ranging 1:0.8 to 1:1.2 in 1 ml disposable UV-Vis cuvettes and vortex for 30 sec. Monitor the reaction by recording the associated UV-Vis spectra of each reaction mixture every 15 min over 1.5 h. The reaction should run to completion during this time, after which the spectra should not change. Repeat the above procedure with different reagent ratios as needed to acquire gold–gold sulfide particles with the desired optical properties. After determining this ratio, the reaction can be scaled linearly to produce larger volumes.

Centrifuging is next required to remove small, gold colloid that is also produced during this reaction. Split the reaction product into 20 ml volumes in 50 ml conical tubes and centrifuge for a total of three rounds for 30 min each at 1900g. After each round, remove the supernatant and resuspend the pellet in an equivalent amount of ultrapure water. Probe sonicate the suspensions at 5 W for 30 sec to redisperse. Monitor the removal of gold colloid via UV-Vis spectroscopy, which should show an increase in the ratio of peak extinction values of gold–gold sulfide nanoparticles to gold colloid. Store the final, purified product at 4 °C. A sample TEM image and the associated UV-Vis spectra of NIR absorbing gold–gold sulfide nanoparticles are shown in Figure 3.

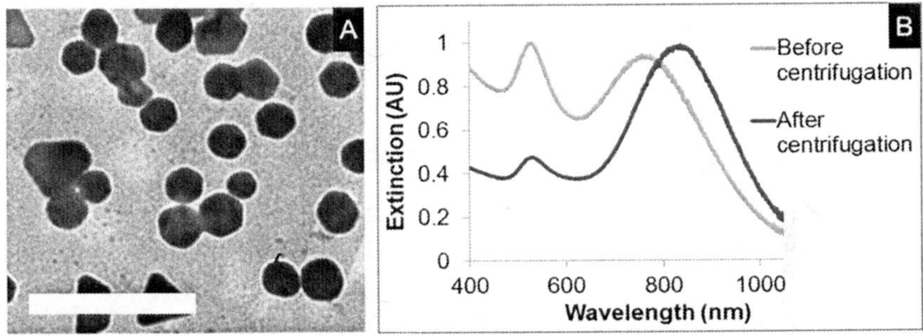

Figure 3. A) Transmission electron micrograph and B) UV-Vis spectra of gold–gold sulfide nanoparticles with maximum extinction within the NIR window. Removal of small, gold colloid that also forms during synthesis is achieved via centrifugation, as indicated by decrease in the extinction peak at 530 nm. Scale bar = 100 nm.

Gold Nanorod Synthesis

Gold nanorods can be fabricated via seed-mediated growth, using the surfactant cetyltrimethylammonium bromide (CTAB) and silver nitrate ($AgNO_3$) to achieve particles with high monodispersity and controlled aspect ratios in aqueous solutions (*17, 34, 35*). In the first step, a seed suspension is made using gold salt, CTAB, and sodium borohydride ($NaBH_4$). This suspension is then added to a growth solution containing CTAB, gold salt, ascorbic acid, and varying amounts of $AgNO_3$ to yield the final product.

Gold Seed Fabrication

Immediately before nanorod synthesis, prepare the precursor solutions of 1 mM $HAuCl_4$ and 200 mM CTAB (Sigma Aldrich) in ultrapure water. The 1 mM $HAuCl_4$ solution can be made by diluting the 1% (29.7 mM) $HAuCl_4$ stock solution as prepared above and will also be used in the growth solution as described below. For seed fabrication, dilute the 1 mM solution 1:2 to prepare a final concentration of 0.5 mM HAuCl$_4$. Add 5 ml of 0.5 mM HAuCl$_4$ to 5 ml of 200 mM CTAB in a glass beaker. Stir at medium speed on a magnetic stir plate for 1 min. The product should appear bright yellow-brown in color. Increase the stirring rate. Quickly add 600 μl of 10 mM NaBH$_4$ (Acros Organics) in ultrapure water that has been pre-chilled to 2 – 8 °C. Thoroughly mix the suspension by rapid stirring for 2 min. The obtained seed suspension should appear pale yellow-brown in color, corresponding to the formation of gold particles less than 5 nm in size. This suspension is stable for a period up to 1 week at room temperature.

Gold Nanorod Growth

To determine the appropriate reagent ratios that produce gold nanorods with desired optical properties, prepare growth solutions containing CTAB, HAuCl$_4$, and ascorbic acid (Fisher Scientific) with varying amounts of 4 mM AgNO$_3$ (Acros Organics) in ultrapure water. First add 5 ml of 1 mM HAuCl$_4$ to 5 ml of 200 mM CTAB in five glass beakers and stir for 30 sec on a magnetic stir plate at medium speed. Next add 50, 100, 150, 200, or 250 μl of 4 mM AgNO$_3$ and stir for an additional 1 min. To these five beakers, add 70 μl of 78.8 mM ascorbic acid and stir again for an additional 1 min. The growth solution should progress from yellow-brown to colorless. Finally, add 12 μl of the seed suspension as prepared in the previous section. Continue mixing for 1 min and then allow the suspensions to sit for 3 h. Color changes that occur as the nanorods form should stabilize within this time period.

Acquire UV-Vis spectra for each of the nanorod suspensions. Nanorods produced with greater amounts of AgNO$_3$ should exhibit greater aspect ratios, corresponding to a red shift in the longitudinal plasmon resonance (Figure 4). Determine the appropriate reagent ratios that yield gold nanorods with the desired optical properties. As with other gold particles described in this chapter, reagent volumes can be scaled linearly to produce larger amounts of nanorods.

Next concentrate particles as needed and remove residual CTAB via two rounds of centrifugation at 3300g for 20 min. Remove the supernatant after each round and resuspend in ultrapure water followed by probe sonication at 5 W for 1 min. Gold nanorod suspensions can be stored at 4 °C until further use.

Protein Conjugation and PEG Passivation

Proteins can be conjugated to gold nanoparticle surfaces by employing a heterobifunctional poly(ethylene glycol) (PEG) linker containing an n-hydroxysuccinimide ester (NHS) group for coupling to amines as well as an ortho-pyridyl disulfide (OPSS) group for attachment to gold. Although tethered to nanoparticle surfaces, the protein is still mobile with the flexible PEG linker in place, facilitating ligand binding. Remaining gold surface area should then be backfilled with PEG-SH to promote particle stability and prevent nonspecific protein adsorption. The following procedures should be performed on ice, unless otherwise noted.

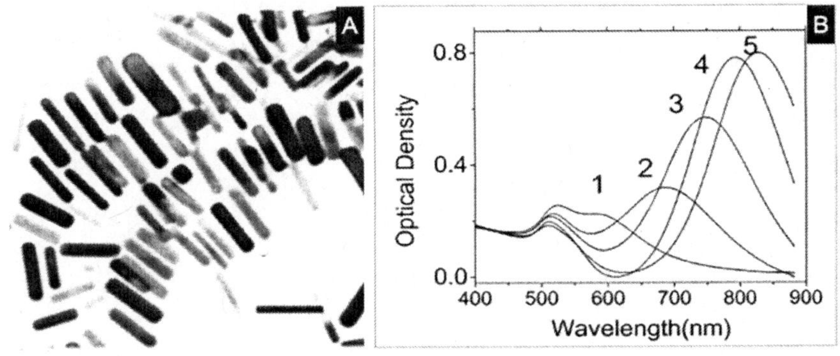

Figure 4. A) Transmission electron micrograph and B) UV-Vis spectra of gold nanorods. Numbers (1 – 5) correspond to increasing amounts of AgNO$_3$ as described in the gold nanorod growth section. Scale bar = 50 nm. (Reprinted with permission from Nikoobakht, B.; El-Sayed, M. A. Chem. Mater. 2003, 15, 1957–1962. Copyright 2003 American Chemical Society) (17).

Protein Conjugation to OPSS-PEG-NHS

Dissolve the targeting protein of interest in a volume of 100 mM NaHCO$_3$ at pH 8.5 to achieve a final protein concentration of 10 – 20 μM. Dissolve OPSS-PEG-NHS (MW = 2 kDa, Creative PEG Works) in 100 mM NaHCO$_3$ at pH 8.5 to achieve a polymer concentration of 25 μM. After dissolving the OPSS-PEG-NHS, work quickly as the hydrolysis half-life of NHS ester groups is on the order of ~10 min at this pH and 4 °C (36). Add the necessary volume of 25 μM OPSS-PEG-NHS solution to the protein solution to react at a 10:1 PEG:protein molar ratio. Vortex the PEG-protein mixture immediately for 1 min, followed by rocking at 4 °C for at least 2 h. Store the OPSS-PEG-protein at -80 °C or proceed with conjugation to gold nanoparticles as outlined below.

Note that the optimal PEG:protein molar ratio will depend upon the number of surface accessible amines on the particular protein. Fewer PEG chains per protein will be needed for proteins with high amine content, which can be determined by the ninhydrin assay if necessary. Additionally, excessive numbers of PEG chains per protein may hinder bioactivity and interfere with its targeting capability. Small scale optimization reactions may be required to determine the minimum number of PEG chains required to achieve efficient protein conjugation. Successful OPSS-PEG-NHS conjugation to proteins can be evaluated using SDS-PAGE (sodium dodecyl sulfate polyacrylamide gel electrophoresis) followed by silver staining (Silver Stain Plus Kit, Bio Rad Laboratories) for the protein. PEG-protein conjugates will appear as a smear of bands of higher molecular weights than the protein alone (because of the polydispersity of PEG products, more bands will also be present with higher PEG:protein ratios).

Nanoparticle Conjugation to Protein and PEG Passivation

Thaw an aliquot of OPSS-PEG-protein on ice. To one 20 ml glass scintillation vial, add 9 ml of gold nanoparticles with a peak optical extinction of 1.5 in ultrapure water. Add the necessary volume of OPSS-PEG-protein to the gold nanoparticle suspension at one of the following protein:nanoparticle ratios: 1500:1 for gold–silica nanoshells, 150:1 for gold–gold sulfide nanoparticles and gold nanorods, and 50:1 for THPC gold colloid. Vortex the nanoparticle-protein mixture for 30 sec, followed by rocking for 1 h at 4 °C.

Next, dissolve PEG-SH (MW = 5 kDa, Laysan Bio) in prechilled ultrapure water and bring to a final concentration of 25 µM if working with the larger gold–silica particles or 250 µM if using the smaller gold–gold sulfide, THPC gold particles, or gold nanorods. Vortex the solution to ensure complete dissolution. Add 1 ml of the PEG-SH solution to the nanoparticle-protein suspension. Vortex the suspension for 30 sec to ensure adequate mixing and then rock the mixture at 4 °C overnight.

Unbound protein and PEG chains can be removed via centrifugation. After removing the supernatant, resuspend the particles in ultrapure water or solvent of choice (phosphate buffered saline, cell culture media, etc.). If storage is required, it is recommended that the particle conjugates be kept at 4 °C, although shelf life will depend on protein stability in the particular solvent.

Gold Nanoparticle ELISA

A variety of techniques exist to characterize nanoparticles, both unfunctionalized and functionalized. After synthesis, gold nanoparticle size and polydispersity should be determined with electron microscopy. UV-Vis spectroscopy can also be used, as mentioned, to assess optical properties both before and after functionalization, as this can help to determine if significant particle aggregation has occurred. Polymer functionalized particles, such as those described herein, will demonstrate a slight red-shift in the wavelength at which plasmon resonance occurs. Additionally, dynamic light scattering (DLS) and zeta potential characterization are excellent evaluation methods and show differences in hydrodynamic diameter and surface charge post functionalization. Below we describe a process similar to an enzyme-linked immunosorbent assay (ELISA) to determine the amount of targeting proteins displayed on nanoparticle surfaces. The methods below describe an indirect ELISA, employing both a primary and secondary antibody-horseradish peroxidase (HRP) conjugate, although a direct ELISA with only an antibody-HRP could also be performed.

Begin by coating six 2 ml centrifuge tubes with 3 wt% bovine serum albumin (BSA) in phosphate buffered saline (PBS), hereafter referred to as PBSA, by rocking at room temperature for 1 h. Next prepare 1.5 ml of gold nanoparticles, functionalized with the targeting protein of choice and backfilled with PEG-SH as outlined above (referred to as "Group 1" henceforth). Particles conjugated to PEG-SH only are recommended as a control (referred to as "Group 2" henceforth). Ensure that the peak optical extinction values are at least 2 and equal for Groups 1 and 2.

Remove the PBSA and add 450 μl of the nanoparticle suspension to each of the three tubes for Groups 1 and 2 to run the assay in triplicate. Thaw the primary antibody on ice and prepare at least 310 μl in PBSA at 100 μg / ml. Add 50 μl of the primary antibody at this concentration to each centrifuge tube of nanoparticles to give a final antibody concentration of 10 μg / ml. Rock all tubes overnight at 4 °C.

Centrifuge the tubes for a total of three rounds to remove unreacted primary antibody. After the first and second round, remove 400 μl of supernatant and add 400 μl of fresh PBSA. After the third round of centrifugation, remove 400 μl of supernatant and add 350 μl of fresh PBSA to give a final volume of 450 μl / tube. Thaw the antibody-horseradish peroxidase conjugate (directed against the primary antibody for an indirect ELISA or against the targeting protein on nanoparticles for a direct ELISA) on ice. Prepare at least 320 μl of antibody-peroxidase conjugate in PBSA at 200 μg / ml. Add 50 μl of the antibody at this concentration to each centrifuge tube of nanoparticle suspension to give a final antibody concentration of 20 μg / ml. Rock all tubes at room temperature for 1 h.

Centrifuge the tubes for a total of three rounds to remove unreacted secondary antibody. After each round, remove 400 μl of supernatant and add 400 μl of fresh PBSA. Upon completion of these washing steps, remove 100 μl from each 2 ml tube and place the particle suspensions in a separate set of tubes (referred to as "nanoparticle samples" henceforth). Next remove another 200 μl from each 2 ml tube and place into labeled 500 μl UV-Vis microcuvettes to be used in nanoparticle quantification discussed later (referred to as "quantification samples" henceforth). Centrifuge the remaining contents of the six 2 ml tubes to completely pellet remaining nanoparticles. Afterwards, remove 100 μl of the supernatant and place it into another set of tubes (referred to as "supernatant samples" henceforth). Take extra care not to disturb the pellet during this step.

Preliminary Enzyme Assay

Now perform the preliminary enzyme assay on one set of the replicates to determine the optimal concentration to run the entire assay across all replicates. Make three 10X serial dilutions (1:10, 1:100, and 1:1000) of this replicate for both the nanoparticle and supernatant samples from Groups 1 and 2. To do this, remove 50 μl and add to 450 μl of PBS in separate tubes labeled according to the dilutions made. Remove 70 μl from each of the three serial dilutions in PBS and transfer to another set of tubes.

Next prepare the phosphate-citrate buffer (Sigma Alrich) by adding one buffer tablet to 100 ml of deionized water according to manufacturer's instructions. Mix the buffer until completely dissolved. This solution must be used fresh, within 30 min of preparation. Add one 3,3′,5,5′-tetramethylbenzidine dihydrochloride tablet (TMB, Sigma Aldrich) to 10 ml of phosphate-citrate buffer (according to manufacturer's instructions) in a foil wrapped conical tube. Vortex at high speed to completely dissolve.

Then add 700 µl of the TMB solution to each of the six tubes containing 70 µl of the three 10X serial dilutions as described above. Vortex each solution to mix followed by rocking at room temperature for 15 min. HRP catalyzed hydrolysis of the TMB substrate is indicated by a blue color change. Stop the reaction by addition of 200 µl 2 M H_2SO_4 to each of the tubes with the TMB solution. Any blue color seen previously should change to yellow upon addition of the acid. Centrifuge each tube to pellet any nanoparticles in the solution that may interfere with absorbance readings later.

Transfer 200 µl of each sample/TMB/acid mixture to designated wells within a 96-well plate. Measure the absorbance at 450 nm and note which serial dilution gives readings above background but below saturation. This dilution will be used for the remainder of the assay.

Full Enzyme Assay and Nanoparticle Quantification

Next prepare the remaining replicates from Groups 1 and 2 of the nanoparticle and supernatant samples at the optimal dilution as determined in the preliminary enzyme assay. Also prepare standard solutions of the antibody-peroxidase conjugate by making dilutions of the original 200 µg / ml solution in PBSA. First, add 10 µl of the original solution to 9.99 ml of PBS to give a final concentration of 0.2 µg / ml (standard 1). Vortex to mix thoroughly. Next perform eight 2X serial dilutions by removing 500 µl and adding to 500 µl PBS, vortexing after each dilution to mix (standards 2 – 9). The last standard solution should be pure PBS (standard 10).

For all samples prepared at the optimal dilution as well as standards 1 – 10, remove 70 µl and place in separate microcentrifuge tubes. Perform the full enzyme assay with each of the 70 µl-loaded tubes as described above in the preliminary enzyme assay, beginning with the preparation of the phosphate-citrate buffer and ending with absorbance readings at 450 nm.

Finally, add 300 µl of ultrapure water to the cuvettes containing the quantification samples mentioned above. Probe sonicate each sample at 5 W for 10 sec to fully redisperse particles. Acquire UV-Vis spectra and record the peak optical extinction. Now calculate the degree of protein conjugation to nanoparticle surfaces. Plot a standard curve, taking care to account for all dilutions, and interpolate to determine the secondary antibody concentrations in samples from Groups 1 and 2. Again, take care to account for all dilutions made (including the 10X serial dilutions). By subtracting the antibody amounts in the supernatant samples from those in the nanoparticle samples, one can then determine the amount of targeting protein in the nanoparticle samples by approximating a 1:1 ratio between targeting protein and antibody-peroxidase conjugate. Using Mie Theory, Beer's Law, and the optical extinction values recorded for the quantification samples, calculate nanoparticle concentrations for each of the samples and then finally the number of targeting proteins per nanoparticle.

Cell Targeting

Before proceeding with *in vitro* targeting studies with protein conjugated nanoparticles, verifying the high expression of target cell surface receptors on a positive control cell line and low expression on a negative control cell line with immunohistochemistry is recommended. For example, the tyrosine kinase receptor EphA2 is overexpressed in a variety of cancer types and by as much as 100 fold in metastatic prostate cancer as compared to non-invasive prostate epithelial cells (*37, 38*). Therefore, EphA2 represents an interesting target for a variety of nanoparticle-based diagnostic and therapeutic platforms by using the receptor's ligand ephrinA1. Two cell types of interest for testing these platforms include EphA2+ PC3 and EphA2– HDF cell lines (*22*). Below we describe a silver staining method for determining the efficacy of gold nanoparticle targeting to a particular cell type.

Begin by preparing a 24-well tissue culture plate with cell lines positive and negative for the targeted receptor of interest, at or near confluency. For this assay, targeted nanoparticles will be incubated with both cell types. Suggested controls include particles conjugated to PEG only as well as cell-only samples, not treated with either nanoparticle type.

Next prepare 4.5 ml of protein and PEG only conjugated nanoparticle suspensions as described above at a desired concentration in a physiologic solvent of choice (PBS, cell culture media, etc.). Appropriate nanoparticle concentrations will require some optimization, depending on the level of receptor or target expression. To start, varying peak extinction values between 2 and 5 is recommended for this *in vitro* assay. These suspensions can also be filtered with 0.45 or 0.80 μm low binding filter membranes if further purity is required.

Aspirate the media from the cells. Add 1 ml of nanoparticle suspension to designated wells and 1 ml of solvent to cell-only controls. Return the well plate to the tissue culture incubator and incubate the particles with cells long enough to allow for sufficient settling time (1 – 2 h for ~150 nm gold–silica nanoshells and several hours or overnight for the smaller THPC gold colloid, gold–gold sulfide nanoparticles, and gold nanorods). After the incubation period, aspirate the nanoparticle suspensions and rinse the cells three times with 2 ml PBS per well. Fix the cells with 2.5% glutaraldehyde in PBS. Add 1 ml per well, incubate at room temperature for 20 min, and then rinse the cells two times with 1 ml PBS per well. Note that as an alternative to incubating nanoparticles with adherent cells, trypsinized cells can be hybridized with particles for 30 min at 37 °C, rinsed via 3 rounds of centrifugation and PBS exchanges, and then seeded onto tissue culture plastic. Cells should be allowed to adhere for one overnight period before assaying nanoparticle targeting.

Next prepare the silver enhancement solution (Sigma Aldrich) per manufacturer's instructions. Add 1 ml of silver solution per well of cells. Monitor the silver deposition under phase contrast microscopy. Approximately 30 min is typically sufficient development time. Note that silver precipitation can begin to stain proteins on cells nonspecifically if development is allowed to occur for an excessive amount of time. When imaging cells during the development process, pay close attention to the cell-only control to limit nonspecific staining as much as possible. Stop the reaction by rinsing cells three times with 1 ml deionized water. After the final rinse, add 1 ml PBS to each well. Image samples with phase contrast microscopy.

Targeted nanoparticles should provide enhanced contrast when incubated with cells positive for the targeted receptor of interest. Targeted nanoparticles incubated with the negative control cell line as well as PEG only conjugated particles should resist cell binding, wash away during the PBS rinses, and provide little to no contrast after silver staining (Figure 5). If available, dark field microscopy is another tool for evaluating binding of some nanoparticle types to cells. In this case, a similar incubation and rinsing protocol can be followed. Cells should be seeded on coverglass instead of well plates and then coverslipped before imaging. No silver stain is required.

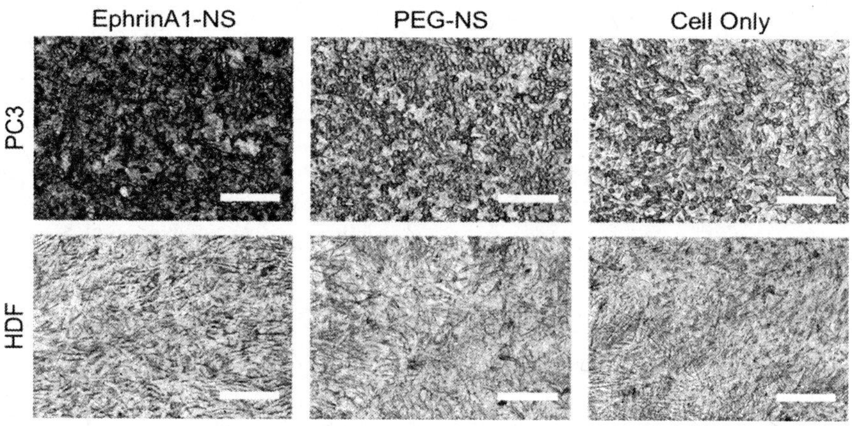

Figure 5. Phase contrast micrographs of silver stained (top row) PC3 prostate cancer cells and (bottom row) human dermal fibroblasts or HDF cells after incubation with gold–silica nanoshells (NS) conjugated to ephrinA1 or PEG only. EphrinA1-NS show robust binding to PC3 cells via the EphA2 receptor. Because HDF cells express very low levels of EphA2, ephrinA1-NS offer little to no contrast. PEG-NS demonstrate little to no binding as well for both cell types. Cell only controls show low background staining. Scale bars = 100 μm.

Conclusions

Nanotechnology poses a powerful tool for the diagnosis and treatment of human cancer. With their established biocompatibility, unique optical properties, and ease of surface functionalization, gold nanoparticles in particular show great promise as contrast agents in molecular cancer imaging and absorptive agents in photothermal ablation. Gold colloid, gold–silica nanoshells, gold–gold sulfide nanoparticles, and gold nanorods have all been employed successfully in these applications. Moreover, targeting these particles with proteins to cancer cell-surface markers has demonstrated enhanced specificity of imaging and therapy *in vitro* thus far, warranting continued study of these protein-particle conjugates *in vivo*.

Acknowledgments

This research is funded in part by the National Science Foundation Center for Biological and Environmental Nanotechnology (NSF CBEN) at Rice University. Andrew Coughlin is also supported by a National Science Foundation Graduate Research Fellowship (NSF GRF, award number 0940902) as well as a National Defense Science and Engineering Graduate (NDSEG) Fellowship through the Air Force Office of Scientific Research (AFOSR). The authors would like to thank Dr. Melissa McHale for assistance with editing.

References

1. American Cancer Society Cancer Facts & Figures 2012.
2. American Cancer Society Global Cancer Facts & Figures, 2nd ed., 2008.
3. Connor, E. E.; Mwamuka, J.; Gole, A.; Murphy, C. J.; Wyatt, M. D. *Small* **2005**, *1*, 325–327.
4. Shukla, R.; Bansal, V.; Chaudhary, M.; Basu, A.; Bhonde, R. R.; Sastry, M. *Langmuir* **2005**, *21*, 10644–10654.
5. von Maltzahn, G.; Park, J.-H.; Agrawal, A.; Bandaru, N. K.; Das, S. K.; Sailor, M. J.; Bhatia, S. N. *Cancer Res.* **2009**, *69*, 3892–3900.
6. Dickerson, E. B.; Dreaden, E. C.; Huang, X.; El-Sayed, I. H.; Chu, H.; Pushpanketh, S.; McDonald, J. F.; El-Sayed, M. A. *Cancer Lett.* **2008**, *269*, 57–66.
7. Chen, J.; Wang, D.; Xi, J.; Au, L.; Siekkinen, A.; Warsen, A.; Li, Z.-Y.; Zhang, H.; Xia, Y.; Li, X. *Nano Lett.* **2007**, *7*, 1318–1322.
8. Skrabalak, S. E.; Chen, J.; Sun, Y.; Lu, X.; Au, L.; Cobley, C. M.; Xia, Y. *Acc. Chem. Res.* **2008**, *41*, 1587–1595.
9. Huang, X.; Qian, W.; El-Sayed, I. H.; El-Sayed, M. A. *Lasers Surg. Med.* **2007**, *39*, 747–753.
10. Javier, D. J.; Nitin, N.; Levy, M.; Ellington, A.; Richards-Kortum, R. *Bioconjugate Chem.* **2008**, *19*, 1309–1312.
11. Gobin, A. M.; Lee, M. H.; Halas, N. J.; James, W. D.; Drezek, R. A.; West, J. L. *Nano Lett.* **2007**, *7*, 1929–1934.
12. Hirsch, L. R.; Stafford, R. J.; Bankson, J. A.; Sershen, S. R.; Rivera, B.; Price, R. E.; Hazle, J. D.; Halas, N. J.; West, J. L. *Proc. Natl. Acad. Sci. U.S.A.* **2003**, *100*, 13549–13554.
13. Huang, X.; El-Sayed, I. H.; Qian, W.; El-Sayed, M. A. *J. Am. Chem. Soc.* **2006**, *128*, 2115–2120.
14. Loo, C.; Lowery, A.; Halas, N.; West, J.; Drezek, R. *Nano Lett.* **2005**, *5*, 709–711.

15. O'Neal, D. P.; Hirsch, L. R.; Halas, N. J.; Payne, J. D.; West, J. L. *Cancer Lett.* **2004**, *209*, 171–176.

16. Weissleder, R. *Nat. Biotechnol.* **2001**, *19*, 316–317.

17. Nikoobakht, B.; El-Sayed, M. A. *Chem. Mater.* **2003**, *15*, 1957–1962.

18. Chen, J.; Wiley, B.; Li, Z. -Y; Campbell, D.; Saeki, F.; Cang, H.; Au, L.; Lee, J.; Li, X.; Xia, Y. *Adv. Mater.* **2005**, *17*, 2255–2261.

19. Oldenburg, S. J.; Averitt, R. D.; Westcott, S. L.; Halas, N. J. *Chem. Phys. Lett.* **1998**, *288*, 243–247.

20. Choi, J.; Yang, J.; Park, J.; Kim, E.; Suh, J.; Huh, Y.; Haam, S. *Adv. Funct. Mater.* **2011**, *21*, 1082–1088.

21. Qian, X.; Peng, X.-H.; Ansari, D. O.; Yin-Goen, Q.; Chen, G. Z.; Shin, D. M.; Yang, L.; Young, A. N.; Wang, M. D.; Nie, S. *Nat. Biotechnol.* **2008**, *26*, 83–90.

22. Gobin, A. M.; Moon, J. J.; West, J. L. *Int. J. Nanomed.* **2008**, *3*, 351–358.

23. Li, X.; Zhou, H.; Yang, L.; Du, G.; Pai-Panandiker, A. S.; Huang, X.; Yan, B. *Biomaterials* **2011**, *32*, 2540–2545.

24. Day, E. S.; Bickford, L. R.; Slater, J. H.; Riggall, N. S.; Drezek, R. A.; West, J. L. *Int. J. Nanomed.* **2010**, *5*, 445–454.

25. Gobin, A. M.; Watkins, E. M.; Quevedo, E.; Colvin, V. L.; West, J. L. *Small* **2010**, *6*, 745–752.

26. Duff, D. G.; Baiker, A.; Edwards, P. P. *Langmuir* **1993**, *9*, 2301–2309.

27. Stöber, W.; Fink, A.; Bohn, E. *J. Colloid Interface Sci.* **1968**, *26*, 62–69.

28. Zhou, H. S.; Honma, I.; Komiyama, H.; Haus, J. W. *Phys. Rev. B: Condens. Matter* **1994**, *50*, 12052–12056.

29. Averitt, R. D.; Sarkar, D.; Halas, N. J. *Phys. Rev. Lett.* **1997**, *78*, 4217–4220.

30. Averitt, R. D.; Westcott, S. L.; Halas, N. J. *J. Opt. Soc. Am. B* **1999**, *16*, 1824–1832.

31. Norman, T. J.; Grant, C. D.; Magana, D.; Zhang, J. Z.; Liu, J.; Cao, D.; Bridges, F.; Van Buuren, A. *J. Phys. Chem. B* **2002**, *106*, 7005–7012.

32. Zhang, J. Z.; Schwartzberg, A. M.; Norman, T., Jr.; Grant, C. D.; Liu, J.; Bridges, F.; Van Buuren, T.; Raschke, G.; Brogl, S.; Susha, A. S.; Rogach, A. L.; Klar, T. A.; Feldmann, J.; Fieres, B.; Petkov, N.; Bein, T.; Nichtl, A.; Kürzinger, K. *Nano Lett.* **2005**, *5*, 809–812.

33. Schwartzberg, A. M.; Grant, C. D.; Van Buuren, T.; Zhang, J. Z. *J. Phys. Chem. C* **2007**, *111*, 8892–8901.

34. Sau, T. K.; Murphy, C. J. *Langmuir* **2004**, *20*, 6414–6420.

35. Rostro-Kohanloo, B. C.; Bickford, L. R.; Payne, C. M.; Day, E. S.; Anderson, L. J. E.; Zhong, M.; Lee, S.; Mayer, K. M.; Zal, T.; Adam, L.; Dinney, C. P. N.; Drezek, R. A.; West, J. L.; Hafner, J. H. *Nanotechnology* **2009**, *20*, 434005.

36. Hermanson, G. T. *Bioconjugate Techniques*, 2nd ed.; Academic Press: New York, 2008.

37. Walker-Daniels, J.; Coffman, K.; Azimi, M.; Rhim, J. S.; Bostwick, D. G.; Snyder, P.; Kerns, B. J.; Waters, D. J.; Kinch, M. S. *Prostate* **1999**, *41*, 275–280.

38. Surawska, H.; Ma, P.; Salgia, R. *Cytokine Growth Factor Rev.* **2004**, *15*, 419–433.

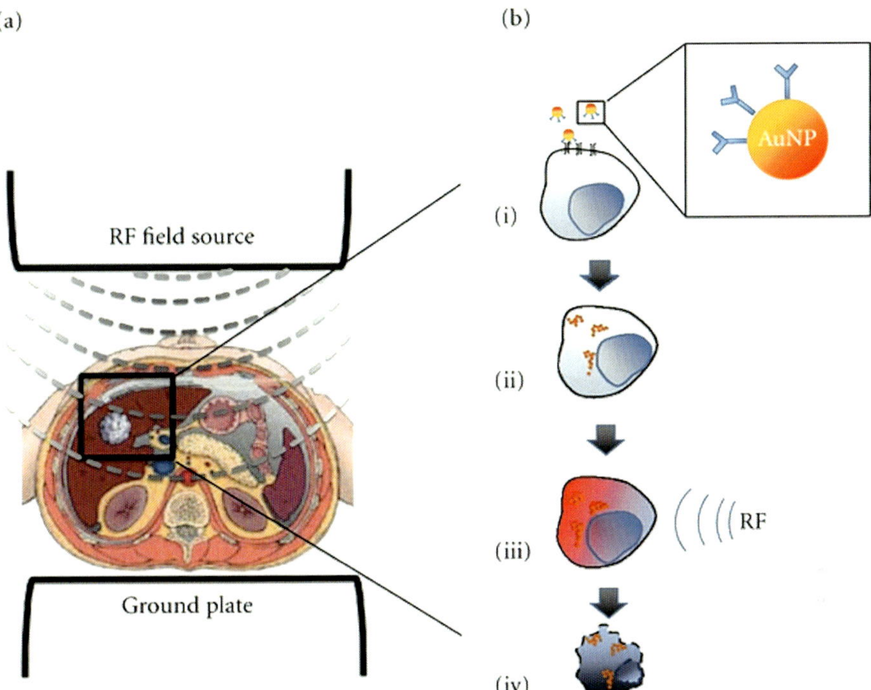

Figure 1. Principle of the noninvasive RF based treatment of hepatocellular carcinoma (HCC). A. RF field source is used to generate a 13.56 MHz electromagnetic field that penetrates tissues and reaches the tumor. B. Nanoparticles that can be thermally activated are conjugated to monoclonal antibodies against known targets expressed on HCC (i), internalized specifically by cancer cells after systemic administration (ii), and upon RF activation release heat (iii) causing targeted cancer cell death. (Chapter 6)

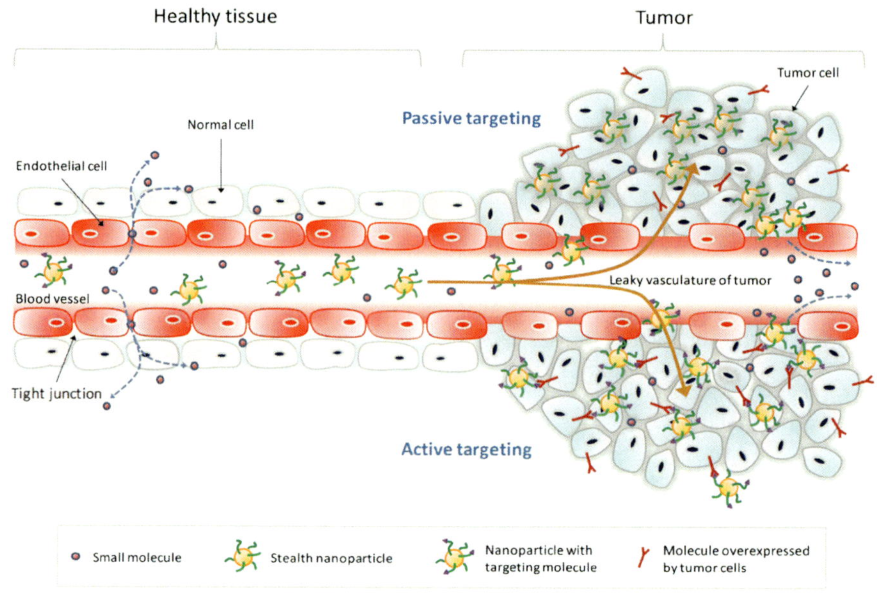

Figure 1. Schematic representation of blood vessels irrigating healthy tissue (left) and tumor (right). (Chapter 7)

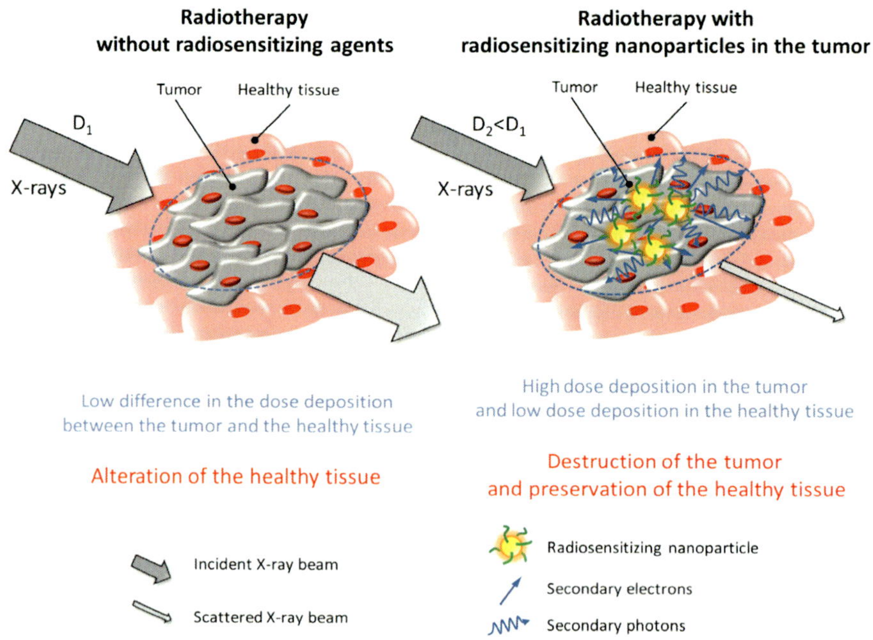

Figure 2. Schematic principle of radiotherapy with and without radiosensitizing nanoparticles. (Chapter 7)

Molecular contrast agents

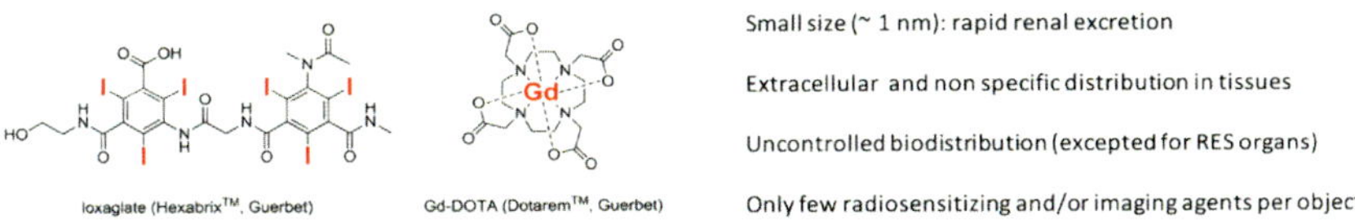

Multifunctional nanoparticles

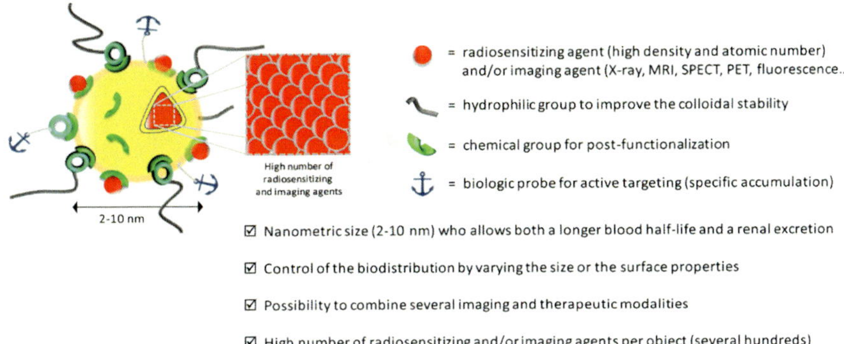

Figure 3. Schematic representation and characteristics of commercially available molecular agents and multifunctional nanoparticles. (Chapter 7)

Organs / Nanoparticles properties	Hydrodynamic Diameter (HD)	Zeta Potential (ζ pot.)	Surface properties and functionalization	
Kidneys / Renal clearance	HD < 6–10nm	ζ pot. >0 fast, 0 mV, <0 slow	Hydrophilic polymers grafting increases body half life and decrease renal clearance	
Liver and spleen / Hepatic filtration SRE uptake	Hepatic filtration if 10nm<HD<100nm; Hepatic uptake if HD > 100nm; Splenic uptake if HD > 200nm	Increased RES uptake; Decreased RES uptake	Hydrophilic polymers grafting increases body half life and decrease RES uptake	Hydrophobic surface increase RES uptake and thus decrease body half life
Tumor / Tumoral retention	10nm–400nm EPR effect	Decreased EPR effect probability due to RES uptake; Increased EPR effect probability due to extended circulation time	Hydrophilic polymers grafting increase body half life and increase EPR effect probability — Passive targeting	Some biomolecules allow specific recognition — Active targeting

Figure 4. Relationship between nanoparticles properties and their biodistribution. (Chapter 7)

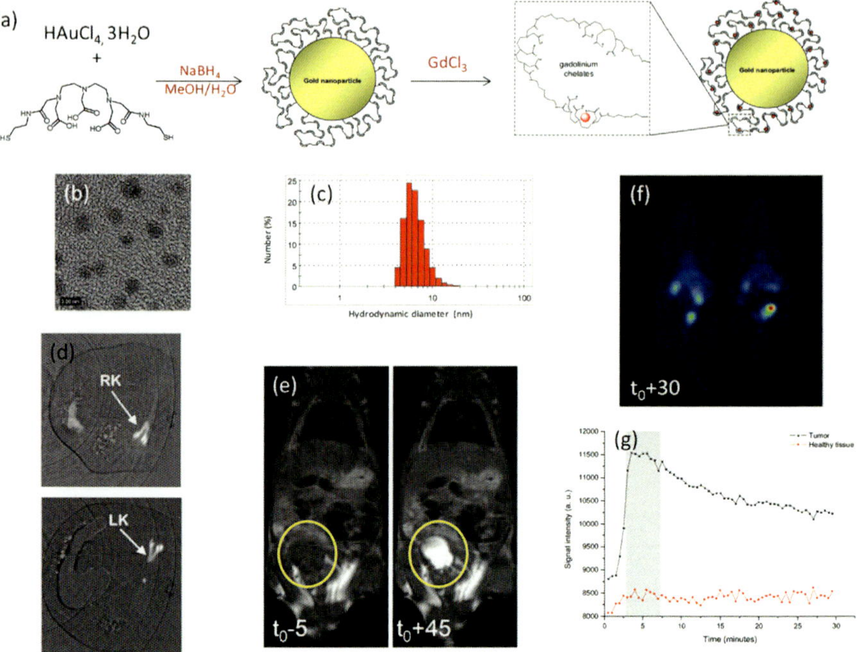

Figure 5. (a) Synthesis of Au@DTDTPA-Gd$_{50}$ nanoparticles. (b) TEM micrograph of Au@DTDTPA nanoparticles. (c) size distribustion of Au@DTDTPA nanoparticles by photon correlation spectroscopy. (d) contribution of the gold core to X-ray imaging (synchrotron radiation computed tomography) (RK and LK for right and left kidney, respectively). Contribution of gadolinium and 99m-technetium chelates to imaging: (e) T$_1$-weighted images before (t$_0$-5) and 45 after (t$_0$+45) intravenous injection of Au@DTDTPA-Gd$_{50}$ nanoparticles to a mouse, (f) γ-scintigraphy image acquired 30 minutes after intravenous injection of Au@DTDTPA-^{99m}Tc nanoparticles to rats and (g) temporal evolution of the MRI signal in tumor and in an equivalent surface in normal tissue in the left hemisphere after injection of Au@DTDTPA-Gd$_{50}$ nanoparticles to gliosarcoma bearing rat. (Chapter 7)

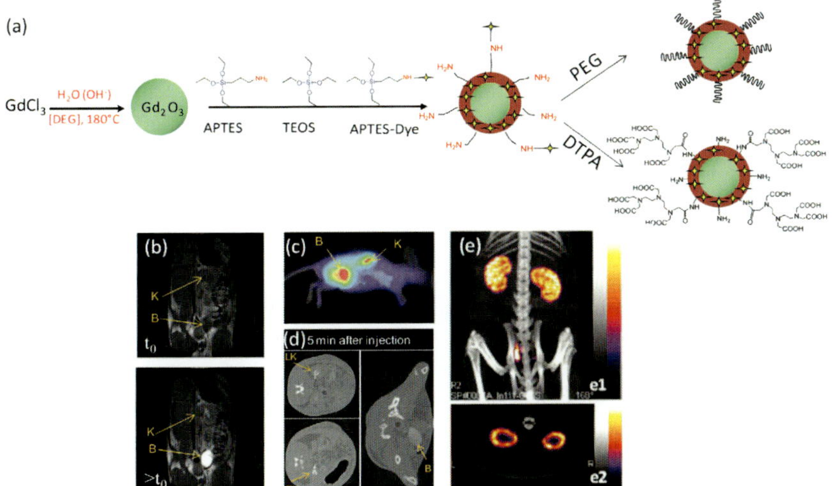

Figure 6. (a) Synthesis of GadoSiPEG and GadoSiDTPA nanoparticles. (b) T_1-weighted images before (t_0) and 32 minutes after ($>t_0$) intravenous injection of GadoSiDTPA nanoparticles to a rat. (c) Fluorescence reflectance imaging of a nude mouse 3 hours after the injection of GadoSiPEG. (d) SRCT images of a series of transverse slices including the right and left kidneys and the bladder (B) of a 9L gliosarcoma-bearing rat acquired 5 min after the intravenous injection of GadoSiDTPA nanoparticles. (e) SPECT/CT image acquired 2 days after intravenous injection of GadoSiDTPA-^{111}In nanoparticles to rats (posterior projection (e1) and transversal slice (e2)). B for bladder; K, RK and LK, for kidney, right and left kidney respectively. (Chapter 7)

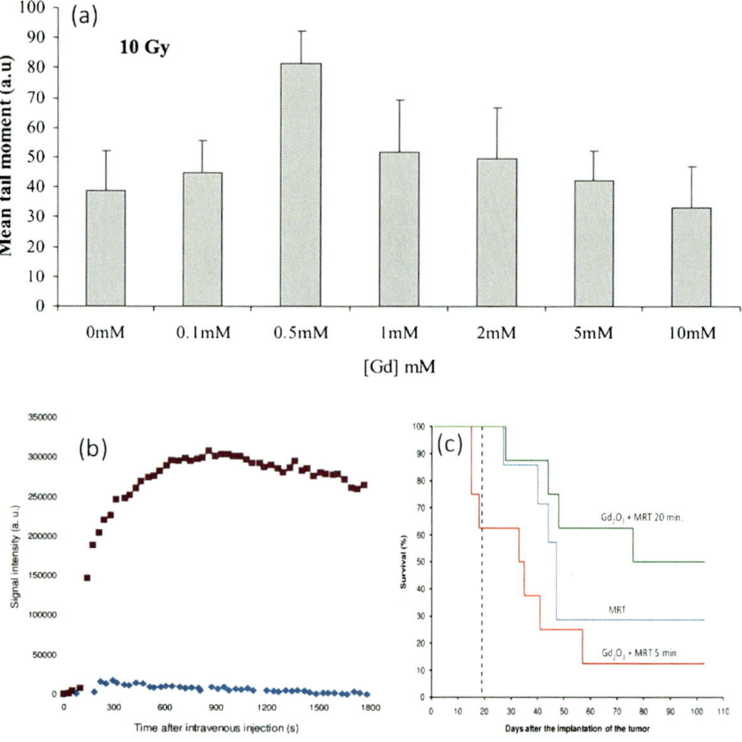

Figure 7. (a) In vitro radiosensitizing effect of gadolinium oxide nanoparticles at 10 Gy using a radiation at 660 keV. (b) Temporal evolution of the MRI signal in tumor (purple curve) and in an equivalent surface in normal tissue in the left hemisphere (blue curve) after injection of GadoSiDTPA nanoparticles to gliosarcoma bearing rat. (c) Survival curve comparison obtained on gliosarcoma bearing rats without treatment (black dashed curve, n = 4 rats), only treated by MRT (blue curve, n = 7), and treated by MRT 5 min (red curve, n = 8) and 20 min (green curve, n = 8) after intravenous injection. (Chapter 7)

Chapter 4

The Application of Peptide Functionalized Gold Nanoparticles

Tao Li,[a] Xiuxia He,[a,b] and Zhenxin Wang[*,a]

[a]State Key Laboratory of Electroanalytical Chemistry,
Changchun Institute of Applied Chemistry,
Chinese Academy of Sciences, Changchun 130022, China
[b]School of Science and Technology,
Changchun University of Science and Technology,
Changchun 130022, China
*E-mail: wangzx@ciac.jl.cn. Fax: (+86) 431-85262243

Peptide enabled synthesis of nanoparticles (NPs) offer the ability to control structures, properties and functionalities. Therefore, a wide variety of peptide functionalized gold nanoparticles (AuNPs) have been synthesized to meet different experimental conditions by the interactions of AuNPs with the primary amino acid sequences and/or secondary structures of peptides. The peptide functionalized AuNPs can provide a range of surface functionalities because peptides provide an invaluable resource to control structures and properties of AuNPs. The peptide functionalized AuNPs have been extensively explored as probes for sensing/imaging various analytes/targets, including ions and molecules, and excellent drug delivery systems for selective target binding as well as therapeutic effects for cancer treatment.

Keywords: Peptide; gold nanoparticle; application; bioanalysis; drug delivery

© 2012 American Chemical Society

Introduction

In the past two decades, gold nanoparticles (AuNPs) have been extensively studied and widely employed as probes for sensing/imaging wide ranges of analytes/targets, and building blocks for fabricating nanostructures and/or nanodevices (*1–8*). For instance, massive AuNP-based colorimetric assays have been developed for detecting different targets including metallic cations, small molecules, nucleic acids, proteins and cells, because of their unique optical properties (known as "surface plasmon resonance" (SPR) or "localized surface plasmon resonance" (LSPR)) (*2–9*). These technological applications normally require that AuNPs have high stability and solubility in aqueous media combined with a controllable chemical reactivity of the ligand shell (*2–11*). One particularly attractive application is the use of peptide functional AuNPs in bioanalytical and biomedical field for addressing biological events/processes in the form of recognition, sensing, assembly, and catalysis, because peptides provide an invaluable resource to control structures and properties of AuNPs (*12–14*). Especially, the versatile properties of peptides as recognition molecules open the route to many applications in biology and nanotechnology. Peptides can be used not only as ligands for targeting biomarkers but also as multidentate agents to stabilize AuNPs. For example, the pentapeptide CALNN and its derivative are able to convert citrate stabilized AuNPs into extremely stable AuNPs with some specific recognition properties (*15*).

In the present review, we will focus on preparation of peptide functionalized spherical AuNPs and applications of the peptide functionalized AuNPs for ions and molecules detection, enzymatic functionality/activity evaluation, targeting delivery and cellular analysis, highlighting some of their technical challenges and the new trends by means of a set of selected recent applications.

Synthesis of Peptide Functionalized Gold Nanoparticles

There are three mainly approaches (i.e. ligand exchange method, chemical reduction method, and chemical conjugation method) for preparing peptide functionalized AuNPs. The ligand exchange method necessitates using pre-synthesized peptide ligands which can strongly interact with AuNPs (*15–24*). Traditionally, gold–cysteine based peptide interactions are preferred to prepare peptide functionalized AuNPs by ligand exchange method since they form locally strong Au-S covalent bonds (*22*). Examples of cysteine based peptides include peptides with a single cysteine (C) residue located at either the N- or C-terminus (e.g., CALNN) or at several locations within the sequence (e.g., GCGGCGGKGGCGGCG) for multidentate binding (*15, 16*). Among the conventional methods of peptide functionalized AuNPs preparation, the ligand exchange method is most generally used and simplest approach to confer specific molecular recognition properties to the nanoparticles by including a defined percentage of a modified peptide in the preparation process.

In the chemical reduction method, the peptides are employed as reducing agents and/or capping reagents to generate the peptide functionalized AuNPs (*12, 25–32*). For instance, peptides contain tyrosine (Y) residues which are normally employed as reducing agents to synthesis the peptide functionalized AuNPs (*27–30*). The Y residues present in the peptides can reduce tetrachloroaurate (III) anions (AuCl$_4^-$) to the gold atoms, which eventually combine to form AuNPs. Lee et al. have used a "bottom-up" approach to formulate a set of rudimentary rules for the size- and shape-controlled peptide synthesis of AuNPs from the properties of the 20 natural α-amino acids for AuCl$_4^-$ reduction and Au^0 binding (*31*). They found that the formation of AuNPs by short peptides is primarily determined by two basic functionalities of the amino acid residues: reducing capability for AuCl$_4^-$ and capping (binding) capability for the AuNPs formed. The peptide sequence-crystal growth relationship could be used to produce AuNPs with prescribed geometrical features depending on the needs of the application. Serizawa et al. found that reductions of AuCl$_4^-$ by HEPES in the presence of adequate amounts of Cys-terminal peptides led to the production of peptide functionalized AuNPs under ambient conditions (*32*). The basic peptides including the RPTR sequence, which are difficult to apply to ligand exchange reactions, can be successfully introduced on nanoparticles by the approach.

Nanoparticles functionalized with molecules (e.g., thiol-containing oligoethylene glycols (PEG or OEG) with a terminated amine or carboxylic group) that provide reactive sites for binding of NH$_2$-, -COOH or biotin groups have also been used for the specific attachment of peptides (*33–37*). There are various chemical conjugation methods available for cross-linking peptide to AuNPs. The most commonly used strategy is so-called EDC/NHS coupling. For example, the KPQPRPLS peptide has been conjugated to nanoparticles of sphere or rodlike shape using EDC/sulfo-NHS coupling reaction by Bartczak and Kanaras (as shown in Figure 1) (*33*). They found that the degree of peptide coupling and the nanoparticle stability are strongly dependent on the experimental conditions. Kumar et al. have also successfully conjugated two peptides, therapeutic (p12) and targeted (CRGDK), on the surface of tiopronin capped AuNPs without changing the properties of peptides by the EDC/NHS coupling strategy (*36*). The chemical conjugation method can be employed to transfer hydrophilic peptide to water soluble AuNP surface, or synthesize small peptide functionalized AuNPs (e.g., 2 nm in diameter).

Applications of Peptide Functionalized Gold Nanoparticles

Because AuNPs have unique optical properties, peptide functionalized AuNPs have been extensively used as detecting probe to develop novel AuNP-based assay for chemical sensing and biomedical imaging applications. These assays promise increased sensitivity and specificity, multiplexing capability, and short turnaround times. In addition, peptide functionalized AuNPs are excellent building blocks for fabricating nanodevices/nanostructures for various technological applications such as optoelectronics, medical diagnostics and drug delivery since the properties of the surface peptide chains can be easily controlled by external stimuli such as pH and solution composition (*38–44*). Here, we provide an overview of recent examples of applications of the peptide functionalized AuNPs involved in assay development, targeting delivery and biomedical imaging.

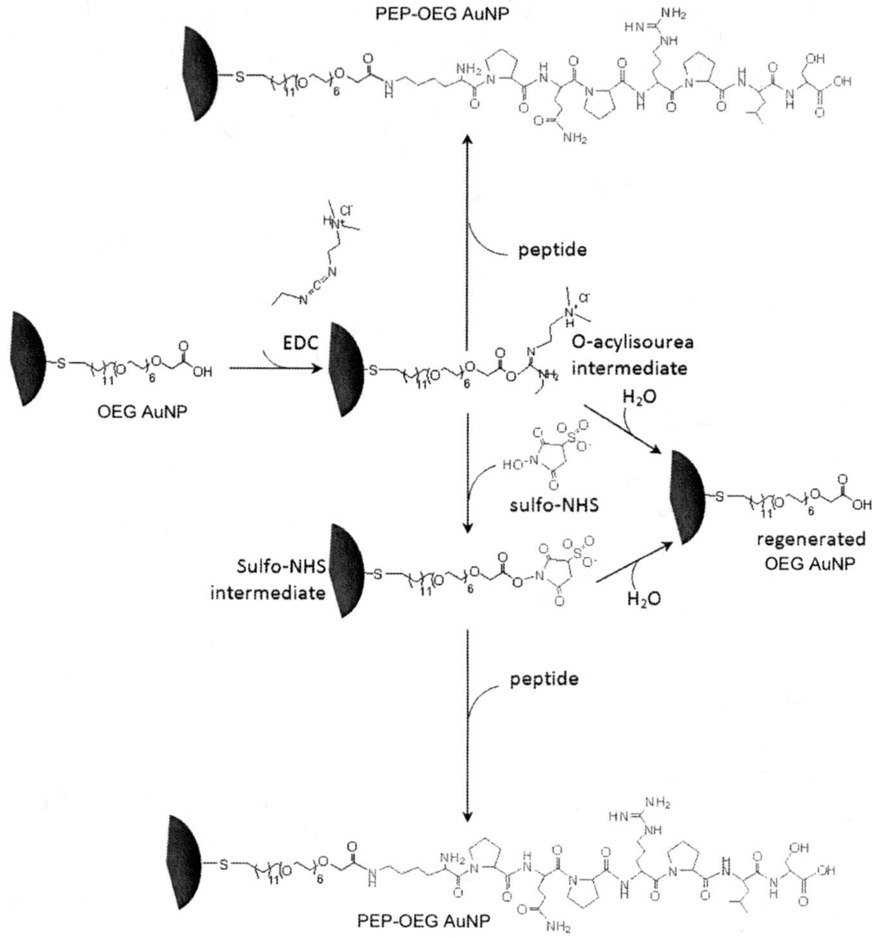

Figure 1. Schematic representation of conjugation of the KPQPRPLS peptide with monocarboxy(1-mercaptoundec-11-yl) hexa(ethylene glycol) (OEG) modified AuNPs by EDC and Sulfo-NHS reaction. Reproduced with permission from reference (33). Copyright 2011 American Chemical Society.

Molecule/Ion Detection

Because many metal ions and molecules can interplay with amino acids or peptides, rational and combinatorial design of peptide capping ligands for AuNPs is fascinating to develop novel colorimetric probes for chemical sensing (*45–54*). For instance, Ye et al. have developed a colorimetric assay for parallel detection of Cd^{2+}, Ni^{2+} and Co^{2+} utilizing functional peptide ligand, CALNNDHHHHHH capped AuNPs as sensing elements (*52*). Upon addition of Cd^{2+}, Ni^{2+} and Co^{2+} (or a combination of them) to the peptide functionalized AuNPs solution, its color would change from red to blue or purple due to the aggregation of peptide functionalized AuNPs induced by the binding between the metal ions and peptide ligands.   The method showed good selectivity for Cd^{2+}, Ni^{2+} and Co^{2+} over other metal ions, and detection limit as low as 0.05 μM Cd^{2+}, 0.3 μM Ni^{2+} or 2 μM Co^{2+}, respectively. Wang et al. have successfully developed a colorimetric assay based on bipeptides (glutathione (GSH) and CALNN) capped AuNPs for monitoring Pb^{2+} in living cell (53). In this case, GSH is recognizing group which enables to react with Pb^{2+}, and CALNN is stabilizing ligand which improves the stability of AuNPs under physiological condition. Panitch et al. have developed a AuNP-based method to examine the interactions between heparin and peptide (54). This method relies on aggregation of the CALNN stabilized AuNPs in the presence of heparin-binding peptides and retardation of their aggregation once the peptides bind to heparin. The binding affinities of heparin with peptides can be distinguished by the method. This label-free and easy-to-implement method is expected to be a good complement to other existing methods in enlightening the heparin-peptide interactions.

The assemblies of peptide functionalized AuNPs can also be used to monitor the pH change in biological systems and biomolecular interactions at interfaces because the structure of peptide assemblies can be controlled by external stimuli such as pH and solution composition (42–44). Liedberg et al. have demonstrated that polypeptide (The design of the polypeptides was based on the SA-42 motif, sequence NAADLEKAIEALEKHLEAKGPVDAAQLEKQLEQAFEAFERAG) capped AuNPs can be controllably aggregated upon exposure to Zn^{2+} (as shown in Figure 2) (44). The aggregation of particles could be completely and repeatedly reversed by removal of the zinc ions using EDTA. The possibility to assemble nanoparticles utilizing these specific and biologically inspired interactions in a controlled and reversible manner provides a highly interesting route for nanoengineering and programming new types of composite materials and devices. The optical and chemical properties of AuNPs also provide an interesting tool for a deeper and more fundamental understanding of specific molecular interactions involved in polypeptide folding.

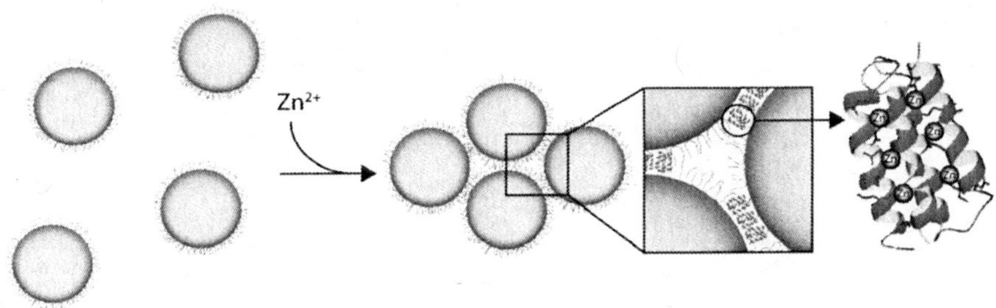

Figure 2. The binding of Zn^{2+} by JR2EC immobilized on AuNPs induces dimerization and folding between peptides located on separate particles resulting in particle aggregation. Reproduced with permission from reference (44). Copyright 2008 American Chemical Society.

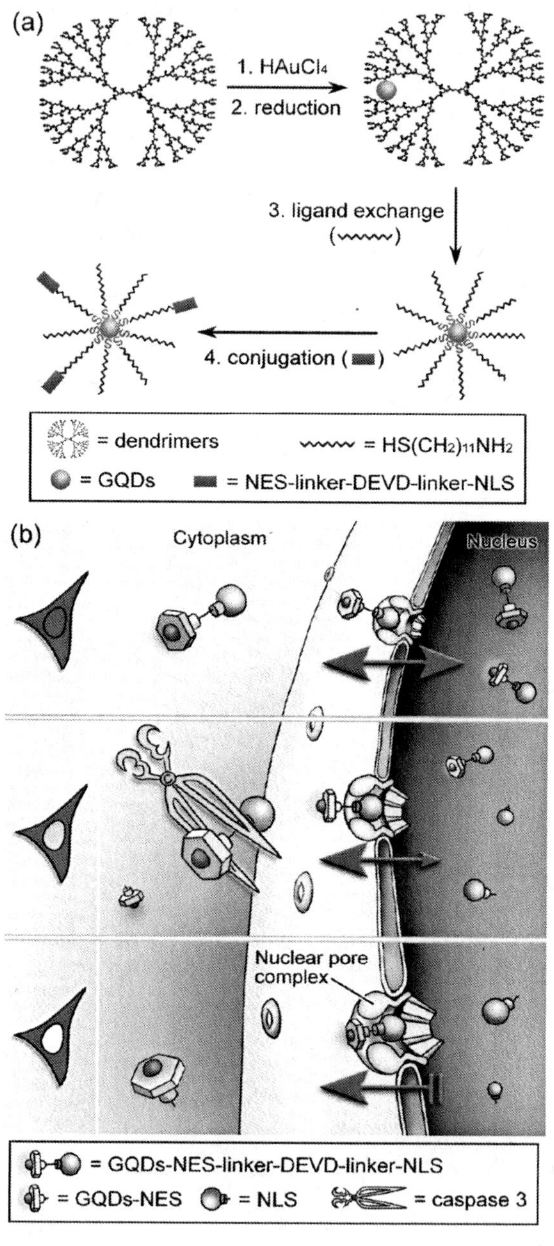

Figure 3. (a) Illustration of GQD synthesis and derivatization. (b) Schematic representation of the nucleus shuttle of GQDs functionalized with the peptide moiety containing NLS, NES and the caspase-3 responsive DEVD that allow monitoring of cell apoptotic progression. Reproduced with permission from reference (86). Copyright 2010 American Chemical Society.

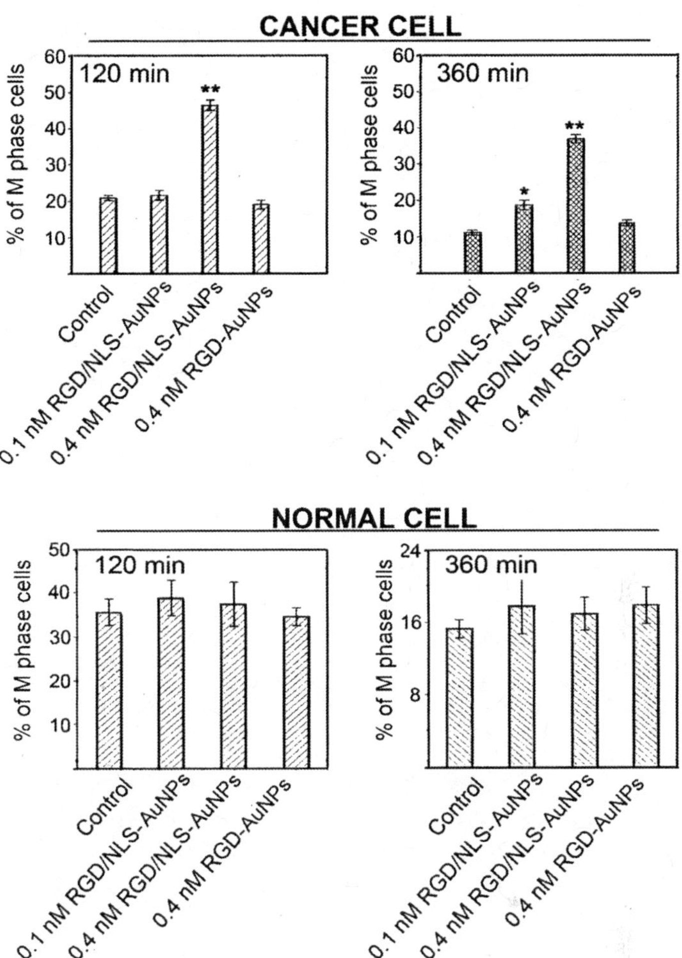

Figure 4. (top) M phase (mitosis phase of cell division) accumulation of cancer cells in the presence of 0.4 nM nuclear-targeting AuNPs (RGD/NLS-AuNPs) suggests complete cell division (cytokinesis) has not taken place. (bottom) Correspondent data for normal cells. Reproduced with permission from reference (96). Copyright 2010 American Chemical Society.

Enzymatic Activity Assay

The detection of functionalities/activities of enzymes (*e.g.*, kinases and proteases) is a central task in proteomics because enzymes play vital regulatory roles in many metabolic pathways and cell communication. Numbers of peptide functionalized AuNP-based assays have been developed for studying enzyme functionality and inhibition (*55–68*). Kim et al. have demonstrated that protein kinase and its inhibition can be assayed with high sensitivity on peptide functionalized AuNPs (*59*). Phosphorylation of peptides on the AuNP-monolayers was detected by using an anti-phosphotyrosine antibody and Cy3-labeled secondary antibody as a probing molecule. In contrast to conventional self-assembled monolayers, the AuNP monolayers generated higher density of peptide substrates and easier accessibility of protein kinase, due to the spherical and three-dimensional geometry on surfaces, resulting in highly sensitive assay of protein kinase functionality and inhibition. Stevens et al. have developed a peptide-modified AuNP-based colorimetric assay for rapid kinase inhibitor screening (*66*). In this assay, peptide-modified AuNPs are enzymatically phosphorylated and aggregated on a surface or in solution by action of phosphor-specific antibodies. Qu et al. successfully establish a homogeneous fluorometric method for the assay of multiple proteinases using AuNPs–peptide–fluorophore conjugates as the substrate (*68*). The assay offers a sensitive and rapid evaluation of proteinases activity operating either in an endpoint or real-time format. Compared with traditional fluorescence-based methods, which generally employ doubly labeled molecular beacons to serve as the substrates, only one terminal of the peptide needs to be labeled with fluorescence moiety for this strategy. The experimental results also demonstrate that the assay format is an ideal platform for high-throughput and/or point-of-care applications.

The Targeting Delivery of Nanoparticles by Basic Peptides

Nuclear targeting of nanoparticles in live cells is generating widespread interest because of the prospect of developing novel diagnostic and therapeutic strategies such as gene therapy. In particular, AuNPs as nanocarriers have tremendous growth in the pharmaceutical field for intracellular drug and gene delivery, mainly due to their large interacting surface, offering indispensible advantages to enhance potency, high specificity and low toxicity (*69, 70*). However, nuclear delivery of the nanoparticles requires bypassing the formidable barriers of the cellular membrane and the nuclear membrane. Many basic peptides (e.g., cell-penetrating peptides (CPPs), nuclear localization signal (NLS) peptides) derive from various virus proteins (*e.g.*, HIV-1 Tat protein, simian virus 40 and flock house virus coat proteins) and antimicrobial peptides (*e.g.*, lactoferricin B, indolicidin and penetratin) have been reported to translocate through the cell membranes and some of them target into the nucleus (*71–76*). Therefore, CPP functionalized AuNPs have emerged as attractive candidates for delivery of various payloads into their targets (*77–86*). As early as 2003, Tkachenko et al. have demonstrated nuclear entry of 20 nm AuNPs indirectly conjugated to various nuclear localization signal (NLS) peptides through a shell of bovine serum albumin (BSA) protein (*77*). Wang et al. found that arginine-rich peptide (CALNNR$_8$) functionalized AuNPs exhibit the specific recognition properties for the nucleus and endoplasmic reticulum targeting (*21*). They also demonstrated that the translocation is effected by the size of nanoparticle and CALNNR$_8$ surface coverage, i.e., large nanoparticle or AuNPs with low CALNNR$_8$ surface coverage has weak ability of the transmembrane. Oh et al. studied the cellular uptake of a series of cell penetrating peptide functionalized AuNPs ranging in diameter from

2.4 to 89 nm (*81*). They also found that AuNP cellular uptake is directly dependent on the surface display of the CPP and that the ultimate intracellular destination is further determined by AuNP diameter. This suggests the potential for controlling cellular uptake and specific delivery of biologically active cargos to subcellular targets just by altering the size of the central AuNP nanoplatform. Recently, the evolution of the intracellular distribution of TAT-functionalized AuNPs with time has been studied by Brust et al. (*82*). They found that the particles appear to negotiate intracellular membrane barriers quite freely, including the possibility of direct membrane transfer. This finding provides support for the hypothesis that cell-penetrating peptides can enable small objects to negotiate membrane barriers also in the absence of dedicated transport mechanisms. More importantly, co-functionalization of basic peptides with other specific biomolecules (e.g. oligonucleotides, peptides, proteins) on AuNP surfaces could pave the way toward creating new generations of heterofunctionalized nanomaterials for a variety of diagnostic and therapeutic applications (*83, 84*). For instance, Mirkin et al. have designed a heterofunctionalized AuNP consisting of a 13-nm AuNP containing both thiolated antisense oligonucleotides and cysteine-terminated basic peptides (*85*). The heterofunctionalized AuNPs are prepared easily and show perinuclear localization and an enhanced gene regulation activity when tested in a cellular model. Yang et al. have synthesized a kind of subnanometer photoluminescent gold quantum dots (GQDs) which are functionalized with a peptide moiety that contains nuclear export signal (NES), nuclear localization signal (NLS) and capsase-3 recognition (DEVD) sequences (*86*). The NES-linker-DEVD-linker-NLS peptide functionalized GQDs can be employed as molecular probes for the real-time monitoring cellular apoptosis (as shown in Figure 3).

The Targeting Delivery of Nanoparticles by Interactions of Peptides with Receptors

Because receptors for peptides are highly expressed on a variety of neoplastic and non-neoplastic cells, surface modification of AuNPs with specific peptides can improve cell type uptake and ensure that the AuNPs reach the desired target (*87–89*). Furthermore, integrin $\alpha_v\beta_3$, an important biomarker overexpressed in sprouting tumor vessels and most tumor cells, plays a critical role in regulating tumor growth, metastasis and tumor angiogenesis (*90, 91*). The RGD (arginine-glycine-aspartic acid) short peptides can specifically bind with integrin $\alpha_v\beta_3$ (*92*). And the peptides containing an RGD sequence have shown a high level of internalization within tumor cells via receptor-mediated endocytosis. These attractive physical properties coupled with their smaller size make peptides ideal candidates for synthesizing tumor-specific AuNPs which can be used to develop novel method for tumor diagnosis and therapy (*93–97*). Porta et al. have demonstrated that RGD-(GC)$_2$ functionalized AuNPs are accumulated in the cancer cells by receptor-mediated entrance and that their entrance pathway most likely drives them into the endosome (*95*). El-Sayed et al. have synthesized 30 nm RGD and NLS functionalized AuNPs which exhibit cancer cell nucleus-specific targeting (*96*). They found that that nuclear targeting of

AuNPs in cancer cells cause cytokinesis arrest, leading to the failure of complete cell division and thereby resulting in apoptosis (as shown in Figure 4). The result indicates that the RGD and NLS functionalized AuNPs can be used alone as an anticancer therapeutic material. Nam et al. have demonstrated that radioactive iodine-labeled, RGD-PEGylated AuNPs can target tumor selectively and be taken up by tumor cells via integrin $\alpha_v\beta_3$ -receptor-mediated endocytosis with no cytotoxicity (*97*). These promising results show that radioactive-iodine-labeled RGD-PEGylated AuNPs have potential for highly specific and sensitive tumor imaging or for use as angiogenesis-targeted SPECT/CT imaging probes. Kanaras et al. have investigated a range of intensities in which laser treatment of endothelial cells (HUVECs) in the presence of peptide KPQPRPLS functionalized AuNPs is used to control the cellular response (*98*). The peptide KPQPRPLS functionalized AuNPs preferentially bind to the vascular endothelial growth factor receptors (VEGFR-1).The experimental show that plasmon-mediated mild laser treatment, combined with specific targeting of cellular membranes, enables new routes for controlling cell permeability and gene regulation in endothelial cells.

Conclusion and Outlook

The surface and core properties of peptide functionalized AuNPs can be engineered for individual and multifold applications, including chemical sensing, biomedical imaging, clinical diagnosis and therapy. However, there are a number of critical issues that require addressing, such as recognition mechanism and 3-dimensional structures of the peptide ligands on the AuNP surfaces and long-term health effects of the nanomaterials. It will result in generating a diverse array of novel nanomaterials for better sensing and recognition of targets, and development of smart nanocarriers for drug delivery and new nanomedicines for selective tumor therapy if the peptide structure and orientation on nanoparticle surfaces can be predicted and precise controlled.

Acknowledgments

The authors thank National Basic Research Program of China (No. 2011CB935800), NSFC (Grant No. 21075118) and Jilin Provincial Science and Technology Department (Grant No. 20100701) for financial support.

References

1. Daniel, M. C.; Astruc, D. *Chem. Rev.* **2004**, *104*, 293–346.
2. Katz, E.; Willner, I. *Angew. Chem. Int. Ed.* **2004**, *43*, 6042–6108.
3. Rosi, N. L.; Mirkin, C. A. *Chem. Rev.* **2005**, *105*, 1547–1562.
4. Myroshnychenko, V.; Rodríguez-Fernández, J.; Pastoriza-Santos, I.; Funston, A. M.; Novo, C.; Mulvaney, P.; Liz-Marzán, L. M.; García de Abajo, F. J. *Chem. Soc. Rev.* **2008**, *37*, 1792–1805.
5. Wang, Z. X.; Ma, L. N. *Coord. Chem. Rev.* **2009**, *253*, 1607–1618.
6. de la Rica, R.; Matsui, H. *Chem. Soc. Rev.* **2010**, *39*, 3499–3509.

7. Aili, D.; Stevens, M. M. *Chem. Soc. Rev.* **2010**, *39*, 3358–3370.

8. Moyano, D. F.; Rotello, V. M. *Langmuir* **2011**, *27*, 10376–10385.

9. Liu, D.; Wang, Z.; Jiang, X. *Nanoscale* **2011**, *3*, 1421–1433.

10. Lee, K. S.; El-Sayed, M. A. *J. Phys. Chem. B* **2006**, *110*, 19220–19225.

11. Connor, E. E.; Mwamuka, J.; Gole, A.; Murphy, C. J.; Wyatt, M. D. *Small* **2005**, *1*, 325–327.

12. Slocik, J. M.; Naik, R. R. *Chem. Soc. Rev.* **2010**, *39*, 3454–3463.

13. Dickerson, M. B.; Sandhage, K. H.; Naik, R. R. *Chem. Rev.* **2008**, *108*, 4935–4978.

14. Chen, C. L.; Rosi, N. L. *Angew. Chem., Int. Ed.* **2010**, *49*, 1924–1942.

15. Lévy, R.; Thanh, N. T. K.; Doty, R. C.; Hussain, I.; Nichols, R. J.; Schiffrin, D. J.; Brust, M.; Fernig, D. G. *J. Am. Chem. Soc.* **2004**, *126*, 10076–10084.

16. Krpetic, Z.; Nativo, P.; Porta, F.; Brust, M. *Bioconjugate Chem.* **2009**, *20*, 619–624.

17. Heinz, H.; Farmer, B. L.; Pandey, R. B.; Slocik, J. M.; Patnaik, S. S.; Pachter, R.; Naik, R. R. *J. Am. Chem. Soc.* **2009**, *131*, 9704–9714.

18. Liu, Y.; Shipton, M. K.; Ryan, J.; Kaufman, E. D.; Franzen, S.; Feldheim, D. L. *Anal. Chem.* **2007**, *79*, 2221–2229.

19. Xie, H.; Tkachenko, A. G.; Glomm, W. R.; Ryan, J. A.; Brennaman, M. K.; Papanikolas, J. M.; Franzen, S.; Feldheim, D. L. *Anal. Chem.* **2003**, *75*, 5797–5805.

20. Guerrero, A. R.; Caballero, L.; Adeva, A.; Melo, F.; Kogan, M. J. *Langmuir* **2010**, *26*, 12026–12032.

21. Sun, L. L.; Liu, D. J.; Wang, Z. X. *Langmuir* **2008**, *24*, 10293–10297.

22. Dubois, L. H.; Zegarski, B. R.; Nuzzo, R. G. *J. Am. Chem. Soc.* **1990**, *112*, 570–579.

23. Wang, Z. X.; Lévy, R.; Fernig, D. G.; Brust, M. *Bioconjugate Chem.* **2005**, *16*, 497–500.

24. Lévy, R. *ChemBioChem* **2006**, *7*, 1141–1145.

25. Higashi, N.; Kawahara, J.; Niwa, M. *J. Colloid Interface Sci.* **2005**, *288*, 83–87.

26. Pietersen, L. K.; Govender, P.; Kruger, H. G.; Maguire, G. E. M.; Wesley-Smith, J.; Govender, T. *Int. J. Pept. Res. Ther.* **2010**, *16*, 291–295.

27. Si, S.; Tarun, K.; Mandal, T. K. *Chem.-Eur. J.* **2007**, *13*, 3160–3168.

28. Bhattacharjee, R. R.; Das, A. K.; Haldar, D.; Si, S.; Banerjee, A.; Mandal, T. K. *J. Nanosci. Nanotechnol.* **2005**, *5*, 1141–1147.

29. Si, S.; Bhattac, S. D.; Sastry, M. *J. Colloid Interface Sci.* **2004**, *269*, 97–102.

30. Toroz, D.; Corni, S. *Nano Lett.* **2011**, *11*, 1313–1318.

31. Tan, Y. N.; Lee, J. Y.; Wang, D. I. C. *J. Am. Chem. Soc.* **2010**, *132*, 5677–5686.

32. Serizawa, T.; Hirai, Y.; Aizawa, M. *Langmuir* **2009**, *25*, 12229–12234.

33. Bartczak, D.; Kanaras, A. G. *Langmuir* **2011**, *27*, 10119–10123.

34. de la Fuente, J. M.; Berry, C. C. *Bioconjugate Chem.* **2005**, *16*, 1176–1180.

35. Fan, J.; Chen, S.; Gao, Y. *Colloid Surf., B* **2003**, *28*, 199–207.

36. Kumar, A.; Ma, H.; Zhang, X.; Huang, K.; Jin, S.; Liu, J.; Wei, T.; Cao, W.; Zou, G.; Liang, X. *Biomaterials* **2012**, *33*, 1180–1189.

37. Sun, L. L.; Wang, J. E.; Wang, Z. X. *Nanoscale* **2010**, *2*, 269–276.

38. Duchesne, L.; Wells, G.; Fernig, D. G.; Harris, S. A.; Lévy, R. *ChemBioChem* **2008**, *9*, 2127–2134.

39. Samit, G.; Arindam, B. *Macromol. Chem. Phys.* **2009**, *210*, 1422–1432.

40. Coomber, D.; Bartczak, D.; Gerrard, S. R.; Tyas, S.; Kanaras, A. G.; Stulz, E. *Langmuir* **2010**, *26*, 13760–13762.

41. Hwang, L.; Chen, C. L.; Rosi, N. L. *Chem. Commun.* **2011**, *47*, 185–187.

42. Si, S.; Mandal, T. K. *Langmuir* **2010**, *26*, 13760–13762.

43. Higuchi, M.; Ushiba, K.; Kawaguchi, M. *J. Colloid. Interf. Sci.* **2007**, *308*, 356–363.

44. Aili, D.; Enander, K.; Rydberg, J.; Nesterenko, I.; Björefors, F.; Baltzer, L.; Liedberg, B. *J. Am. Chem. Soc.* **2008**, *130*, 5780–5788.

45. Sigel, A.; Sigel, H.; Sigel, R. K. O. *Metal Ions in Life Sciences*, 2nd ed; John Wiley & Sons: New York, 2008; ISBN: 0470513241.

46. Seal, B. L.; Panitch, A. *Biomacromolecules* **2003**, *4*, 1572–1582.

47. Si, S.; Kotal, A.; Mandal, T. K. *J. Phys. Chem. C* **2007**, *111*, 1248–1255.

48. Li, X. K.; Wang, J. E.; Sun, L. L.; Wang, Z. X. *Chem. Commun.* **2010**, *46*, 988–990.

49. Savage, A. C.; Pikramenou, Z. *Chem. Commun.* **2011**, *47*, 6431–6433.

50. Cortez, J.; Vorobieva, E.; Gralheira, D.; Osorio, I.; Soares, L.; Vale, N.; Pereira, E.; Gomes, P.; Franco, R. *J. Nanopart. Res.* **2011**, *13*, 1101–1113.

51. Si, S.; Raula, M.; Paira, T. K.; Mandal, T. K. *ChemPhysChem* **2008**, *9*, 1578–1584.

52. Zhang, M.; Liu, Y.-Q.; Ye, B.-C. *Analyst* **2012**, *137*, 601–607.

53. Zhu, D.; Li, X. K.; Liu, X.; Wang, J. E.; Wang, Z. X. *Biosens. Bioelectron.* **2012**, *31*, 505–509.

54. Jeong, K. J.; Butterfield, K.; Panitch, A. *Langmuir* **2008**, *24*, 8794–8800.

55. Wang, Z. X.; Lévy, R.; Fernig, D. G.; Brust, M. *J. Am. Chem. Soc.* **2006**, *128*, 2214–2215.

56. Guarise, C.; Pasquato, L.; Filippis, V. De; Scrimin, P. *Proc. Natl. Acad. Sci. U.S.A.* **2006**, *103*, 3978–3982.

57. Pengo, P.; Baltzer, L.; Pasquato, L.; Scrimin, P. *Angew. Chem., Int. Ed.* **2007**, *46*, 400–404.

58. Zhen, S. J.; Li, Y. F.; Huang, C. Z.; Long, Y. F. *Talanta* **2008**, *76*, 230–232.

59. Kim, Y. P.; Oh, Y.-H.; Kim, H.-S. *Biosens. Bioelectron.* **2008**, *23*, 980–986.

60. harjee, R. R.; Banerjee, A.; Mandal, T. K. *Chem.-Eur. J.* **2006**, *12*, 1256–1265.

61. Selvakannan, P. R.; Mandal, S.; Phadtare, S.; Gole, A.; Pasricha, R.; Adyanthaya, S. D.; Sastry, M. *J. Colloid Interface Sci.* **2004**, *269*, 97–102.

62. Free, P.; Shaw, C. P.; Levy, R. *Chem. Commun.* **2009**, *50*, 5009–5011.

63. Yin, B. C.; Zhang, M.; Tan, W.; Ye, B. C. *ChemBioChem* **2010**, *11*, 494–497.

64. Kang, J.-H.; Asami, Y.; Murata, M.; Kitazaki, H.; Sadanaga, N.; Tokunaga, E.; Shiotani, S.; Okada, S.; Maehara, Y.; Niidome, T.; Hashizume, M.; Mori, T.; Katayama, Y. *Biosens. Bioelectron.* **2010**, *25*, 1869–1874.

65. Zhang, M.; Yin, B. C.; Wang, X. F.; Ye, B. C. *Chem. Commun.* **2011**, *47*, 2399–2401.

66. Gupta, S.; Andresen, H.; Stevens, M. M. *Chem. Commun.* **2011**, *47*, 2249–2251.

67. Yang, X.-C.; Mo, Z.-H. *Chin. J. Anal. Chem.* **2010**, *38*, 1333–1336.

68. Wang, X.; Geng, J.; Miyoshi, D.; Ren, J.; Sugimoto, N.; Qu, X. *Biosens. Bioelectron.* **2010**, *26*, 743–747.

69. Torchilin, V. P. *Biopolymers* **2008**, *90*, 604–610.

70. Giljohann, D. A.; Seferos, D. S.; Daniel, W. L.; Massich, M. D.; Patel, P. C.; Mirkin, C. A. *Angew. Chem., Int. Ed.* **2010**, *49*, 3280–3294.

71. Mitchell, D. J.; Kim, D. T.; Steinman, L.; Fathman, C. G.; Rothbard, J. B. *J. Pept. Res.* **2000**, *56*, 318–325.

72. Voge, H. J.; Schibli, D. J.; Jing, W.; Lohmeier-Vogel, E. M.; Epand, R. F.; Epand, R. M. *Biochem. Cell Biol.* **2002**, *80*, 49–63.

73. Schibli, D. J.; Epand, R. F.; Vogel, H. J.; Epand, R. M. *Biochem. Cell Biol.* **2002**, *80*, 667–677.

74. Guerrero, S.; Araya, E.; Fiedler, J. L.; Arias, J. I.; Adura, C.; Albericio, F.; Giralt, E.; Arias, J. L.; Fernández, M. S.; Kogan, M. *Int. J. Nanomedicine* **2010**, *5*, 897–913.

75. Pujals, S.; Bastús, N. G.; Pereiro, E.; López-Iglesias, C.; Puntes, V. F.; Kogan, M. J.; Giralt, E. *ChemBioChem* **2009**, *10*, 1025–1031.

76. Bartczak, D.; Nitti, S.; Millar, T. M.; Kanaras, A. G. *Nano Today* **2011**, *6*, 478–492.

77. Tkachenko, A. G.; Xie, H.; Coleman, D.; Glomm, W.; Ryan, J.; Anderson, M. F.; Franzen, S.; Feldheim, D. L. *J. Am. Chem. Soc.* **2003**, *125*, 4700–4701.

78. Nativo, P.; Prior, I. A.; Brust, M. *ACS Nano* **2008**, *2*, 1639–1644.

79. Maus, L.; Dick, O.; Bading, H.; Spatz, J. P.; Fiammengo, R. *ACS Nano* **2010**, *4*, 6617–6628.

80. Yang, H.; Fung, S. Y.; Liu, M. *Angew. Chem., Int. Ed.* **2011**, *50*, 9643–9646.

81. Oh, E.; Delehanty, J. B.; Sapsford, K. E.; Susumu, K.; Goswami, R.; Blanco-Canosa, J. B.; Dawson, P. E.; Granek, J.; Shoff, M.; Zhang, Q.; Goering, P. L.; Huston, A.; Medintz, I. L. *ACS Nano* **2011**, *5*, 6434–6448.

82. Krpetic, Z.; Saleemi, S.; Prior, I. A.; See, V.; Qureshi, R.; Brust, M. *ACS Nano* **2011**, *5*, 5195–5201.

83. Pietersen, L. K.; Govender, P.; Kruger, H. G.; Maguire, G. E. M.; Govender, T. *J. Nanosci. Nanotechnol.* **2011**, *11*, 3075–3083.

84. Mandal, D.; Maran, A.; Yaszemski, M. J.; Bolander, M. E.; Sarkar, G. *J. Mater. Sci.: Mater. Med.* **2009**, *20*, 347–350.

85. Patel, P. C.; Giljohann, D. A.; Seferos, D. S.; Mirkin, C. A. *Proc. Natl. Acad. Sci. U.S.A.* **2008**, *105*, 17222–17226.

86. Lin, S.-Y.; Chen, N.-T.; Sun, S.-P.; Chang, J. C.; Wang, Y.-C.; Yang, C.-S.; Lo, L.-W. *J. Am. Chem. Soc.* **2010**, *132*, 8309–8315.

87. Bartholdi, M. F.; Wu, J. M.; Pu, H. F.; Troncoso, P.; Eden, P. A.; Feldman, R. I. *Int. J. Cancer* **1998**, *79*, 82–90.

88. Markwalder, R.; Reubi, J. C. *Cancer Res.* **1999**, *59*, 1152–1159.

89. Bartczak, D.; Sanchez-Elsner, T.; Louaf, F.; Milla, T. M.; Kanara, A. G. *Small* **2011**, *7*, 388–394.

90. Allman, R.; Cowburn, P.; Mason, M. *Eur. J. Cancer* **2000**, *36*, 410–422.

91. Hood, J.; Cheresh, D. *Nat. Rev. Cancer* **2002**, *2*, 91–100.

92. Pierschbacher, M.; Ruoslahti, E. *Nature* **1984**, *309*, 30–33.

93. Hosta-Rigau, L.; Olmedo, I.; Arbiol, J.; Cruz, L. J.; Kogan, M. J.; Albericio, F. *Bioconjugate Chem.* **2010**, *21*, 1070–1078.

94. Arosio, D.; Manzoni, L.; Araldi, E. M. V.; Scolastico, C. *Bioconjugate Chem.* **2011**, *22*, 664–672.

95. Scarì, G.; Porta, F.; Fascio, U.; Avvakumova, S.; Santo, V. D.; Simone, M. D.; Saviano, M.; Leone, M.; Gatto, A. D.; Pedone, C.; Zaccaro, L. *Bioconjugate Chem.* **2012**, *23*, 340–349.

96. Kang, B.; Mackey, M. A.; El-Sayed, M. A. *J. Am. Chem. Soc.* **2010**, *132*, 1517–1519.

97. Kim, Y. H.; Jeon, J.; Hong, S. H.; Rhim, W. K.; Lee, Y. S.; Youn, H.; Chung, J. K.; Lee, M. C.; Lee, D. S.; Kang, K. W.; Nam, J. M. *Small* **2011**, *7*, 2052–2060.

98. Bartczak, D.; Muskens, O. L.; Millar, T. M.; Sanchez-Elsner, T.; Kanaras, A. G. *Nano Lett* **2011**, *11*, 1358–1363.

Chapter 5

Gold Nanoparticles for the Development of Transdermal Delivery Systems

Dakrong Pissuwan[1,§] and Takuro Niidome[*,1,2,3]

[1]Department of Applied Chemistry, Faculty of Engineering, Kyushu University, 744 Motooka, Nishi-ku, Fukuoka 819-0395, Japan
[2]Center for Future Chemistry, Kyushu University, 744 Motooka, Nishi-ku, Fukuoka 819-0395, Japan
[3]International Research Center for Molecular System, Kyushu University, 744 Motooka, Nishi-ku, Fukuoka 819-0395, Japan
§Current address: Immunology Frontier Research Center, Osaka University, 3-1 Yamadaoka, Suita, 565-0871, Osaka, Japan
***E-mail: niidome.takuro.655@m.kyushu-u.ac.jp**

The transdermal delivery is an attractive approach for the delivery of therapeutic molecules, which provides a lot of benefits to patients. However, the delivery of high-molecular-weight drugs (especially hydrophilic agents) is problematic because the difficulty in skin penetration of these molecules. In recent years, gold nanoparticles have been used in several biomedical applications. They could be candidates for transdermal delivery systems. Here we review the possibility of using gold nanoparticles to develop treatment methods *via* the skin. The penetration pathways, applications, and toxicity effects of gold nanoparticles in transdermal delivery systems are discussed.

Introduction

In recent years, gold nanoparticles (GNPs) have received considerable attention due to their unique physical and chemical properties. Their surfaces are easy for chemical modification, and can be modified by many bioactive molecules (*1, 2*). These properties enable them to be used extensively in various biological

© 2012 American Chemical Society

applications: photothermal therapy (*3*), delivery of drugs and genes (*4*, *5*) and bioimaging (*6*). Several studies have shown how GNPs can be used to target cancer cells (*7*, *8*), bacteria (*9*), and parasites (*10*).

The transdermal route has shown promising potential for the delivery of drugs and therapeutic materials. However, the barrier property of skin causes difficulties for the creation of transdermal delivery systems (TDSs). Because of the many benefits of using GNPs in the applications mentioned above, they could be used to enhance the efficiency of TDSs.

This chapter focuses on a general overview of recent studies using GNPs to improve TDSs. The properties of GNPs and their mechanism of penetration are discussed. Additionally, the applications of GNPs in TDSs as well as their toxicity are detailed.

The Skin as a Route for Delivery Systems

Transdermal delivery is a non-invasive route for the delivery of therapeutic materials from the intact skin to the systemic circulation. Every square centimeter of the surface of the human skin contains ≈10–70 hair follicles and ≈200–250 sweat glands (*11*). Hence, the skin surface could be a conduit for the delivery of therapeutic substances from the intact skin to inside the body. Additional benefits of using TDSs are pain-free delivery, few side effects, providing a low-frequency of drug administration, and avoidance of the first pass-effect in the liver. Moreover, the steady permeation of drugs/therapeutic materials transversing the skin allows constant levels of the drug in plasma, so drugs can be maintained for a prolonged period in the circulation (*11*).

Nevertheless, the problem that must be overcome in the delivery of all therapeutic substances passing through the skin is the stratum corneum (SC). This barrier has a complex structural arrangement, and the major constituent is lipids (*12*). To overcome this barrier, studies have been carried out to find a way to enhance substance delivery across the skin.

In general, a TDS comprises two main processes: percutaneous absorption and penetration of therapeutic substances. In the case of percutaneous absorption, substances are absorbed from outside to underneath the skin. Because of these two main processes of transdermal delivery and the potent benefits of GNPs, studying the ability of GNPs to penetrate or pass through intact skin is useful. Such studies could help to elicit further benefits for TDSs, including the treatment of skin diseases.

Absorption and Penetration of GNPs through the Skin

The interactions of GNPs with various mammalian cells have been studied in the past few years (*13*, *14*). Recently, the uptake of gold particles across the skin barrier has attracted appreciable interest. The combination between GNPs and deoxyribonucleic acid (DNA) vaccines was reviewed by Dean et al. (*15*). That review showed that the DNA-coated gold particles could enhance the penetration of vaccines from the outer layer of the skin to Langerhans cells (which have an important role in the immune system of the skin). This approach is known as "particle-mediated vaccine for epidermal delivery" (Figure 1), which has become an innovative approach to improve vaccine delivery. This research area could lead to the development of TDSs using gold as nanosize particles.

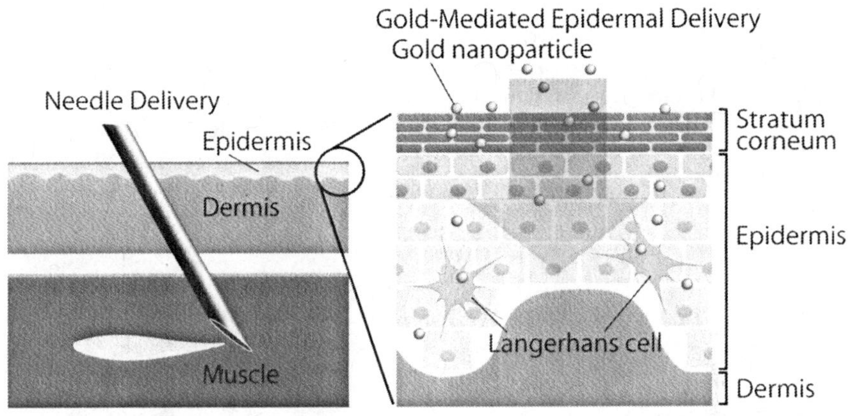

Figure 1. Gold particle-mediated DNA vaccine for epidermal delivery. Reproduced with permission from (15). Copyright (2005) (Elsevier).

It has been shown that cellular uptake of GNPs is dependent upon their size, shape, and chemical surface (*14, 16, 17*). It appears that these properties also affect the absorption and penetration of GNPs into the skin. It has been recently reported that GNPs have a high potential to "overwhelm" the skin barrier. *In vitro* permeation of GNPs at different sizes (approximately 15, 102, 198 nm) through rat skin was demonstrated by Sonavane et al. (*18*). Their study showed that the permeation of GNPs through the skin was highest if GNPs of diameter 15 nm were applied to the rat skin. They observed that, the larger the size of GNPs, the lower the permeation of gold particles through the rat skin. Additionally, they found that a small size of GNPs could pass more rapidly from the outer layer to the deeper area of the skin compared with larger-sized GNPs (*18*). Another study by Huang et al. (*19*) reported that GNPs of diameter ≈5 nm could permeate rapidly through the epidermis in mouse skin. However, only a small amount of GNPs was observed in the dermis. The penetrative ability of small-sized GNPs could arise from their induction of lipid modulation.

In 2011, Labouta et al. (*20*) investigated the skin penetration of thiol-coated GNPs (diameter, ≈6 nm) in human skin using multiphoton imaging-pixel analysis. They showed that thiol-coated GNPs of diameter 6 nm penetrated into the SC and later migrated to deeper layers of the skin. The same research team subsequently studied the penetration of GNPs at different surface modifications, sizes, vehicles, and concentrations through the human skin. Non-polar (dodecanethiol-coated gold of size ≈6 nm; cetrimide-coated gold of size ≈15 nm) and negatively charged (lecithin-coated gold of size ≈6 nm; citrate-coated gold of size ≈15 nm) GNPs were used for skin penetration. After applying these GNPs to the skin and incubating for 24 h, all types of gold could penetrate the SC and migrate to the deeper skin layer excluding citrate-coated GNPs. An important point to note here is that citrate-coated GNPs of size 15 nm were dispersed in water. However, cetrimide-coated GNPs of size 15 nm were dispersed in toluene, and this type of particle could be detected in the deeper layer of the skin (Figure 2). This implies that not only size can affect skin penetration but also the dispersion solvent and charge of gold particles (*21*). In the study by Sonavane et al. (*18*) mentioned above, they showed that 15-nm citrate–GNPs could penetrate to a deeper layer of rat skin, but the same type of gold particles could not penetrate into human skin (*21*). This could be because different structures and components of skin provide different skin permeability for particles.

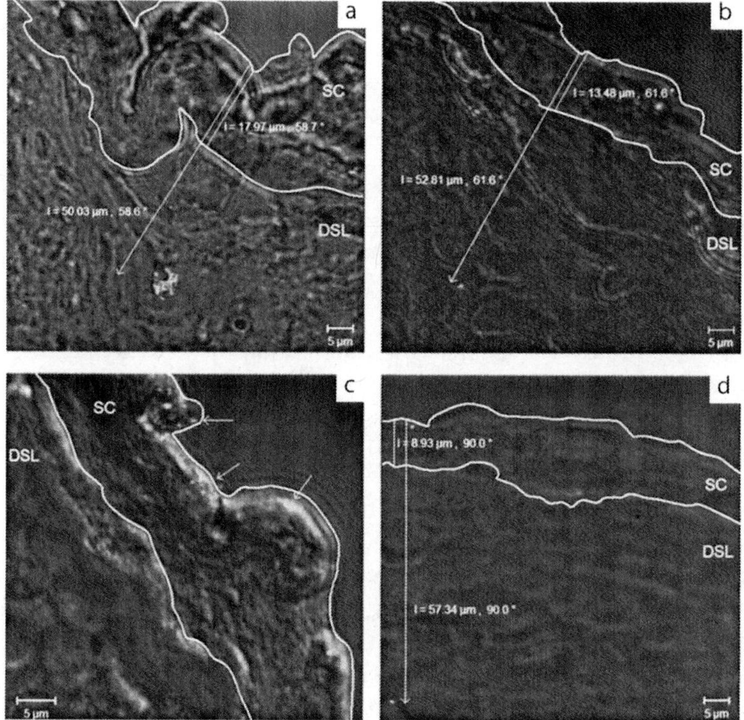

Figure 2. Penetration of different surface coatings and sizes of gold nanoparticles (GNPs) from the outer layer to the deeper inner layer of human skin. (a) 6-nm dodecanethiol-coated GNPs in toluene, (b) 6-nm lecithin-coated GNPs in water, (c) 15-nm citrate-coated GNPs in water, and (d) 15-nm cetrimide-coated GNPs in toluene. The green color represents gold nanoparticles; SC denotes the stratum corneum; and DSL is the deeper skin layer. Reproduced with permission from (21). Copyright (2011) (The Royal Society of Chemistry).

The reason why 15-nm cetrimide-coated GNPs could penetrate to deeper skin layers is probably due to their hydrophobic properties. The SC consists of lipids in intercellular spaces (*22*). Therefore, the hydrophobic molecules of cetrimide-coated gold particles (which are also lipophilic) could be enhanced to permeate and then penetrate through the skin rather than hydrophilic molecules (*23, 24*). The same direction of permeation was reported when 6-nm hydrophilic (lecithin-coated gold) and hydrophobic (dodecanethiol-coated gold) particles were applied to the skin. The permeation and penetration of hydrophilic nanoparticles could occur *via* hydrophilic transepidermal pores of skin, which can open naturally to a maximum width of ≈20–30 nm (*25*).

The same authors also tried to investigate the aggregation of gold particles inside the skin by measuring the absorption of GNPs after exposing them to the skin. They found that the absorption peak of 15-nm citrate-gold nanoparticles was red shift and broaden and this means that the aggregation of gold nanoparticles occurred inside the skin. Interestingly, other forms of GNPs used in their study did not cause significant aggregation after exposure to skin for 24 h. The cause of aggregation may be the reaction between citrate molecules on the surface of gold particles and biomolecules such as lipids and proteins in the skin (*21*).

All studies mentioned above show that the features of GNPs that could aid permeation and penetration could be the size of particles, the surface properties of gold particles, and the compatibility of gold particles with lipids in the SC. GNPs could penetrate *via* the intercellular lipid spaces between SC corneocytes. Penetration *via* this route is called as the "intercellular route". Another route by which GNPs penetrate the skin can be the "appendage route" (through the sebaceous glands, sweat glands, and hair follicles). Both routes are common routes for molecules (including nanoparticles) to penetrate the skin (*26, 27*).

It is well known that compounds with a molecular mass >500 Da cannot cross the skin barrier (*28*). Therefore, various approaches have been developed to overwhelm the skin barrier and enhance the absorption and penetration of high-molecular-weight drugs cross the skin. For instance, this can involve the use of chemical enhancers to change the structure of skin lipids (*29*). It was reported that preparations of nanoparticles in organic solvents such as toluene, cyclohexane, and chloroform increased the penetrative efficacy of nanoparticles (*30, 31*). Other approaches, such as electrophoresis and photomechanical waves, have been used to improve the efficiency of transdermal delivery (*32, 33*). Laser microablation was also reported to enhance the permeability of nanoparticles into the skin dermis (*34*).

GNPs have shown a capability to penetrate into the skin. Therefore, they are considered to be good candidates for TDSs, which are discussed in the next section.

Application of GNPs in Transdermal Delivery

As discussed above, several approaches have been used to improve TDSs. Unfortunately, TDSs have disadvantages such as difficulties in the delivery of hydrophilic macromolecules and side effects from the use of chemical enhancers.

Considering the various applications of GNPs in biomedical applications and their properties, GNPs have become attractive materials for improving TDSs.

Azarbayjani et al. (*35*) described the use of GNPs in cosmetic facial masks. In their study, GNPs were loaded into nanofiber face masks containing polyvinyl alcohol and randomly methylated β-cyclodextrin. The SC of adult abdominal skin was slightly detached. More fragments of the SC and expansion of inter-keratinocyte spaces were found after exposing gold-loaded nanofibers to the skin for 24 h. Another example of a study using GNPs in cosmetic applications was initiated by Pornpattananangkul et al. (*36*), who bound carboxyl-modified GNPs to cationic liposomes. Liposomes have been used in transdermal formulations for several years. However, the SC still limits their stability and permeability (*37*). The binding of GNPs to liposomes could stabilize liposomes and increase the efficiency of using liposomes for transdermal delivery.

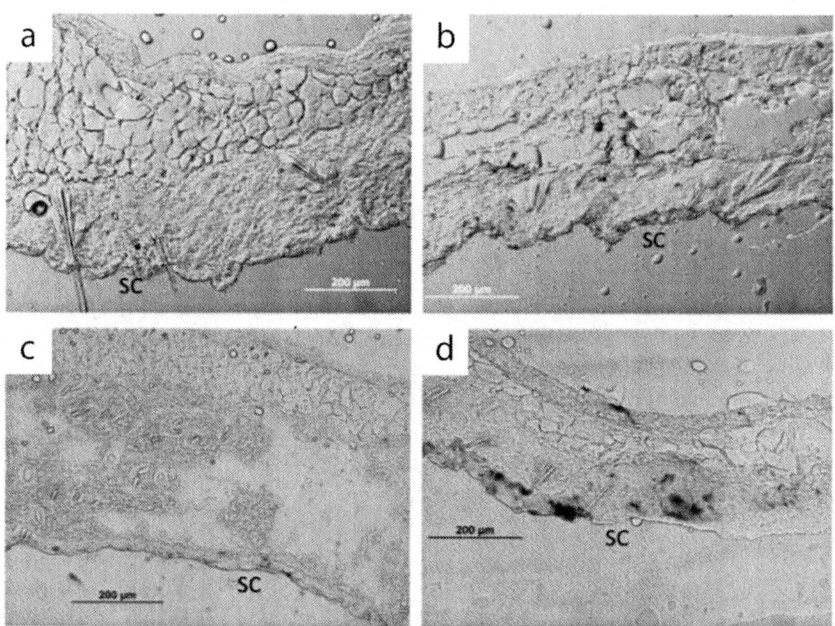

Figure 3. Penetration of the proteins horseradish peroxidase and β-galactosidase with and without gold nanoparticles in mouse skin. (a) Horseradish peroxidase, (b) horseradish peroxidase/gold, (c) β-galactosidase, and (d) β-galactosidase/gold. Reproduced with permission from (19). Copyright (2012) (Elsevier).

The use of GNPs for percutaneous delivery of "protein drugs" was reported by Huang et el. (*19*). GNPs of mean size 5 nm and a negative zeta potential (−18.3 mV) were employed. GNPs were mixed with two molecular-weight proteins (45-kDa horseradish peroxidase; 460-kDa β-galactosidase) as co-administered protein drugs. After exposure of mouse skin to the gold–protein mixture for 2 h, both types of proteins co-administered with GNPs could penetrate the SC. Two proteins were mainly detected in the epidermis and smaller amounts were observed in the dermis (Figure 3). In contrast, the penetration of both proteins without GNPs was very low. This finding implies that GNPs can induce skin permeability as a result of the penetration of high-molecular-weight proteins into the skin.

Transcutaneous immunization using GNPs has also been investigated. GNPs were mixed with ovalbumin (OVA) antigen (OVA/GNP) and applied to mouse skin. After two boosters of OVA/GNP at the second and the fourth week, the concentration of antibody that responded to OVA antigen was increased. However, mice applied with OVA without GNPs showed very low concentrations of antibody response.

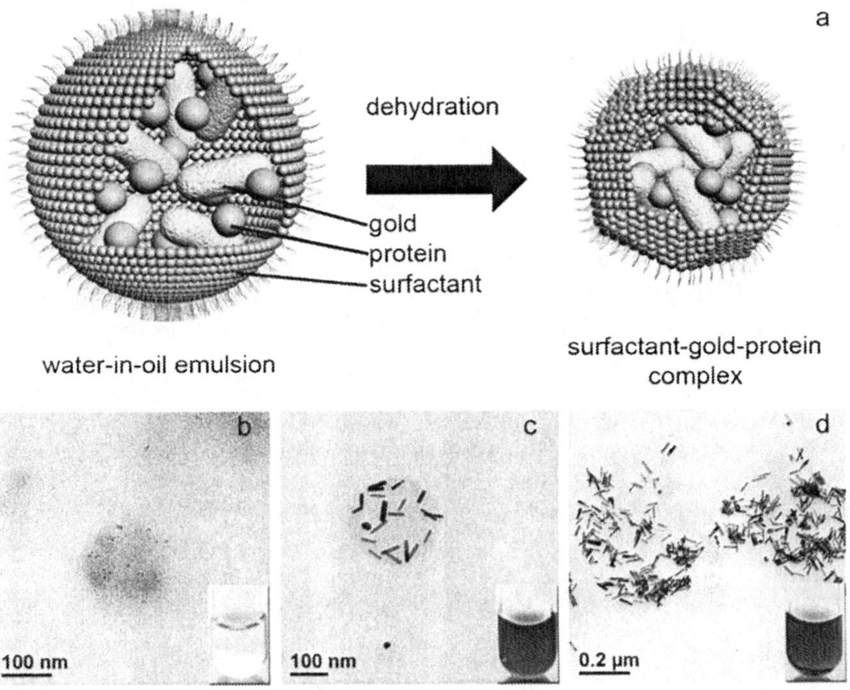

Figure 4. Transmission electron microscopy (TEM) images showing formulation of a solid-in-oil dispersion of gold nanorods combined with protein. (a) Complex formation (schematic), (b) TEM image of the formulation without gold nanorods, (c) TEM image of the formulation with gold nanorods, and (d) TEM image of the formulation without proteins. Reproduced with permission from (41). Copyright (2011) (John Wiley and Sons).

As mentioned in earlier section, a laser light can be used to enhance skin permeability. However, by using lasers to enhance transdermal delivery, heat generation may become difficult to control. Using gold nanorods for photothermal therapeutic applications has recently been reported (*38–40*). The main reason for these applications is because gold nanorods have longitudinal plasmon resonances which can be "tuned" into the near-infrared biological window. Our research team was the first to report the possibility of using gold nanorods for transdermal delivery. We prepared a combination of methoxy(polyethylene glycol)-thiol-coated gold nanorods with OVA as an oil-based formulation called "solid-in-oil dispersion of gold nanorods" (Figure 4). The complex was applied to the dorsal skin of mice *in vivo*. Mice were then irradiated with a light from a xenon lamp for 10 min at a power density of $\approx$6.0 W/cm^2. After two transcutaneous immunization boosters, the induction of antibodies in mice was significantly higher than that observed in mice treated with the same formulation and method but without gold nanorods (*41*). It seems that our method provided a significant increase in antibody production in 2 weeks compared with the method reported by Huang et el. (*19*) (who needed 4 weeks to induce antibody production).

After we successfully used our novel approach for transcutaneous immunization we then used a similar formulation for the transdermal delivery of insulin to mice with type-1 diabetes. An oil-based complex of gold nanorods with insulin and surfactant could significantly decrease blood glucose levels in diabetic mice after applying the complex to the dorsal skin of mice *in vivo* and then irradiating them with near-infrared light. Our approach presents a higher efficiency for insulin delivery than that observed for subcutaneous injection (*42*).

Besides transdermal delivery, GNPs could be used to enhance local heating in the treatment of skin diseases such as skin cancer (*43*). The schematic summary of using gold nanoparticles for transdermal delivery systems are shown in Figure 5.

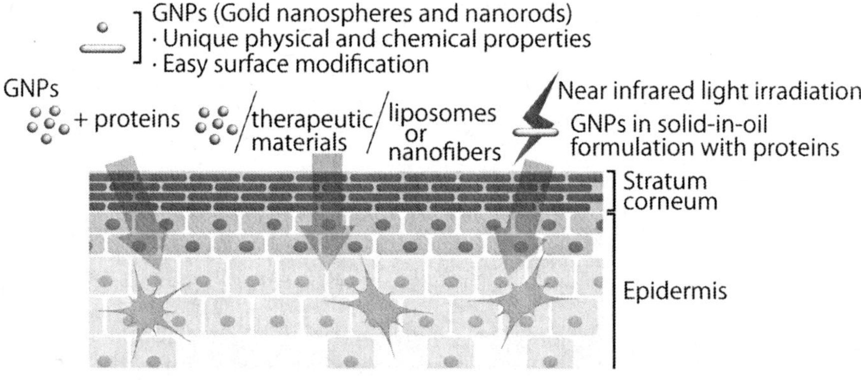

Figure 5. Schematic summary of using gold nanoparticles for transdermal delivery systems.

Toxicity of GNPs

GNPs have very useful properties for transdermal delivery and also have low toxicity. Nevertheless, investigating the toxicity of GNPs is important. It is well known that gold nanorods were coated with a cationic surfactant called cetyltrimethylammonium bromide (CTAB) to control the size and shape of gold nanorods. However, CTAB is toxic to mammalian cells. Thus, surface modification of gold nanorods is applied to overcome toxicity. One example is coating gold nanorods with the which is a biocompatible polymer poly(ethyleneglycol) (PEG) (*7*). A recent report showed that 25 μg/mL PEG-coated gold nanorods inhibited the proliferation of human HaCat keratinocyte cells (*44*). The response of reactive oxygen species (ROS) and disruption of the membrane potential of mitochondria was detected in HaCat cells treated with PEG-coated gold nanorods. Both effects were also detected when cells were treated with spherical GNPs (mercaptopropanesulfonate-coated GNPs), but these effects were less pronounced than those in skin treated with PEG-coated gold nanorods. This phenomenon could be because gold nanorods (dimensions, $\approx$16.7 $\times$43.8 nm) had a higher surface area to interact with cell membranes than spherical GNPs (diameter, $\approx$20 nm) (*44*). The effect of 13-nm citrate-GNPs on the growth, shape, and cell mechanism of human dermal fibroblasts has also been reported (*45*).

It seems that the long-term effects and side effects of applying GNPs *in vivo via* the skin have not been reported extensively. The toxicity mechanism of GNPs is unclear and controversial. However, GNPs seem to provide more benefits for improving TDSs compared with other methods such as chemical enhancers.

Conclusions

GNPs can interact with the skin barrier, leading to increases in skin permeability and enhancement of the delivery of high-molecular-weight therapeutic materials (especially hydrophilic molecules) through the skin. Therefore, GNPs could be good candidates for skin immunization and improving TDSs. The toxicity associated with the application of GNPs for TDSs needs further investigation.

Acknowledgments

This work was supported by a Grant-in-Aid for Scientific Research (B) (No. 22300158) from the Japan Society for the Promotion of Science.

References

1. Kumar, A.; Ma, H.; Zhang, X.; Huang, K.; Jin, S.; Liu, J.; Wei, T.; Cao, W.; Zou, G.; Liang, X.-J. Gold nanoparticles functionalized with therapeutic and targeted peptides for cancer treatment. *Biomaterials* **2012**, *33* (4), 1180–1189.

2.   Tiwari, P.; Vig, K.; Dennis, V.; Singh, S. Functionalized goldnanoparticles and their biomedical applications. *Nanomaterials* **2011**, *1* (1), 31–63.

3.   Pissuwan, D.; Valenzuela, S. M.; Cortie, M. B. Therapeutic possibilities of plasmonically heated gold nanoparticles. *Trends Biotechnol.* **2006**, *24* (2), 62–67.

4.   Dreaden, E. C.; Alkilany, A. M.; Huang, X.; Murphy, C. J.; El-Sayed, M. A. The golden age: gold nanoparticles for biomedicine. *Chem. Soc. Rev.* **2012**, *41* (7), 2740–2779.

5.   Pissuwan, D.; Niidome, T.; Cortie, M. B. The forthcoming applications of gold nanoparticles in drug and gene delivery systems. *J. Controlled Release* **2011**, *149* (1), 65–71.

6.   Hutter, E.; Maysinger, D. Gold nanoparticles and quantum dots for bioimaging. *Microsc. Res. Tech.* **2011**, *74* (7), 592–604.

7.   Niidome, T.; Yamagata, M.; Okamoto, Y.; Akiyama, Y.; Takahashi, H.; Kawano, T.; Katayama, Y.; Niidome, Y. PEG-modified gold nanorods with a stealth character for in vivo applications. *J. Controlled Release* **2006**, *114* (3), 343–347.

8.   Huang, X.; El-Sayed, I. H.; Qian, W.; El-Sayed, M. A. Cancer cells assemble and align gold nanorods conjugated to antibodies to produce highly enhanced, sharp, and polarized surface Raman spectra: a potential cancer diagnostic marker. *Nano Lett.* **2007**, *7* (6), 1591–1597.

9.   Pissuwan, D.; Cortie, C. H.; Valenzuela, S. M.; Cortie, M. B. Functionalised gold nanoparticles for controlling pathogenic bacteria. *Trends Biotechnol.* **2010**, *28* (4), 207–213.

10. Pissuwan, D.; Valenzuela, S. M.; Miller, C. M.; Killingsworth, M. C.; Cortie, M. B. Destruction and control of *Toxoplasmagondii Tachyzoites* using gold nanosphere/antibody conjugates. *Small* **2009**, *5* (9), 1030–1034.

11. Keleb, E.; Sharma, R. K.; Mosa, E. B.; Aljahwi, A.-a. Z. Transdermal drug delivery system- design and evaluation. *Int. J. Adv. Pharm. Sci.* **2011**, *1* (3), 201–211.

12. Brisson, P. Percutaneous absorption. *Can. Med. Assoc. J.* **1974**, *110* (10), 1182–1185.

13. Bartczak, D.; Muskens, O. L.; Nitti, S.; Sanchez-Elsner, T.; Millar, T. M.; Kanaras, A. G. Interactions of human endothelial cells with gold nanoparticles of different morphologies. *Small* **2012**, *8* (1), 122–130.

14. Chithrani, B. D.; Ghazani, A. A.; Chan, W. C. W. Determining the size and shape dependence of gold nanoparticle uptake into mammalian cells. *Nano Lett.* **2006**, *6* (4), 662–668.

15. Dean, H. J.; Haynes, J.; Schmaljohn, C. The role of particle-mediated DNA vaccines in biodefense preparedness. *Adv. Drug Delivery Rev.* **2005**, *57* (9), 1315–1342.

16. Chithrani, D. B.; Dunne, M.; Stewart, J.; Allen, C.; Jaffray, D. A. Cellular uptake and transport of gold nanoparticles incorporated in a liposomal carrier. *Nanomed.: Nanotechnol., Biol., Med.* **2010**, *6* (1), 161–169.

17. Nativo, P.; Prior, I. A.; Brust, M. Uptake and intracellular fate of surface-modified gold nanoparticles. *ACS Nano* **2008**, *2* (8), 1639–1644.

18. Sonavane, G.; Tomoda, K.; Sano, A.; Ohshima, H.; Terada, H.; Makino, K. In vitro permeation of gold nanoparticles through rat skin and rat intestine: Effect of particle size. *Colloids Surf. B* **2008**, *65* (1), 1–10.

19. Huang, Y.; Yu, F.; Park, Y.-S.; Wang, J.; Shin, M.-C.; Chung, H. S.; Yang, V. C. Co-administration of protein drugs with gold nanoparticles to enable percutaneous delivery. *Biomaterials* **2010**, *31* (34), 9086–9091.

20. Labouta, H. I.; Kraus, T.; El-Khordagui, L. K.; Schneider, M. Combined multiphoton imaging-pixel analysis for semiquantitation of skin penetration of gold nanoparticles. *Int. J. Pharm.* **2011**, *413* (1–2), 279–282.

21. Labouta, H. I.; El-Khordagui, L. K.; Kraus, T.; Schneider, M. Mechanism and determinants of nanoparticle penetration through human skin. *Nanoscale* **2011**, *3* (12), 4989–4999.

22. Wertz, P. W. Stratum corneum lipids and water. *Exog. Dermatol.* **2004**, *3* (2), 53–56.

23. Küchler, S.; Herrmann, W.; Panek-Minkin, G.; Blaschke, T.; Zoschke, C.; Kramer, K. D.; Bittl, R.; Schäfer-Korting, M. SLN for topical application in skin diseases—Characterization of drug–carrier and carrier–target interactions. *Int. J. Pharm.* **2010**, *390* (2), 225–233.

24. Desai, P.; Patlolla, R. R.; M., S. Interaction of nanoparticles and cell-penetrating peptides with skin for transdermal drug delivery. *Mol. Membr. Biol.* **2010**, *27* (7), 247–259.

25. SchÄtzlein, A.; Cevc, G. Non-uniform cellular packing of the stratum corneum and permeability barrier function of intact skin: a high-resolution confocal laser scanning microscopy study using highly deformable vesicles (Transfersomes). *Br. J. Dermatol.* **1998**, *138* (4), 583–592.

26. Schneider, M.; Stracke, F.; Hansen, S.; Schaefer, F. U. Nanoparticles and their interactions with the dermal barrier. *Derm.-Endocrinol.* **2009**, *1* (4), 109–206.

27. Zhang, L. W.; Yu, W. W.; Colvin, V. L.; Monteiro-Riviere, N. A. Biological interactions of quantum dot nanoparticles in skin and in human epidermal keratinocytes. *Toxicol. Appl. Pharmacol.* **2008**, *228* (2), 200–211.

28. Kumar, R.; Philip, A. Modified transdermal technologies: breaking the barriers of drug permeation via the skin. *Trop. J. Pharm. Res.* **2007**, *6* (1), 633–644.

29. Pathan, I. B.; Setty, C. M. Chemical penetration enhancers for transdermal drug delivery systems. *Trop. J. Pharm. Res.* **2009**, *8* (2), 173–179.

30. Labouta, H. I.; Liu, D. C.; Lin, L. L.; Butler, M. K.; Grice, J. E.; Raphael, A. P.; Kraus, T.; El-Khordagui, L. K.; Soyer, H. P.; Roberts, M. S.; Schneider, M.; Prow, T. W. Gold nanoparticle penetration and reduced metabolism in human skin by toluene. *Pharm Res.* **2011**, *28* (11), 2931–2944.

31. Xia, X. R.; Monteiro-Riviere, N. A.; Riviere, J. E. Skin penetration and kinetics of pristine fullerenes (C60) topically exposed in industrial organic solvents. *Toxicol. Appl. Pharmacol.* **2010**, *242* (1), 29–37.

32. Denet, A.-R.; Vanbever, R.; Préat, V. Skin electroporation for transdermal and topical delivery. *Adv. Drug Delivery Rev.* **2004**, *56* (5), 659–674.

33. Lee, S.; Mulholland, S. E.; Doukas, A. G. Photomechanical transdermal delivery: the effect of laser confinement. *Lasers Surg. Med.* **2001**, *28* (4), 344–347.

34. Genina, E. A.; Dolotov, L. E.; Bashkatov, A. N.; Terentyuk, G. S.; Maslyakova, G. N.; Zubkina, E. A.; Tuchin, V. V.; Yaroslavsky, I. V.; Altshuler, G. B. Fractional laser microablation of skin aimed at enhancing its permeability for nanoparticles. *Quantum Electron.* **2011**, *41* (5), 396–401.

35. Fathi-Azarbayjani, A.; Qun, L.; Chan, Y.; Chan, S. Novel vitamin and gold-loaded nanofiber facial mask for topical delivery. *AAPS PharmSciTech* **2010**, *11* (3), 1164–1170.

36. Pornpattananangkul, D.; Olson, S.; Aryal, S.; Sartor, M.; Huang, C.-M.; Vecchio, K.; Zhang, L. Stimuli-responsive liposome fusion mediated by gold nanoparticles. *ACS Nano* **2010**, *4* (4), 1935–1942.

37. Zhang, L.; Granick, S. How to stabilize phospholipid liposomes (using nanoparticles). *Nano Lett.* **2006**, *6* (4), 694–698.

38. Pissuwan, D.; Valenzuela, S. M.; Cortie, M. B. Prospects for gold nanorod particles in diagnostic and therapeutic application. *Biotechnol. Genet. Eng. Rev.* **2008**, *25* (1), 93–112.

39. Huang, X.; Neretina, S.; El-Sayed, M. A. Gold Nanorods: From synthesis and properties to biological and biomedical applications. *Adv. Mater.* **2009**, *21* (48), 4880–4910.

40. C.S, R.; Kumar, J.; V, R.; M, V.; Abraham, A. Laser immunotherapy with gold nanorods causes selective killing of tumour cells. *Pharmacol. Res.* **2012**, *65* (2), 261–269.

41. Pissuwan, D.; Nose, K.; Kurihara, R.; Kaneko, K.; Tahara, Y.; Kamiya, N.; Goto, M.; Katayama, Y.; Niidome, T. A solid-in-oil dispersion of gold nanorods can enhance transdermal protein delivery and skin vaccination. *Small* **2011**, *7* (2), 215–220.

42. Nose, K.; Pissuwan, D.; Goto, M.; Katayama, Y.; Niidome, T. Gold nanorods in an oil-base formulation for transdermal treatment of Type 1 diabetes in mice. *Nanoscale* **2012**.

43. Salas-García, I.; Fanjul-Vélez, F.; Ortega-Quijano, N.; Lavín-Castanedo, A.; Mingo-Ortega, P.; Arce-Diego, J. L. Effect of gold nanoparticles in the local heating of skin tumors induced by phototherapy. *Proc. SPIE* **2011**, *8092*, 809204.

44. Schaeublin, N. M.; Braydich-Stolle, L. K.; Maurer, E. I.; Park, K.; MacCuspie, R. I.; Afrooz, A. R. M. N.; Vaia, R. A.; Saleh, N. B.; Hussain, S. M. Does shape matter? Bioeffects of gold nanomaterials in a human skin cell model. *Langmuir* **2012**, *28* (6), 3248–3258.

45. Pernodet, N.; Fang, X.; Sun, Y.; Bakhtina, A.; Ramakrishnan, A.; Sokolov, J.; Ulman, A.; Rafailovich, M. Adverse effects of citrate/gold nanoparticles on human dermal fibroblasts. *Small* **2006**, *2* (6), 766–773.

Chapter 6

Nanoparticles for Noninvasive Radiofrequency-Induced Cancer Hyperthermia

Stuart J. Corr,[1,2,3] Mustafa Raoof,[1] Lon J. Wilson,[2,3] and Steven A. Curley[*,1,4]

[1]Dept. of Surgical Oncology, M.D. Anderson Cancer Center, University of Texas
[2]Dept. of Chemistry, William Marsh Rice University
[3]Richard E. Smalley Institute for Nanoscale Science & Technology, Rice University
[4]Dept. of Mechanical Engineering and Materials Science, William Marsh Rice University
[*]E-mail: scurley@mdanderson.org. Fax: (+1) 713-745-2436

Effective cancer therapies that are less invasive and less toxic are highly desirable. Shortwave radio-frequency (RF) electrical fields have excellent depth of penetration into biologic tissues with minimal toxicity. Study of the release of heat by targeted nanoparticles exposed to RF fields has produced interesting results and led to the possibility of using nanoparticles internalized into cancer cells to produce targeted thermal injury to cancer cells while minimizing toxicity in normal cells.

Introduction

Thermal interactions between nanomaterials and radio waves in the high-frequency radio spectrum (13.56 MHz) are currently being investigated for new methods of noninvasive cancer therapy. It is envisioned that nanoparticles (NPs) can act as radio-frequency-induced heat sources that initiate hyperthermia and necrosis once internalized within cancer cells. This cancer treatment has the potential to be completely *noninvasive* due to the relatively long (22 m) full-body penetrating wavelengths. These propagating waves gradually raise the

© 2012 American Chemical Society

body temperature over time due to the nonzero dielectric loss associated with biological organs and tissue. However, by allowing targeted NPs to accumulate within cancer cells over time, preferential heating of cancer is favored.

(a)

(b)

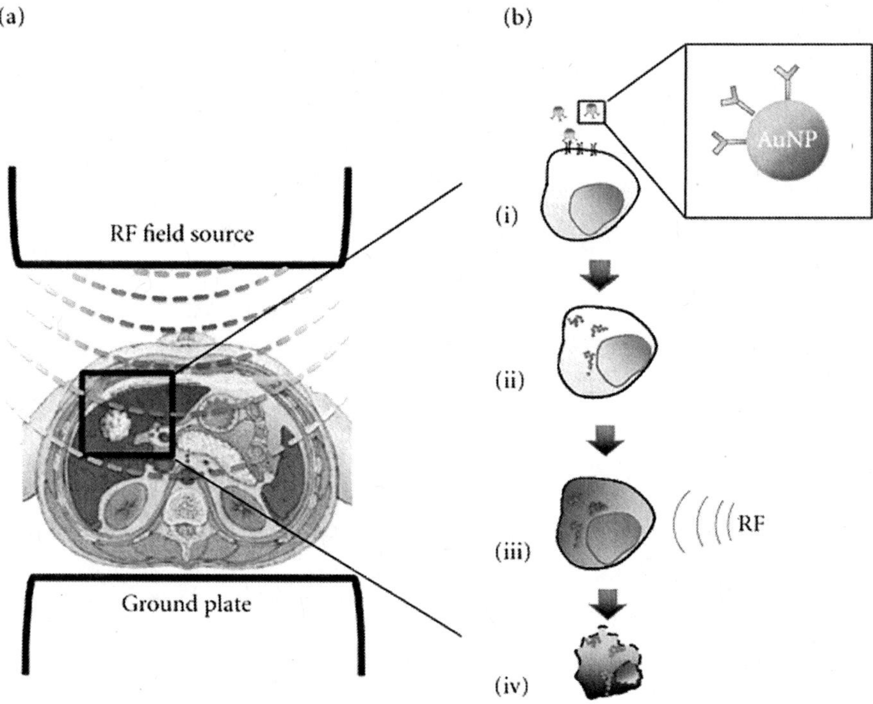

Figure 1. Principle of the noninvasive RF based treatment of hepatocellular carcinoma (HCC). A. RF field source is used to generate a 13.56 MHz electromagnetic field that penetrates tissues and reaches the tumor. B. Nanoparticles that can be thermally activated are conjugated to monoclonal antibodies against known targets expressed on HCC (i), internalized specifically by cancer cells after systemic administration (ii), and upon RF activation release heat (iii) causing targeted cancer cell death. (see color insert)

By functionalizing NPs with cancer-specific targeting agents such as anti-bodies and peptides, it is hypothesized that thermal energy can propagate within cancer cells that have internalized these nanomaterials, leaving neighboring healthy cells intact. This principle is highlighted in Figure 1. In order to fully develop and optimize this treatment several fundamental steps must first be accomplished. Understanding of the exact heating mechanism of NPs when exposed to radio waves as well as their behavior within cancer cells are two basic science questions which need to be fully examined for complete system development.

So far, gold NPs (*1–4*), carbon nanotubes (*5*), and quantum dots (*6*) have all shown exciting properties when used for both *in vitro* and *in vivo* RF applications. More recently, an extensive set of key experiments within our laboratories (*7, 8*) have helped gain insight into the mechanism of heat production from gold NPs (AuNPs) within a radiofrequency (RF) field. These findings were subsequently supported with *in vitro* studies which showed critical importance on NP stability within endolysomal cellular compartments for optimized treatment. To build upon previous review articles and book chapters (*9–12*) the proceeding sections in this chapter aim to give the reader an overview of the most *recent* advancements of the use of AuNPs for noninvasive RF cancer hyperthermia, as well as a brief overview of previous results utilizing both quantum dots and carbon nanotubes.

Hyperthermia and Targeted Noninvasive RF Cancer Therapy

The history of elevating body temperature above normal core temperature (a.k.a. hyperthermia) to treat cancers is more than a century old. The earliest observations from Busch demonstrated regression of a sarcoma by fever resulting from erysipelas, a type of bacterial skin infection (*13*). Later, Coley purposely administered the bacterial toxins from Streptococcus group A (etiological agent of erysipelas) to reproduce these effects. A variety of cellular and biological processes in cancer cells are preferentially and adversely affected by hyperthermia (41-43°C) which may explain the effects observed by Coley and are still not fully understood. While there is biologic plausibility of using hyperthermia as a potent modality against cancer, the translation of this modality to the clinic has been an ongoing challenge.

A multitude of approaches have been utilized in the past to achieve uniformly elevated temperature in the cancer tissue with mediocre success. Currently, the majority of hyperthermia systems employ electromagnetic radiation to deliver noninvasive locoregional thermal therapy. Electromagnetic radiation below the visible spectrum is considered to be nonionizing and hence relatively safe for hyperthermia applications. An important parameter for interaction of electromagnetic radiation with matter is the penetration depth. This is defined as the depth at which the intensity of the radiation inside the material is attenuated to about 37% of the incident radiation. At infrared frequencies (400 THz), the penetration depth is less than a few centimeters while at high radiofrequencies (13.56 MHz), the penetration depth can be more than 30 cm. Therefore, the penetration depth favors the use of high frequency RF fields in the design of hyperthermia systems, however microwave and infrared systems have also been used for noninvasive superficial, interstitial or endocavitary applications.

There are at least three systems that were developed to exploit RF fields for thermal therapy that have been investigated in humans. The first was a capacitively-coupled device developed by LeVeen in the 1960s that utilized radiofrequencies in the range of 8-27 MHz (*14–19*). A similar capacitively-coupled system that utilized 8 MHz RF fields was developed in Japan and is currently in use (*20–38*). By changing the relative size of electrodes, the depth of hyperthermia in the tissue could be adjusted providing some control over tissue heating with this system. Given the inherent limitation of nonselective hyperthermia in the previous devices, a focused phase-array system has recently been developed in Europe (*39–48*). Multiple antennae are arranged circumferentially around the body as dipole antenna pairs. These antennae allow for "focusing" of electromagnetic energy to the intended tumor tissue by constructive interference and help minimize by-stander damage. When coupled with real-time thermography such as magnetic resonance thermography, this system promises to be a significant advance in safe and effective thermal therapy of deep-seated cancers. Despite these advances concerns over by-stander normal tissue damage and limitations in noninvasive thermography have prevented wide-spread use of RF hyperthermia systems for cancer therapy.

With the advent of nanotechnology, there has been a renewed interest in exploring interactions of nanomaterials with electromagnetic radiation for thermal therapy of cancers. Initial reports of plasmonic heating of AuNPs under near-infrared exposure provided a significant breakthrough and are currently under development for translation to the clinic (*49*). Since then, numerous nanomaterials have been shown to be effective in photothermal therapy of tumors in pre-clinical studies (*50–52*). As mentioned previously, photothermal therapy is not suited for deep-seated tumors because of its limited penetration depth. For that purpose, the interactions of nanomaterials with RF fields have been explored, as well as other promising techniques such as focused ultra-sound (*53–58*).

Remotely heating nanomaterials using radio waves allows for the possibility of achieving tumor-targeted hyperthermia even for deep-seated tumors. The premise of this targeted thermal therapy is based in the notion that NPs can be designed for optimal delivery to the tumor tissue through active (i.e. bioconjugation with targeting moieties) or passive mechanisms (i.e. shape, size, stiffness, etc.). Once delivered to the tumor tissue, exposure to radio waves can remotely activate the NPs to generate heat that is dissipated only within the tumor cells without damage to adjacent or distant normal tissues. Combining molecular targeting with nanoscale hyperthermia represents the future for safe and effective thermal therapy of cancer.

In the following sections we discuss the relevant physical and biological studies published in the past decade towards the clinical development of this modality. All studies discussed here utilize a capacitively-coupled RF system (13.56 MHz) - the Kanzius external RF generator. While several studies have reported the *in vitro* efficacy of NPs in combination with radio waves (*59, 60*), comparison across the studies are precluded because of differences in experimental set-up and difficulty in determination of incident electric field intensity.

Heat Production from Gold Nanoparticles

Factors Limiting Heat Production

Previous work within our laboratories has shown that both AuNPs and carbon nanotubes can promote effective thermal ablation of hepatocellular, colorectal, and pancreatic cancer - all of which have exceptionally high morbidity rates upon diagnosis. There have been several publications however which state that these heating effects arise primarily from Joule heating of the background buffers which the NPs are suspended in, and not the NPs themselves. In particular, some work has both theoretically (*61, 62*) and experimentally (*60*) dismissed the use of AuNPs as RF thermal heat sources.

Taking these considerations into account, an extensive set of experiments were initiated within our laboratories to validate and quantify thermal interactions between citrate-capped AuNPs and RF energy. The results of these experiments, currently under review, suggest that AuNP heating rates are dependent on size, concentration, and surface area and that all background buffers must be eliminated to allow for full analysis of RF-AuNP interactions. Also, colloidal stability seems to be an important factor in heat production. As recently shown by Raoof *et al.* (*7*) hyperthermia treatment can be optimized by modulating the pH of the cellular endolysomal nanoenvironment to prevent AuNP aggregation (this is explained in more detail in the proceeding sections). At this stage, it is clear that further work needs to be concentrated on elucidating the exact heating mechanism to allow for full optimization of NP-targeted cancer hyperthermia treatment.

Gold Nanoparticle Heating Mechanism

Previous models by Hanson *et al.* (*61, 62*) have included classical Mie theory for the absorption of electromagnetic energy by small spheres (including absorbing AuNP coatings), as well as electronic absorption occurring at the surface of the AuNPs due to electron spill-out, surface roughness, and effects of surface phonons. In his work (*62*), he demonstrated that the maximum cross-sectional absorption area for uncoated AuNPs of diameters 200 nm at 10 MHz would be ~2.67×10^{-26} m^2, and is dominated by the magnetic dipole contribution. The cross-sectional absorption area could be increased to 2.1×10^{-21} m^2 by coating the AuNP with a thin (1 nm) dielectric layer of conductivity 0.0063 $S.m^{-1}$. Furthermore, for the same diameter, the maximum possible cross-sectional absorption area was found to be 1.6×10^{-20} m^2 for an optimized spherical NP of conductivity 0.078 $S.m^{-1}$, which is far from the conductivity of gold (4.6×10^7 $S.m^{-1}$) and is strongly dominated by the electric dipole contribution.

The most recent theoretical thermodynamic evaluation by Hanson *et al.* concluded that at RF frequencies the absorption of metallic AuNPs is too small, even for unit volume fractions, to enhance absorption and that any observed heating is directly due to the water itself (*61*). In their work, they investigated AuNPs of diameter 40 nm (of volume fraction 10^{-6} with a coating of conductivity 0.01 $S.m^{-1}$) suspended in 1.5 ml of DI water (in a quartz cuvette) under ambient open-air conditions. These parameters are similar to the experimental setup used in our recent investigations (*4, 6, 7, 11*).

To achieve a noticeable increase in temperature by tens of degrees it was estimated that the power absorbed per NP would need to be increased by a factor of 10^7 for noticeable heating effects to take place. Using parameters from our previous work, they assumed that the electric-field intensity outside the cuvette was on the lower end of the $kV.m^{-1}$ scale. However, as we discovered through use of refined E-field probes (data under review), our RF E-field strength is actually ~770 $kV.m^{-1}$ rather than the ~12.5 $kV.m^{-1}$ previously thought. Considering that the recent work of Raoof *et al.* (*7*) has observed heat generation from purified AuNPs, where the background bulk heating rates of the buffer suspension have been subtracted, we feel these theoretical models should be re-evaluated.

The most recent theoretical models published concerning the absorption of RF energy by AuNP suspensions looked at both Mie theory and Maxwell–Wagner effective medium theory (with the effect of the counter-ion relaxation and electric-double layer also taken into account), as well as dielectric loss resulting from electrophoretic particle acceleration (*63*). In this study, similar to the conclusions of Hanson *et al.*, it was concluded that the Mie absorption cross-section of individual AuNPs is very small and cannot account for heating effects in either a weak electrolyte solution or tissues rich in water.

Alternatively, using Maxwell-Garner theory, the effect upon the attenuation coefficient (α) of a propagating electromagnetic wave was also investigated and found to be ~50 times smaller for a colloidal suspension of 10 nm diameter AuNPs (volume fraction 2×10^{-2}) when compared to the attenuation induced by the host medium alone. As the results of these two models were essentially equivalent, it was suggested that dielectric losses are primarily dominated by the host medium with negligible contribution from the AuNPs themselves (*63*). The effect of the electric-double layer in the context of Maxwell-Wagner theory was also investigated and found to do little to enhance the absorption properties of AuNPs (although the electric double-layer did affect the conductivity of the NP).

The key conclusions by Sassaroli *et al.* were that both the ionic contribution and the motion of charged particles are potentially important factors in increased absorption of RF energy by AuNP suspensions. If we compare our recent experimental data (under review) with their numerical simulations of electrophoretic movement of charged NPs in a water medium, direct parallels can be drawn as they demonstrated that dielectric loss was proportional to NP surface charge and concentration, yet inversely proportional to size (for a fixed volume fraction). Having found similar trends in our own recent work we speculate that further theoretical investigations should be initiated.

Biological Studies with Biofunctional Nanoparticles

Gold Nanoparticles

Gold nanoparticles are favored as hyperthermia targeting agents because of their excellent biocompatibility. They have been used for their medicinal value in a number of diseases and have a proven track record of safety in humans. They can be synthesized to precise dimensions using simple methods, and can be modified using gold-thiol chemistry for multi-functionality. Given these advantages, it is not surprising that AuNPs have been investigated as a radiowave susceptor in a number of cell lines including liver, pancreatic and prostate cancer cells.

Gannon *et al.* exposed human liver cancer cells (Hep3B) and pancreatic cancer cells (Panc-1) to 5 nm AuNPs (*2*). They found that these NPs were internalized in cancer cells but had negligible cytotoxicity when used alone. Further study demonstrated that 1-5 min RF exposure of cancer cells that had internalized AuNPs (exposed at 67 μM) induced significant cell death compared to controls that were only treated with RF exposure. Heating of AuNPs in human liver cancer cells (HepG2) was also reported by Cardinal *et al.* (*64*). They utilized 13 nm citrate-capped AuNPs at a concentration of 4nM for an exposure of 4 hours. Immediately after, cell culture media containing AuNPs was replaced with fresh media and the cells were exposed to RF. They noted that the bulk media of AuNP treated-cells heated faster than the bulk media of control cells. Although the determination of heat source was not made, increased cell kill was noted only in cells treated with AuNPs with subsequent RF exposure. The findings were further validated in the rat hepatoma (JM-1) model where intra-tumoral injection of AuNPs followed by RF resulted in significant necrosis compared to tumors treated with RF alone. These two independent studies provided the proof-of-concept that supported use of AuNPs for RF-based hyperthermia.

Several studies have focused on investigating antibody-conjugated AuNPs for RF-based hyperthermia with the ultimate aim of delivering molecularly targeted hyperthermia at a single cancer cell-level. Curley *et al.* functionalized ~5 nm AuNPs with an FDA approved anti-epidermal growth factor receptor (EGFR-1) monoclonal antibody (also known as cetuximab, C225) (*65*). They demonstrated cancer cell selective internalization of AuNPs after targeting which was not seen with AuNPs conjugated with nonspecific IgG antibodies. They found that the internalization was rapid and the AuNPs localized to cytoplasmic vesicles within cancer cells. Minimal toxicity was noted with these conjugates alone (62.5-500 µM) in the four human cancer cell lines tested (Panc-1, Difi, Cama-1 and SN12PM6). When cells were exposed to an RF field, after first washing the cells with PBS to remove excess AuNPs, cancer cells targeted with C225-conjugated AuNPs underwent enhanced apoptosis and necrosis compared to nonspecific control AuNPs. This thermal cytotoxicity could be attenuated when internalization was blocked by pre-incubation with excess C225 demonstrating selectivity of thermal cytotoxicity.

Further experiments demonstrated that the cancer cell-selective toxicity was only seen in cells that over-expressed EGFR-1 (Panc-1 and Difi) but not in cells with low expression of EGFR-1 (Cama-1 and SN12PM6). These findings were further confirmed by Glazer *et al.* (*66*). It was demonstrated that concentrations of C225-AuNPs as low as 100 nM were sufficient to cause internalization and significant thermal toxicity after RF exposure selectively in EGFR-expressing Panc-1 cells but not in low EGFR-1 expressing Cama-1 cells. They noted a 20-fold increase in necrotic population that was attributed to RF-induced targeted hyperthermia using C225-AuNP. While these studies demonstrated tumor cell selective hyperthermia, noncovalent functionalization of AuNPs is not optimal and may cause desorption of targeting antibody by nonspecific serum proteins (*67*).

To address this possibility the authors modified the conjugation protocol in their subsequent study, where noncovalent conjugation of antibody to AuNPs was replaced with pseudo-covalent conjugation using gold-thiol chemistry by means of a linker. In this study, again, internalization and tumor cell-selective RF-triggered cytotoxicity could be demonstrated in Panc-1 cells albeit at a higher concentration of 500 µM (*68*). The conjugates were also administered *in vivo* in an ectopic mouse model of human pancreatic cancer at a dose of 10 mg/kg/wk. The majority of gold nanoconjugates accumulated in liver and spleen with some accumulation in pancreatic tumors. The mice that were treated with gold nanoconjugates followed by RF exposure once weekly for 7 weeks demonstrated significant retardation of tumor growth compared to untreated, RF-only, or C225-AuNP-only control group mice. In order to validate these findings the authors repeated these studies with another pancreatic cancer specific target MUC1, using PAM4 antibody with similar results. Interestingly, thermal cytotoxicity was not noted in liver and spleen where maximum accumulation of AuNPs was observed: the reason for which remains unclear.

Taken together these studies provide *in vitro* and *in vivo* evidence of tumor cell specific effects of AuNPs targeted to these cancer cells and activated with RF field exposure. Contrary to these findings, lack of theoretical plausibility of AuNP heating, as described in the previous sections, and at least some experimental evidence negating RF absorption by AuNPs has questioned these findings. A fundamental limitation in resolving this discrepancy stems from our inability to perform single cell thermography. In addition, a mechanistic understanding of heat generation from AuNPs after RF activation within cancer cells is lacking. In an attempt to address those questions we have recently evaluated the effects of AuNP aggregation and RF-triggered thermal therapy.

We (among others) have noted that AuNP aggregation attenuates heating in a non-biological context (*69, 70*). It is also known that AuNPs localize to membrane-bound cytoplasmic vesicles, the endolysosomes, after internalization where the acidic pH of these vesicles and activation of proteases induces aggregation of AuNPs (*69*). Our findings demonstrated that by modulating pH of these vesicles, intracellular AuNP aggregation could be prevented. Furthermore, RF-triggered thermal cytotoxicity was higher for nonaggregated AuNPs in comparison to aggregated AuNPs. These results are currently being translated into *in vivo* tumor models.

Through our efforts to perform thermography at a cellular level, we have recently developed an assay that allows quantification of intracellular thermal dose up to 100 cumulative equivalent minutes at 43°C (CEM43) (*71*). Using this assay we were able to demonstrate that within our experimental conditions, cancer cells treated with C225-AuNPs and exposed to RF accumulated thermal dose 6-fold higher than control cells treated with RF alone after 4 minutes of RF activation. The assay utilizes thermal denaturation of firefly luciferase as a surrogate for thermal dose inside cancer cells. Luciferin is converted to oxyluciferin by luciferase and as a consequence light is emitted. By quantifying photons emitted, this assay allows retrospective thermal dose quantification but is labor-intensive and has a limited dynamic range. Newer techniques need to be developed that would allow thermal dose quantifications at a single cells level in real-time and confirm the experimental findings reported in prior studies.

Quantum Dots

Antibody-conjugated quantum dots were investigated by Glazer *et al* (*72*). RF heating of unconjugated quantum dots was not discussed in that study. Two types of quantum dots (indium-gallium-phosphide and cadmium-selenide) were evaluated after conjugation to C225. These C225-conjugated quantum dots were selectively internalized by EGFR-1 over-expressing pancreatic cancer (Panc-1) cells but not by cells with minimal expression of EGFR (Cama-1), at a dose of 2-200 nM. At these concentrations the quantum dots were found to be minimally toxic by themselves (viability 82% vs. 99% in untreated controls). However, after exposure to a sub-toxic dose of RF, significant toxicity was observed in cells that had internalized quantum dots as opposed to cells that were treated with RF only (viability 47.5% vs. 95%). This was further confirmed in co-culture experiments where low and high EGFR-1 expressing cells were cultured together and exposed to C225-conjugated quantum dots. The authors noted that RF exposure has no effect on Cama-1 cells but significant toxicity was observed in Panc-1 cells, which was attributed to targeted heating of quantum dots. In addition, C225-conjugated 10 nm AuNPs were also compared in the same manner and performed slightly better than the same dose of C225-conjugated quantum dots. As with AuNPs, the mechanism of cell death was found to be a combination of apoptosis and necrosis.

Single-Walled Carbon Nanotubes

Gannon *et al.* evaluated CoMoCAT single-walled carbon nanotube (SWNT) cytotoxicity in the RF field (*5*). These SWNTs were functionalized using a noncovalent interaction with a biocompatible polymer, Kentera. The authors demonstrated that human liver (Hep3B or HepG2) or pancreatic (Panc-1) cells were able to internalize these SWNTs despite their larger size and unusual aspect ratio and showed minimal toxicity after incubation at a dose of 5-500 mg.L^{-1} for 24 hours. However, a brief 2-minute RF exposure resulted in a RF dose- and SWNT dose-dependent toxicity. For proof-of-principle, the authors injected these SWNTs in a VX2 rabbit model of liver cancer. After intra-tumoral injection, the animals were exposed to RF field. On histological examination, necrosis was seen at the site of SWNT deposition whereas the surrounding normal and distant tumor tissue remained viable. Further work on full-length SWNTs is proceeding with caution based on numerous studies raising concerns for potential toxicity in pre-clinical studies. Ultra-short single-walled carbon nanotubes (US-SWNTs) and fullerenes may provide a much safer alternative and are under investigation. The studies discussed here clearly demonstrate that metallic or semi-conducting NPs can be used for targeted toxicity in cancer cells after RF activation.

Conclusion

This RF-based nano-heat system has the potential to be a useful modality in cancer therapy but several questions need to be addressed before translation of this modality to the clinic. At the cellular level, while heating of NPs and subsequent dissipation of heat to cancer cells is the most likely explanation for the RF-triggered toxicity, direct evidence of this has been lacking. It is also not clear how many NPs in a cell are needed to kill that cancer cell without harming the adjacent normal cell. Future advances in cellular thermography and bio-mathematical modeling can help address these questions and also entertain the possibility of nonthermal cytotoxicity. At the organism level, targeting NPs to cancerous tissue remains a challenge. Despite surface modification with targeting moieties, most NPs eventually localize to Kupffer cells and resident phagocytes in the liver and spleen, respectively. While no toxicity to these organs has been reported in mouse studies thus far, future studies need to provide an explanation for this phenomenon. In addition, newer targeting moieties (antibodies, aptamers, peptides etc.) may facilitate active targeting while additional surface modification may improve immune evasion for better delivery of the NPs to cancer cells. Last but not least, advances in *in vivo* thermography will have to be incorporated in the design of RF systems to allow for thermal dosimetry, real-time feedback and safety monitoring to make this a realistic treatment option in cancer patients.

References

1. Curley, S. A.; Cherukuri, P.; Briggs, K.; Patra, C. R.; Upton, M.; Dolson, E.; Mukherjee, P. J. Noninvasive radiofrequency field-induced hyperthermic cytotoxicity in human cancer cells using cetuximab-targeted gold nanoparticles. *J. Exp. Ther. Oncol.* **2008**, *7*, 313.
2. Gannon, C. J.; Patra, C. R.; Bhattacharya, R.; Mukherjee, P.; Curley, S. A. Intracellular gold nanoparticles enhance non-invasive radiofrequency thermal destruction of human gastrointestinal cancer cells. *J. Nanobiotechnol.* **2008**, *6*, 2.
3. Moran, C. H.; et al. Size-Dependent Joule Heating of Gold Nanoparticles Using Capacitively Coupled Radiofrequency Fields. *Nano Res.* **2009**, *2*, 400.
4. Glazer, E. S.; Bs, K. L. M.; Zhu, C. H.; Curley, S. A. Pancreatic carcinoma cells are susceptible to noninvasive radio frequency fields after treatment with targeted gold nanoparticles. *Surgery* **2010**, *148*, 319.
5. Gannon, C. J.; et al. Carbon nanotube-enhanced thermal destruction of cancer cells in a noninvasive radiofrequency field. *Cancer* **2007**, *110*, 2654.
6. Glazer, E. S.; Curley, S. A. Radiofrequency field-induced thermal cytotoxicity in cancer cells treated with fluorescent nanoparticles. *Cancer* **2010**, *116*, 3285.
7. Raoof, M.; et al. Stability of antibody-conjugated gold nanoparticles in the endolysosomal nanoenvironment: implications for noninvasive radiofrequency-based cancer therapy. *Nanomedicine* **2012**Feb17.
8. Raoof, M.; Zhu, C.; Kaluarachchi, W. D.; Curley, S. A. Luciferase-based protein denaturation assay for quantification of radiofrequency field-induced targeted hyperthermia: Developing an intracellular thermometer. *Int. J. Hyperthermia* **2012**, *28*, 202.
9. Cherukuri, P.; Curley, S. A. Use of nanoparticles for targeted, noninvasive thermal destruction of malignant cells. *Methods Mol. Biol.* **2010**, *624*, 359.
10. Cherukuri, P.; Glazer, E. S.; Curley, S. A. Targeted hyperthermia using metal nanoparticles. *Adv. Drug Delivery Rev.* **2010**, *62*, 339.

11. Glazer, E. S.; Curley, S. A. Non-invasive radiofrequency ablation of malignancies mediated by quantum dots, gold nanoparticles and carbon nanotubes. *Ther. Delivery* **2011**, *2*, 1325.

12. Raoof, M.; Curley, S. A. Non-invasive radiofrequency-induced targeted hyperthermia for the treatment of hepatocellular carcinoma. *Int. J. Hepatology* **2011**, *2011*, 676957.

13. Field, S. B.; Bleehen, N. M. Hyperthermia in the treatment of cancer. *Cancer Treat Rev.* **1979**, *6*, 63.

14. LeVeen, H. H.; et al. Radiofrequency thermotherapy, local chemotherapy, and arterial occlusion in the treatment of nonresectable cancer. *Am. Surg.* **1984**, *50*, 61.

15. Armitage, D. W.; LeVeen, H. H.; Pethig, R. Radiofrequency-induced hyperthermia: computer simulation of specific absorption rate distributions using realistic anatomical models. *Phys. Med. Biol.* **1983**, *28*, 31.

16. LeVeen, H. H.; Pontiggia, P. Radiofrequency thermotherapy (RFTT) in the treatment of cancer. *Prog. Clin. Biol. Res.* **1982**, *107*, 731.

17. LeVeen, H. H.; Obrien, P.; Wallace, K. M. Radiofrequency thermo therapy for cancer. *J S C Med. Assoc.* **1980**, *76*, 5.

18. Sugaar, S.; LeVeen, H. H. A histopathologic study on the effects of radiofrequency thermotherapy on malignant tumors of the lung. *Cancer* **1979**, *43*, 767.

19. LeVeen, H. H.; Wapnick, S.; Piccone, V.; Falk, G.; Ahmed, N. Tumor eradication by radiofrequency therapy. Responses in 21 patients. *JAMA* **1976**, *235*, 2198.

20. Ohguri, T.; et al. Deep regional hyperthermia for the whole thoracic region using 8 MHz radiofrequency-capacitive heating device: relationship between the radiofrequency-output power and the intra-oesophageal temperature and predictive factors for a good heating in 59 patients. *Int. J. Hyperthermia* **2011**, *27*, 20.

21. Ohguri, T.; et al. Radiotherapy with 8-MHz radiofrequency-capacitive regional hyperthermia for stage III non-small-cell lung cancer: the radiofrequency-output power correlates with the intraesophageal temperature and clinical outcomes. *Int. J. Radiat. Oncol., Biol., Phys.* **2009**, *73*, 128.

22. Ohguri, T.; et al. Radiotherapy with 8 MHz radiofrequency-capacitive regional hyperthermia for pain relief of unresectable and recurrent colorectal cancer. *Int. J. Hyperthermia* **2006**, *22*, 1.

23. Ohguri, T.; et al. Effect of 8-MHz radiofrequency-capacitive regional hyperthermia with strong superficial cooling for unresectable or recurrent colorectal cancer. *Int. J. Hyperthermia* **2004**, *20*, 465.

24. Yamamoto, K.; Tanaka, Y. Radiofrequency capacitive hyperthermia for unresectable hepatic cancers. *J. Gastroenterol.* **1997**, *32*, 361.

25. Nagata, Y.; et al. Clinical results of radiofrequency hyperthermia for malignant liver tumors. *Int. J. Radiat. Oncol., Biol., Phys.* **1997**, *38*, 359.

26. Uchibayashi, T.; et al. Radiofrequency capacitive hyperthermia combined with irradiation or chemotherapy for patients with invasive bladder-cancer. *Oncol. Rep.* **1995**, *2*, 773.

27. Urata, K.; et al. Radiofrequency hyperthermia for malignant liver tumors: the clinical results of seven patients. *Hepatogastroenterology* **1995**, *42*, 492.

28. Uchibayashi, T.; Yamamoto, H.; Kunimi, K.; Koshida, K.; Nakajima, K. Radiofrequency capacitive hyperthermia combined with irradiation or chemotherapy for patients with invasive bladder cancers. *Int. Urol. Nephrol.* **1995**, *27*, 735.

29. Uchibayashi, T.; et al. Combined treatment of radiofrequency capacitive hyperthermia for urological malignancies. *Oncol. Rep.* **1994**, *1*, 937.

30. Fujimura, T.; et al. Radiofrequency capacitive hyperthermia for superficial malignant-tumors. *Int. J. Oncol.* **1993**, *2*, 1017.

31. Uchibayashi, T.; et al. Studies of temperature rise in bladder cancer and surrounding tissues during radiofrequency hyperthermia. *Eur. Urol.* **1992**, *21*, 299.

32. Hamazoe, R.; Maeta, M.; Murakami, A.; Yamashiro, H.; Kaibara, N. Heating efficiency of radiofrequency capacitive hyperthermia for treatment of deep-seated tumors in the peritoneal cavity. *J. Surg. Oncol.* **1991**, *48*, 176.

33. Hashimoto, T.; Hisazumi, H.; Nakajima, K.; Matsubara, F. Studies on endocrine changes induced by 8 MHz local radiofrequency hyperthermia in patients with bladder cancer. *Int. J. Hyperthermia* **1991**, *7*, 551.

34. Kato, H.; et al. Control of specific absorption rate distribution using capacitive electrodes and inductive aperture-type applicators: implications for radiofrequency hyperthermia. *IEEE Trans. Biomed. Eng.* **1991**, *38*, 644.

35. Hamazoe, R.; et al. Attempt to induce total-body hyperthermia by whole-abdominal hyperthermia using a radiofrequency capacitive-heating system: an experimental study in dogs. *Int. J. Hyperthermia* **1991**, *7*, 385.

36. Nishimura, Y.; et al. Radiofrequency (RF) capacitive hyperthermia combined with radiotherapy in the treatment of abdominal and pelvic deep-seated tumors. *Radiother. Oncol.* **1989**, *16*, 139.

37. Tanaka, R.; Kim, C. H.; Yamada, N.; Saito, Y. Radiofrequency hyperthermia for malignant brain tumors: preliminary results of clinical trials. *Neurosurgery* **1987**, *21*, 478.

38. Hiraoka, M.; et al. Radiofrequency capacitive hyperthermia for deep-seated tumors. I. Studies on thermometry. *Cancer* **1987**, *60*, 121.

39. Lee, W. M.; Ameziane, A.; van den Biggelaar, A. M.; Rietveld, P. J.; van Rhoon, G. C. Stability and accuracy of power and phase measurements of a VVM system designed for online quality control of the BSD-2000 (-3D) DHT system. *Int. J. Hyperthermia* **2003**, *19*, 74.

40. Lamprecht, U.; Gromoll, C.; Hehr, T.; Buchgeister, M.; Bamberg, M. An on-line phase measurement system for quality assurance of the BSD 2000. Part II: results of the phase measurement system. *Int. J. Hyperthermia* **2000**, *16*, 365.

41. Gromoll, C.; Lamprecht, U.; Hehr, T.; Buchgeister, M.; Bamberg, M. An on-line phase measurement system for quality assurance of the BSD 2000. Part I: technical description of the measurement system. *Int. J. Hyperthermia* **2000**, *16*, 355.

42. Turner, P. F. Comparisons between the BSD-2000 Quad Amplifier and a new prototype solid state amplifier for deep regional hyperthermia. *Int. J. Hyperthermia* **1999**, *15*, 339.

43. Raskmark, P.; Larsen, T.; Hornsleth, S. N. BSD 2000 multi-applicator hyperthermia system using scattering- or S-parameters. *Int. J. Hyperthermia* **1992**Jul-Aug, *8*, 555.

44. Myerson, R. J.; et al. Phantom studies and preliminary clinical experience with the BSD 2000. *Int. J. Hyperthermia* **1991**, *7*, 937.

45. Wust, P.; et al. Determinant factors and disturbances in controlling power distribution patterns by the hyperthermia-ring system BSD-2000. 2. Measuring techniques and analysis. *Strahlenther. Onkol.* **1991**, *167*, 172.

46. Notter, M.; Schwegler, N.; Burkard, W. Thermoradiotherapy of pelvic tumours with the BSD 2000: first clinical results. *Adv. Exp. Med. Biol.* **1990**, *267*, 439.

47. Turner, P. F.; Tumeh, A.; Schaefermeyer, T. BSD-2000 approach for deep local and regional hyperthermia: physics and technology. *Strahlenther. Onkol.* **1989**, *165*, 738.

48. Turner, P. F.; Schaefermeyer, T. BSD-2000 approach for deep local and regional hyperthermia: clinical utility. *Strahlenther. Onkol.* **1989**, *165*, 700.

49. Link, S.; El-Sayed, M. A. Shape and size dependence of radiative, non-radiative and photothermal properties of gold nanocrystals. *Int. Rev. Phys. Chem.* **2000**, *19*, 409.

50. Lal, S.; Clare, S. E.; Halas, N. J. Nanoshell-Enabled Photothermal Cancer Therapy: Impending Clinical Impact. *Acc. Chem. Res.* **2008**, *41*, 1842.

51. O'neal, D. P.; Hirsch, L. R.; Halas, N. J.; Payne, J. D.; West, J. L. Photothermal tumor ablation in mice using near infrared-absorbing nanoparticles. *Cancer Lett.* **2004**, *209*, 171.

52. Loo, C.; Lowery, A.; Halas, N. J.; West, J.; Drezek, R. Immunotargeted nanoshells for integrated cancer imaging and therapy. *Nano Lett.* **2005**, *5*, 709.

53. Kim, Y. S.; et al. Cancer treatment using an optically inert Rose Bengal derivative combined with pulsed focused ultrasound. *J. Controlled Release* **2011**, *156*, 315.

54. O'Neill, B. E.; Karmonik, C.; Li, K. C. An optimum method for pulsed high intensity focused ultrasound treatment of large volumes using the InSightec ExAblate(R) 2000 system. *Phys. Med. Biol.* **2010**, *55*, 6395.

55. O'Neill, B. E.; Karmonik, C.; Sassaroli, E.; Li, K. C. Estimation of thermal dose from MR thermometry during application of nonablative pulsed high intensity focused ultrasound. *J. Magn. Reson. Imaging* **2012**, *35*, 1169.

56. O'Neill, B. E.; Li, K. C. Augmentation of targeted delivery with pulsed high intensity focused ultrasound. *Int. J. Hyperthermia* **2008**, *24*, 506.

57. O'Neill, B. E.; et al. Pulsed high intensity focused ultrasound mediated nanoparticle delivery: mechanisms and efficacy in murine muscle. *Ultrasound Med. Biol.* **2009**, *35*, 416.

58. Sassaroli, E.; Li, K. C.; O'Neill, B. E. Modeling focused ultrasound exposure for the optimal control of thermal dose distribution. *TheScientificWorldJournal* **2012**, *2012*, 252741.

59. Kruse, D. E.; et al. A Radio-Frequency Coupling Network for Heating of Citrate-Coated Gold Nanoparticles for Cancer Therapy: Design and Analysis. *IEEE Trans. Biomed. Eng.* **2011**, *58*, 10.

60. Li, D.; et al. Negligible absorption of radiofrequency radiation by colloidal gold nanoparticles. *J. Colloid Interface Sci.* **2011**, *358*, 47.

61. Hanson, G. W.; Monreal, R. C.; Apell, S. P. Electromagnetic absorption mechanisms in metal nanospheres: Bulk and surface effects in radiofrequency-terahertz heating of nanoparticles. *J. Appl. Phys.* **2011**, *109*, 124306.

62. Hanson, G. W.; Patch, S. K. Optimum electromagnetic heating of nanoparticle thermal contrast agents at rf frequencies. *J. Appl. Phys.* **2009**, *106*, 054309.

63. Sassaroli, E.; Li, K. C. P.; O'Neill, B. E. Radio frequency absorption in gold nanoparticle suspensions: a phenomenological study. *J. Phys. D: Appl. Phys.* **2012**, *45*, 075303.

64. Abbott, D. E.; et al. Resection of liver metastases from breast cancer: Estrogen receptor status and response to chemotherapy before metastasectomy define outcome. *Surgery* **2012**, *151*, 710.

65. Curley, S. A.; et al. Noninvasive radiofrequency field-induced hyperthermic cytotoxicity in human cancer cells using cetuximab-targeted gold nanoparticles. *J. Exp. Ther. Oncol.* **2008**, *7*, 313.

66. Glazer, E. S.; Massey, K. L.; Zhu, C.; Curley, S. A. Pancreatic carcinoma cells are susceptible to noninvasive radio frequency fields after treatment with targeted gold nanoparticles. *Surgery* **2010**, *148*, 319.

67. Kumar, S.; Aaron, J.; Sokolov, K. Directional conjugation of antibodies to nanoparticles for synthesis of multiplexed optical contrast agents with both delivery and targeting moieties. *Nat. Protoc.* **2008**, *3*, 314.

68. Glazer, E. S.; et al. Noninvasive radiofrequency field destruction of pancreatic adenocarcinoma xenografts treated with targeted gold nanoparticles. *Clin. Cancer Res.* **2010**, *16*, 5712.

69. Raoof, M.; et al. Stability of antibody-conjugated gold nanoparticles in the endolysosomal nanoenvironment: implications for noninvasive radiofrequency-based cancer therapy. *Nanomedicine* **2012**Feb17.

70. Li, D.; et al. Negligible absorption of radiofrequency radiation by colloidal gold nanoparticles. *J. Colloid Interface Sci.* **2011**, *358*, 47.

71. Raoof, M.; Zhu, C.; Kaluarachchi, W. D.; Curley, S. A. Luciferase-based protein denaturation assay for quantification of radiofrequency field-induced targeted hyperthermia: Developing an intracellular thermometer. *Int. J. Hyperthermia* **2012**, *28*, 202.

72. Glazer, E. S.; Curley, S. A. Radiofrequency field-induced thermal cytotoxicity in cancer cells treated with fluorescent nanoparticles. *Cancer* **2010**, *116*, 3285.

Chapter 7

The Design of Hybrid Nanoparticles for Image-Guided Radiotherapy

Christophe Alric,[a] Rana Bazzi,[b] François Lux,[a] Gautier Laurent,[b] Matteo Martini,[a] Marie Dutreix,[c] Géraldine Le Duc,[d] Pascal Perriat,[e] Stéphane Roux,[*,b] and Olivier Tillement[a]

[a]Laboratoire de Physico-Chimie des Matériaux Luminescents, UMR 5620 CNRS, Université Claude Bernard Lyon 1, 69622 Villeurbanne Cedex, France
[b]Institut UTINAM, UMR 6213 CNRS - Université de Franche-Comté, 16 route de Gray, 25030 Besançon, France
[c]Institut Curie, Equipe, "Recombinaison et instabilité des génomes", UMR 2027, Orsay, France
[d]European Synchrotron Radiation Facility, ID 17 Biomedical Beamline, Polygone Scientifique Louis Néel, 6 rue Jules Horowitz, 38000 Grenoble, France
[e]Matériaux Ingénierie et Science, UMR 5510 CNRS, INSA de Lyon, 69621 Villeurbanne Cedex, France
*E-mail: stephane.roux@univ-fcomte.fr

Many studies revealed the high potential of multifunctional nanoparticles for biomedical applications. Since these nanoparticles can be designed for combining imaging and remotely controlled therapeutic activity, image-guided therapy which rests on the induction of nanoparticles toxicity by external stimulus when the nanoparticles content is both high in the diseased zone and low in the healthy tissue can be envisaged, especially for fighting cancer (one of the most important cause of mortality in several countries). Image-guided therapy should lead to valuable improvements in radiation-based therapy provided that the multifunctional radiosensitizing nanoparticles developed for increasing the selectivity of the radiotherapy (and therefore the efficiency) meet the criteria imposed by *in vivo* applications and by the physical principles of the interaction between radiation and the matter.

© 2012 American Chemical Society

Diagnosis and Treatment of Cancer

Cancer

The term cancer includes a set of diseases characterized by unlimited proliferation of cells that, following the alteration of their genetic heritage, escape the normal mechanisms of differentiation and regulation of their proliferation. This increase in surplus populations of cells destroys the surrounding tissue and may be accompanied by a dispersion throughout the body. In many cases, this can cause death (*1*).

There are many different types of cancer (more than a hundred) that can affect any part of the body. The most common cancers in women are (in descending order of frequency) breast cancer, colorectal cancer and cancers of the cervix and lungs. In men, lung cancer and prostate cancer are most common, followed by colorectal and stomach cancer (*2*). An estimated eleven million cancers are diagnosed worldwide each year and between seven and nine million people die, which is about 13% of global mortality. According to estimates by international health agencies, the annual number of cancer deaths is expected to rise to twelve million in 2030 (*3*).

Induction and Growth

It is now accepted that the development of cancer can be outlined in three phases, which can take several years in total (*1*):

- The initiation, which corresponds to irreversible damage to the DNA of a cell (either spontaneously or after exposure to a chemical, physical, or biological carcinogen...). The "initiated" cell multiplies;
- The promotion, which corresponds to prolonged exposure (repeated or continuous) to a substance that maintains and stabilizes the lesion through the proliferation of "initiated" cells;
- The progression, which corresponds to the acquisition of the properties of uncontrolled proliferation, acquisition of independence, loss of differentiation, local and metastatic invasion.

Cancer cells are fundamentally different from "normal" cells. They have the ability to escape the regulatory mechanisms of the cell population (homeostasis) and thus gain "immortality" allowing them to divide again and again (*1, 4*). Moreover, they can implement various mechanisms to avoid detection or destruction by the immune system (tumor escape) (*5*). This allows them to multiply rapidly and without any control.

Growth of a malignant tumor is then very fast and leads to the invasion and destruction of surrounding healthy tissue (tumor invasion). Through the secretion of specific enzymes, tumor cells degrade the extracellular matrix, making it more loose and disorganized. Loss of adhesion between the cells and the displacement acquisition capacity (cell migration) result in the access of tumor cells to vessels (*1, 4*). The latter are of fundamental importance for tumor growth since they ensure the delivery of nutriments and oxygen to the tumor.

Angiogenesis and Metastasis

In normal physiological conditions, the creation of new blood capillaries from preexisting vessels (angiogenesis) is finely controlled by complex molecular mechanisms (balance between induction and inhibition of the angiogenesis) by taking into account the physiological needs (*6*). Certain stimuli (hypoxia, pressure generated by a large proliferation of cells) can lead to a shift in the balance regulating the formation of new blood vessels. This change, designated as the "angiogenic switch", is reflected by the increase in the concentration of angiogenesis inducing molecules which is concomitant with the decrease in the concentration of inhibiting molecules (*7*).

Up to a volume of about 1 mm^3, oxygen and nutrients are delivered to a growing solid tumor by diffusion through the surrounding interstitial spaces. Beyond this critical size, the tumor is exposed to hypoxia (*8*). Cancer cells secrete many factors inducing the angiogenic switch. As a result, blood vessels are created from preexisting vasculature (*9, 10*). These factors, such as vascular endothelial growth factor (VEGF) or fibroblast growth factors (FGFs) are involved in proliferation and arrangement of endothelial cells necessary for the formation of new vessels which will then bring oxygen and nutrients for tumor growth. The latter begins to grow exponentially. The formation of lymphatic neo-vessels was also initiated by the secretion of growth factors in tumor development (*11*).

The formation of new vessels (blood and lymph) induced by the growth of a solid tumor appears to be the essential phenomenon for the dissemination of cancer cells in the body (*11, 12*). They are able to break into new vessels (intravasion), to travel and then to extract (extravasation) to reach another place. Cancer cells can then establish distant colonies (metastases). They may then induce neovascularization within their location in order to grow and spread in the body (*13, 14*). Metastases are by far the largest cause of mortality from cancer (*15*).

The establishment of neovascularization is a sequenced process essential for tumor growth and metastasis (*10, 16*). The resulting new vessels which are made haphazardly, constitute a prime target for many innovative treatments (*17, 18*).

EPR Effect

Solid tumors generally have a different physiology from that of healthy tissue (*19, 20*). These different specificities, which depend on the types of cancers, can be exploited in order to only treat tumors while sparing the surrounding healthy tissue.

One of the most remarkable illustrations of the physiology of solid tumors is their hyper-vascularization, which is structurally and functionally abnormal in most tumors (*7, 21*). Because of their rapid growth which is continually induced by growth factors, newly formed blood vessels are often disorganized, tortuous and have non-constant diameter (*7, 22, 23*). Endothelial cells that compose them are often not contiguous. The fenestration of the vessels which irrigate the tumors is characterized by void spaces between cells of hundreds of nanometers (*24–26*), whereas in the case of continuous healthy endothelium of blood capillaries, the distance between endothelial cells is of the order of 3 to 6 nm, allowing the passage of ions and small molecules necessary for the microcirculation (*27*) (Figure 1).

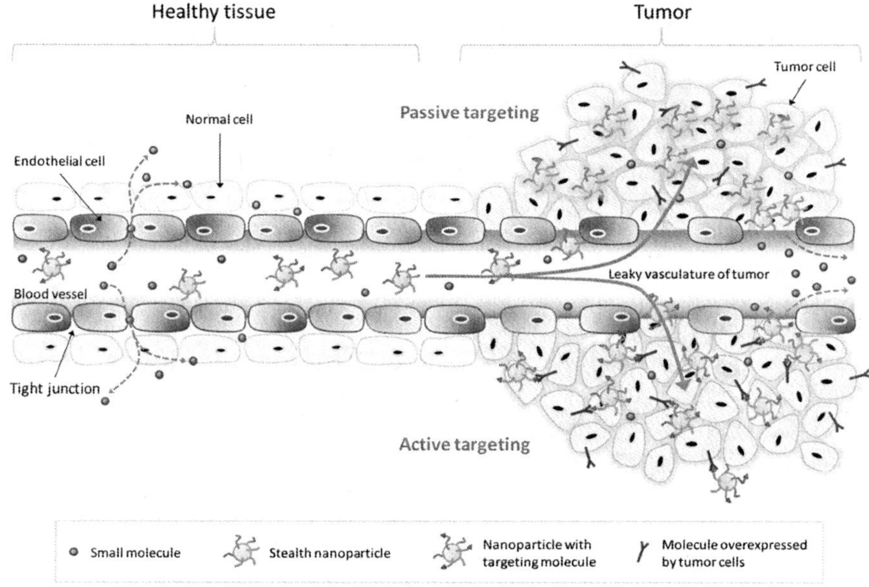

Figure 1. Schematic representation of blood vessels irrigating healthy tissue (left) and tumor (right). (see color insert)

Blood vessels from neoangiogenesis are often porous and permeable (*28, 29*). Another notable feature of the environment of solid tumors is the absence of fully functional lymphatic vessels, despite the lymphogenesis initiated by growth factors (*7, 30*). Associated with permeable hyper-vascularization of tumor, the lack of lymphatic drainage may result in accumulation of macromolecules in the interstitial spaces, that can be exploited to target the tumor environment (*31*). This phenomenon, known as the EPR effect (Enhanced Permeability and Retention Effect), allows a "passive" targeting of highly vascularized solid tumors, using macromolecules or nano-objects (Figure 1). The EPR effect is now one of the major pathways for the detection and treatment of solid tumors (*32*).

Moreover in the case of solid tumors, most of the cells (both cancerous and stroma cells) overexpress on their surface, because of extensive metabolism, different molecules which are less present on normal cells. These molecules are mostly membrane receptors (proteins) involved in the growth and survival of cancer cells. Their expression differs among cancer types. For example, receptors for growth factors, such as EGFR (Epidermal Growth Factor Receptor), are overexpressed in some cancers including colorectal or HER2 (Human Epidermal Growth Factor Receptor-2), from certain types of breast cancer. The overexpression of these markers gives sometimes an estimation of the virulence of the disease and permits to assess the prognosis (*33, 34*). Many ligands have been developed to allow specific targeting of cancerous cell surface receptors, thus paving the way for targeted therapies (*35–37*) (Figure 1).

In most cases, active targeting is possible only if passive targeting is observed. The attractive potential of active targeting does not rest only on the increase of accumulated therapeutic agents in the tumor but also in the increase of the residence time owing to the interaction with cancerous cell receptors. The return in bloodstream of the therapeutic agents occurs indeed more belatedly than in the case of passive targeting.

Another promising strategy is the targeting of receptors expressed by stroma cells and tumor vascularization, particularly the neoangiogenic endothelial cells (*38, 39*). The vascularization plays an important nutritional role in the survival and growth of solid tumors (intake of oxygen and nutrients) but also allows its metastasis. It is therefore an essential target in anticancer strategy (*40, 41*). The inhibition of neoangiogenesis or the selective destruction of new vessels intended to indirectly causes the death of cancerous cells by depriving them of nutrient and oxygen they need to live and multiply. The antiangiogenic and antivascular therapies are based on the overexpression of membrane markers on the neoangiogenic endothelial cells (as compared to endothelial cells of healthy tissues).

Solid tumors have remarkable physiological characteristics (permeable and uncontrolled vascularization, overexpression of membrane receptors by tumor cells or stroma cells) for different types of cancers and can evolve during their progression. These features can be exploited to detect as early as possible and treat selectively the tumor areas. Currently, medical imaging remains one of the major tools for cancer detection (*1, 42*).

Medical Imaging for the Visualization of Solid Tumors

Medical imaging is now one of the pillars of the fight against cancer. Its contribution in this field includes the detection of lesions before the onset of clinical signs, the evaluation of a stage of disease progression and of the response to treatment and/or the planning of a therapeutic protocol. Medical imaging plays an important role in defining target volumes for radiation treatment (*43, 44*).

Born at the end of 19th century with the discovery and use of X-rays to visualize bone structures, medical imaging today consists of several techniques for noninvasively visualizing within a living organism. These techniques rely on different physical principles that determine their resolution, sensitivity and type of tissue or biological phenomena that they render visible (*45, 46*).

The X-ray imaging is an indispensable technique in medical imaging. Since the discovery of X-rays by Roentgen in 1895, it was the only imaging technique available until the mid-twentieth century. The progress in computer science contributed to the improvements of X-ray imaging.

The X-ray imaging is based on the differential attenuation of a beam of incident X-rays (25–150 keV) by the body tissues. A X-ray beam passes through a patient and is collected on a photographic plate or a photon detector before being converted into digital signals. Tissues absorb radiation depending on their composition (atomic number and density) and thickness. The invention of X-ray computed tomography (or "CT scanner") by Hounsfield in 1972 revolutionized the X-ray imaging. It is based on the acquisition of multiple projections of the patient at different angles through a system of rotation of the X-ray tube (*47*). CT scanner quickly provides three-dimensional snapshots of the living organism anatomy, with excellent spatial resolution (~ 50 microns). This technique is also able to image in a single session deep tissues (whole body). X-ray computed tomography is now one of the major tools in medical imaging (*48*). The main limitation of the X-ray imaging is the use of ionizing radiation, which limits the number of examinations per patient (*48*). This technique is also hampered by a poor resolution of soft tissue and a low sensitivity (*42*). The injection of contrast agents allows in some extent improving the contrast of soft tissues, but the sensitivity of the X-ray imaging remains below that of other imaging techniques.

The majority of contrast agents for X-ray imaging are iodinated molecules. Iodine increases the contrast of soft tissues (where contrast agents are) by the photoelectric effect, thanks to its high atomic number ($Z(I) = 53$). The contrast agents for X-ray imaging are derivatives of 1,3,5-triiodobenzene (*49*). Once injected into the bloodstream, iodinated contrast agents (ICAs) are spread throughout the extracellular space due to their low molecular weight (less than 2000 g.mol^{-1}) and are rapidly cleared from the body, mainly by renal excretion (*50*). They allow both better visualization of soft tissue and a transient contrast enhancement between normal tissue and diseased tissues (including tumor hypervascularization). The limitations of molecular ICAs are their rapid elimination from the body, their non-specific distribution and high viscosity due to very high concentrations of iodine (from 100 to 370 g.L^{-1}).

Magnetic Resonance Imaging (MRI)

The magnetic resonance imaging (MRI) is a medical imaging technique which provides highly resolved three-dimensional images of living bodies using nuclear magnetic resonance (NMR). MRI exploits the magnetic properties of protons (constituting about 85% of biological tissue) to image inside a body, without using ionizing radiation (*51*). Parameters at the origin of contrast in MRI images are the relaxation time of the tissue magnetization (T_1 and T_2)

and the density of protons ρ in these tissues. The relaxation times are identical for the same type of tissue and therefore allow to distinguish different tissue types. The relaxation times are sensitive to the physicochemical environment of protons in imaged tissues. MRI allows obtaining anatomical images of soft tissue (containing many protons) of very good quality and is not limited in depth acquisition. It is the reference technique for the study of central nervous system, both for anatomical and functional study (*e.g.* brain). MRI is currently one of the most commonly used imaging techniques, especially for the detection of tumors.

However, MRI has significant limitations, which include a long acquisition time and the inability to image large areas (*e.g.* the whole body) in a single acquisition. Another weakness of this technique is its low sensitivity which sometimes makes difficult the distinction between normal and pathological tissues. For more reliable medical diagnosis, the administration of contrast agents to patients is commonly used to enhance the native contrast between different tissues. The use of magnetic contrast agents can enhance the natural contrast between different tissue types in decreasing the relaxation time of protons they contain. The ability of a contrast agent to modify the relaxation time of protons that surround it is quantitatively represented by its relaxivity (longitudinal r_1 or transversal r_2 relaxivity).

The contrast agents for MRI are divided into two types: superparamagnetic agents and paramagnetic compounds (*52*). Even if they act on both longitudinal (T_1) and transverse (T_2) relaxation times of protons, their effect is only predominant on one out of two. The superparamagnetic agents (T_2 or negative contrast agents) are iron oxide nanoparticles (5 to 200 nm) coated with a layer of hydrophilic polymer (*e.g.* dextran). Depending on their size, they are called SPIOs (SuperParamagnetic Iron Oxides, > 50 nm) or USPIOs (Ultrasmall SuperParamagnetic Iron Oxides, <50 nm). On MRI images, their presence results in a darkening of the image in areas where they are distributed. Nanoparticles of iron oxides are mainly used for the detection of lesions of the reticuloendothelial system (liver, spleen and lymph nodes), because of their significant retention in these organs when they are healthy. They therefore darken the healthy areas in images while diseased zones (devoid of superparamagnetic contrast agents) appear bright (*52*).

The paramagnetic contrast agents (T_1 or positive contrast agents) cause an increase in the signal. The zones containing the paramagnetic agents appear therefore bright. They are mainly composed of gadolinium(III) chelates (*52*). Complexation of the gadolinium(III) by linear or macrocyclic polyaminocarboxylate ligands ensures a safe use (free gadolinium ion is toxic), while allowing a transfer of magnetization to the protons of the surrounding tissues (*53*). The paramagnetic complexes are primarily intravenously injected at concentrations of 0.5 to 1 mol.L^{-1}. Their biodistribution depends on their size and their hydrophilic/hydrophobic character. After injection, gadolinium(III) chelates of the first generation (Gd-DTPA (DTPA: diethylenentriaminepentaacetic acid), Gd-DOTA (DOTA: 1,4,7,10-tetraazacyclododecane-1,4,7,10-tetraacetic acid) and their non-ionic derivatives) rapidly escape from the bloodstream and are distributed in the extracellular medium, because of their low molecular weight (less than 1000 g.mol^{-1}) (Scheme 1).

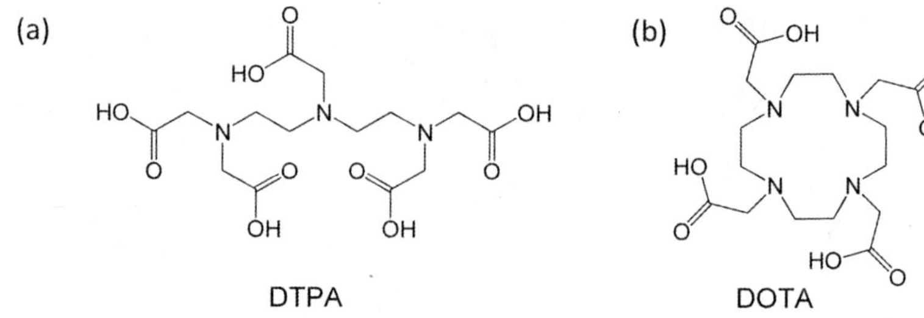

Scheme 1. The structures of the commercially available ligands (a) DTPA, (b) DOTA

Their presence in the tumor due to its permeable hyper-vascularization rapidly and transiently enhances the contrast. However they are rapidly eliminated through the urinary system. It must be pointed out that these agents cross the blood-brain barrier (BBB) when it is injured, for example in the case of a brain tumor (*52, 54*). The low blood half-life of these agents reduces the acquisition time window and decreases the likelihood of durably achieving a passive targeting of pathological areas. If the commercially available contrast agents for MRI have an important role in the detection and the therapy monitoring in oncology (*55*), they are however handicapped by their rapid elimination and by their lack of specificity. Moreover, the molecular character of these gadolinium(III) chelates limits their ability to induce a change in the longitudinal relaxation time of protons because of their low rotational correlation time which is inherent of their small size (*56*).

Nuclear Imaging

Nuclear imaging rests on the biodistribution of radioisotopes administered to the patient to achieve functional imaging of the body. It is divided into two techniques depending on the nature of the emitters: γ-scintigraphy and Single Photon Emission Computed Tomography, SPECT, (using γ emitters) and Positron Emission Tomography, PET, (using positron (β^+) emitters) (*57*). Unlike the γ-scintigraphy, SPECT and PET provide images of the three-dimensional radionuclide distribution in an organism. The radionuclide may be administered either as a salt or complex and eventually coupled to a vector exhibiting a biological activity. Different radionuclides can be used depending on the type of clinical examination (Table 1).

Table 1. Characteristics of radionuclides used for scintigraphy (γ emitters) and PET (β⁺ emitters)

Radionuclide	Half-life (h)	Energies (keV)
Technetium-99m (^{99m}Tc)	6.03	140.5 (88.5 %)
Indium-111 (^{111}In)	67.32	171.3 (90.6 %); 245.4 (94.1 %)
Iodine-123 (^{123}I)	13.22	159.0 (83.25 %)
Fluorine-18 (^{18}F)	109.8	633.5 (96.9 %)
Gallium-68 (^{68}Ga)	67.7	1899 (87.9 %)

The nuclear imaging techniques (primarily SPECT) can quite quickly realize functional imaging of an organism with an excellent sensitivity. They also allow precisely quantifying the activity in each organ and monitoring the biodistribution of a radiotracer over time (dynamic acquisition).

One of the main limitations of nuclear imaging techniques stems from their ionizing character, although the radiopharmaceutical compounds are used at very low concentrations. The low spatial resolution is another major limitation of these techniques, making it problematic for example the precise location of a tumor. The current trend is therefore to couple the SPECT or PET with a CT scanner to take advantage of the excellent spatial resolution of the latter and the high sensitivity of nuclear imaging techniques. The superposition of two types of images is attractive since it provides both anatomical and functional information (*58, 59*).

Fluorescence Imaging

The fluorescence imaging consists in visualizing the photons of visible light emitted by certain tissues (autofluorescence) or exogenous fluorophores after being excited by photons of higher energy. This imaging technique is well suited for obtaining morphological and functional information. Currently well developed in the *in vitro* and *ex vivo* studies of cellular mechanisms (*60, 61*), fluorescence imaging is increasingly used in preclinical studies on small animals, particularly in the field of drug development (*62*). However, application to humans of this type of imaging can be problematic because of:

- The autofluorescence of the tissue, which decreases the signal to noise ratio (and thus the sensitivity);
- The diffusion and the absorption of photons by biological tissues, which alter the spatial resolution and confine this technique to the imaging of the superficial tissues.

Table 2. Characteristics of the main medical imaging techniques

Technique	Detection	Resolution	Acquisition time	Sensitivity[a] $(mol.L^{-1}\ IA)$	Depth	Imaging agents[b]	Comments
CT scanner	X-ray photons	50–200 µm	Minutes	10^{-3} M	no limit	Iodinated molecules	Ionizing technique; low sensitivity but high resolution
MRI	electromagnetic variations	10–100 µm	Minutes - hour	10^{-3}–10^{-5} M	no limit	Gd^{3+} chelates; iron oxides	Non-ionizing technique; low sensitivity but high resolution
SPECT	γ-ray photons	1–2 mm	Minutes	10^{-10}–10^{-11} M	no limit	^{99m}Tc, ^{111}In,	Ionizing technique; low resolution but high sensitivity (precise quantification)
PET	γ-ray photons (511 keV)	1–2 mm	10 secondes - minutes	10^{-11}–10^{-12} M	no limit	^{18}F, ^{64}Cu,	Ionizing technique; low resolution but high sensitivity (precise quantification)
Fluorescence	NIR photons	1–3 mm	Secondes - minutes	10^{-9}–10^{-12} M	< 2 cm (2D) 2-6 cm (3D)	NIR organic dyes	Non-ionizing technique; low resolution but high sensitivity (semi-quantification for 2D)

[a] IA : Imaging agents (including the contrast agents which improve the sensitivity of the technique (MRI, X-ray imaging) and the agents which are required for the image acquisition (radionuclides for SPECT and PET or NIR dyes for fluorescence imaging)). [b] Only few examples are given.

Biological tissues strongly absorb UV and visible photons (up to 650 nm), because of the presence of certain components (*e.g.* water and hemoglobin) (*63*). Some of these constituents behave as fluorophores, responsible for the tissue autofluorescence (*64*). The use of photons having a wavelength between 650 and 900 nm (in the red and near infrared region (NIR)) can limit the phenomena of absorption by the tissues. This range of wavelength (650–900 nm) defines the "optical window" that allows a deeper exploration (few centimeters) and ensures the safety of the imaging method. Furthermore, the tissue autofluorescence is minimized in the NIR region (*65*). Many fluorophores whose excitation wavelength is located in the NIR region have been recently developed (*66*) and successfully used to visualize *in vivo* various biological phenomena (*67*). The fluorescence imaging is distinguished by its high sensitivity, its easy implementation and low cost but is hampered by a relatively low penetration depth (< 2 cm). Anatomical or functional images can quickly be performed without using ionizing radiation. The two-dimensional images provide semi-quantitative information but quantitative data to greater depths (2–6 cm) and with higher spatial resolution should be expected with the development of three-dimensional methods (*68*).

Towards a Multimodal Imaging

The imaging techniques described above exploit different physical principles, which determine the characteristics of these modalities. All imaging techniques do not provide the same information and their advantages and disadvantages are not the same (Table 2) (*69–71*).

Thus, CT and MRI can be used to evaluate the anatomy of a patient with a sub-millimeter spatial resolution and to locate precisely certain lesions. Both methods also provide a satisfactory tissue contrast – especially when using contrast agents – and can differentiate diseased tissue from healthy tissue by revealing structural differences or abnormalities of perfusion. However, these methods are unable to discriminate tissues which are anatomically normal but with impaired functioning. In contrast, nuclear imaging methods (γ-scintigraphy, SPECT and PET) permit to assess the functioning of tissues or organs with excellent sensitivity, but with insufficient spatial resolution. This lack of anatomical landmarks which is inherent in the use of radionuclides can not pinpoint the areas where are the tissues with altered function.

To access complementary information from different imaging techniques, the combination of several imaging modalities becomes increasingly common in clinical practice. Obtaining, for the same area, morphological and functional information can be achieved either by merging images acquired by complementary techniques or by using imaging devices combining several modalities (hybrid or multimodal imaging) (*72*). In both cases, the goal is to get a synergistic effect by combining the strengths of complementary techniques. Among the many possible combinations, the equipment integrating the modalities of positron emission tomography and CT scanner (PET/CT) are most common today (*73*), but devices combining single photon emission tomography and a CT scanner

(SPECT/CT) have recently been developed. Devices PET/CT and SPECT/CT allow the simultaneous acquisition of anatomical and functional information and allow a better localization of pathological areas while increasing the sensitivity and image quality compared to traditional PET and SPECT devices (*74*).

One of the favorite fields of these multimodal imaging is oncology, both for the diagnostic step (*75*) and for the definition of target volumes to be irradiated in the course of radiotherapy treatment (*76*). To minimize patient exposure and visualize the soft tissue with both excellent spatial resolution and high sensitivity, hybrid devices PET/MRI (*77*) and SPECT/MRI (*78*) have recently been described. Other combinations are being developed: optical imaging/(PET or SPECT), CT/MRI or optical imaging/MRI (*79*). These new combinations aim at maximizing the benefits of each modality (sensitivity, spatial resolution, anatomic or functional imaging, etc.). The goal is to detect and locate earlier lesions (including tumors) to allow the follow-up of the patient more quickly and correctly. Alongside the development of multimodal techniques, an intense research effort is made in the field of multimodal probes, which has spawned a proliferation of new objects that can be monitored simultaneously by multiple imaging modalities (*80, 81*).

Radiotherapy

An estimated 50% of patients with cancer should receive radiotherapy, either alone or in combination with other types of treatment (surgery and/or chemotherapy) (*82*). The cost to benefit ratio of radiotherapy is very favorable as compared to other techniques applied in oncology. This largely explains its extensive use (*83*). Cancer treatment by radiation therapy consists in an energy deposition in the largest possible area to eradicate tumor while sparing the surrounding healthy tissue. Different radiation types can be used as part of radiotherapy. The main ones are the X-ray photons of high energy (4–25 MeV) from linear accelerators. Depending on circumstances, electrons (8–30 MeV) or rarely protons, ions or neutrons can still be used. Whatever the type of ionizing radiation used, the principle of action remains the same. The energy deposition is performed by interactions between high energy X-ray beam and the constituents (mainly electrons) of the irradiated medium. The energy is deposited through the tumor by a succession of physical events leading to the formation of highly reactive species which cause the alteration of proteins and genetic material of cancerous cells. This may lead to the cell death and, under certain conditions, to the destruction of tumor. These events can be grouped into several phases occurring at characteristic time scales (*84*):

- The initial physical step, corresponding to the transfer of radiation energy in the medium through which it passes, by creation of ionization or excitation;
- Physicochemical step, during which are formed unstable and highly reactive species (radicals, peroxides);
- Molecular stage, during which these unstable species cause alterations in the different molecules constituting cells, mainly DNA;

- Cell stage, during which the cellular repair mechanisms try to remedy the various chemical alterations of cellular constituents. This step determines the survival or cell death.

Physical Processes: Interaction between Ionizing Radiation and Matter

Radiation is considered as ionizing when it produces ionization in the material through which it passes by transferring energy to its components. As part of the anticancer radiotherapy, ionizing radiation consists of high energy X-ray photons. This is indirectly ionizing radiation: ionizing character stems mainly from charged particles (electrons) which are set in motion during the interactions. These secondary electrons will in turn cause energy deposition by ionization or excitation in the medium. The high-energy X-rays used in radiotherapy interact with matter (mainly electrons) according to three predominant processes: the photoelectric effect, Compton effect and pair production (*85*).

The photoelectric effect involves the absorption of the total energy of a photon by an electron of the inner electron shells (K and L mainly) of an atom. The incident photon disappears and transfers all of its energy to an electron of inner shell, which is expelled from the atom. This mode of interaction occurs only when the incident photon energy is close by higher value of the binding energy of the electron. This condition is only fulfilled for electron of inner shells when using the X-ray beam. A cascade of electronic rearrangements and emission of fluorescence photons or Auger electrons takes place to fill holes in the electronic shells. The photoelectric effect is predominant for low photon energies and for dense materials with a high atomic number.

The Compton effect is an inelastic scattering which can be considered as an elastic collision between an incident photon and an electron. The incident photon energy is distributed between the scattered photon energy and kinetic energy of the electron. The portion of the energy transmitted to the Compton electron increases with incident photon energy. The Compton effect is predominant in the energy range commonly used in radiotherapy (4–25 MeV) (*86*). When the energy decreases, the Compton effect is predominant for the lighter elements (biological tissues).

When a photon passes close to the intense electric field prevailing in the vicinity of a nucleus, it can give rise to a pair (electron-positron). The positron, antiparticle of the electron, is identical to the latter but has an opposite charge. Energy equal to 2×511 keV is spent to materialize the two particles. The effect of materialization can take place only if the photon has a minimum energy of 1.02 MeV, corresponding to the rest mass of two particles. It becomes dominant for very high energies. The positron from the phenomenon of materialization will, once thermalized, annihilate when meeting with an electron of the medium. Two γ photons of energy equal to 511 keV are then sent in opposite directions.

According to its energy and the characteristics of the medium, a photon passing through matter has different probabilities of undergoing one of three types of interactions (*85*). The photoelectric effect is predominant at low energy, for materials having a high atomic number and at moderate energy when the atomic number Z is high. The Compton effect is predominant over a wide energy range, mainly for light elements (Z <40). This is the main mode of interaction of X-ray photons applied in the treatment with conventional radiotherapy. The pair becomes dominant only for very high energies. The interaction between incident photons and the component of matter exposed to the radiation, produces secondary photons and electrons (photoelectrons, Auger electrons). The latter will then creates cascades of ionization in the medium until to transfer the energy of the incident photons. The secondary photons (from Compton scattering) are also involved in energy deposition, *via* photoelectric and Compton interactions.

Physicochemical and Molecular Processes

Primary and secondary radiation (electrons, photons) from physical processes (described in the preceding paragraph) deposit their energy in cells by ionizing or exciting the different cell constituent (water, protein, nucleic acids, etc.). This leads to the formation of molecular ions or unstable radicals. Chemical alterations of biomolecules can be caused by two different mechanisms (*84*) :

- By indirect action, *via* the formation of reactive species from the radiolysis of water.
- By direct action, due to the excitation or ionization of biomolecules themselves.

During the indirect process, the water, which is the main cellular component (70–80 wt %), is excited or ionized by the radiation generated by the physical step (photons, electrons). These reactions led to the rapid formation (in nanoseconds) of highly reactive radical species. Few tens of nanoseconds after irradiation, the main species existing in a biological medium after irradiation are: $HO^{\bullet}$, $O_2^{\bullet-}$ and e^-_{aq} (radical species) together with H_2O_2, H_2 and H^+ (non-radical species) (*84*). All these species are toxic because they alter the cell components after chemical reaction.

However, in the direct process, biomolecules are directly ionized or excited by ionizing radiation (*via* the primary radiation or due to the secondary electrons or photons). The unstable biomolecules can get rid of their excess energy by the emission of Auger electron or fluorescence photons. This secondary radiation can in its turn interact with cellular components in the vicinity. The ionization and excitation of biomolecules can also cause the appearance of unstable species and highly reactive ionic radicals. These species characterized by a short lifetime can react with each other or with unstable species produced in the vicinity (oxygen, hydrogen peroxide). These reactions can cause the denaturation of biomolecules. All cellular components (plasma membrane, cytoplasm, nucleus) and thus all the molecules that compose them (phospholipids (the main constituents of membranes), proteins, sugars, enzymes, nucleic acids, water, etc.) may be the target of ionizing radiation, directly or indirectly. The alteration of these biomolecules (*e.g.* the peroxidation of phospholipids) may cause cell death (*84*). The DNA molecule is however considered to be the critical target of ionizing radiation. The DNA molecule contains the genetic information necessary for the synthesis of proteins which ensure the full functioning of cells. DNA plays therefore a central role and even minor damage can lead to cell death. Like all biomolecules, the DNA molecule can be damaged directly or indirectly by ionizing radiation. In the case of conventional radiation therapy (using high energy X-ray photons), the DNA damage is mainly due to the indirect action of reactive oxygen species (including the hydroxyl radical HO$^\bullet$), which contribute to about 60–70% of lesions (*84*). Recently, the influence of very low energy electrons (<20 eV) on the damage to the DNA molecule has been shown (*87*). There are several types of DNA damage by ionizing radiation (Table 3).

Table 3. Estimation of the mean amount of radiation-induced damages of DNA in mammalian cells for a dose of 1 Gy

Damages	*Number of radiation-induced damages*
Single-strand breaks	~ 1000
Loss of bases	~ 2000
Damages of bases	~ 2000
Damages of sugar	~ 1200
Double-strand breaks	~ 40
DNA-DNA crosslinks	~ 30
DNA-proteins crosslinks	~ 150

The consequences of the lesions on the cell survival depend on the type of lesions, their number, their concentration and their management by the various mechanisms of DNA repair. While most radiation-induced damages are very common (single-strand breaks, loss or change of bases), it now seems well established that double-strand breaks in DNA, although less numerous, play a key-role in the radiation-induced death (*88*). However, the physical and biological effects of ionizing radiation affect both tumor cells and healthy cells. The success of radiation therapy is therefore based on achieving the best compromise possible between (*89*):

- The deposit of the total prescribed dose (often high) in a volume exactly equal to the target volume in order to destroy the tumor and prevent recurrence;
- The preservation of surrounding healthy tissue, in order to avoid early and/or late side effects (*90*). Irradiation of sensitive tissues or organs can indeed induce more or less severe and disabling lesions.

The success of cancer treatment by radiotherapy does not rest only on the total eradication of the tumor but also on the preservation of healthy tissue included in the irradiated volume. This differential effect, allowing the eradication of cancer lesion without surrounding healthy tissues are harmed, can be optimized by taking into account the knowledge in three important fields (radiobiology, technology, pharmacology) according to the type and grade of solid tumors (*84*):

- Radiobiology should provide tools for optimizing the small differences biological behavior between cancerous tissue and healthy tissue. Temporal splitting of the prescribed dose and increasing the dose rate are examples;
- Technology of irradiation (ballistics, nature of radiation), has made great advances which allow the deposition of a maximum dose in the tumor and a minimum dose in the surrounding healthy tissue. Implementation of innovative radiotherapy techniques, such as "conformational" radiotherapy with or without intensity modulation, the image-guided radiotherapy, proton and particle therapy fits into this framework (*91–94*);
- Pharmacology is an active domain for the development of drugs or other exogenous agents which are able to increase cell sensitivity towards the irradiation. The combination of radiotherapy and of molecules which inhibit the repair of radiation-induced molecular lesions is an interesting example (*95*).

In this chapter, only the pharmacological modulation of irradiation using heavy elements is addressed.

Radiosensitization of Tumors by the Use of Heavy Elements

Principle of Radiosensitization

Due to its mode of action, radiotherapy is a technique which rarely discriminates tumor and healthy tissue. This lack of specificity is a source of many recurrences of cancer because the preservation of healthy tissue requires the deposition of low dose which is tolerable for healthy tissue but is too weak for the eradication of tumor. To increase the selectivity of the radiotherapy, the use of radiosensitizing agents which are composed of heavy elements has been proposed (*96*). The propensity of elements with a high atomic number (Z) to absorb X-rays permit to focus the dose deposition in the area where they are concentrated. Furthermore, this absorption is followed by the emission of secondary radiation whose physical and chemical effects can contribute to damage biological tissue containing radiosensitizing agents. The specific accumulation of such agents within tumors should improve the selectivity of radiotherapy and thus increase efficiency by concentrating the effects of dose deposition in the tumor, thereby helping to protect the surrounding healthy tissue (Figure 2).

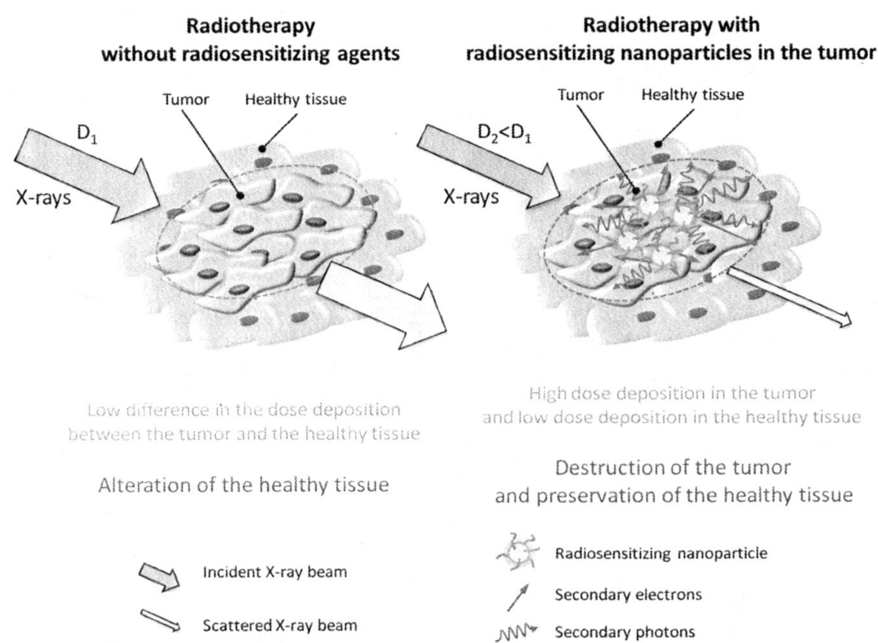

Figure 2. Schematic principle of radiotherapy with and without radiosensitizing nanoparticles. (see color insert)

To achieve an increase of local dose deposition, their irradiation can be performed by using various types of radiation (particles or photons) or different energies (poly- or monochromatic beams, broad beam or microbeams). The choice of the radiosensitizer and radiation to be used to reach an optimal differential effect is currently the subject of numerous investigations. While most studies are designed to optimize nonradiative de-excitation (Auger electron emission) which occurs after the photoactivation of heavy atoms, the mechanisms by which they act as radiosensitizers are not yet fully understood and seem to depend on the nature of the high Z element of the radiosensitizers (*97*).

The Radiosensitization

Iodinated Compounds

Due to their availability and daily clinical use, contrast agents for X-ray imaging (iodinated molecules) appear as promising candidates to achieve the increased dose deposition by photoelectric effect. The relatively high atomic number of iodine ($Z(I) = 53$) favors the attenuation of a beam of X-ray photons. This property has been exploited for increasing the dose deposition. Several authors have shown encouraging results of the radiosensitizing effect of iodine on cells, preclinical models of animal tumors and ultimately in humans, using kilovoltage X-ray CT scanner (*98–101*). However, the iodinated molecules used are not internalized by the cells. They remain therefore away from the cell nucleus and its components (such as DNA). In order to plentifully exploit the emission of Auger electrons, whose mean free path ranges from several nanometers and several micrometers (depending on their energy) (*102*), iodinated molecules should be incorporated into the cell nucleus. An iodinated analogue of thymidine (5-iodo-2′-deoxyuridine, IUdR) has been used with relative success as radiosensitizer due to its ability to incorporate within the DNA molecule (*103*). The irradiation of V79 hamster cells labeled with IUdR at an energy equal to 33.4 keV (200 eV above the iodine K-edge) was 1.4 times more effective in reducing the clonogenic survival than irradiation at 32.9 keV (200 eV below the iodine K-edge) (*104*). This result seems to confirm that the irradiation of a heavy element with a monochromatic beam energy just above the K-edge generate a localized dose deposition probably due to the Auger electron cascades. However, in a more recent study, Corde *et al.* have shown that an optimal biological effect (clonogenic cell survival) was obtaining after irradiation in the presence of iodinated species (ICAs or IUdR) when the difference in absorption between iodine and biological matter was the largest (*i.e.* for an energy of 50 keV). When radioresistant SQ20B human cells are irradiated by synchrotron beam (at 50 keV) in the presence of ICAs (outside the cell nucleus) and IUdR (within the cell nucleus), radiosensitization factors are 2.03 and 2.60, respectively (*105*). In other words, iodine in nucleus exerts a greater radiosensitizing effect than outside, since the radiosensitization factor is higher in the case of IUdR. The proximity of the radiosensitizers with DNA seems therefore important for an increased efficiency of radiotherapy. Due to the short action range of Auger electrons, the presence of the radiosensitizer in vicinity of DNA (i.e. in the nucleus) is required for benefiting of the Auger electron shower. However, this study shows that heavy elements (*i.e.* with high *Z)* exert a radiosensitizing effect even if they are not located within cells (ICAs). The *in vivo* application of radiosensitizers which are not able to cross the cell membranes can therefore be envisaged.

The Platinum Salts

Compounds with platinum (*e.g.* the complex ion cis-diaminedichloroplatine (II), better known as cisplatin) are currently used in chemotherapy which exploits their ability to intercalate into the DNA double helix. Another therapeutic use of platinum compounds can be envisaged since the high atomic number of platinum (Z(Pt) = 78) makes it very attractive for radiosensitization. Irradiation with monochromatic synchrotron beam at 78.8 keV (400 eV above the platinium K-edge) led to a greater production of double-strand breaks (DSB) and a slower DNA repair than in the case of an irradiation at 78.0 keV (400 eV below the platinium K-edge). This result suggests an increase in the dose deposition in the vicinity of the DNA, in particular via the photoelectrons and Auger electrons emission in the case of irradiation 78.8 keV. However, irradiation of rats bearing F98 glioma after an intratumorous injection of cisplatin led to similar increased survival for both energies (34% survival at one year). This unexpected result highlights the difficulty to transpose results from *in vitro* to *in vivo* studies (*106*).

However, irradiation with a monochromatic beam at an energy slightly above the platinum K-edge (78.8 keV) induced more DNA double-strand breaks and better survival of glioma bearing rats than in the case of a polychromatic beam irradiation (6 MeV) when cisplatin is used (*in vitro* and *in vivo*). The choice of the energy is an important parameter to consider for the effectiveness of radiosensitization by heavy elements (*107*).

The various studies on the use of radiosensitizing agents to improve the effectiveness of radiotherapy illustrate the importance of the pair (radiosensitizer + irradiation beam). In practice, the choice is often the result of a compromise between several factors, including the following:

- The difference in the absorption of radiation between the radiosensitizer and biological tissues (high for energies below a few hundred keV) ;
- Preference for non-radiative de-excitation (Auger electrons) which induces an additional localized dose deposition (*i.e.* preference for monochromatic beam, just above the absorption edge which is still far from being readily available);
- The probability of nonradiative de-excitation, connected to the fluorescence yield (for the heavy elements, Auger electron emission is more frequent after ionization of L and M-shells) (*108*) ;
- Penetration depth of the radiation (an energy below 30 keV is difficult to envisage in clinic even if it allows a better understanding of the phenomena involved in the radiosensitization of DNA or cells);
- The availability of irradiation equipment (CT scanner and the conventional megavoltage radiotherapy equipments are widespread while the synchrotrons are rare but "compact" synchrotron sources are under development);
- The administration mode of the radiosensitizers (intravenous injection is often preferred to intratumorous injection);
- The safe behavior of the radiosensitizer (not toxic without activation);
- The appropriate biodistribution of the radiosensitizer (preferential accumulation in tumor tissues and clearance from normal tissues).

A broad consensus seems to be formed about the optimum energy range to obtain a *in vivo* radiosensitizing effect. It ranges from a few tens of keV to a few hundred of keV (depending on the element used) to maximize the absorption of X or γ-radiation. Most comparative studies have clearly demonstrated the benefits of the use of low energy, (*98, 109*), despite their lower penetration in tissues.

The choice of the element to be used remains an open question. Although they are daily applied for radiographic examinations, the iodinated contrast agents (ICAs) do not seem to be ideal candidates. Indeed, their extracellular biodistribution and rapid elimination from the body after intravenous injection prevent a sufficient accumulation in the tumor. Their intratumorous injection seems also relatively difficult for clinical implementation, especially in the case of tumors located in extremely sensitive organs like the brain. Platinum complexes, which accumulate in the DNA of tumor cells, could represent an interesting choice, provided that these salts are present only at very low levels in normal tissues because of their many well-known side effects when used for chemotherapy (*110*). The use of radiosensitizers whose toxicity is induced by the external radiation only when accumulation in tumor and renal clearance of the excess are confirmed by medical imaging seems *a priori* more appropriate.

The Ideal Radiosensitizer

The ideal radisosensitizer is an agent which renders the radiotherapy more efficient because its presence will induce a higher and more controlled dose deposition. Consequently, the destruction of a solid tumor by radiotherapy without alteration of healthy tissue can be envisaged provided that the radiosensitizers are only present in the zone to be eradicated. The ideal radiosensitizer should therefore overcome the lack of selectivity of radiotherapy which constitutes one of the main drawback of this therapeutic technique.

For designing the ideal radiosensitizer, it is necessary to take into account the physiological differences between healthy tissue and tumor, the characteristics of the interaction of the ionizing radiation with high Z elements and with the elements of biological matter and the advantage to monitor the biodistribution by medical imaging. This radiosensitizer must be composed of a large amount of high Z elements (for the radiosensitization and for X-ray imaging), exhibit a safe behavior after intravenous injection (no toxicity in absence of radiation, preferential accumulation in solid tumors but renal elimination of the excess, *i.e.* the radiosenstizers must be sufficiently large for avoiding the extravasation from healthy vessels but sufficiently small for allowing the release from leaky vessels which irrigate the tumor) and be designed for a follow-up by medical imaging in order to initiate the irradiation at the most opportune moment (high content in the tumor and low content in the surrounding healthy tissues).

Since the iodinated compounds, the platinum salts and more generally the molecular radiosenstizers do not meet all these criteria (*vide supra*), the development of more efficient radiosensitizers is required. The intense research activity devoted to the nanoscience and the nanotechnology shows that the multifunctional nanoparticles are better suited for image-guided therapy than the molecules.

Nanoparticles for the Detection and Destruction of Cancer Cells

Nanoparticles

Nanoparticles are assemblies of a few tens to several thousands of atoms. Moreover, at least one of the dimensions of nanoparticles is less than 100 nm. This small size confers to the nanoobjects original properties which are different from properties observed for the bulk parent material. One of the most representative examples of the change in properties at the nanoscale is the color of colloidal suspensions of gold or silver nanoparticles, different from that of bulk materials. The optical properties of these nanoparticles are governed by localized surface plasmon resonance (LSPR) which is the collective oscillation of the electrons in the conduction band of metallic nanoparticles that are excited by light (*111*). As a result, an intense absorption band is observed for specific wavelengthes. The resonance wavelength, which defines the color of the colloids, is dependent on the shape and size of the nanoparticles but also on the dielectric constant of the medium and the distance between nanoparticles. This property confers to colloidal suspensions of gold and silver nanoparticles colors ranging, respectively, from red to violet and from yellow to green depending on conditions. It has been particularly exploited by Mirkin's group for a simple colorimetric detection of oligonucleotides strands (target). The hybridization of the target with complementary oligonucleotides (probes) which are tethered on gold nanoparticles induces agglomeration of nanoparticles. As a result, the color of the gold colloid changes from red to blue (*112*).

Another classic illustration of the emergence of novel properties at the nanoscale is the fluorescence of semiconducting nanocrystals (or quantum dots, *e.g.* CdSe), whose emission wavelength depends on the size. Despite remaining questions about their potential toxicity (*113*), the development of semiconducting nanocrystals has led to many applications based on *in vitro* and *in vivo* fluorenscence imaging (*114*).

Nanotechnology has thus become an important research scope for twenty years. This scope is extremely broad and has allowed major scientific breakthroughs in domains as wide as the electronics, composites, catalysis or medicine. In this chapter we will present some examples of promising applications of nanoparticles in the biomedical field.

Biomedical Application of Nanoparticles

The advantage of using nanoparticles in biomedicine lies primarily in their nanometer size, which enables them to explore the interior of living organisms, especially at the cellular or even molecular scale. The size of the nanoparticles is much lower than that of the cells and similar to the size of biological molecules (*115*). The reduced size is at a origin of a tremendous variety of applications in the biomedical field. Only a few will be mentioned.

In addition to a size suitable for the exploration of living organisms, another advantage of nanoparticles is that they can gather different properties in a very small object (Figure 3). The characteristics of the nanoparticles depend on the components, the size and the shape of the nanoparticles. Through an accurate control of these parameters, various multifunctional nanoparticles can be tailored for a specific application. One of the best demonstrations of the potential of multifunctional nanomaterials was carried out by the research group of Weissleder. They managed to follow up cells by MRI, fluorescence imaging and γ-scintigraphy after their labeling by multifunctional dextran coated iron oxide nanoparticles. The follow-up by MRI was possible thanks to the superparamagnetic character of the iron oxide while the detection by fluorescence and scintigraphy results from the post-functionalization of the dextran shell by fluorescent TAT peptide and by ^{111}In-chelates. In addition to ensure the post-functionalization of the iron oxide core (*i.e.* to enlarge the palette of properties), the biocompatible dextran shell improves the colloidal stability of the nanoparticles in a biological medium. However the encapsulation of the superparamagnetic core by dextran is accompanied by a size increase, from 5 to 45 nm (*116*). These multifunctional nanoparticles thus allow tracking labeled cells with three types of complementary medical imaging (MRI, scintigraphy and fluorescence).

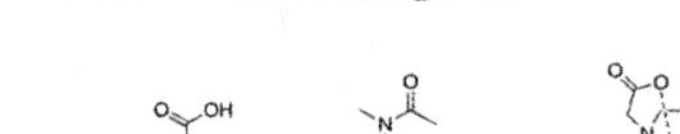

Molecular contrast agents

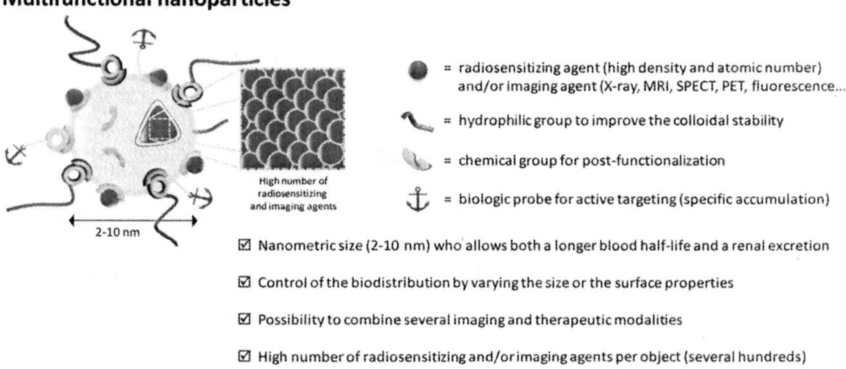

Multifunctional nanoparticles

Figure 3. Schematic representation and characteristics of commercially available molecular agents and multifunctional nanoparticles. (see color insert)

Constraints on the Use of Nanoparticles in the Body

The administration of nanoparticles to a living organism requires that they have physicochemical properties compatible with their use in a biological environment. Knowledge of the constraints imposed by the biological environment is essential to develop nanoparticles for biomedical applications.

The preferred route of administration in the body is the intravenous injection, which reaches all the organs, even those that are inaccessible by local injection. The *in vivo* application of the nanoparticles requires a high chemical and colloidal stability in physiological conditions which are mainly characterized by a pH of about 7.4, high salinity (equivalent to an aqueous solution of 150 mM NaCl), a temperature of about 37°C and a high concentration of proteins. If the stability is essential for any *in vivo* use, the efficacy of nanoparticles (whatever their purpose:

imaging and/or therapy) rests on their ability to reach their target (organ, tumor, etc.). To achieve this, the nanoparticles must remain in the bloodstream long enough. This condition is mainly governed by two interrelated phenomena:

- The elimination of the nanoparticles from bloodstream and the body by the immune system, which reduces their effectiveness and limits the probability of being in contact with the target;
- A safe biodistribution of the nanoparticles, which ensures a suitable concentration of the nanoparticles in the targeted zone.

Elimination by the Immune System

Once introduced into the bloodstream, nano-objects can trigger a cascade of immune reactions which lead to the rapid removal from the bloodpool and from the body (*117*). These complex mechanisms are mostly initiated by the adhesion of plasma proteins (opsonins) on the nanoparticle surface. This "opsonization" facilitates the capture of nanoparticles and their phagocytosis by cells of the reticuloendothelial system (RES), including the resident macrophages of the liver (Kupffer cells) or spleen (*118*). The capture of nanoparticles by the RES reduces their blood half-life. The probability of interaction with their target is therefore minimized. Opsonization is a phenomenon dependent on the physicochemical properties of nanoparticles (hydrophilic/hydrophobic character, size, shape, charge and surface chemistry). The influence of these parameters on the opsonization and uptake by the RES (liver, spleen) depends on the types of nanoparticles (Figure 4).

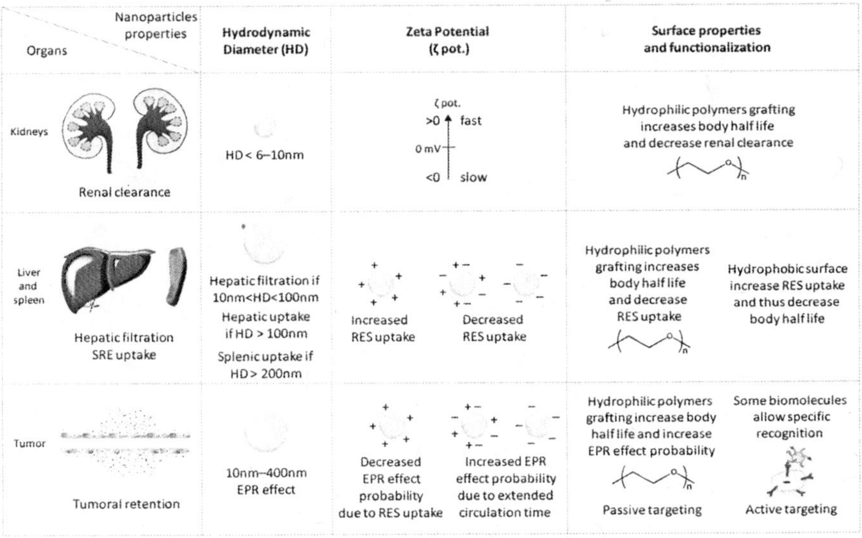

Organs / Nanoparticles properties	Hydrodynamic Diameter (HD)	Zeta Potential (ζ pot.)	Surface properties and functionalization	
Kidneys — Renal clearance	HD < 6–10nm	ζ pot. >0 ↑ fast; 0 mV; <0 slow	Hydrophilic polymers grafting increases body half life and decrease renal clearance	
Liver and spleen — Hepatic filtration, SRE uptake	Hepatic filtration if 10nm<HD<100nm; Hepatic uptake if HD > 100nm; Splenic uptake if HD > 200nm	Increased RES uptake / Decreased RES uptake	Hydrophilic polymers grafting increases body half life and decrease RES uptake	Hydrophobic surface increase RES uptake and thus decrease body half life
Tumor — Tumoral retention	10nm–400nm EPR effect	Decreased EPR effect probability due to RES uptake / Increased EPR effect probability due to extended circulation time	Hydrophilic polymers grafting increase body half life and increase EPR effect probability — Passive targeting	Some biomolecules allow specific recognition — Active targeting

Figure 4. Relationship between nanoparticles properties and their biodistribution. (see color insert)

However, some trends are emerging. The hydrophobic particles are very rapidly extracted from the bloodstream because they are captured by the RES, while the hydrophilic particles have a longer blood half-life (*119*). The functionalization of the nanoparticles by hydrophilic polymers (typically polyethylene glycol, PEG) is a common strategy for increasing their blood half-life (*120, 121*). On the other hand, the size and charge of nanoparticles are important parameters affecting the adhesion of opsonins. The reduction in size can limit the adhesion. The presence of negative charges or the absence of charge onto the nanoparticles also impedes the opsonization (*122*). On the contrary, the adhesion is likely to occur when the nanoparticles are large and/or they are positively charged. Moreover the presence of positive charges is often associated with some toxicity (*117, 122*). Once the nanoparticles sequestered by the RES (phagocytes), they are then digested and eliminated by the organism when they are biodegradable (liposomes, polymers). Otherwise, they are stored and accumulate in the phagocytes. Their future is quite uncertain and this accumulation raises the question of their long-term toxicity (*117*).

Control of the Biodistribution of the Nanoparticles

Several parameters influence the biodistribution of nanoparticles after intravenous injection, which can be grouped into two groups: (i) physiological factors (bloodstream, endothelial permeability of vessels), (ii) the physicochemical properties of nanoparticles. After injection, the nanoparticles are distributed mainly in the organs where bloodstream is high but their distribution is also related to the physiology of the organs, in particular the appearance of the endothelium (*122*). According to the organs, the endothelium may be continuous (arteries, blood vessels, lungs), fenestrated (glands, gastrointestinal mucosa, kidneys: pores of about 60 nm) or discontinuous (liver: pores of 50–100 nm). The porosity of the endothelium exerts a strong influence on the biodistribution of nanoparticles (depending on size) because it determines their clearance (*vide infra*). Owing to their immaturity, vessels from tumor neoangiogenesis are permeable and tortuous. The fenestration of the endothelium of tumors, which can amount to several hundred nanometers, is at the origin of the EPR effect. Besides the absence of lymphatic drainage, it facilitates the accumulation of nanoparticles and their distribution in the tumor interstitial space. Such a fenestration allows the passive targeting of the tumor (*123, 124*). The realization of passive targeting is obviously facilitated by a high residence time of nanoparticles in the bloodstream. This implies that the nanoparticles are designed to avoid the opsonization which leads to the uptake by RES (Figures 1 and 4). The preferential accumulation of nanoparticles in tumors can also be achieved by active targeting which takes advantage of the overexpression of certain markers by tumor cells or endothelial cells of the neovessels (*125, 126*) (Figures 1 and 4).

The physicochemical properties (size, charge and surface chemistry) of nanoparticles also affect their distribution (Figure 4). As previously shown, a positive charge is often associated with hepatic uptake which decreases the blood half-life of the nanoparticles. On the contrary, neutral or negatively charged nanoparticles usually circulate for a longer time and, if their size is adapted, are removed by renal excretion (*122*). The size (or rather the hydrodynamic diameter) plays a role in the biodistribution of nanoparticles because it determines their mode of elimination (*127*) (Figure 4). A hydrodynamic diameter greater than 8–10 nm is associated with hepatic elimination, after rapid opsonization, but the latter can nevertheless be deferred by an appropriate functionalization (hydrophilic polymer). However, the large particles (> 8–10 nm) will inevitably be captured by the liver, either by hepatocytes (through biliary excretion in feces) or Kupffer cells. If the nanoparticles are biodegradable they will be degraded, otherwise they will be stored (*127*). This long-term sequestration could have deleterious effect when using metal nanoparticles or semiconducting nanocrystals (quantum dots). Moreover undesirable interferences with further medical examinations can be expected. The need for a complete elimination for this type of particles has recently been emphasized by Choi *et al.* (*128*).

It is now recognized that a hydrodynamic diameter of less than 5–6 nm leads to renal elimination whatever their surface charge (*127, 128*). An intermediate hydrodynamic diameter (*i.e.* 6–10 nm) leads to renal elimination, but the surface charge then plays a role; the elimination is faster for positive particles than for negative or neutral particles (*127*). While the renal clearance minimizes the residence time of nanoparticles in the organism, it nonetheless seems to be the most appropriate way for the removal of non-biodegradable particles (*129*).

Remotely Controlled Therapy

The research activity devoted to the development of multimodal contrast agents and nano-objects for therapy led to the emergence of multifunctional nanoparticles combining both imaging and therapy (*130, 131*) (Figure 3). Thanks to these new agents, image-guided therapy can be envisaged. This concept is based on the remotely induction of the therapeutic activity of non-toxic nanoparticles at the most opportune moment determined by various medical imaging techniques. In other words, harmless nanoparticles become toxic under the effect of physical or chemical stimulus induced when the content of the nanoparticles is both high in the tumor and very low in the surrounding healthy tissue. Recent work has shown that the release of chemotherapy drugs could be induced by chemical changes in the environment of tumor (pH or the presence of enzymes) (*132, 133*). However, the application of external physical stimuli should *a priori* permit a better control of the therapeutic activity of the multifunctional nanoparticles, because the activation is spatially well controlled (laser, beam of neutrons or X-rays). This type of application requires therefore to have multifunctional nanoparticles combining imaging and physically triggered therapy.

Ideally, this type of nano-objects must be able to reach the zone of interest and accumulate preferentially, by taking advantage of the characteristics of the targeted tissue (*e.g.* solid tumor) without being detected by the immune system. It must also be followed up by several imaging techniques to be precisely localized and become toxic due to the activation by an external stimulation when its concentration (estimated from imaging data) is both high in the diseased zone and low in the surrounding healthy tissue. The therapeutic effect can be remotely triggered by the interaction with a focused physical stimulus (visible light in the case of photodynamic therapy, neutron beam in the case of neutron capture therapy,…) (*134, 135*). A cascade of physical reactions then generates localized damage around the nanoparticles.

The idea of using nanoparticles that can both induce a toxic effect after interaction with a physical stimulation and be visualized in real time through one or more imaging techniques seems undeniably attractive for focusing the damage in the tumor while preserving the surrounding healthy tissue. Among the physical trigger, X-ray beam seems very well suited because of their widespread application in therapy.

Multifunctional Nanoparticles for Imaging-Guided Radiotherapy

Introduction

During the last century, the field of cancer treatment has benefited from major advances. Improved surgical techniques, combined with the radiotherapy (*136*), chemotherapy (*137*) and targeted therapies (immunotherapy) (*138*) have increased significantly both the cure rate and the quality of life for patients. Despite the therapeutic arsenal currently available, the prognosis of certain cancers remains bleak, particularly because of their infiltrative nature, their location or their resistance to the current treatments. A particularly demonstrative example is the case of glioblastoma, which is the most aggressive and common brain tumor (*139*). Despite a multidisciplinary approach (surgery combined with radiochemotherapy (*140*)), the median survival of patients with these tumors is extremely low (10–15 months) (*139*). This can be explained by its infiltrative nature which renders difficult the complete excision of the tumor. In addition, glioblastoma is a particularly radioresistant tumor in a radiosensitive organ. The dose that the brain can tolerate is generally insufficient for the eradication of the tumor. Moreover the lack of specificity of the radiotherapy prevents the selective destruction of cancerous cells which infiltrate into the healthy tissue of the brain. The efforts to improve the effects of radiotherapy which however remains a promising treatment for glioblastoma have not shown significant influence on the lifespan of patients, which remains very low (*141, 142*). The treatment of glioblastoma has reached a therapeutic impasse. For this reason, new strategies should be considered.

Since the brain-blood barrier (BBB), which constitutes a quasi impenetrable barrier, becomes highly porous when a tumor grows in the brain, the radiosensitization with multifunctional nanoparticles seems an attractive route for breaking the deadlock.

Functional Gold Nanoparticles for the Treatment of Cancer

Characteristics of Gold Nanoparticles

Owing to its properties at the nanometer scale, gold appears as one of the most interesting materials for the development of multifunctional nanoparticles. Moreover gold exhibits a high chemical stability (oxidation resistance) and is probably the most biocompatible metal, even at the nanoscale (*143–145*). In addition, gold has a high atomic number ($Z(Au) = 79$) and a high density (19.3). These features confer a large absorption cross section of X-ray photons which makes gold nanoparticles particularly attractive for use combining X-ray imaging and radiotherapy. The extensive study of the tunable optical properties of gold nanoparticles which depend on their size, shape, composition and environment led to the development of various multifunctional gold nanostructures (*146, 147*). The spherical gold nanoparticles are commonly synthesized by applying or adapting the protocols developed by Frens and Brust. The method of Frens (*148*) consists in the reduction of a gold salt ($HAuCl_4$) with a solution of sodium citrate (mild reducing agent). This method yields gold nanoparticles coated with citrate ions in aqueous suspension. Gold nanoparticles, whose size is in the range 10–150 nm (depending on gold to reducing agent ratio) can be functionalized with thiolated or aminated molecules by displacement of weakly adsorbed citrate ions. The main drawbacks of this synthesis method are the large size and low concentration of the nanoparticles. The method described by Brust (*149*) allows the production of concentrated colloidal suspensions by reducing a gold salt with sodium borohydride ($NaBH_4$, strong reducing agent) in the presence of thiol ligands, either in biphasic solvents mixture (toluene or chloroform/water) or in monophasic aquoalcoholic mixture. During the reaction, the thiol ligands provide control of the size of the nanoparticles and allow the colloidal stability after the reaction. Besides a greater concentration of the colloids in gold nanoparticles, the Brust method leads to the formation of smaller nanoparticles (1.5–30 nm) as compared with the Frens method. The use of appropriate thiolated ligands allows the production of functionalized "ready-for-use" nanoparticles without further chemical treatment (except purification). A post-functionalization may however be performed by ligand exchange or by using any reactive chemical functions eventually carried by thiolated ligands (*e.g.* carboxylic acid functions (*150*)).

Because of the plasmon resonance phenomenon, the spherical gold nanoparticles are characterized by a high absorption of light in the visible range (520–550 nm). The wavelength value of plasmon band depends on the size, the distance between the nanoparticles and the dielectric constant of the medium.

Apart from the spherical gold nanoparticles (the most common gold nanoobjects), gold nanostructures can exhibit different shapes. We distinguish, for example, gold nanorods, characterized by their aspect ratio (length/diameter) (*151*), the nanoshells (silica nanoparticles coated with a gold layer) (*152*), and the gold nanocages (hollow and porous gold cubes) (*153*). These gold nanostructures are characterized by original optical properties. Their tuneable optical properties (which can be exploited for imaging and photothermal therapy), their propensity to absorb X-ray photons (which opens the door to X-ray imaging and dose enhancement in radiotherapy) and the ease to functionalize them (which allows to adapt the behavior of the nanoparticles in biological media) render the gold nanostructures very attractive for the image-guided therapy, particularly in the field of cancer (*154*).

Gold Nanoparticles for Imaging and Cancer Therapy

Many studies on multifunctional gold nanoparticles combining several imaging techniques and remotely controlled therapy have been described. Only a few examples will be discussed to illustrate the immense potential of this type of nanoparticles.

The optical properties of gold nanostructures can be adjusted appropriately in order to exploit them for both imaging and therapy. In the case of nanoshells (consisting of silica beads coated with a layer of gold) and nanorods, the wavelength corresponding to the plasmon resonance may be shifted toward the longer wavelengths respectively by playing with the thickness of the gold layer (and/or with the silica core diameter) and the aspect ratio (*155*). These gold nanostructures can therefore be tailored for scattering and absorbing the light in the NIR region which corresponds to the "transparency window" of the biological tissue (*156*). These two properties (scattering and absorption of the NIR light) have been exploited respectively for imaging and photothermal therapy. Using gold nanoshells coated with antibodies for specifically targeting cancerous cells, the research groups of Halas and Drezek was able to image and destroy cancerous cells (*157*). Promising results in photothermal therapy were also obtained *in vivo* (*158*). Meanwhile, the team of El-Sayed was able to obtain similar results (imaging and therapy) using gold nanorods, on cells (*159*) or *in vivo* (*160*). Since this therapeutic strategy is based on NIR light whose penetration depth does not exceed 2 cm, clinical application is limited to the treatment of superficial tumors. The high potential of gold nanoshells for photothermal therapy is currently tested in a clinical trial (*161*).

However the gold nanoshells and nanorods are handicapped by a large size which is not compatible with an intravenous injection and by a relatively complex synthesis (as compared with spherical nanoparticles). Despite their large size, the targeting of tumor by gold nanoshells was successfully performed by using monocyte cells as cargo.

The ease to functionalize the gold nanoparticles has paved the way for drug delivery, a research field where the Rotello's group is particularly enterprising (*162–164*). The delivery of chemotherapeutic agents by gold nanoparticles (*165*) raises a growing interest, but remains at the stage of tests on cells (*166, 167*). Other molecules having a destructive effect on cancer cells can be used. For example, PEGylated gold nanoparticles carrying TNF-α, a cytokine known for its antitumor activity, have recently been described (*168*). Following promising tests in animals, this type of gold nanoparticle began a phase of clinical trials (*169*).

Another interesting property of gold nanoparticles is their propensity to absorb X-rays, thanks to the physical properties of the gold element (high atomic number and high density). This property was originally exploited for enhancing the effect of the radiotherapy. The works of Herold *et al.* (*170*) have shown that gold beads induce a radiosensitizing effect. However the particle size (1.5–3.0 microns in diameter) was too important to allow a safe administration and an accumulation in the solid tumors. Based on this observation, Hainfeld *et al.* performed the first *in vivo* study of radiosensitization by gold nanoparticles (in mice) (*171*). Because of their small size (diameter of the gold core: about 1.9 nm), these particles diffuse rapidly in the body and accumulate transiently in the tumor because of its large blood supply. The injection of nanoparticles combined with X-ray irradiation (250 kVp) causes a decrease in the average size of tumors and significantly improves the survival of mice bearing a subcutaneous EMT–6 tumor (*171*).

Hainfeld *et al.* then showed that the same gold nanoparticles could be used as contrast agents to better visualize tumors in X-ray imaging, showing the ability of these objects to combine imaging and therapy (*172*). This innovative project has generated enormous enthusiasm among several teams. Numerous studies on the use of nanoparticles as radiosensitizers (*173–175*) or as contrast agents for imaging X (*176–178*) have been completed and confirmed the previous results. However, and surprisingly, no other study was conducted on the combined use of these two properties (imaging/radiation therapy), despite the promising applications of gold nanoparticles for therapy guided by medical imaging.

Gadolinium Chelates Coated Gold Nanoparticles: Au@DTDTPA-Gd

If the opportunity to image the accumulation by X-ray before the radiotherapy with the same X-ray facility is very attractive, the repetition of X-ray imaging sessions is not recommended because of the harmfulness of X-ray. For a safer *in vivo* monitoring of the gold nanoparticles, the synthesis of gadolinium chelate coated nanoparticles has been proposed. Molecular gadolinium chelates, such as DTPA-Gd or DOTA-Gd, are widely used as positive contrast agents for MRI (Figure 3) (*179*). Moreover this technique contributes to improve the comfort of the patient because it is non-invasive, rapid and avoids the use of radiochemicals (*180*). MRI is consequently one of the most powerful and most widely used medical imaging techniques.

The *in vivo* application of gold nanoparticles requires a high colloidal stability in physiological media. The latter are characterized by an elevated ionic strength and the presence of proteins. Both can induce for different reasons the undesirable agglomeration of the nanoparticles. Since imaging and therapy applications require the injection of colloids with a high concentration of gold nanoparticles, Brust's protocol seems more suited for yielding multifunctional gold nanoparticles with a size smaller than 5 nm (*149*). This protocol is based on the reduction of gold salt by sodium borohydride ($NaBH_4$) in methanol-water mixture containing thiolated ligands. The adsorption of the ligands onto the nanoparticles during their synthesis allows the control of the nanoparticle growth and avoids the agglomeration. Our strategy for the synthesis of gold nanoparticles combining radiotherapy with MRI and X-ray imaging techniques consisted first in a screening of ligands able to ensure three important roles: (i) the improvement of the colloidal stability in salty medium even at high concentration of gold nanoparticles, (ii) the improvement of the grafting of ligands onto the particles in order to avoid the desorption and (iii) the possibility to perform a post-functionalization of the gold nanoparticles for enlarging the application field (complementary imaging technique, targeting).

The improvement of the grafting is a crucial issue that must be addressed. Even if sulfur is the element among all those of the periodic table that exhibits the highest affinity for gold and establishes the most intense interaction, stronger linkage between ligands and gold nanoparticles are needed for ensuring a lasting immobilization of the functional molecules. The increase of temperature and the competition for the adsorption onto gold nanoparticles in physiological media between ligands and molecules containing at least one sulfur and/or nitrogen atoms favor the desorption of the ligands. Such an undesirable modification of the chemical composition of the nanoparticle surface can lead either to the precipitation of the nanoparticles or to an unexpected biodistribution if the desorbed ligands are replaced by biomolecules present in the biological medium. The detrimental desorption phenomenon can be limited when the ligands carry two thiols, as shown with dihydrolipoic acid (DHLA). K-edge XANES experiments (*150, 181*) demonstrated that DHLA is anchored onto gold nanoparticles by both sulfur ends. Moreover the desorption of DHLA is very restricted (<5%) when the nanoparticles are redispersed in aqueous solution. For this reason, a dithiolated derivative of diethylenetriaminepentaacetic acid (DTDTPA) was used for the synthesis of gold nanoparticles (Au@DTDTPA) by adapting Brust's protocol. DTDTPA was chosen for coating gold nanoparticles because this ligand will favor the immobilization of gadolinium ions onto the gold nanoparticles. Indeed, DTPA and its bis-amide derivatives (like DTDTPA) are able to form highly stable gadolinium(III) chelates. In contrast to DHLA, DTDTPA is anchored onto gold core by only one sulfur end. However the second one allows the formation of a disulfide bridge with the sulfur end of a neighboring DTDTPA. As a result, Au@DTDTPA nanoparticles which are synthesized by the application of a modified Brust's protocol are composed of a gold core (2.4±0.5 nm) embedded in a multilayered shell of DTDTPA (hydrodynamic diameter : 6.6±1.8 nm) (Figures 5a-c). These particles are characterized by a high colloidal stability in a large pH range (between 2 and 14, including the pH of physiological media (7.4)) owing to the electrostatic repulsion between each negatively charged particle. Au@DTDTPA nanoparticles can entrap 150 gadolinium(III) ions per particle (Au@DTDTPA-Gd$_x$, with 0<x<150) (*182*). Such a structure, which contains a gold (Z(Au)=79) core with high absorption cross section of X-ray photons due to atomic number of gold and a paramagnetic layer of DTDTPA-Gd, exhibit a great potential for X-ray-imaging, MRI and radiotherapy. Although the presence of 150 Gd^{3+} ions is recommended for inducing a contrast of MR images as strong as possible, such a quantity is however detrimental for the colloidal stability of concentrated aqueous suspensions of gold nanoparticles (>1g.L^{-1}) because the positive charges of these ions compensate the negative charge of DTDTPA and weaken consequently the electrostatic repulsion between Au@DTDTPA-Gd nanoparticles.

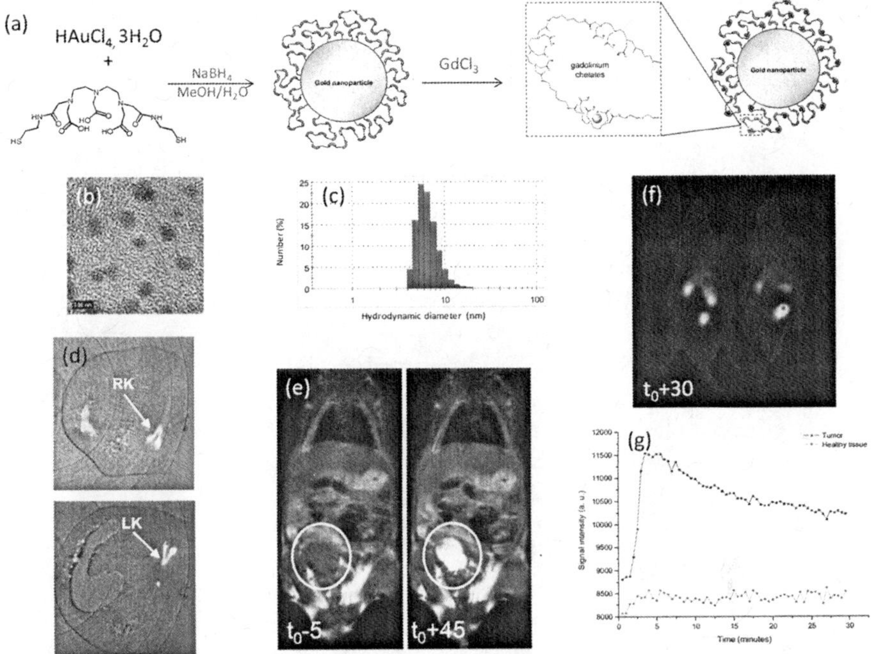

Figure 5. (a) Synthesis of Au@DTDTPA-Gd$_{50}$ nanoparticles. (b) TEM micrograph of Au@DTDTPA nanoparticles. (c) size distribustion of Au@DTDTPA nanoparticles by photon correlation spectroscopy. (d) contribution of the gold core to X-ray imaging (synchrotron radiation computed tomography) (RK and LK for right and left kidney, respectively). Contribution of gadolinium and 99m-technetium chelates to imaging: (e) T_1-weighted images before (t_0-5) and 45 after (t_0+45) intravenous injection of Au@DTDTPA-Gd$_{50}$ nanoparticles to a mouse, (f) γ-scintigraphy image acquired 30 minutes after intravenous injection of Au@DTDTPA-^{99m}Tc nanoparticles to rats and (g) temporal evolution of the MRI signal in tumor and in an equivalent surface in normal tissue in the left hemisphere after injection of Au@DTDTPA-Gd$_{50}$ nanoparticles to gliosarcoma bearing rat. (see color insert)

For a gold content of 10 g.L^{-1}, the immobilization of 50 Gd^{3+} ions constitutes a good compromise between colloidal stability and contrast enhancement for X-ray imaging and MRI (Figures 5d-e) (*183*). Thanks to their design, Au@DTDTPA-Gd$_{50}$ nanoparticles can be followed up by X-ray imaging (in transmission and in tomography (Synchrotron Radiation Computed Tomography, SRCT) mode performed at the biomedical beamline (ID17) of the European Synchrotron Radiation Facility, ESRF) and by MRI after intravenous injection of an aqueous colloid (10 g Au.L^{-1}, pH = 7.4 and [NaCl] = 150 mM) to rats and mice (Figure 5). These particles exhibit pharmacokinetics which is well suited for applications combining imaging and therapy. When they are injected to healthy animals, they circulate freely without undesirable accumulation in liver, lungs and spleen and are rather quickly removed from body by urine (*183*). This observation which was expected since the hydrodynamic diameter is inferior to 8 nm (Figure 4) was confirmed by elemental analysis of the organs but also by scintigraphy. Indeed, the replacement of Gd^{3+} by ^{111}In or ^{99m}Tc allows monitoring the distribution of the gold nanoparticles by scintigraphy and SPECT (Figure 5f). The *ex vivo* γ-counting was consistent with the ICP-OES analysis. At least 70% of the gold nanoparticles were removed by urine and about 10% by feces. No gold nanoparticle was detected in brain but the intravenous injection of Au@DTDTPA-Gd$_{50}$ to brain tumor (gliosarcome 9L) bearing rats led to a positive contrast enhancement of MR images in the periphery of the tumor (*184*). The enhancement of contrast which reflects the presence of the nanoparticles in the tumor zone can be explained by a denser vasculature around the tumor. This peculiarity which allows a preferential distribution around the tumor and a better distinction between healthy and diseased tissue was exploited for radiotherapy. After injection of Au@DTDTPA nanoparticles, the MRI signal in tumor quickly increases, reaches a maximum 4–5 minutes after the injection and decreases (Figure 5g). The comparison with the curve of signal intensity recorded in healthy tissue shows that the most opportune moment (high gold content in tumor and low gold content in healthy tissue) for the irradiation is ranging between 3 and 7 minutes after the intravenous injection of Au@DTDTPA nanoparticles (Figure 5g). While the median survival of rats receiving no treatment (group 1, control) is equal to 22.5 days, the median survival of rats treated with microbeam radiation therapy (MRT at ID17, ESRF) alone (group 2) is increased to 72.5 days (Table 4).

Table 4. Mean and median survival times and increase of lifespan (ILS) of 9L gliosarcoma bearing rats after different treatments

Groups	*Survival (days)*		*ILS (%)*
	Mean	*Median*	
(1) Control	22.5	22.5	/
(2) MRT only	106.1	72.5	222.2
(3) Au@DTDTPA 50 mM 0,7 mL + MRT 5 min.	154.0	129.0	473.3

This significant increase in survival is comparable to that obtained by Bouchet *et al.* using similar irradiation conditions (median survival equal to 65 days; ILS = 225%) (*185*). Remarkably, three of the eight rats treated with MRT are still alive 100 days after implantation. These results demonstrate the potential of using the technique of microbeam to treat aggressive brain tumors (median survival being multiplied by a factor approximately 3.2). The median survival of rats is increased to 129 days when they are treated by MRT five minutes after intravenous injection of 0.7 mL of a colloidal suspension of nanoparticles Au@DTDTPA ([Au] = 50 mM, group 3), which corresponds to an increase of survival of 473% as compared with the control group (group 1). This type of treatment can extend the lifespan of rats compared to treatment consisting only of a bidirectional radiation by MRT (group 2; ILS = 78%). In addition, the combined use of intravenous injection of gold nanoparticles and irradiation allows the survival of some animals for a long time. Thus, five of the eight animals of the group 3 are still alive one hundred days after irradiation and two survived more than three hundred days. These results confirm the existence of a *in vivo* radiosensitizing effect of gold nanoparticles Au@DTDTPA, which increases the lifespan of animals.

Multifonctional Gadolinium Oxide Nanoparticles

If gold nanoparticles received much attention owing to their well known potential for biomedical applications, lanthanides based nanoparticles were rarely studied whereas they exhibit a large range of properties which deserve to be exploited for imaging and therapy.

Gadolinium(III) Containing Crystalline Nanoparticles: Promising MRI Contrast Agents

The rare earth gadolinium(III) ion is the most widely used paramagnetic elements for the elaboration of contrast agents for MRI due to its seven unpaired electrons and relatively long electronic relaxation. As free gadolinium is extremely toxic, gadolinium ions are buried within chelate structures, DTPA (diethylenetriaminepentaacetic acid, Magnevist®), DOTA (1,4,7,10-tetraazacyclododecane-1,4,7,10-tetraacetic acid, Dotarem®), etc... (Scheme 1) (*53, 179, 180*). This caging limits the number of coordinated water molecules associated with each gadolinium(III) ion, and hence limits the modification of signal intensity that could be induced with such paramagnetic element. Another limitation is due to the small size (<1 nm) of clinically used Gd-complexes, which leads to short rotational correlation time, hence limiting the modification of proton relaxivity (Figure 3). Their small size also reduced the blood half-life of Gd-complexes, and bloodpool imaging is only accessible during the first minutes after intraveneous injection of the contrast agent. Moreover, as they are quickly eliminated from the body, both passive (by taking advantage of EPR effect in tumoral tissues) and active targeting strategies (by grafting relevant target-specific ligand on the surface of the contrast agents) are limited. The development of gadolinium(III) based nanoparticles aims therefore at affording positive contrast agents for MRI with a stronger ability of contrast enhancement (due to both the increase of gadolinium(III) number and the decrease of rotational motion), better pharmacokinetic parameters, a better control of biodistribution and a larger palette of properties than molecular gadolinium complexes did. To achieve this goal, three strategies were explored: (i) the functionalization of nanomaterials by gadolinium chelates (within or onto the nanostructures (gadolinium complexes loaded liposomes (*186*), gadolinium ions entrapped in zeolites (*187*) and in mesoporous silica nanoparticles (*188*), gadolinium chelates immobilized on quantum dots (*189, 190*), on lipid particles (*191*), and on gold nanoparticles (*182, 183*))); (ii) the entrapment of gadolinium (III) salts in carbon nanostructures (fullerenes (*192–195*), carbon nanotubes (*196, 197*)); (iii) the synthesis of multifunctional gadolinium containing crystalline nanoparticles. They are all characterized by an increase of the molecular weight and of the amount of Gd(III) ions per contrast agent. As a consequence of their structure, some of them were easily functionalized by biotargeting groups and/or fluorescent molecules conferring them additional attractive features (*186, 189–191, 198–203*). Besides the paramagnetic character of Gd^{3+} ion which is very useful for enhancing the contrast of MR images, gadolinium element exhibits two interesting properties for therapeutic applications. First, gadolinium element is characterized by a relatively high atomic number (Z(Gd) = 57). A radiosensitizing effect is therefore expected for compounds containing gadolinium. Moreover two isotopes of this element (^{155}Gd and ^{157}Gd) are distinguished by a huge neutron capture cross section which can be exploited for neutron capture therapy. Nanoparticles containing a large amount of gadolinium(III) ions seem well suited for image-guided therapy since the treatment can be remotely induced by the exposure to X-ray or thermal neutron beam when a favorable distribution is observed by MRI. Among the different classes of nanomaterials containing gadolinium(III), crystalline gadolinium oxide appears very attractive for the preparation of nanoparticles combining imaging and therapy because a very high gadolinium content can be obtained even in ultrasmall nanoparticles, *i.e.* in nanoparticles designed for renal clearance (*204–206*).

Synthesis and Functionalization of Gadolinium Oxide Nanoparticles

Pioneering studies devoted to gadolinium oxide nanoparticles were performed by Roberts and Watkins (*204, 205*). They revealed the potential of gadolinium oxide nanoparticles to be applied as contrast agents for MRI. However these preliminary works suffered from the availability of well defined hydrophilic gadolinium oxide nanoparticles. The exploitation of the attractive characteristics of gadolinium oxide nanoparticles for MRI became conceivable only when a reproducible and efficient synthesis protocol of naked gadolinium oxide nanoparticles and their functionalization were developed. A reproducible and reliable synthesis of gadolinium oxide nanoparticles was reported by our group (*207, 208*). These particles were synthesized by applying with some modifications the polyol routes which opened the door to the synthesis of a large range of inorganic crystalline nanoparticles. The synthesis of gadolinium oxide nanoparticles is based on the alkaline hydrolysis of gadolinium chloride in diethylene glycol (DEG), a high boiling diol (Figure 6a). Diethylene glycol plays an important role. It allows carrying out the synthesis at 180°C and to accurately control the size of gadolinium oxide nanoparticles (between 1 and 5 nm according to the experimental conditions) because the high viscosity of DEG and its adsorption on growing particles limit their growth and avoid the agglomeration.

The strategy developed by our group for achieving a high colloidal stability in biological fluid (pH 7.4, high ionic strength, high concentration of biomolecules, T ~ 37°C) rests on the encapsulation of the gadolinium oxide cores in a polysiloxane shell (*209*). The most striking feature of the polysiloxane shell results from its chemical nature which can provide new properties to the particles. The polysiloxane shell was indeed obtained by hydrolysis-condensation of tetraethyl orthosilicate (TEOS) and by aminopropyltriethoxysilane (APTES). The use of APTES allows the sequential functionalization of the polysiloxane shell: before and after the hydrolysis-condensation step, *i.e.* before and after the formation of the polysiloxane shell because each APTES carries an amine (NH_2) function. Before the hydrolysis condensation, APTES can be conjugated to organic dyes thanks to the coupling between amine group of APTES and the isothiocyanate or NHS ester group of the organic fluorophores while the polysiloxane shell can be post-functionalized by the covalent grafting of hydrophilic PEG chains (GadoSiPEG) (*121, 210*), polyaminocarboxylate moieties (GadoSiDTPA) (*211*) and/or of biotargeting groups (*209, 212*) (Figure 6a). The encapsulation of gadolinium oxide cores enlarges the properties range of these particles because the inner part of the polysiloxane shell is functionalized by organic dyes and the outer part is derivatized by hydrophilic molecules. As expected, these gadolinium oxide nanoparticles encapsulated in fluorescent and hydrophilic polysiloxane shell can be followed up after intravenous injection to mice and rats by fluorescence imaging (due to Cyanine 5 (NIR organic dye) covalently bound to the polysiloxane network), by X-ray imaging (due to relatively high Z(Gd)), by MRI (due to the presence of gadolinium(III) in the core of each particle, r_1 = 8.8 mM^{-1}.s^{-1} and r_2/r_1 = 1.3 for GadoSiPEG (hydrodynamic diameter: 11–14 nm) and r_1 = 9.4 mM^{-1}.s^{-1} and r_2/r_1 = 1.13 for GadoSiDTPA (hydrodynamic diameter: 2 nm) and by SPECT (thanks to the immobilization of ^{111}In when the polyaminocarboxylate ligands are tethered instead of PEG chains) (Figures 6b-e).

The potential of gadolinium oxide cores for MRI was confirmed thanks to the studies performed by a research group of Linköping University (Sweden) (*206*). Their works are very interesting and complementary to the one done by our group. Indeed, they propose an alternative way for derivatizing the gadolinium oxide cores and therefore for exploiting their paramagnetic character in MRI experiments. Their strategy consists in the capping of gadolinium oxide nanoparticles which were synthesized according to the polyol route (*207, 208*) by organic acids or a by the grafting of PEG-silane molecules. After their synthesis in DEG, the resulting gadolinium oxide cores are coated by DEG molecules which can be replaced in hot DEG by citric acid (CA) or dimercaptosuccinic acid (DMSA) (*213, 214*). This layer of organic acid molecules acts as a primer for further functionalization and in peculiar for the PEGylation which is required for biological applications (*215*). The PEGylation can also be performed using PEG-silane since alkoxysilyl end-group of PEG-silane favors its grafting on oxide nanoparticles through the condensation between the alkoxysilyl groups and hydroxyl (OH) groups present on the surface of the oxide materials (Gd_2O_3@PEG) (*216*). The grafting of nonimmunogenic and nonantigenic PEG is expected to render the gadolinium oxide cores more robust, to improve the colloidal stability in biological media, to favor the uptake by cells and to enhance the blood retention.

These PEGylated oxide nanoparticles which are prepared by the research group of Linköping University exhibit, as expected, relatively high longitudinal relaxivity r_1. The relaxivity r_1 values of Gd_2O_3@PEG, Gd_2O_3@DMSA-PEG, and Gd_2O_3@CA-PEG (based on the amount of gadolinium(III) ions in each sample) were 12.0, 14.2, and 7.8 $mM^{-1}.s^{-1}$, respectively (*217*). It must however be pointed out that the different strategies for the functionalization of gadolinium oxide cores provide similar performance in terms of positive contrast enhancement for MRI. Unfortunately their behavior after intravenous injection in small animals (rats, mice) cannot be compared because no *in vivo* imaging experiment was reported with the PEGylated organic acid capped gadolinium oxide cores in contrast to GadoSiPEG and GadoSiDTPA. Although the encapsulation of the gadolinium oxide cores by polysiloxane shell seems more difficult to control, it allows nevertheless enlarging more easily the palette of properties of the contrast agents since post-functionalization of gadolinium oxide cores can be sequentially realized, in contrast to the encapsulation by organic acids.

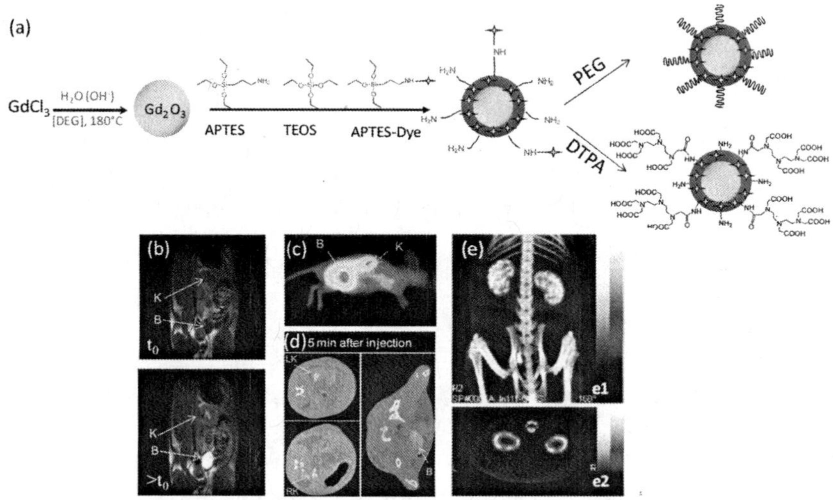

Figure 6. (a) Synthesis of GadoSiPEG and GadoSiDTPA nanoparticles. (b) T_1-weighted images before (t_0) and 32 minutes after (>t_0) intravenous injection of GadoSiDTPA nanoparticles to a rat. (c) Fluorescence reflectance imaging of a nude mouse 3 hours after the injection of GadoSiPEG. (d) SRCT images of a series of transverse slices including the right and left kidneys and the bladder (B) of a 9L gliosarcoma-bearing rat acquired 5 min after the intravenous injection of GadoSiDTPA nanoparticles. (e) SPECT/CT image acquired 2 days after intravenous injection of GadoSiDTPA-^{111}In nanoparticles to rats (posterior projection (e1) and transversal slice (e2)). B for bladder; K, RK and LK, for kidney, right and left kidney respectively (see color insert)

Other routes for the preparation of gadolinium oxide based contrast agents have been recently developed. Gadolinium oxide nanoparticles in carbon nanotubes (*218*) and in the protein apoferritin (*219*), porous and hollow gadolinium oxide nanoparticles (*220*) were synthesized for MRI application. Although the enhancement induced by these nanostructures based on Gd_2O_3 appears sufficient for MRI, their application as *in vivo* contrast agent seems to be limited due to the lack of colloidal stability in biological fluid and/or a large size.

The in Vitro Application of GadoSiPEG and GadoSiDTPA for Image-Guided Radiotherapy

Due to the presence of gadolinium, the GadoSiPEG and GadoSiDTPA nanoparticles with imaging multimodality (MRI, X-ray imaging, NIR fluorescence imaging and eventually SPECT) are expected to enhance the dose effect of neutron and X-ray beams. These particles are therefore designed for image-guided therapy (neutron capture therapy and radiotherapy). In order to confirm their potential, *in vitro* experiments combining imaging and irradiation were performed.

As widely mentioned in the literature, objects containing gadolinium carry the promise to replace advantageously the boron based compounds for neutron capture therapy. The internalization of GadoSiPEG nanoparticles in bioluminescent murine lymphoma cells (EL4-luc) which was monitored by fluorescence (Rhodamine B isothiocyanate was covalently bound to the polysiloxane shell), MRI and elemental analysis (ICP) generates no cytotoxicity and no alteration of the cell proliferation until $[Gd]_{incubation} = 0.3$ mM. Unloaded cells are also not affected by the exposure to a thermal neutron beam if the delivered dose is equal to 3Gy (or less). But the irradiation of EL4-luc cells after incubation in presence of GadoSiPEG ($[Gd]_{incubation} = 0.05$ mM) by harmless thermal neutron beam (3 Gy) generates a great killing effect since all cells are destroyed (*221*). The GadoSiPEG nanoparticles exhibit therefore a great potential for neutron capture therapy. However this therapeutic strategy remains essentially restricted to Japan and to the United States of America. Neutron capture therapy is obviously less common than the radiotherapy.

Owing to the relatively high Z of the gadolinium element, it was demonstrated that GadoSiDTPA strongly absorb the X-ray photons of high flux synchrotron beam since they induce a contrast enhancement of X-ray images (Figure 6d) (*211*). This property can also be exploited for radiotherapy since DNA strand breaks were observed when radioresistant U87 cells (human glioblastoma) were irradiated after incubation with gadolinium oxide nanoparticles (comet assays). For $[Gd]_{incubation} = 0.5$ mM, a maximum number of DNA breaks was generated at 10 Gy and after this concentration, the number of DNA breaks decreases and becomes stable because of particles aggregation that limits the efficiency of therapy (Figure 7a). Moreover the survival of the cells after irradiation decreases in presence of GadoSiDTPA nanoparticles while the internalization of these nanoparticles within the U87 cells does not lead to any mortality (*222*). This study opened the door to *in vivo* application of these gadolinium oxide-based nanoparticles since they can induce a radiosensitizing effect in an energy range easily used for radiotherapy.

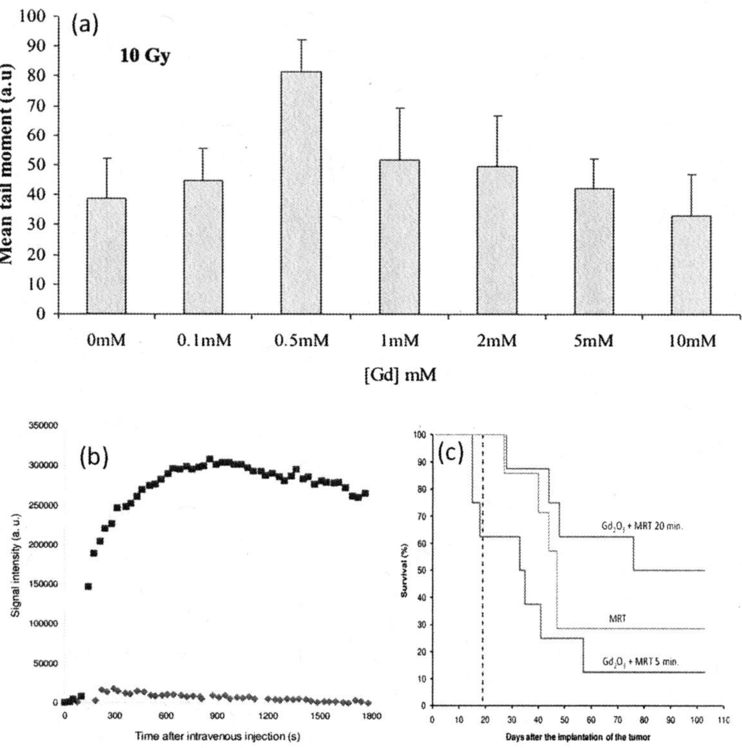

Figure 7. (a) In vitro radiosensitizing effect of gadolinium oxide nanoparticles at 10 Gy using a radiation at 660 keV. (b) Temporal evolution of the MRI signal in tumor (purple curve) and in an equivalent surface in normal tissue in the left hemisphere (blue curve) after injection of GadoSiDTPA nanoparticles to gliosarcoma bearing rat. (c) Survival curve comparison obtained on gliosarcoma bearing rats without treatment (black dashed curve, n = 4 rats), only treated by MRT (blue curve, n = 7), and treated by MRT 5 min (red curve, n = 8) and 20 min (green curve, n = 8) after intravenous injection. (see color insert)

The in Vivo Application of GadoSiDTPA for Image-Guided Radiotherapy

The *in vivo* imaging experiments after intravenous injection of GadoSiPEG revealed that the particles exhibit attractive biodistribution and pharmacokinetics parameters which are accurately controlled by the length of the PEG chains and the nature of the end group (*121, 210*). GadoSiDTPA and GadoSiPEG (with a short

PEG chain ended by COOH group) freely circulate in the bloodstream without undesirable non-specific accumulation in liver, spleen and lungs (Figure 6b-e). Moreover these nanoparticles are rather quickly removed from body essentially by renal excretion (Figure 6b-e) (*121, 210, 223*).

All these observations denote a safe behavior of nanoparticles when they are intravenously injected to small animals. But the accumulation of GadoSiDTPA nanoparticles was observed after intravenous injection to gliosarcoma (9L) bearing rats (murine brain tumor). This was revealed by the variation of the intensity of the signal in the tumor and in the healthy tissue and the obvious delineation of the tumor due to the enhancement of the positive contrast induced by the presence of the particles in the tumor region (Figure 7b). This accumulation which can be explained by the EPR effect was exploited for radiotherapy since GadoSiDTPA nanoparticles exert a radiosenstizing effect as revealed by *in vitro* experiments (*222*). As expected, the treatment of brain tumor bearing rats by radiotherapy after intravenous injection of GadoSiDTPA led to a longer survival than the one of diseased animals treated only by radiotherapy but only when hybrid gadolinium oxide nanoparticles was administered at least 20 minutes before the exposure to therapeutic X-ray beam (*211*) (Figure 7c). It must be pointed out that a shorter increase in lifespan (ILS) is observed when the irradiation was performed 5 minutes after the injection. These contrasted observations can be explained from the data collected by MRI which obviously showed a temporal evolution of the biodistribution of the radiosensitizing nanoparticles in the brain. This was confirmed by elemental analysis of both brain hemispheres (9L glisarcoma is implanted in the right hemisphere). The amount of gadolinium based nanoparticles decreased in course of time but the decrease is more important in the healthy tissue than in the tumor. As a result, the difference in gadolinium content between tumor and surrounding healthy tissue is higher 20 minutes than 5 minutes after the injection of GadoSiDTPA nanoparticles. Since the ILS is more important when the irradiation was performed 20 minutes after the intravenous injection of the radiosensitizers whereas the content of gadolinium in the tumor was smaller, it can be deduced that the efficiency of the treatment which rests on the interaction between radiosensitizers and X-ray beam mainly depends on the difference of gadolinium content between tumor and surrounding healthy tissue. In other words, the radiosensitizng effect was probably stronger when irradiation was performed 5 minutes after injection but healthy tissues were also seriously altered due to the dose enhancement of X-rays induced by the presence of particles. This study reveals therefore that the most opportune moment for the irradiation is not when the gadolinium content in tumor is the highest but the moment when both the difference of gadolinium content between tumor and healthy tissue is high and the amount of gadolinium is low in the surrounding healthy tissue. This study clearly indicates the benefit that can be derived from therapeutic agents which are also designed for medical imaging because the most opportune moment for remotely inducing their therapeutic activity can be determined thanks to the monitoring of their biodistribution by medical imaging. The GadoSiDTPA nanoparticles which combine imaging multimodality and radiosensitization are well suited for image-guided radiotherapy.

Conclusion

As highlighted by the studies focused on Au@DTDTPA-Gd and GadoSiDTPA nanoparticles, an appropriate design of multifunctional nanoparticles can lead to the increase of survival of rats bearing radioresistant aggressive tumor (9L gliosarcoma) in radiosensitive organ (brain) when the diseased animals were treated by radiotherapy after intravenous injection of the nanoparticles. For achieving such encouraging results, some criteria imposed by biological (nanoparticles biodistribution, accumulation in solid tumor, clearance) and physical (lack of selectivity of radiotherapy, interaction between radiation and biological matter) constraints must be met by the nanoparticles designed for image-guided therapy. In the case of non-biodegradable nanoparticles (like metallic nanoparticles), the size is the main key-parameter. The success of the therapy using multifunctional nanoparticles rests on a paradox since the nanoparticles must be sufficiently large for a preferential accumulation in tumor (at least by passive targeting) and sufficiently small for renal clearance. Their hydrodynamic diameter does therefore not exceed 6–10 nm which constitutes the limit for renal clearance. This mode of elimination (by urine) must indeed be preferred since it implies no degradation of the nanoparticles in contrast to the hepatobiliary mode (by feces). Although the size imposed by the renal clearance is very reduced, the incorporation in a single nanoobject of a sufficient amount of high Z elements (Au, Gd) for efficient X-ray imaging and radiosensitization is possible. Moreover, other imaging modalities (MRI, fluorescence imaging, scintigraphy) can be integrated, thanks to the functionalization of the nanoparticles. The possibility to monitor the biodistribution by several complementary medical imaging techniques constitutes a valuable asset since more reliable data are obtained by combining the high sensitivity of a technique (e.g. fluorescence imaging, nuclear imaging) to the high resolution of another one (e.g. MRI, CT). The follow-up of radiosensitizing nanoparticles which are designed for meeting the criteria imposed by *in vivo* application is essential for improving the effects of radiotherapy (complete eradication of the tumor and preservation of the surrounding healthy tissues) since the irradiation is induced when the content of the nanoparticles is high in the tumor and very low in the healthy tissue. It must be pointed out that the improvement of the selectivity (and finally of the efficiency) of radiotherapy when radiosensitizing nanoparticles are used mainly results from the difference in high Z element content between the tumor and the healthy tissue. The localization by medical imaging of the radiosensitizers is therefore very important. In the case of the treatment of gliosarcoma, the accumulation of the radiosensitizing nanoparticles (Au@DTDTPA and GadoSiDTPA) by passive targeting was sufficient for inducing a great increase in lifespan.

Thanks to the biodistribution monitoring, radiotherapy can be guided by medical imaging. This opens the door to the personalized therapy which aims at adapting the treatment to each patient. However the treatment of cancer is not confined to the eradication of the tumor but also, in many cases, to the destruction of metastases which remains an absolute challenge. This implies that either the treatment is applied before the appearance of the metastases or all metastases are targeted by radiosensitizing agents. The early detection of tumor and the destruction of metastases are crucial issues that should be addressed. For this reason, the strategy based on active targeting of multifunctional nanoparticles designed for image-guided therapy deserves to be explored.

References

1. *Cancer Medicine*, 8th ed.; Hong, W. K., Bast, R. C., Hait, W. N., Kufe, D. W., Pollock, R. E., Weichselbaum, R. R., Holland, J. F., Frei, E., Eds.; People's Medical Publishing House: Shelton, 2010.
2. *Cancer Incidence and Mortality Worldwide in 2008 (GloboCan 2008) Home Page, International Agency for Research on Cancer*; http://globocan.iarc.fr/, accessed Jun 10, 2010.
3. *World Health Organization Home Page*; http://www.who.int, accessed Jun 10, 2010.
4. Hanahan, D.; Weinberg, R. A. *Cell* **2000**, *100*, 57–70.
5. Igney, F. H.; Krammer, P. H. *J. Leukocyte Biol.* **2002**, *71*, 907–920.
6. Carmeliet, P. *Nature* **2005**, *438*, 932–936.
7. Carmeliet, P.; Jain, R. K. *Nature* **2000**, *407*, 249–257.
8. Sherwood, L. M.; Parris, E. E.; Folkman, J. *N. Engl. J. Med.* **1971**, *285*, 1182–1186.
9. Hanahan, D.; Folkman, J. *Cell* **1996**, *86*, 353–364.
10. Bergers, G.; Benjamin, L. E. *Nat. Rev. Cancer* **2003**, *3*, 401–410.
11. Stacker, S. A.; Achen, M. G.; Jussila, L.; Baldwin, M. E.; Alitalo, K. *Nat. Rev. Cancer* **2002**, *2*, 573–583.
12. Alitalo, K.; Tammela, T.; Petrova, T. V. *Nature* **2005**, *438*, 946–953.
13. Chambers, A. F.; Groom, A. C.; MacDonald, I. C. *Nat. Rev. Cancer* **2002**, *2*, 563–572.
14. Friedl, P.; Wolf, K. *Nat. Rev. Cancer* **2003**, *3*, 362–374.
15. Mehlen, P.; Puisieux, A. *Nat. Rev. Cancer* **2006**, *6*, 449–458.
16. Verheul, H.; Voest, E.; Schlingemann, R. *J. Pathol.* **2004**, *202*, 5–13.
17. Kerbel, R.; Folkman, J. *Nat. Rev. Cancer* **2002**, *2*, 727–739.
18. Jain, R. K. *Science* **2005**, *307*, 58–62.
19. Brown, J. M.; Giaccia, A. J. *Cancer Res.* **1998**, *58*, 1408–1416.
20. Kuszyk, B. S.; Corl, F. M.; Franano, F. N.; Bluemke, D. A.; Hofmann, L. V.; Fortman, B. J.; Fishman, E. K. *Am. J. Roentgenol.* **2001**, *177*, 747–753.
21. Weis, S. M.; Cheresh, D. A. *Nature* **2005**, *437*, 497–504.
22. McDonald, D. M.; Choyke, P. L. *Nat. Med.* **2003**, *9*, 713–725.
23. Ribatti, D.; Nico, B.; Crivellato, E.; Vacca, A. *Cancer Lett.* **2007**, *248*, 18–23.
24. Yuan, F.; Dellian, M.; Fukumura, D.; Leunig, M.; Berk, D. A.; Torchilin, V. P.; Jain, R. K. *Cancer Res.* **1995**, *55*, 3752–3756.
25. Hobbs, S. K.; Monsky, W. L.; Yuan, F.; Roberts, W. G.; Griffith, L.; Torchilin, V. P.; Jain, R. K. *Proc. Natl. Acad. Sci. U.S.A.* **1998**, *95*, 4607–4612.

26. Hashizume, H.; Baluk, P.; Morikawa, S.; McLean, J. W.; Thurston, G.; Roberge, S.; Jain, R. K.; McDonald, D. M. *Am. J. Pathol.* **2000**, *156*, 1363–1380.

27. Bell, D. R. In *Medical Physiology: Principles for Clinical Medicine*; Rhoades, R. A., Bell, D. R., Eds.; Lippincott Williams & Wilkins: Baltimore, 2009; Chapter 15, pp 275-290.

28. Roberts, W. G.; Palade, G. E. *Cancer Res.* **1997**, *57*, 765–772.

29. Maeda, H.; Bharate, G.; Daruwalla, J. *Eur. J. Pharm. Biopharm.* **2009**, *71*, 409–419.

30. Padera, T. P.; Kadambi, A.; di Tomaso, E.; Carreira, C. M.; Brown, E. B.; Boucher, Y.; Choi, N. C.; Mathisen, D.; Wain, J.; Mark, E. J.; Munn, L. L.; Jain, R. K. *Science* **2002**, *296*, 1883–1886.

31. Fang, J.; Nakamura, H.; Maeda, H. *Adv. Drug Delivery Rev.* **2011**, *63*, 136–151.

32. Wang, X.; Wang, Y.; Chen, Z. G.; Shin, D. M. *Cancer Res. Treat.* **2009**, *41*, 1–11.

33. Rubin, I.; Yarden, Y. *Ann. Oncol.* **2001**, *12*, S3–S8.

34. Ludwig, J. A.; Weinstein, J. N. *Nat. Rev. Cancer* **2005**, *5*, 845–856.

35. Allen, T. M. *Nat. Rev. Cancer* **2002**, *2*, 750–763.

36. Imai, K.; Takaoka, A. *Nat. Rev. Cancer* **2006**, *6*, 714–727.

37. Sharkey, R. M.; Goldenberg, D. M. *Ca-Cancer J. Clin.* **2006**, *56*, 226–243.

38. Hofmeister, V.; Schrama, D.; Becker, J. *Cancer Immunol. Immun.* **2008**, *57*, 1–17.

39. Neri, D.; Bicknell, R. *Nat. Rev. Cancer* **2005**, *5*, 436–446.

40. Thorpe, P. E. *Clin. Cancer Res.* **2004**, *10*, 415–427.

41. Ferrara, N.; Kerbel, R. S. *Nature* **2005**, *438*, 967–974.

42. Fass, L. *Mol. Oncol.* **2008**, *2*, 115–152.

43. Benaron, D. A. *Cancer Metast. Rev.*, *21*, 45–78.

44. Brindle, K. *Nat. Rev. Cancer* **2008**, *8*, 94–107.

45. Lentle, B.; Aldrich, J. *Lancet* **1997**, *350*, 280–285.

46. Elliott, A. *Nucl. Instrum. Methods, Sect. A* **2005**, *546*, 1–13.

47. Michael, G. *Phys. Educ.* **2001**, *36*, 442–451.

48. Brenner, D. J.; Hall, E. J. *New Engl. J. Med.* **2007**, *357*, 2277–2284.

49. Krause, W.; Schneider, P. *Top. Curr. Chem.* **2002**, *222*, 107–150.

50. Idée, J.-M.; Nachman, I.; Port, M.; Petta, M.; Le Lem, G.; Le Greneur, S.; Dencausse, A.; Meyer, D.; Corot, C. *Top. Curr. Chem.* **2002**, *222*, 151–171.

51. Pooley, R. A. *Radiographics* **2005**, *25*, 1087–1099.

52. Geraldes, C. F. G. C.; Laurent, S. *Contrast Media Mol. Imaging* **2009**, *4*, 1–23.

53. Caravan, P.; Ellison, J.; McMurry, T.; Lauffer, R. *Chem. Rev.* **1999**, *99*, 2293–2352.

54. Gries, H. *Top. Curr. Chem.* **2002**, *221*, 1–24.

55. Bellin, M.-F. *Eur. J. Radiol.* **2006**, *60*, 314–323.

56. Caravan, P. *Accounts Chem. Res.* **2009**, *42*, 851–862.

57. Badawi, R. D. *Phys. Educ.* **2001**, *36*, 452–459.

58. Bybel, B.; Brunken, R. C.; DiFilippo, F. P.; Neumann, D. R.; Wu, G.; Cerqueira, M. D. *Radiographics* **2008**, *28*, 1097–1113.

59. Blodgett, T. M.; Meltzer, C. C.; Townsend, D. W. *Radiology* **2007**, *242*, 360–385.

60. Lichtman, J. W.; Conchello, J.-A. *Nat. Methods* **2005**, *2*, 910–919.

61. Giepmans, B. N. G.; Adams, S. R.; Ellisman, M. H.; Tsien, R. Y. *Science* **2006**, *312*, 217–224.

62. Dufort, S.; Sancey, L.; Wenk, C.; Josserand, V.; Coll, J.-L. *Biochim. Biophys. Acta, Biomembr.* **2010**, *1798*, 2266–2273.

63. Licha, K.; Olbrich, C. *Adv. Drug Delivery Rev.* **2005**, *57*, 1087–1108.

64. Wagnieres, G. A.; Star, W. M.; Wilson, B. C. *Photochem. Photobiol.* **1998**, *68*, 603–632.

65. Frangioni, J. V. *Curr. Opin. Chem. Biol.* **2003**, *7*, 626–634.

66. Licha, K. *Top. Curr. Chem.* **2002**, *222*, 1–29.

67. Rao, J.; Dragulescu-Andrasi, A.; Yao, H. *Curr. Opin. Biotechnol.* **2007**, *18*, 17–25.

68. Weissleder, R. *Nat. Rev. Cancer* **2002**, *2*, 11–18.

69. Massoud, T. F.; Gambhir, S. S. *Gene. Dev.* **2003**, *17*, 545–580.

70. Weissleder, R.; Pittet, M. J. *Nature* **2008**, *452*, 580–589.

71. Baker, M. *Nature* **2010**, *463*, 977–980.

72. Townsend, D.; Cherry, S. *Eur. Radiol.* **2001**, *11*, 1968–1974.

73. von Schulthess, G. K.; Steinert, H. C.; Hany, T. F. *Radiology* **2006**, *238*, 405–422.

74. Even-Sapir, E.; Keidar, Z.; Bar-Shalom, R. *Semin. Nucl. Med.* **2009**, *39*, 264–275.

75. Bockisch, A.; Freudenberg, L. S.; Schmidt, D.; Kuwert, T. *Semin. Nucl. Med.* **2009**, *39*, 276–289.

76. Dawson, L. A.; Menard, C. *Oncologist* **2010**, *15*, 338–349.

77. Pichler, B. J.; Wehrl, H. F.; Kolb, A.; Judenhofer, M. S. *Semin. Nucl. Med.* **2008**, *38*, 199–208.

78. Goetz, C.; Breton, E.; Choquet, P.; Israel-Jost, V.; Constantinesco, A. *J. Nucl. Med.* **2008**, *49*, 88–93.

79. Cherry, S. R. *Semin. Nucl. Med.* **2009**, *39*, 348–353.

80. Jennings, L. E.; Long, N. J. *Chem. Commun.* **2009**, 3511–3524.

81. Louie, A. *Chem. Rev.* **2010**, *110*, 3146–3195.

82. Delaney, G.; Jacob, S.; Featherstone, C.; Barton, M. *Cancer* **2005**, *104*, 1129–1137.

83. Baskar, R.; Lee, K. A.; Yeo, R.; Yeoh, K. W. *Int. J. Med. Sci.* **2012**, *9*, 193–199.

84. Tubiana, M.; Wambersie, A.; Dutreix, J. *Introduction to Radiobiology*; Taylor & Francis: London, 1990.

85. *Radiation Oncology Physics: A Handbook for Teachers and Students*; Podgorsak, E. B., Ed.; International Atomic Energy Agency: Vienna, 2005.

86. Hubbell, J. H. *Phys. Med. Biol.* **1999**, *44*, R1–R22.

87. Boudaïffa, B.; Cloutier, P.; Hunting, D.; Huels, M. A.; Sanche, L. *Science* **2000**, *287*, 1658–1660.

88. *Basic Clinical Radiobiology*, 4th ed.; Joiner, M., van der Kogel, A., Eds.; Hodder Arnold: London, 2009.

89. Beyzadeoglu, M.; Ozyigit, G.; Ebruli, C. *Basic Radiation Oncology*; Springer-Verlag: Berlin, 2010.

90. Stone, H. B.; Coleman, C. N.; Anscher, M. S.; McBride, W. H. *Lancet Oncol.* **2003**, *4*, 529–536.

91. Dearnaley, D. P.; Khoo, V. S.; Norman, A. R.; Meyer, L.; Nahum, A.; Tait, D.; Yarnold, J.; Horwich, A. *Lancet* **1999**, *353*, 267–272.

92. Hong, T. S.; Ritter, M. A.; Tome, W. A.; Harari, P. M. *Br. J. Cancer* **2005**, *92*, 1819–1824.

93. Verellen, D.; Ridder, M. D.; Linthout, N.; Tournel, K.; Soete, G.; Storme, G. *Nat. Rev. Cancer* **2007**, *7*, 949–960.

94. Schulz-Ertner, D.; Tsujii, H. *J. Clin. Oncol.* **2007**, *25*, 953–964.

95. Wilson, G. D.; Bentzen, S. M.; Harari, P. M. *Semin. Radiat. Oncol.* **2006**, *16*, 2–9.

96. Kobayashi, K.; Usami, N.; Porcel, E.; Lacombe, S.; Sech, C. L. *Mutat. Res., Rev. Mutat. Res.* **2010**, *704*, 123–131.

97. Pignol, J.-P.; Rakovitch, E.; Beachey, D.; Le Sech, C. *Int. J. Radiat. Oncol. Biol. Phys.* **2003**, *55*, 1082–1091.

98. Santos Mello, R.; H., C.; Winter, J.; Kagan, A. R.; Norman, A. *Med. Phys.* **1983**, *10*, 75–78.

99. Iwamoto, K. S.; Norman, A.; Kagan, A. R.; Wollin, M.; Olch, A.; Bellotti, J.; Ingram, M.; Skillen, R. G. *Radiother. Oncol.* **1990**, *19*, 337–343.

100. Norman, A.; Ingram, M.; Skillen, R.; Freshwater, D.; Iwamoto, K.; Solberg, T. *Radiat. Oncol. Invest.* **1997**, *5*, 8–14.

101. Rose, J. H.; Norman, A.; Ingram, M.; Aoki, C.; Solberg, T.; Mesa, A. *Int. J. Radiat. Oncol. Biol. Phys.* **1999**, *45*, 1127–1132.

102. Kassis, A. I. *Int. J. Radiat. Biol.* **2004**, *80*, 789–803.

103. Miller, R. W.; Degraff, W.; Kinsella, T. J.; Mitchell, J. B. *Int. J. Radiat. Oncol. Biol. Phys.* **1987**, *13*, 1193–1197.

104. Laster, B. H.; Thomlinson, W. C.; Fairchild, R. G. *Radiat. Res.* **1993**, *133*, 219–224.

105. Corde, S.; Joubert, A.; Adam, J.-F.; Charvet, J.-F.; Le Bas, A.-M.; Estève, F.; Elleaume, H.; Balosso, J. *Br. J. Cancer* **2004**, *91*, 544–551.

106. Biston, M.-C.; Joubert, A.; Adam, J.-F.; Elleaume, H.; Bohic, S.; Charvet, A.-M.; Estève, F.; Foray, N.; Balosso, J. *Cancer Res.* **2004**, *64*, 2317–2323.

107. Biston, M.-C.; Joubert, A.; Charvet, A.-M.; Balosso, J.; Foray, N. *Radiat. Res.* **2009**, *172*, 348–358.

108. Hubbell, J. H.; Trehan, P. N.; Singh, N.; Chand, B.; Mehta, D.; Garg, M. L.; Garg, R. R.; Singh, S.; Puri, S. *J. Phys. Chem. Ref. Data* **1994**, *23*, 339–364.

109. Nath, R.; Bongiorni, P.; Rossi, P. I.; Rockwell, S. *Int. J. Radiat. Oncol. Biol. Phys.* **1990**, *18*, 1377–1385.

110. Florea, A.-M.; Büsselberg, D. *Cancers* **2011**, *3*, 1351–1371.

111. Kelly, K. L.; Coronado, E.; Zhao, L. L.; Schatz, G. C. *J. Phys. Chem. B* **2003**, *107*, 668–677.

112. Elghanian, R.; Storhoff, J. J.; Mucic, R. C.; Letsinger, R. L.; Mirkin, C. A. *Science* **1997**, *277*, 1078–1081.

113. Bottrill, M.; Green, M. *Chem. Commun.* **2011**, *47*, 7039–7050.

114. Smith, A. M.; Duan, H.; Mohs, A. M.; Nie, S. *Adv. Drug Delivery. Rev.* **2008**, *60*, 1226–1240.

115. Scheinberg, D. A.; Villa, C. H.; Escorcia, F. E.; McDevitt, M. R. *Nat. Rev. Clin. Oncol.* **2010**, *7*, 266–276.

116. Lewin, M.; Carlesso, N.; Tung, C.-H.; Tang, X.-W.; Cory, D.; Scadden, D. T.; Weissleder, R. *Nat. Biotechnol.* **2000**, *18*, 410–414.

117. Dobrovolskaia, M. A.; Aggarwal, P.; Hall, J. B.; McNeil, S. E. *Mol. Pharmaceutics* **2008**, *5*, 487–495.

118. Vonarbourg, A.; Passirani, C.; Saulnier, P.; Benoit, J.-P. *Biomaterials* **2006**, *27*, 4356–4373.

119. Moghimi, S. M.; Szebeni, J. *Prog. Lipid Res.* **2003**, *42*, 463–478.

120. Otsuka, H.; Nagasaki, Y.; Kataoka, K. *Adv. Drug Delivery. Rev.* **2003**, *55*, 403–419.

121. Faure, A.-C.; Dufort, S.; Josserand, V.; Perriat, P.; Coll, J.-L.; Roux, S.; Tillement, O. *Small* **2009**, *5*, 2565–2575.

122. Alexis, F.; Pridgen, E.; Molnar, L. K.; Farokhzad, O. C. *Mol. Pharmaceutics* **2008**, *5*, 505–515.

123. Moghimi, S. M. In *Nanotechnology for Cancer Therapy*; Amiji, M. M., Ed.; CRC Press: Boca Raton, 2007; Chapter 2, pp 11–18.

124. Torchilin, V. *Adv. Drug Delivery Rev.* **2010**, *63*, 131–135.

125. Juliano, R. L.; Alam, R.; Dixit, V.; Kang, H. M. *WIREs: Nanomed. Nanobiotechnol.* **2009**, *1*, 324–335.

126. Ruoslahti, E.; Bhatia, S. N.; Sailor, M. J. *J. Cell Biol.* **2010**, *188*, 759–768.

127. Longmire, M.; Choyke, P. L.; Kobayashi, H. *Nanomedicine* **2008**, *3*, 703–717.

128. Choi, H. S.; Liu, W.; Misra, P.; Tanaka, E.; Zimmer, J. P.; Ipe, B. I.; Bawendi, M. G.; Frangioni, J. V. *Nat. Biotechnol.* **2007**, *25*, 1165–1170.

129. Choi, H. S.; Frangioni, J. V. *Mol. Imaging* **2010**, *9*, 291–310.

130. Kim, J.; Piao, Y.; Hyeon, T. *Chem. Soc. Rev.* **2009**, *38*, 372–390.

131. Park, K.; Lee, S.; Kang, E.; Kim, K.; Choi, K.; Kwon, I. C. *Adv. Funct. Mater.* **2009**, *19*, 1553–1566.

132. Gullotti, E.; Yeo, Y. *Mol. Pharm.* **2009**, *6*, 1041–1051.

133. Lee, E. S.; Gao, Z.; Bae, Y. H. *J. Controlled Release* **2008**, *132*, 164–170.

134. Olivo, M.; Bhuvaneswari, R.; Lucky, S. S.; Dendukuri, N.; Soo-Ping Thong, P. *Pharmaceuticals* **2010**, *3*, 1507–1529.

135. Barth, R. F.; Coderre, J. A.; Vicente, M. G. H.; Blue, T. E. *Clin. Cancer Res.* **2005**, *11*, 3987–4002.

136. Bernier, J.; Hall, E. J.; Giaccia, A. *Nat. Rev. Cancer* **2004**, *4*, 737–747.

137. Chabner, B. A.; Roberts, T. G. *Nat. Rev. Cancer* **2005**, *5*, 65–72.

138. Arruebo, M.; Vilaboa, N.; Sáez-Gutierrez, B.; Lambea, J.; Tres, A.; Valladares, M.; González-Fernández, Á. *Cancers* **2011**, *3*, 3279–3330.

139. Wen, P. Y.; Kesari, S. *N. Engl. J. Med.* **2008**, *359*, 492–507.

140. Stupp, R.; Roila, F. *Ann. Oncol.* **2009**, *20*, iv126–iv128.

141. DeAngelis, L. M. *New Engl. J. Med.* **2001**, *344*, 114–123.

142. Chan, J. L.; Lee, S. W.; Fraass, B. A.; Normolle, D. P.; Greenberg, H. S.; Junck, L. R.; Gebarski, S. S.; Sandler, H. M. *J. Clin. Oncol.* **2002**, *20*, 1635–1642.

143. Connor, E.; Mwamuka, J.; Gole, A.; Murphy, C.; Wyatt, M. *Small* **2005**, *1*, 325–327.

144. Villiers, C.; Freitas, H.; Couderc, R.; Villiers, M.-B.; Marche, P. *J. Nanopart. Res.* **2010**, *12*, 55–60.

145. Alkilany, A.; Murphy, C. *J. Nanopart. Res.* **2010**, *12*, 2313–2333.

146. Schmid, G.; Corain, B. *Eur. J. Inorg. Chem.* **2003**, *2003*, 3081–3098.

147. Daniel, M.-C.; Astruc, D. *Chem. Rev.* **2004**, *104*, 293–346.

148. Frens, G. *Nat. Phys. Sci.* **1973**, *241*, 20–22.

149. Brust, M.; Fink, J.; Bethell, D.; Schiffrin, D. J.; Kiely, C. *J. Chem. Soc., Chem. Commun.* **1995**, 1655–1656.

150. Roux, S.; Garcia, B.; Bridot, J.-L.; Salome, M.; Marquette, C.; Lemelle, L.; Gillet, P.; Blum, L.; Perriat, P.; Tillement, O. *Langmuir* **2005**, *21*, 2526–2536.

151. Murphy, C. J.; Sau, T. K.; Gole, A. M.; Orendorff, C. J.; Gao, J.; Gou, L.; Hunyadi, S. E.; Li, T. *J. Phys. Chem. B* **2005**, *109*, 13857–13870.

152. Loo, C.; Lin, A.; Hirsch, L.; Lee, M.-H.; Barton, J.; Halas, N.; West, J.; Drezek, R. *Technol. Cancer Res. Treat.* **2004**, *3*, 33–40.

153. Yavuz, M. S.; Cheng, Y.; Chen, J.; Cobley, C. M.; Zhang, Q.; Rycenga, M.; Xie, J.; Kim, C.; Song, K. H.; Schwartz, A. G.; Wang, L. V.; Xia, Y. *Nat. Mater.* **2009**, *8*, 935–939.

154. Jelveh, S.; Chithrani, D. B. *Cancers* **2011**, *3*, 1081–1110.

155. Jain, P. K.; Huang, X.; El-Sayed, I. H.; El-Sayed, M. A. *Acc. Chem. Res.* **2008**, *41*, 1578–1586.

156. Weissleder, R. *Nat. Biotechnol.* **2001**, *19*, 316–317.

157. Loo, C.; Lowery, A.; Halas, N.; West, J.; Drezek, R. *Nano Lett.* **2005**, *5*, 709–711.

158. Hirsch, L. R.; Stafford, R. J.; Bankson, J. A.; Sershen, S. R.; Rivera, B.; Price, R. E.; Hazle, J. D.; Halas, N. J.; West, J. L. *Proc. Natl. Acad. Sci. U.S.A.* **2003**, *100*, 13549–13554.

159. Huang, X.; El-Sayed, I.; Qian, W.; El-Sayed, M. *J. Am. Chem. Soc.* **2006**, *128*, 2115–2120.

160. Dickerson, E. B.; Dreaden, E. C.; Huang, X.; El-Sayed, I. H.; Chu, H.; Pushpanketh, S.; McDonald, J. F.; El-Sayed, M. A. *Cancer Lett.* **2008**, *269*, 57–66.

161. Kim, B. Y.; Rutka, J. T.; Chan, W. C. *New Engl. J. Med.* **2010**, *363*, 2434–2443.

162. Kim, C. K.; Ghosh, P.; Pagliuca, C.; Zhu, Z.-J.; Menichetti, S.; Rotello, V. M. *J. Am. Chem. Soc.* **2009**, *131*, 1360–1361.

163. Kim, C. K.; Ghosh, P.; Rotello, V. M. *Nanoscale* **2009**, *1*, 61–67.

164. Duncan, B.; Kim, C.; Rotello, V. M. *J. Controlled Release* **2010**, *148*, 122–127.

165. Gibson, J.; Khanal, B.; Zubarev, E. *J. Am. Chem. Soc.* **2007**, *129*, 11653–11661.

166. Chen, Y.-H.; Tsai, C.-Y.; Huang, P.-Y.; Chang, M.-Y.; Cheng, P.-C.; Chou, C.-H.; Chen, D.-H.; Wang, C.-R.; Shiau, A.-L.; Wu, C.-L. *Mol. Pharmaceutics* **2007**, *4*, 713–722.

167. Brown, S. D.; Nativo, P.; Smith, J.-A.; Stirling, D.; Edwards, P. R.; Venugopal, B.; Flint, D. J.; Plumb, J. A.; Graham, D.; Wheate, N. J. *J. Am. Chem. Soc.* **2010**, *132*, 4678–4684.

168. Paciotti, G. F.; Myer, L.; Weinreich, D.; Goia, D.; Pavel, N.; McLaughlin, R. E.; Tamarkin, L. *Drug Delivery* **2004**, *11*, 169–183.

169. Libutti, S. K.; Paciotti, G. F.; Byrnes, A. A.; Alexander, H. R.; Gannon, W. E.; Walker, M.; Seidel, G. D.; Yuldasheva, N.; Tamarkin, L. *Clin. Cancer Res.* **2010**, *16*, 6139–6149.

170. Herold, M.; Das, I. J.; Stobbe, C. C.; Iyer, R. V.; Chapman, J. D. *Int. J. Radiat. Biol.* **2000**, *76*, 1357–1364.

171. Hainfeld, J. F.; Slatkin, D. N.; Smilowitz, H. M. *Phys. Med. Biol.* **2004**, *49*, N309–N315.

172. Hainfeld, J. F.; Slatkin, D. N.; Focella, T. M.; Smilowitz, H. M. *Br. J. Radiol.* **2006**, *79*, 248–253.

173. Kong, T.; Zeng, J.; Wang, X.; Yang, X.; Yang, J.; McQuarrie, S.; McEwan, A.; Roa, W.; Chen, J.; Xing, J. Z. *Small* **2008**, *4*, 1537–1543.

174. Rahman, W. N.; Bishara, N.; Ackerly, T.; He, C. F.; Jackson, P.; Wong, C.; Davidson, R.; Geso, M. *Nanomed.-Nanotechnol. Biol. Med.* **2009**, *5*, 136–142.

175. Chithrani, D. B.; Jelveh, S.; Jalali, F.; van Prooijen, M.; Allen, C.; Bristow, R. G.; Hill, R. P.; Jaffray, D. A. *Radiat. Res.* **2010**, *173*, 719–728.

176. Cai, Q.; Kim, S. H.; Choi, K. S.; Kim, S. Y.; Byun, S. J.; Kim, K. W.; Park, S. H.; Juhng, S. K.; Yoon, K.-H. *Invest. Radiol.* **2007**, *42*, 797–806.

177. Kattumuri, V.; Katti, K.; Bhaskaran, S.; Boote, E.; Casteel, S.; Fent, G.; Robertson, D.; Chandrasekhar, M.; Kannan, R.; Katti, K. *Small* **2007**, *3*, 333–341.

178. Sun, I.-C.; et al. *Chem.-Eur. J.* **2009**, *15*, 13341–13347.

179. Caravan, P. *Chem. Soc. Rev.* **2006**, *35*, 512–523.

180. Bottrill, M.; Kwok, L.; Long, N. J. *Chem. Soc. Rev.* **2006**, *35*, 557–571.

181. Garcia, B.; Salome, M.; Lemelle, L.; Bridot, J.-L.; Gillet, P.; Perriat, P.; Roux, S.; Tillement, O. *Chem. Commun.* **2005**, 369–371.

182. Debouttière, P.-J.; Roux, S.; Vocanson, F.; Billotey, C.; Beuf, O.; Favre-Réguillon, A.; Lin, Y.; Pellet-Rostaing, S.; Lamartine, R.; Perriat, P.; Tillement, O. *Adv. Funct. Mater.* **2006**, *16*, 2330–2339.

183. Alric, C.; Taleb, J.; Le Duc, G.; Mandon, C.; Billotey, C.; Le Meur-Herland, A.; Brochard, T.; Vocanson, F.; Janier, M.; Perriat, P.; Roux, S.; Tillement, O. *J. Am. Chem. Soc.* **2008**, *130*, 5908–5915.

184. Alric, C.; Serduc, R.; Mandon, C.; Taleb, J.; Le Duc, G.; Le Meur-Herland, A.; Billotey, C.; Perriat, P.; Roux, S.; Tillement, O. *Gold Bull.* **2008**, *41*, 90–97.

185. Bouchet, A.; Lemasson, B.; Duc, G. L.; Maisin, C.; Bräuer-Krisch, E.; Siegbahn, E. A.; Renaud, L.; Khalil, E.; Rémy, C.; Poillot, C.; Bravin, A.; Laissue, J. A.; Barbier, E. L.; Serduc, R. *Int. J. Radiat. Oncol. Biol. Phys.* **2010**, *78*, 1503–1512.

186. Mulder, W. J. M.; Strijkers, G. J.; Griffioen, A. W.; van Bloois, L.; Molema, G.; Storm, G.; Koning, G. A.; Nicolay, K. *Bioconjugate Chem.* **2004**, *15*, 799–806.

187. Platas-Iglesias, C.; Vander Elst, L.; Zhou, W.; Muller, R. N.; Geraldes, C. F. G. C.; Maschmeyer, T.; Peters, J. A. *Chem.-Eur. J.* **2002**, *8*, 5121–5131.

188. Lin, Y.-S.; Hung, Y.; Su, J.-K.; Lee, R.; Chang, C.; Lin, M.-L.; Mou, C.-Y. *J. Phys. Chem. B* **2004**, *108*, 15608–15611.

189. Mulder, W. J. M.; Koole, R.; Brandwijk, R. J.; Storm, G.; Chin, P. T. K.; Strijkers, G. J.; de Mello Donega, C.; Nicolay, K.; Griffioen, A. W. *Nano Lett.* **2006**, *6*, 1–6.

190. van Tilborg, G. A. F.; Mulder, W. J. M.; Chin, P. T. K.; Storm, G.; Reutelingsperger, C. P.; Nicolay, K.; Strijkers, G. J. *Bioconjugate Chem.* **2006**, *17*, 865–868.

191. Vuu, K.; Xie, J.; McDonald, M. A.; Bernardo, M.; Hunter, F.; Zhang, Y.; Li, K.; Bednarski, M.; Guccione, S. *Bioconjugate Chem.* **2005**, *16*, 995–999.

192. Bolskar, R. D.; Benedetto, A. F.; Husebo, L. O.; Price, R. E.; Jackson, E. F.; Wallace, S.; Wilson, L. J.; Alford, J. M. *J. Am. Chem. Soc.* **2003**, *125*, 5471–5478.

193. Mikawa, M.; Kato, H.; Okumura, M.; Narazaki, M.; Kanazawa, Y.; Miwa, N.; Shinohara, H. *Bioconjugate Chem.* **2001**, *12*, 510–514.

194. Tóth, E.; Bolskar, R. D.; Borel, A.; González, G.; Helm, L.; Merbach, A. E.; Sitharaman, B.; Wilson, L. J. *J. Am. Chem. Soc.* **2005**, *127*, 799–805.

195. Sitharaman, B.; Bolskar, R. D.; Rusakova, I.; Wilson, L. J. *Nano Lett.* **2004**, *4*, 2373–2378.

196. Sitharaman, B.; Kissell, K. R.; Hartman, K. B.; Tran, L. A.; Baikalov, A.; Rusakova, I.; Sun, Y.; Khant, H. A.; Ludtke, S. J.; Chiu, W.; Laus, S.; Tóth, E.; Helm, L.; Merbach, A. E.; Wilson, L. J. *Chem. Commun.* **2005**, 3915–3917.

197. Hartman, K. B.; Laus, S.; Bolskar, R. D.; Muthupillai, R.; Helm, L.; Tóth, E.; Merbach, A. E.; Wilson, L. J. *Nano Lett.* **2008**, *8*, 415–419.

198. Frias, J. C.; Williams, K. J.; Fisher, E. A.; Fayad, Z. A. *J. Am. Chem. Soc.* **2004**, *126*, 16316–16317.

199. Frias, J. C.; Ma, Y.; Williams, K. J.; Fayad, Z. A.; Fisher, E. A. *Nano Lett.* **2006**, *6*, 2220–2224.

200. Hüber, M. M.; Staubli, A. B.; Kustedjo, K.; Gray, M. H. B.; Shih, J.; Fraser, S. E.; Jacobs, R. E.; Meade, T. J. *Bioconjugate Chem.* **1998**, *9*, 242–249.

201. Anderson, E. A.; Isaacman, S.; Peabody, D. S.; Wang, E. Y.; Canary, J. W.; Kirshenbaum, K. *Nano Lett.* **2006**, *6*, 1160–1164.

202. Langereis, S.; de Lussanet, Q. G.; van Genderen, M. H. P.; Backes, W. H.; Meijer, E. W. *Macromolecules* **2004**, *37*, 3084–3091.

203. Talanov, V. S.; Regino, C. A. S.; Kobayashi, H.; Bernardo, M.; Choyke, P. L.; Brechbiel, M. W. *Nano Lett.* **2006**, *6*, 1459–1463.

204. Roberts, D.; Zhu, W. L.; Frommen, C. M.; Rosenzweig, Z. *J. Appl. Phys.* **2000**, *87*, 6208–6210.

205. McDonald, M. A.; Watkin, K. L. *Invest. Radiol.* **2003**, *38*, 305–310.

206. Engström, M. M; Klasson, A.; Pedersen, H.; Vahlberg, C.; Käll, P.-O.; Uvdal, K. *Magn. Reson. Mater. Phys., Biol. Med.* **2006**, *19*, 180–186.

207. Bazzi, R.; Flores-Gonzalez, M.; Louis, C.; Lebbou, K.; Dujardin, C.; Brenier, A.; Zhang, W.; Tillement, O.; Bernstein, E.; Perriat, P. *J. Lumin.* **2003**, *102–103*, 445–450.

208. Bazzi, R.; Flores, M.; Louis, C.; Lebbou, K.; Zhang, W.; Dujardin, C.; Roux, S.; Mercier, B.; Ledoux, G.; Bernstein, E.; Perriat, P.; Tillement, O. *J. Colloid Interface Sci.* **2004**, *273*, 191–197.

209. Louis, C.; Bazzi, R.; Marquette, C.; Bridot, J.-L.; Roux, S.; Ledoux, G.; Mercier, B.; Blum, L.; Perriat, P.; Tillement, O. *Chem. Mater.* **2005**, *17*, 1673–1682.

210. Bridot, J.-L.; Faure, A.-C.; Laurent, S.; Rivière, C.; Billotey, C.; Hiba, B.; Janier, M.; Josserand, V.; Coll, J.-L.; Van der Elst, L.; Muller, R.; Roux, S.; Perriat, P.; Tillement, O. *J. Am. Chem. Soc.* **2007**, *129*, 5076–5084.

211. Le Duc, G.; Miladi, I.; Alric, C.; Mowat, P.; Bräuer-Krisch, E.; Bouchet, A.; Khalil, E.; Billotey, C.; Janier, M.; Lux, F.; Epicier, T.; Perriat, P.; Roux, S.; Tillement, O. *ACS Nano* **2011**, *5*, 9566–74.

212. Faure, A.-C.; Hoffmann, C.; Bazzi, R.; Goubard, F.; Pauthe, E.; Marquette, C. A.; Blum, L. J.; Perriat, P.; Roux, S.; Tillement, O. *ACS Nano* **2008**, *2*, 2273–2282.

213. Söderlind, F.; Pedersen, H.; Petoral, R. M., Jr.; Käll, P.-O.; Uvdal, K. *J. Colloid Interface Sci.* **2005**, *288*, 140–148.

214. Pedersen, H.; Söderlind, F.; Petoral, R. M., Jr.; Uvdal, K.; Käll, P.-O.; Ojamäe, L. *Surf. Sci.* **2005**, *592*, 124–140.

215. Hermanson, G. T. *Bioconjugate Techniques*, 2nd ed.; Academic Press: London, 2008.

216. Fortin, M.-A.; Jr, R. M. P.; Söderlind, F.; Klasson, A.; Engström, M.; Veres, T.; Käll, P.-O.; Uvdal, K. *Nanotechnology* **2007**, *18*, 395501.

217. Petoral, R. M.; Söderlind, F.; Klasson, A.; Suska, A.; Fortin, M. A.; Abrikossova, N.; Selegård, L.; Käll, P.-O.; Engström, M.; Uvdal, K. *J. Phys. Chem. C* **2009**, *113*, 6913–6920.

218. Miyawaki, J.; Yudasaka, M.; Imai, H.; Yorimitsu, H.; Isobe, H.; Nakamura, E.; Iijima, S. *J. Phys. Chem. B* **2006**, *110*, 5179–5181.

219. Sánchez, P.; Valero, E.; Gálvez, N.; Dominguez-Vera, J. M.; Marinone, M.; Poletti, G.; Corti, M.; Lascialfari, A. *Dalton Trans.* **2009**, 800–804.

220. Huang, C.-C.; Liu, T.-Y.; Su, C.-H.; Lo, Y.-W.; Chen, J.-H.; Yeh, C.-S. *Chem. Mater.* **2008**, *20*, 3840–3848.

221. Bridot, J.-L.; Dayde, D.; Rivière, C.; Mandon, C.; Billotey, C.; Lerondel, S.; Sabattier, R.; Cartron, G.; Le Pape, A.; Blondiaux, G.; Janier, M.; Perriat, P.; Roux, S.; Tillement, O. *J. Mater. Chem.* **2009**, *19*, 2328–2335.

222. Mowat, P.; et al. *J. Nanosci. Nanotechnol.* **2011**, *11*, 7833–7839.

223. Kryza, D.; Taleb, J.; Janier, M.; Marmuse, L.; Miladi, I.; Bonazza, P.; Louis, C.; Perriat, P.; Roux, S.; Tillement, O.; Billotey, C. *Bioconjugate Chem.* **2011**, *22*, 1145–1152.

Chapter 8

Functional Nanoparticle-Based Bioelectronic Devices

Tadeusz Hepel[*]

Institute of Nanotechnology, Potsdam, New York 13676, U.S.A.
[*]E-mail: elchema@verizon.net

In this review of bioelectronic devices, the progress achieved recently in device performance and miniaturization due to the application of nanotechnology has been presented and critically evaluated. Functional nanoparticles embedded in films of various biodevices act as to enhance the electron transfer between large biomolecules, such as redox proteins, enzymes, or DNA, and the electrode surface and provide new means for the analytical signal transduction due to their unique optical and electronic properties. Metal, semiconductor, and carbon nanoparticles, including nanorods, nanowires and nanotubes, have been reviewed. The carbon nanotubes (CNT), in addition to serving as excellent electrocatalyst, can provide a molecular support (scaffold) to enhance rigidity and robustness and enable device implantation. The integration of biofuel-cells, sensors, and silicon microelectronics creates a framework of a portable or implantable device which can serve in a theranostic treatment of chronic diseases. The advantages and biocompatibility issues of implantable devices and examples of bioelectronic Boolean logic gates enhanced by nanoparticles are discussed for future development of efficient, robust, and safe bioelectronic devices.

Keywords: Au nanoparticles (AuNP); carbon nanotubes (CNT); enzyme biosensors; immunosensors; molecularly imprinted sensors; implantable biosensors; logic gates; BioFET

© 2012 American Chemical Society

Introduction

Since the emergence of nanotechnology, the effects of various nanoparticles on the operational mechanism and performance of biodevices have been extensively investigated (*1–6*).   In this study, we present progress in the design and application of biosensors and bioelectronic devices made possible by incorporating functional nanoparticles composed of different materials, including metals, semiconductors, isolators, polymers, and liposomes, to a bioactive film.   The nanoparticles with different optical and electronic properties modulated by the size and shape, including spherical particles, nanorods, nanotubes, nanoshells, and others, are considered.   We present the nanoparticle-enhanced piezoimmunosensors, molecularly templated polymer films for monolayer-protected gold-nanoparticle analytes, implantable biosensors for continuous disease control, portable microfluidics-based nanowired microsensors for glucose analysis in blood for diabetics, and biogenic nanoelectronic devices, such as the biogenic field-effect transistor, BioFET and biomolecule-based logic gates. The microsensor arrays embedding functionalized nanoparticles are also discussed.

There are many diseases that require constant monitoring of the level of certain biomarkers in blood or body fluids to adjust the medication dosing and reduce the deleterious effects of chemical imbalances caused by the illness. Therefore, considerable efforts have been paid to design and test portable and implantable biosensors for these biomarkers. For instance, the estimated number of diabetic patients is approximately 110 million (*7*) and this number is expected to grow considerably in the nearest decades. Since diabetes requires continuous adjustment of insulin injections to control the blood sugar level, the portable and implantable devices are the best choices for monitoring glycemia, thereby improving the quality of patients' life (*8*). The nanotechnology has entered here with enzymatic sensors for glucose in form of enzyme *nanowiring* with carbon nanotubes (CNT) as developed by Willner and coworkers (*9*).

In terms of the nomenclature, in this Chapter we follow the definition of biosensors and biodevices as those containing biomolecules or biorecognition molecules, such as immunoglobulins, receptor proteins, enzymes, DNA, RNA, cells, or organelles, as well as artificial biomimetic recognition elements such as the molecularly templated polymer films. Moreover, to indicate the involvement of nanoparticles of any composure, we add the prefix *nano-* to form new entities: nanobiosensors, nanobiodevices. Hence, this Chapter deals with nanobiosensors and nanobiodevices and refers the progress achieved in performance, scalability, and sensitivity to the standard macrosystems, represented by biosensors and biodevices.

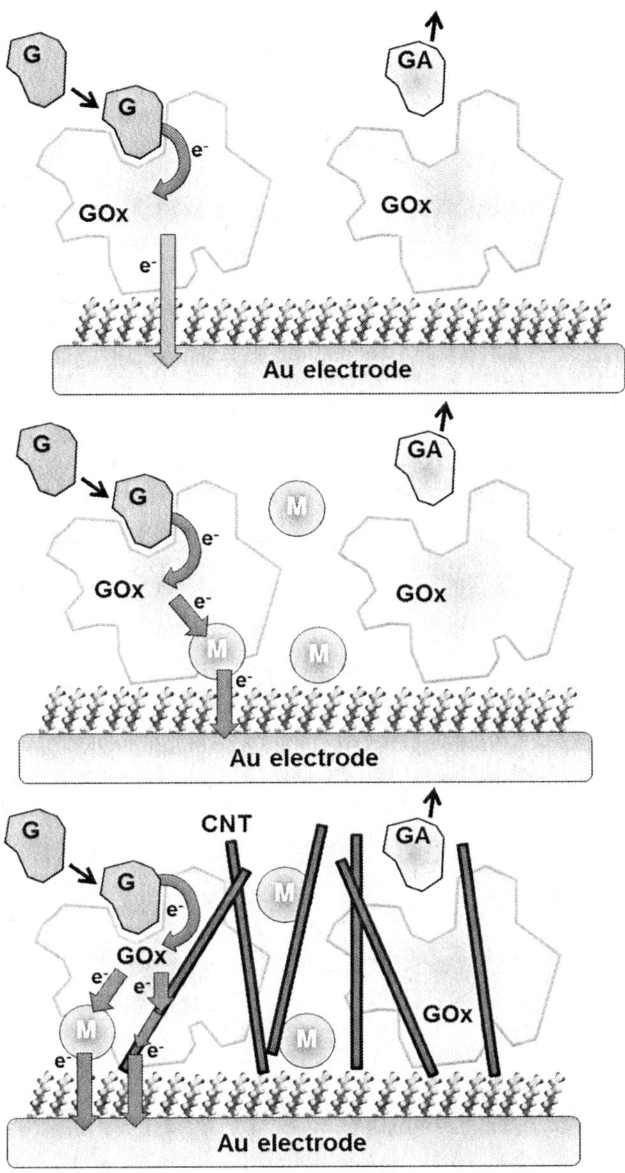

Figure 1. Enzymatic glucose biosensors: (A) generation I sensor with GOx immobilized on a SAM basal film; (B) generation II enhanced sensor with electron-transfer mediator M; (C) generation III carbon nanotube CNT-nanowired enhanced sensor. G- glucose, GOx - glucose oxidase, GA gluconic acid.

Implantable Amperometric Enzyme Biosensors

Principles of Enzyme-Based Biosensing

The enzyme-based biosensors rely on enzymatic reactions to oxidize (or reduce) analyte molecules which themselves exhibit a slow kinetics and cannot be easily quantified by the analytical sensors. Basically, all of the earlier enzyme sensors have utilized hydrogen peroxide or other indicators produced in the enzymatic electrodic reaction that can be readily determined using electrochemical voltammetric techniques. An appropriate enzyme, specific to the analyte to be determined, is immobilized in a sensory film on an electrode surface in such a way that the enzyme activity is preserved. The indicator is generated in the sensory film and/or on its surface in proportion to the analyte concentration. The most popular of analytical signal transduction techniques is the amperometric technique where the sensor is maintained at a constant potential while the indicator current is being measured. The principle of operation of an amperometric enzymatic biosensor is presented in Figure 1.

Typical reactions of enzyme biosensors are illustrated by glucose oxidase (GOx) catalyzed glucose oxidation:

$$\beta\text{-D-glucose} \xrightarrow[\quad -2e \quad]{GOx} \text{D-glucono-1,5- lactone} + 2H^+ \tag{1}$$

$$O_2 + 2H^+ + 2e^- \rightarrow H_2O_2 \tag{2}$$

Reaction (1) is slow and the transfer of electrons to the electrode is strongly hindered. Hence, the electrochemical voltammetric currents recorded for glucose oxidation are very small and difficult to deconvolute from the capacitive contribution and other background currents.

Considerable enhancement of enzymatic biosensors has been achieved in generation II biosensors by utilizing electron mediators that promote transfer of electrons between immobilized-enzyme cofactors and the electrode surface (Fig. 1B). The following molecular mediators have been investigated: ferrocene and its derivative (e.g. poly(vinylferrocene)), polymers based on Os^{2+}/Os^{3+} and bipyridyl (bpy) (e.g. Os(4,4'-dimethyl-2.2'-bipyridyl)2L) (*10*), as well as the conductive polymers: polypyrrole, polyaniline, polythiophene (*11*) which enabled lowering the electroactivation overvoltage for the electron transfer step (Scheme 1).

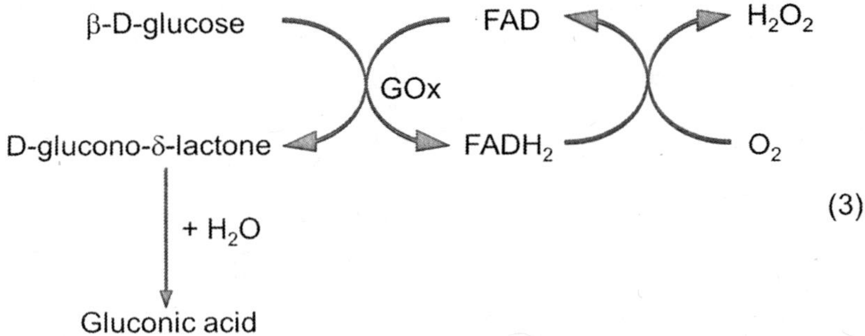

$$(3)$$

Scheme 1. Oxidation of glucose to form gluconolactone, catalyzed by glucose oxidase (GOx). This is accompanied by reduction of FAD to form FADH$_2$ which is then oxidized by molecular oxygen to give back FAD and hydrogen peroxide

Another example of generation II enzyme biosensors is illustrated below (Scheme 2) with a lactate dehydrogenase-based biosensor reaction:

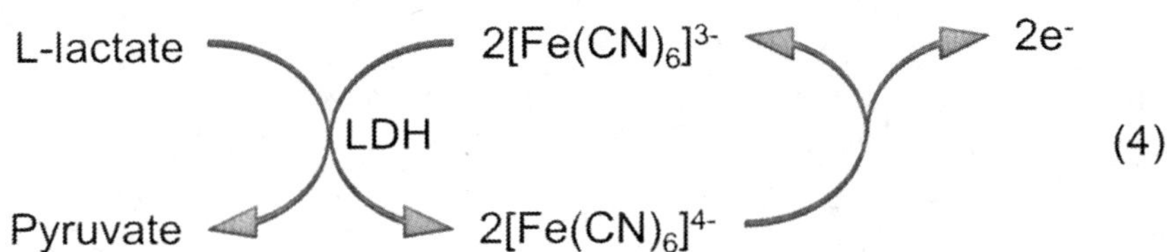

$$(4)$$

Scheme 2. Reactions involved in the enzymatic oxidation of L-lactate, catalyzed by lactate dehydrogenase (LDH) in the presence of ferricyanide. In the last stage, the reduced mediator is oxidized directly on the electrode surface (Pt electrode, E = +250 mV vs. Ag/AgCl reference)

In reactions (1)-(4), glucose (or lactate) is oxidized first by the enzyme with transfer of 2 electrons to the enzyme. In reactions (1)-(2), the enzyme is oxidized back to its oxidized state by oxygen, while in reactions (3)-(4), a mediator, FAD or $Fe(CN)_6^{3-}$, oxidizes the enzyme back to the initial oxidized state. Now, in reactions (3)-(4), the last stage is also different. In reaction (3), the reduced mediator FADH$_2$ is reoxidized back to FAD by oxygen, producing H_2O_2 which can be detected amperometrically. But, in reaction (4), the reduced mediator, $2Fe(CN)_6^{4-}$, is reoxidized directly at the electrode surface giving up 2 electrons to the electrode and generating an analytical signal.

Enzyme Nanobiosensors

Further improvement of enzymatic biosensors was brought by the advent of nanotechnology. Very high direct charge-transfer speed between glucose oxidase and the electrode surface through AuNP-modified FAD was achieved by Willner group (*12*). The enzyme "nanowiring" to the electrode by means of AuNP/FAD network has been a remarkable achievement. In generation III biosensors, highly active electrocatalysts, *viz.* carbon nanotubes (CNT), are embedded in sensory films (*13–15*) providing direct means for the electron transfer from the redox-active oxidoreductase-enzyme cofactor and CNT to the electrode surface (*14–18*). The high conductance of carbon fibers enables fast electron exchange between the reaction centers and the electrode surface (Fig. 1C). Wang and coworkers (*19*) have developed an ultra-sensitive multilayer enzyme-CNT biosensor able to detect down to 80 copies of DNA, making it the most sensitive enzyme detector yet. The layer-by-layer (LBL) assembly based on electrostatic attraction forces consisted of depositing a positively charged polyelectrolyte poly(diallyldimethylammonium chloride) (PDDA) on the oxidized CNT surface containing deprotonated carboxylate groups, followed by assembling a negatively charged enzyme alkaline phosphatase (ALP). They have found that the detector sensitivity increases with the number of PDDA/ALP layers. The LBL stages of the assembly are illustrated in Figure 2. The final layer was then added, PDDA, followed by negatively charged streptavidin (SA). A biotinylated target probe oligonucleotide could then be attached to the $CNT/(PDDA/ALP)_n PDDA/SA$ units. The biotinylated probe oligonucleotide was bound to SA-coated magnetic beads. After mixing the modified CNTs with modified magnetic beads, the complementary target and probe DNA strands hybridize. The duplexes can be easily separated magnetically. The test for complementary DNA strands is done by adding α-naphthyl phosphate which is enzymatically converted to α-naphthol detectable electrochemically using square wave voltammetry. If target oligonucleotide is not complementary to the probe DNA, there is no current of α-naphthol oxidation detected. In this way, very small number of DNA strands can be detected, as little as 80 DNA copies corresponding to 5.4 aM.

Liu and Lin designed a flow-injection enzyme biosensor for detection of glucose using a GOx-modified CNT wiring to a glassy carbon electrode (*20*). Similar biosensor arrangement has been applied to sensors based on other enzymes, choline oxidase (ChO), acetyl cholinesterase (AChE), and double-enzyme, ChO-HRP (*21*).

A dual glucose/insulin biosensor constructed in a 14-gage stainless-steel needle has been developed by Wang and Zhang (*22*). The glucose microsensor was based on GOx enzyme electrode and insulin sensor relied on the electrocatalytic properties of ruthenium dioxide nanoparticles. This approach can be extended to multiple-analyte sensing.

The CNT can serve also as an effective scaffold for immobilizing proteins in a sensory film due to exceptionally strong mechanical properties. The effects of covalent attachment of redox proteins to CNT have been investigated by Guiseppi-Elie et al. (*23*) and others (*24, 25*). Implantable enzyme biosensors are widely used for controlling blood sugar level and other disease biomarkers

(*26–28*). Vasylieva et al. (*28*) have developed covalently attached enzymes to poly(ethylen glycol) diglycidyl ether (PEGDE) on microelectrode biosensors which enabled them to monitor modulated glucose level in the brain by sequential administration of glucose and insulin.

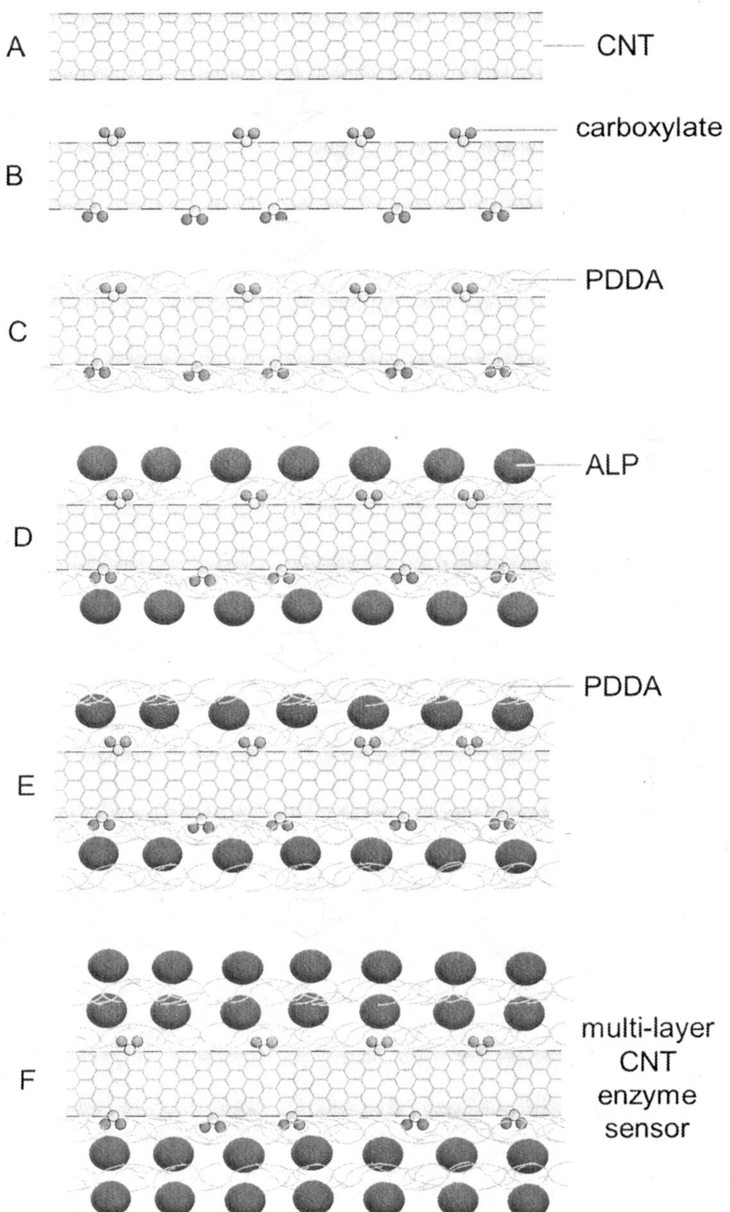

Figure 2. Layer-by-layer assembly of a CNT-wired enzyme sensor. (Reproduced with permission from Ref. (19). Copyright 2005 American Chemical Society.)

The progress in the development of glucose-oxidase biosensors, taking into account nanoparticle role and their effect on sensor performance has been recently discussed by Heller and Feldman (*29*) and Wang (*30*). Note that CNTs have been found to cause dermal toxicity through oxidative stress and inflammatory response (*31*). The latter is already induced by the implantation process itself. Hence, the implanted devices should be equipped with anti-inflammatory measures.

Selecting Enzymes for Bioactive Sensory Films

Table I. Examples of enzymatic reactions driven by oxygen consumption and hydrogen peroxide production

Analyte	*Enzymatic reaction*	*Enzyme*	*Ref*
cholesterol	cholesterol + O_2 → cholestenone + H_2O_2	cholesterol oxidase	(*32*)
creatinine	creatinine + H_2O → creatine creatine + H_2O → urea + sarcosine sarcosine + H_2O → glycine + formaldehyde + H_2O_2	creatininase creatinase sarcosine oxidase	(*33*) (*34*) (*35*)
glucose	β-D-glucose + H_2O → D-glucono-δ-lactone + H_2O_2 D-glucono-δ-lactone + H_2O → gluconic acid	glucose oxidase	(*36*)
glutamate	D-glutamate + O_2 → α-ketoglutarate + NH_3 + H_2O_2	glutamate oxidase	(*37*)
lactate	L-lactate + O_2 → pyruvate + H_2O_2	L-lactate oxidase	(*38*)
lactose	lactose + H_2O → β-D-glucose + β-D-galactose β-D-glucose + H_2O → D-glucono-δ-lactone + H_2O_2	galactosidase glucose oxidase	(*39*)
maltose	maltose + H_2O → 2 D-glucose D-glucose + O_2 → 2-keto-D-glucose + H_2O_2	α-glucosidase pyranose oxidase	(*40*)
pyruvate	pyruvate + HPO_4^{2-} + O_2 → acetyl phosphate + CO_2 + H_2O_2	pyruvate oxidase	(*41*)
sucrose	sucrose + H_2O → α-D-glucose + β-D-fructose α-D-glucose → β-D-glucose β-D-glucose + O_2 → D-glucono-δ-lactone + H_2O_2	invertase mutarotase glucose oxidase	(*42*)

Different enzymes can often be selected in the design of an enzymatic sensor for a given analyte. Several common enzyme-based biosensors employ oxidases. In Table I, examples of enzymatic reactions driven by oxygen consumption and hydrogen peroxide production, based on various oxidases, are presented. The Table includes the analyte, enzyme used, and the enzymatic reactions.

However, instead of oxidases, the dehydrogenases can also be used. There are several hundred dehydrogenases (DH) that are presently known which catalyze different redox reactions in the presence of some coenzymes, for instance, nicotinamide adenine dinucleotide (NAD):

$$\text{substrate} + \text{NAD}^+ \xrightarrow{\text{DH}} \text{product} + \text{NADH} + \text{H}^+ \tag{5}$$

where NAD+ and NADH are the oxidized and reduced forms of the coenzyme, respectively. During the enzymatic reaction, the substrate is oxidized to give the product and NADH plus H$^+$. Although, the direct electrochemical reoxidation of NADH is possible, the high potential of this reaction (> 800mV vs. Ag/AgCl) is impractical. Again, to solve this problem, redox mediators can be used to reoxidize the NADH to NAD$^+$ and the reduced mediator is electrochemically oxidized with the oxidation current being proportional to the substrate (analyte) concentration.

Portable and Implantable Enzyme Nanobiosensors

Semiquantitative, portable glucose testers for diabetes patients are now available on the market. The main advantage of these devices is the ease of use, safety, and immediate display of the sugar level on the screen. The miniature sensors and their rapid response are possible due to the extensive research done by many groups around the world to tackle the diabetes problem.

In Figure 3, portable glucose tester (D), manufactured by Nipro Diagnostics, Inc. (Ft. Lauderdale, FL, U.S.A.), is presented. The sensor strips (A)-(C) contain a microfluidic channel and a GOx-based glucose sensor. The newer strips (B)-(C) require only 0.5 μL of blood and feature patented quad-electrode design enabling to recognize if a blood sample has been delivered. The microfluidic sample tip T is 1×3 mm^2 and the sensor itself (S) is 100 μm in diameter.

A small implantable chip MDEA 5037 with dual-cell microdisc electrode array manufactured by ABTECH Scientific, Inc. (Richmond, VA, U.S.A.) is presented in Figure 4. The width of the chip is 2 mm making it very suitable for easy implantation.

The development of implantable biosensors for simultaneous monitoring of glucose and lactate in tissue is important for trauma management (*45–48*). However, serious difficulties in achieving this goal need to be dealt with (*49–51*), including poor device biocompatibility leading to sensor encapsulation that diminishes the glucose and lactate availability at the sensor surface (*52–54*), as illustrated in Figure 5.

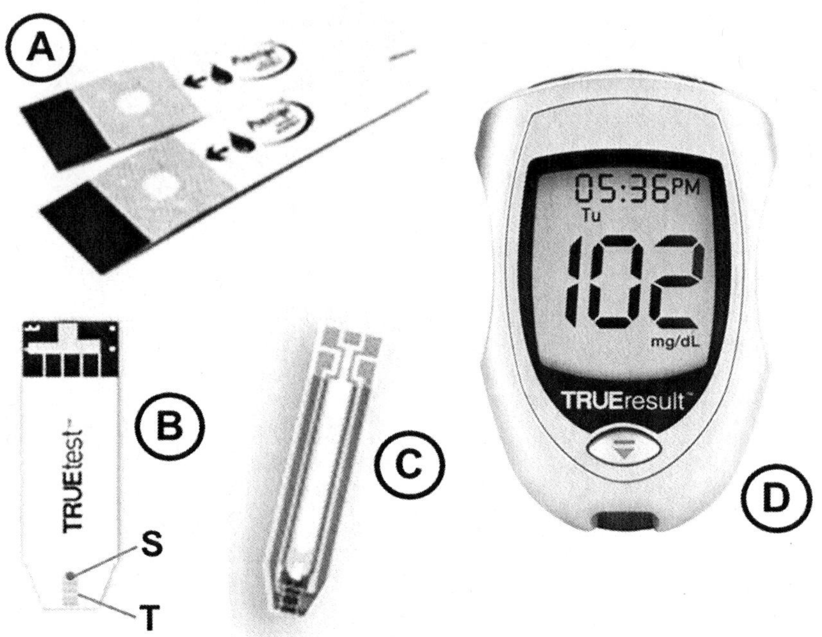

Figure 3. Portable glucose tester (D), manufactured by Nipro Diagnostics, Inc. (Ft. Lauderdale, FL, U.S.A.), and sensor strips (A)-(C). The newer strips (B)-(C) require only 0.5 µL of blood and feature patented quad-electrode design. The sample tip T is 1×3 mm² and the sensor S itself is 100 µm in diameter. (Reprinted with permission from Ref. (43). Copyright 2012 Nipro Diagnostics, Inc.)

The improvement in LOD, sensitivity, and dynamic range are also required. The latter hurdles should be dealt with simultaneously with overcoming pro-inflammatory response of the tissue. The application of nanotechnology for fabrication of implantable biosensors offers outstanding advantages, such as the reduced bioresponse and tissue inflammation, low power consumption, easiness of implantation, and enhanced signal-to-noise figure (*55*). The convenient size of implants is in 1-2 mm range. For instance, the width of implantable chips MDEA 5037 is only 2 mm and the microelectrodes in the array are 100 µm in diameter with interelectrode spacing of 50 µm.

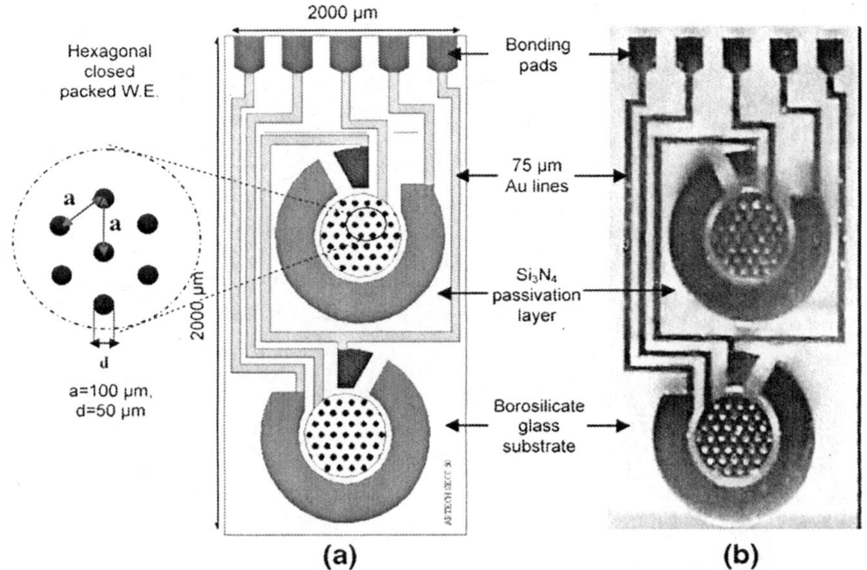

Figure 4. (a) Illustration of MDEA 5037 dual-cell µ-disc electrode array manufactured by ABTECH Scientific, Inc. (b) Photograph of the microfabricated MDEA 5037 chip. The single electrode diameter is 50 µm, interelectrode spacing: 100 µm, and the entire chip dimensions are: 2×5 mm². (Reprinted with permission from Ref. (44). Copyright 2009 Springer.)

To diminish inflammatory tissue response, the implanted devices can be plated with nanostructured silver. The tiopronin-coated AgNP have anti-inflammatory properties (*56–58*). An anti-inflammatory bactericidal AgNP coating has also been considered for implanted plastic catheters (*59*) to prevent endocarditis caused by *Staphylococcus aureus* build-up and entry into the blood stream. Also an anti-inflammatory bactericidal AgNP-based gel for external use on wounds and burns has successfully been tested (*60*). Anti-bacterial activity of biomimetically-capped AgNP has been investigated recently by Amato et al. (*61*). The effect of AuNP in the suppression of inflammatory cytokines has been investigated by evaluating the expression of mRNA, IL-12, and TNF-α (*62*).

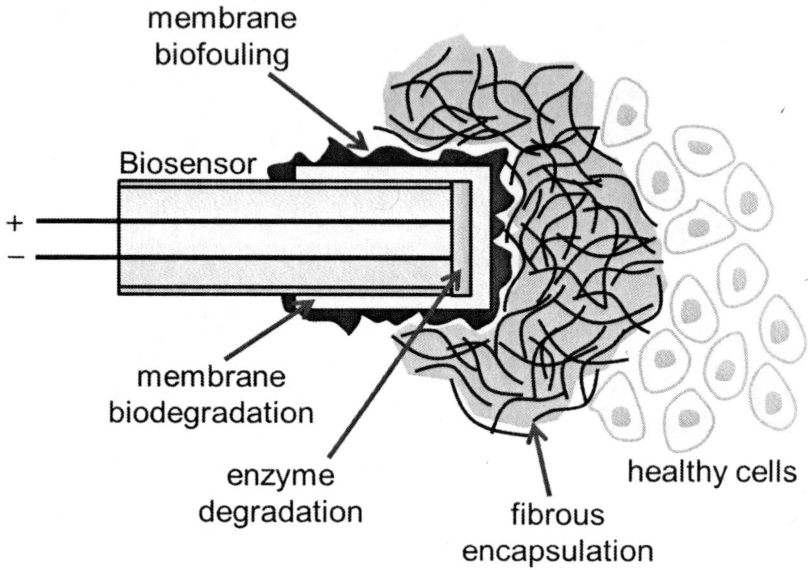

Figure 5. Effects of implantation on the performance of subcutaneous glucose sensor due to the biological response of the host: fibrous encapsulation, membrane degradation, and membrane biofouling. (Reproduced with permission from Ref. (54). Copyright 2000 Springer.)

Nanoparticle-Enhanced Immunosensors

The immunosensors based on antigen-antibody affinity have been applied widely in medicine (*63, 64*), environmental control (*65–67*), and food industry (*68*). The effect of buried potential-barrier in label-less electrochemical immunodetection of glutathione and glutathione-capped gold nanoparticles has been studied by Stobiecka and Hepel (*69*) (Figures 6 and 7). On the basis of model calculations, they have formulated conditions for best sensory film design to limit fouling and electrolyte exchange with the film. Amperometric immunosensors based on a nanostructured gold film have been developed by Tiefenauer et al. (*70*). Other groups have also reported successful development of label-less electrochemical immunosensors based on impedance and/or voltammetric measurements (*71–74*).

In sensors designed for the determination of immunoglubulins (IgG), anodic stripping voltammetry of gold has been applied. The immunoglobulin is first labeled with AuNP. After affinity binding of IgG/AuNP to the sensory film, the anodic stripping voltammetry of gold is performed leading to very low limit of detection (LOD) for IgG. Limoges and coworkers (*75*) were able to detect AuNP-labeled immunoglobulin G with LOD = 3 pM. Similar electrochemical immunoassays with AuNP labels were developed by Yu et al. (*76*) and with AgNP labels by Liao and Huang (*77*).

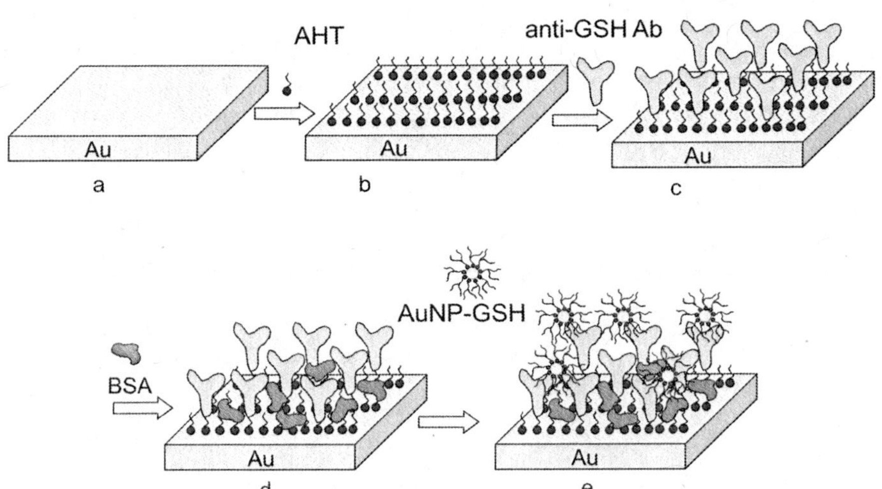

Figure 6. Design of an electrochemical and nanogravimetric immunosensor for the detection of glutathione-capped AuNP; counterions to AHT not shown for clarity. (Reprinted with permission from Ref. (69). Copyright 2011 Elsevier.)

AuNP-enhanced enzyme-labeled immunosensors with $AuNP/TiO_2$ have been developed by the group of Ju (*78–80*), with sol-gel composite (*78*), AuNP embedded in cellulose acetate membrane (*79*), and 3-D sol-gel matrix (*80*).

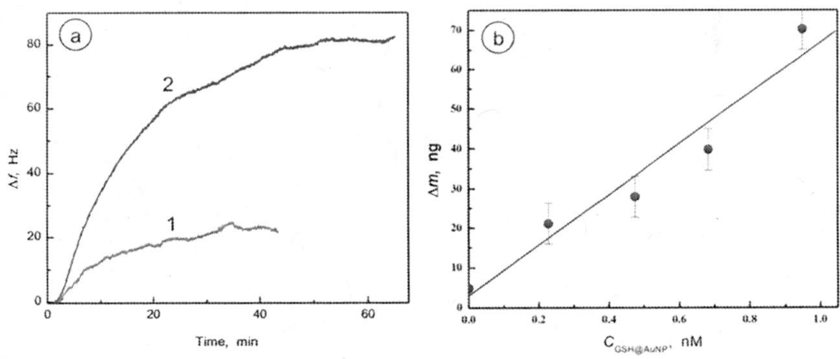

Figure 7. (a) Resonant frequency transients for a QC/Au/AHT/Ab,BSA piezoimmunosensor recorded after addition of: (1) 1.25 mM GSH, (2) 0.95 nM AuNP@GSH; (b) calibration plot of the apparent mass vs. concentration of AuNP@GSH for a QC/Au/AHT/Ab$_{mono}$ sensor in 50 mM PBS, with surface regeneration in 0.2 M glycine solution, pH = 3, after each test. (Reprinted with permission from Ref. (69). Copyright 2011 Elsevier.)

Owing to the outstanding biorecognition propensity of immunoglubulins, the nanoparticle-enhanced immunosensors should be among the best biosensors developed for portable testing and in points-of-care in the near future. The reduction in the sensor size by applying nanofabrication technologies will increase the sensor reliability and substantially reduce the sensor cost which is now relatively high due to the high prices of immunoglobulins.

Biogenic-Gate Field-Effect Transistors (BioFET)

The field-effect transistors (FET) have been utilized in designing sensors since the early works of Bergveld (*81*) on an ion-selective field-effect transistor (ISFET). FET is an electronic semiconductor device in which current flowing between the drain and source electrodes, I_{DS}, can be modulated by coupling the gate-generated electric field to the variable-conductivity channel fabricated between the drain and source inversion-layer regions. The electric field emanating from the gate electrode attracts (or repels) the charge carriers to (from) the conductivity channel causing an increase (decrease) of I_{DS} as a function gate voltage V_G. A typical metal-oxide-semiconductor (MOS) FET device is presented in Figure 8.

Since the charge accumulated on the gate electrode (e.g. due to the applied potential) changes the conductivity of the conductivity channel, the FET can serve as a convenient and very sensitive transducer responding to chemical and physical changes in a sensory film deposited on the gate electrode. This is illustrated in chemical sensing application of FET as an ISFET, presented in Figure 9. Excellent review of ISFETs and, in general, of chemical-sensitive field-effect transistors (CHEMFETs) has been published recently by Janata (*82*).

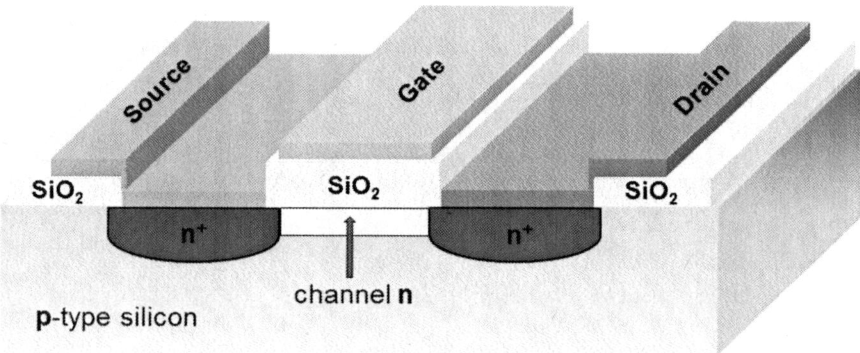

Figure 8. Schematic view of a MOSFET.

To enhance the FET responses and isolate the actual FET microstructure from the sensor area, an extended-gate FET has been proposed (*83*), as illustrated in Figure 10. Kamahori et al. (*84*) have applied the extended-gate field-effect transistor for the detection of DNA hybridization. The characterizations of immobilized DNA–probes and role of applying a superimposed high-frequency voltage onto a reference electrode have been described.

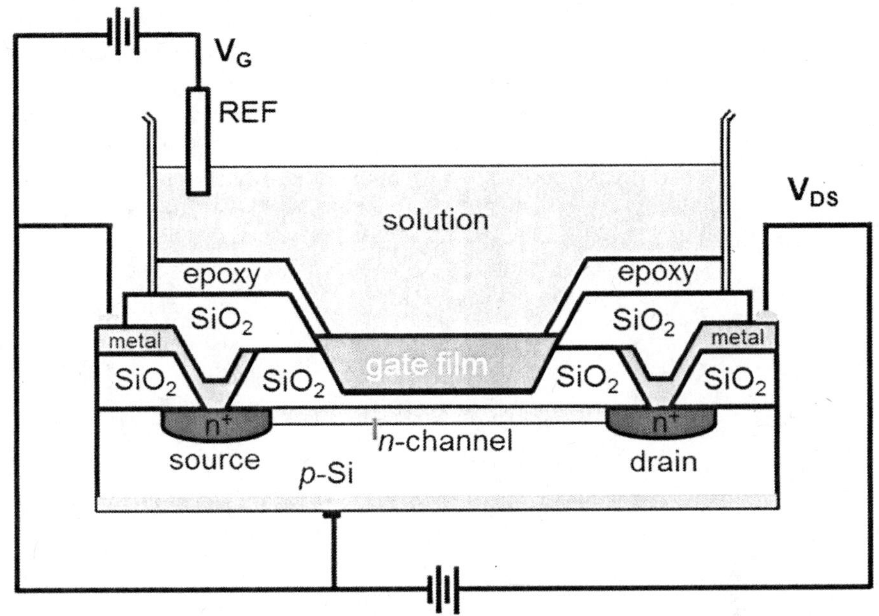

Figure 9. Schematic view of a typical ion-selective field-effect transistor (ISFET).

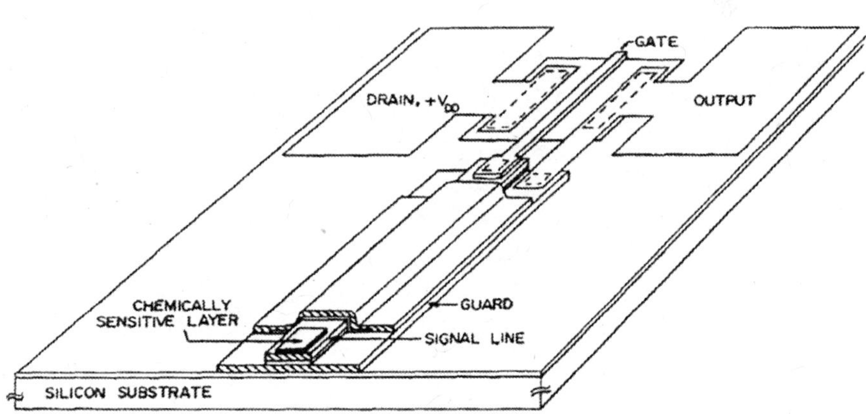

*Figure 10. Schematic diagram of extended gate field effect transistor (83).
(Reprinted with permission from Ref. (83). Copyright 1983 Elsevier.)*

The application of FETs to biosensing is a natural extension of the ISFET development. In Figure 11, a biogenic-gate field-effect transistor (BioFET) device developed by Xu et al. (85) is presented. Here, a multi-layer sensory film, composed of a lactate oxidase enzyme (LOx), MnO_2 nanoparticles, and PDDA polyelectrolyte and assembled using the LBL technique $(PDDA/MnO_2/PDDA/LOD)_n$, has been applied to the FET gate (where PDDA stands for poly(diallyldimethylammonium chloride), a positively charged polyelectrolyte). The H_2O_2 generated in the enzymatic reaction of LOx with lactate analyte was oxidized by MnO_2 nanoparticles producing a local pH change inside the sensory film. The change in pH was sensitively measured by I_{DS} current change.

Highly efficient electron transfer between CNTs and CdSe or CdSe/ZnS quantum dots (Qdots) has been developed by Jeong et al. (86) for a CHEMFET built on a silicon oxide-insulated n-type silicon substrate. They utilized a non-covalently bound pyridine molecule to link a Qdot with a CNT. This provides a low-resistance contact because of the direct coupling of conjugate pyridine system with CNTs and Qdots.

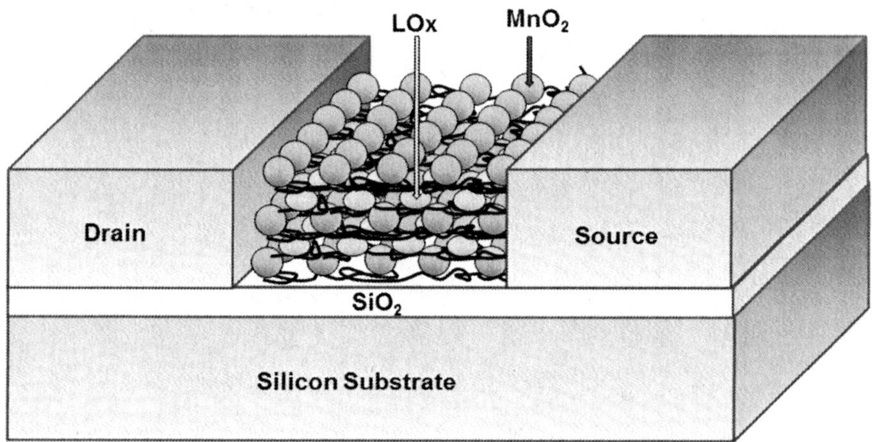

Figure 11. Biogenic-gate field-effect transistor (BioFET) designed by Xu et al. (85), originally called enzyme field-effect transistor (ENFET), based on LBL assembly of lactate oxidase (LOx) enzyme layers with MnO2 nanoparticle layers and polyelectrolyte on the gate electrode in a BioFET.

Oelssner et al. (87) have described the insulation and packaging options available for ISFETs operating under bioenvironment conditions. A microsensor array chip composed of 16×16 ISFETs and incorporating row and column decoders, source and drain follower circuits, switching logic, and ADC has been presented by Milgrewa et al. (88).

Extensive review of the LBL technique in application to functionalization of nanoparticles, designing multi-layer nanoshells, and constructing nanobioelectronic devices has been published by Ariga et al. (89).

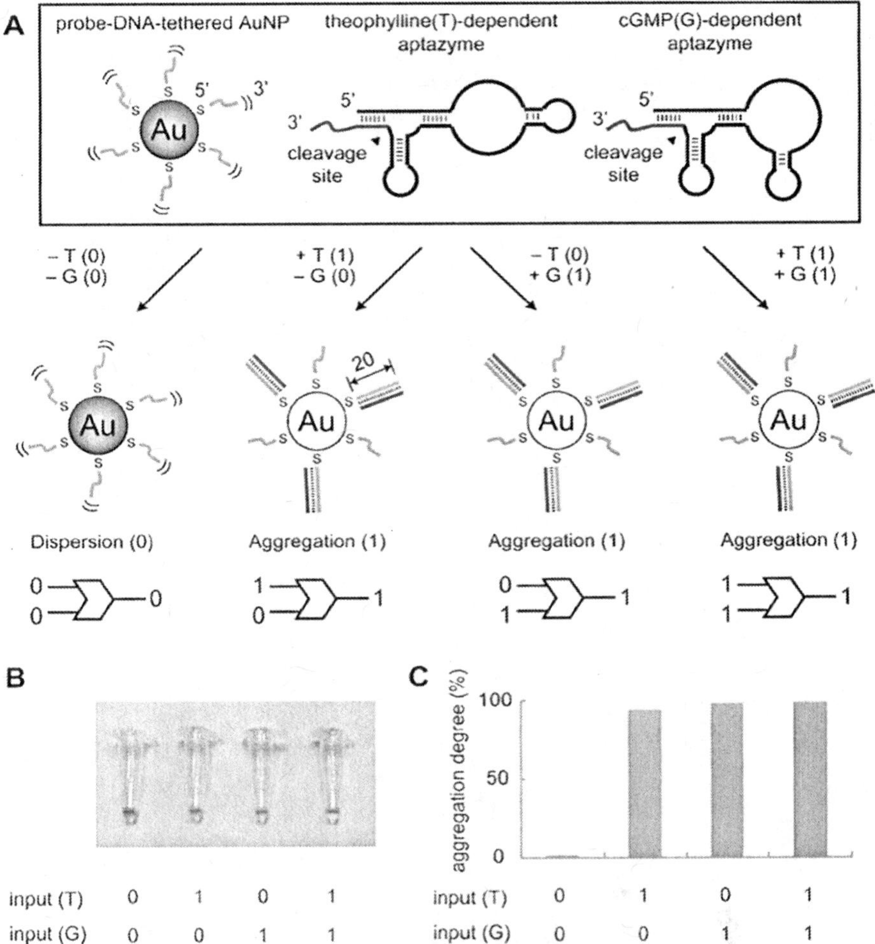

Figure 12. The operational principles of an OR gate designed using aptazymes specific for theophiline (T) and cGMP (G): (A) design strategy; (B, C) Calculation results of the OR gate in the absence (input = 0) or presence (input = 1, 1 mM) of input molecules (T or G) with 5-min reaction; (B) images of AuNP solutions and (C) AuNP aggregation degree. (Reprinted with permission from Ref. (105). Copyright 2009 Royal Society of Chemistry.)

Logic Gates

The extensive homeostasis exploited by living biological organisms for regulation of cellular responses, such as the inflammatory or oxidative stress response, maintenance of biomolecule concentration levels and trace elements, temperature, etc., can be modeled by the equivalent electronic logic gate circuits (*90*). Various biomolecules, including DNA, RNA, enzymes, antibodies, carbohydrates, and also metals, have been utilized to design molecular logic gates mimicking the nature's regulation systems (*91–103*). To increase the gate sensitivity, the number of active sites per gate has been increased by utilizing dendrimers (*104*). Further enhancement can be achieved using nanoparticles. Several examples of bioelectronic Boolean logic gates enhanced by nanoparticles are described below.

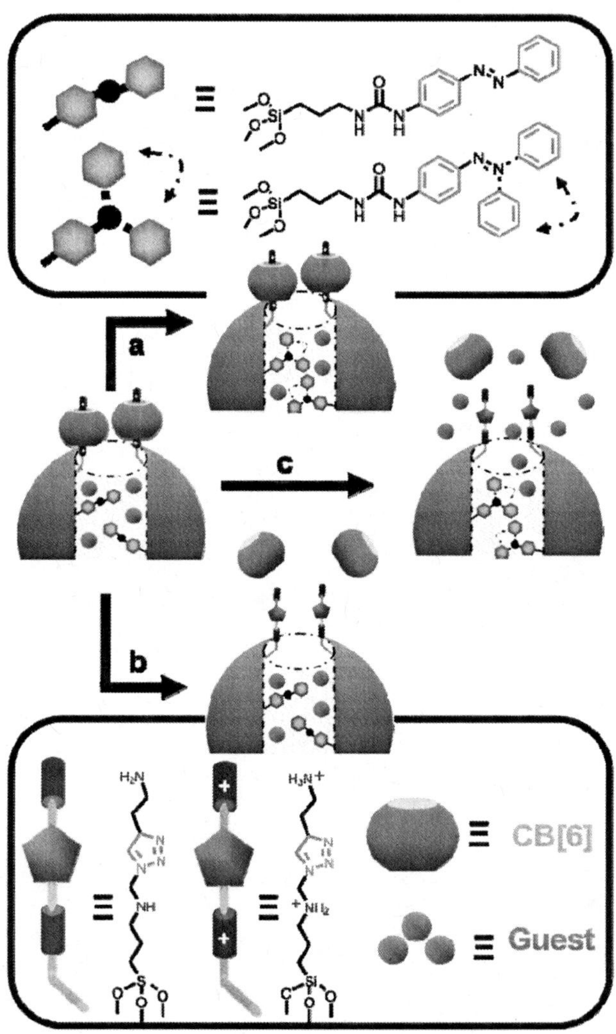

Figure 13. Operation of dual-controlled nanoparticles: (a) Excitation with 448 nm light induces the dynamic wagging motion of the impeller molecules, while the valve molecules remain closed and the drugs (guests) are contained in channels. (b) Addition of NaOH opens the valves, but the impellers are static and keep the drugs contained. (c) Simultaneous excitation with 448 nm light AND addition of NaOH force the drug out of the channels. (Reprinted with permission from Ref. (106). Copyright 2009 American Chemical Society.)

Ogawa and Maeda (*105*) have used non-crosslinking AuNPs to design logic gates based on self-cleaving cleavase-aptazymes, one dependent on theophiline (T) and one dependent on cGMP (G). When T or G is present in the solution then the respective aptazyme is self-cleaving with the release of short oligonucleotide strand that hybridizes with oligonucleotide probe-modified AuNP causing AuNPs to aggregate (Figure 12).

The aggregation leads to color changes detected with the naked eye. The Authors were able to design several types of logic gates including YES, OR, AND gates. In Figure 12, the principles of an OR logic gate operation are presented. The change in visible light absorbance occurs when either T or G, or both, are present in the solution. The degree of aggregation η used as the gate output signal has been determined from the formula:

$$\eta = 100 \, (A_{529, \text{ref}} - A_{529, \text{obs}})/A_{529\text{ref}} \tag{6}$$

where $A_{529, \text{ref}}$ is the absorbance at 529 nm (corresponding to the surface plasmon peak for AuNPs used) and $A_{529, \text{obs}}$ is the absorbance at 529 nm observed upon aggregation.

The use of DNA and RNA molecules for biocomputation has attracted much interest. Nucleic acids are also useful for logic gate-based control and analysis in biological systems. Lee et al. (*107*) have developed a transduction technique that enables displaying the results of the DNA computing based on a colorimetric change induced by gold nanoparticle aggregation. They have applied it to the Boolean logic detection of biomolecules.

Nanoparticles with dual control exhibiting the AND Boolean logic have been investigated by Angelos et al. (*106*). These mesoporous silica nanoparticles have been functionalized with impeller molecules and valve molecules which enable retention and release of drugs stored in nanochannels of the particles on demand. The principle of operation of this useful functional logic AND gate is depicted in Figure 13. The two inputs to the logic gate are: (i) the excitation with 448 nm light and (ii) the injection of NaOH. When the 448 nm light is ON, the impeller molecules are wagging, but at pH = 7, the valve molecules are closed (OFF) and there si no drug release. The addition of NaOH opens the valve molecules, but without 448 nm light illumination, the drugs remain in channels. Only when both inputs are ON (i.e. 448 nm illumination and NaOH present), the drugs are released from silica nanoparticle logic gate.

Quartz Crystal Nanogravimetric (QCN) Sensors

The quartz crystal piezoresonators have been utilized for biosensor designs due to their high sensitivity to minute mass changes of sensory films, comprising a fraction of monolayer of atoms or molecules (*108*). The piezotransduction is based on Sauerbrey equation relating shift of the resonance oscillation frequency of the quartz crystal lattice vibrations Δf in a thin quartz crystal wafer to the mass change Δm in a film rigidly attached to the surface of the wafer (*109*):

$$\Delta f = \frac{2\Delta m n f_0^2}{A\sqrt{\mu_q \rho_q}} \tag{7}$$

where f_0 is the oscillation frequency in the fundamental mode of the series resonance of an AT-cut quartz crystal oscillating in the thickness-shear mode (TSM), n is the overtone number, A is the piezoelectically active surface area, ρ_q is the density of quartz (ρ_q = 2.648 g cm^{-3}), and μ_q is the shear modulus of quartz for AT-cut quartz crystal resonators (ρ_q = 2.947×10^{11} g cm^{-1} s^{-2}).

The quartz crystal resonator wafer with gold electrodes sputtered on both sides of the wafer in the keyhole pattern is shown in Figure 14. In the oscillations in TSM mode, the top surface a quartz crystal wafer moves in opposite direction than the bottom surface.

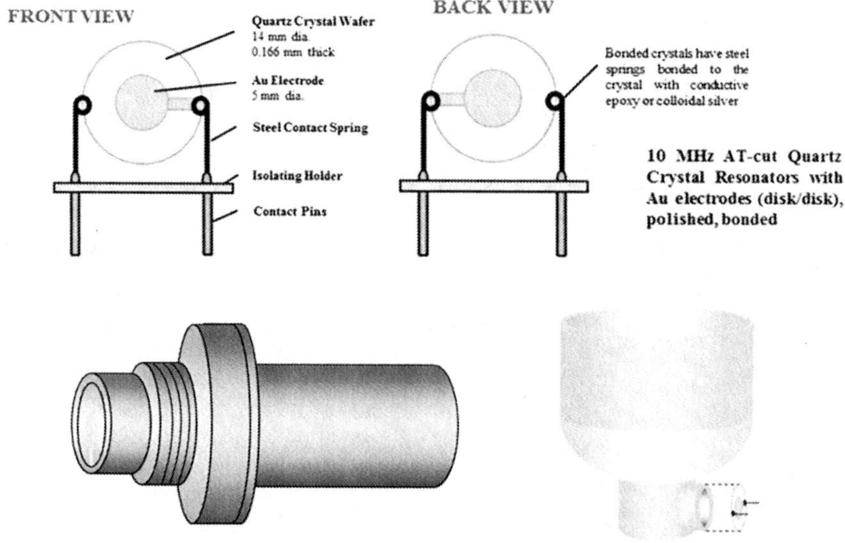

Figure 14. The front and rear view of a quartz crystal resonator wafer (QC) and examples of assembled QC piezoelectrodes, with QC mounted in a submersible tube and a QC sealed to the side opening in EQCN cell.

Note that the amplitude of these oscillations is minute, usually in the range of tens of nanometers. The oscillator should be tuned to the series resonance frequency of working piezoelectrodes to minimize effects due to energy dissipation in viscoelastic films. All experimental variables influencing the resonant frequency (*109*) of the EQCN electrodes such as the temperature, pressure, viscosity and density of the solution, should be kept constant during the apparent mass change measurements. Typical piezoelectrically active (geometrical) surface area of the working Au electrode is 0.2 cm^2 and the real surface area for polished resonators is $A = 0.25$ cm^2 (roughness factor $R = 1.3$). A 100-200 nm thick Au film is usually deposited on a 14-mm diameter, 0.166-mm thick, 10 MHz AT-cut quartz resonator wafer with vacuum evaporated Cr or Ti adhesion interlayer (20 nm thick). The real surface area can be determined for Au-EQCN electrodes by a standard monolayer oxide formation procedure (*109*). For quartz crystals with $f_0 = 10$ MHz, the mass sensitivity is 0.8673 ng/Hz, hence the mass change due to the dissolution/deposition or ingress/egress processes can be easily calculated from the measured frequency shifts. For thick or non-rigid (viscoelastic) films, the acoustic admittance of a quartz plus film system must be measured to correct the apparent mass changes. This can be done using the quartz crystal immittance (QCI) measurements (*109, 110*).

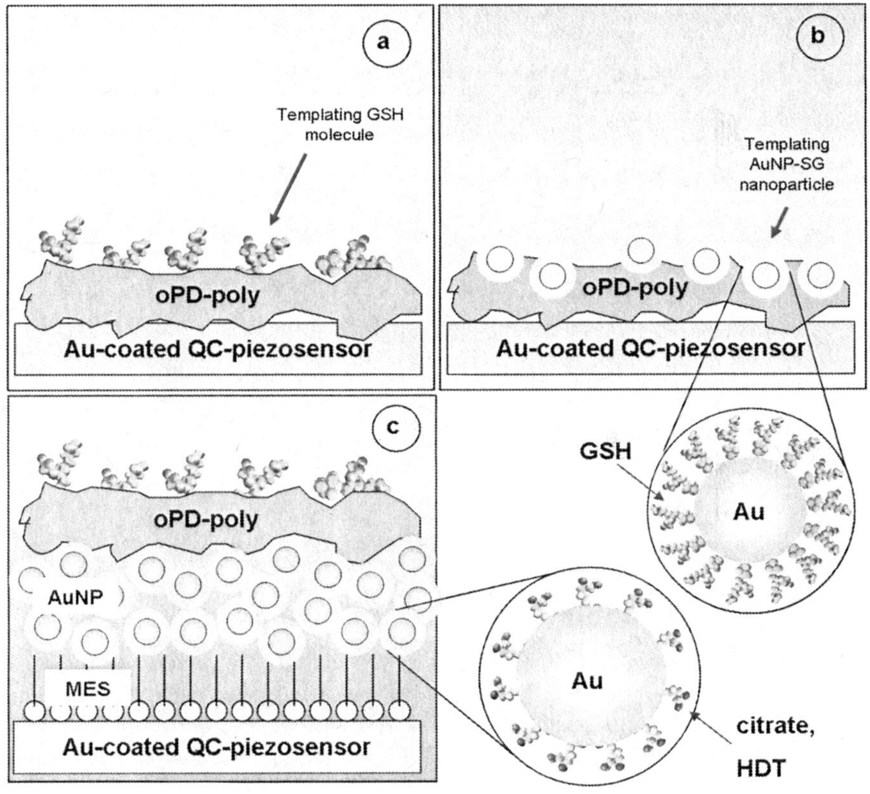

Figure 15. Schematic of GSH-templated sensor designs: (a) GSH embedded in a PoPD polymer film (oPD-poly) on an Au piezoelectrode (sensor T1), (b) AuNP-SG nanoparticle-templated PoPD film on an Au piezoelectrode (sensor T2), and (c) GSH embedded in a PoPD film electrodeposited on a layer of AuNP network assembled on a SAM of MES on an Au piezoelectrode (sensor T3). (Reprinted with permission from Ref. (111). Copyright 2009 Electrochemical Society.)

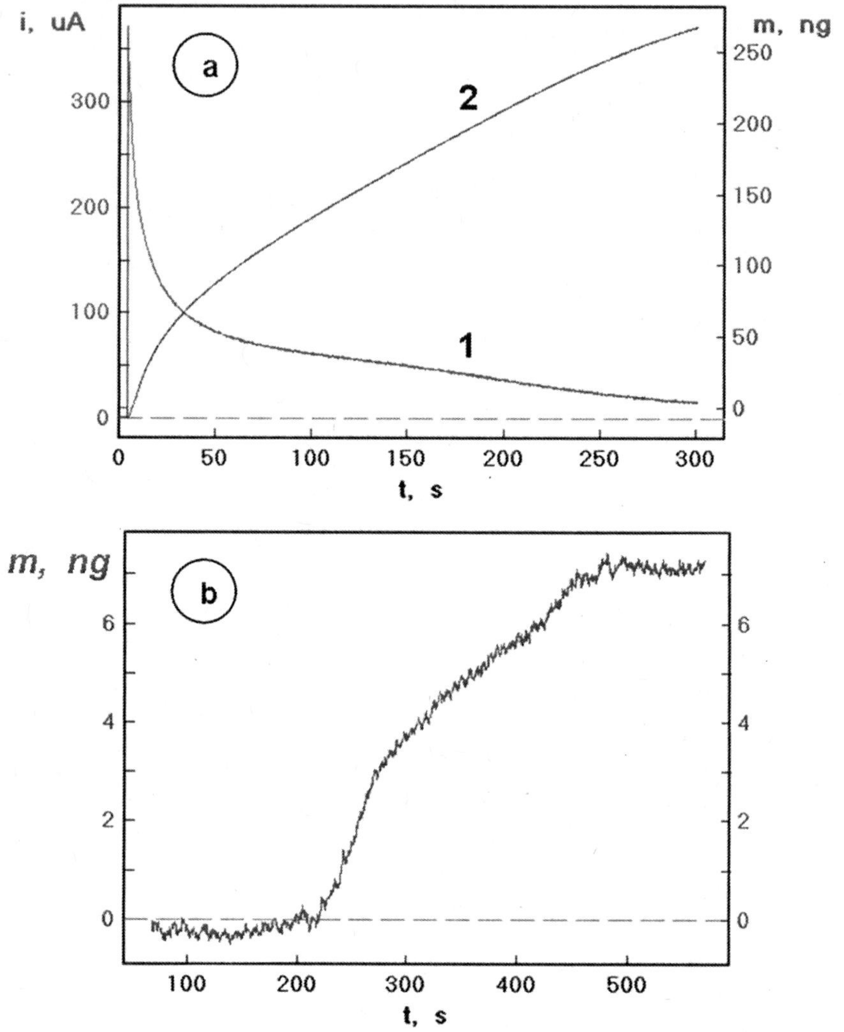

Figure 16. (a) Chronoamperometric (1) and chronogravimetric (2) transients for the initial electropolymerization of a GSH-templated poly(oPD) film T3 on a layer of AuNP-network assembled on a SAM of MES on a QC/Au electrode, from 5 mM oPD + 10 mM GSH + 10 mM HClO$_4$ solution by a potential step from E_1 = 0 to E_2 = +0.8 V vs Ag/AgCl. (b) Apparent mass vs. time response of a T3-type sensor after injection of a free GSH solution (5 mM). (Reprinted with permission from Ref. (111). Copyright 2009 Electrochemical Society.)

The electrochemical quartz crystal nanobalance (EQCN) technique enables simultaneous monitoring of voltamperometric and nanogravimetric characteristics.

The metal nanoparticles used as the analyte enhance the frequency shift due to the high mass of the particles. Recent studies of molecularly-templated conductive-polymer biosensors with glutathione-capped AuNP have shown higher sensitivity than those with glutathione alone (Figures 15 and 16) (*111*). Similar results have been obtained with piezoimmunosensors (*69*).

Scalable Bio-Fuel Cells

Bio-fuel cells are devices that convert chemical energy stored in biogenic fuel (e.g. glucose) and oxidant (e.g. oxygen) into electricity. They act as batteries, except that in fuel cells, the fuel and oxidant can be continuously supplied so that fuel cells can work indefinitely. Although a large scale bio-fuel cells can also be designed, we will present here the progress in developing small scale cells that are suitable to use as the attached or implanted devices. Particular attention will be given to nanoparticle-enhanced bio-fuel cells (*112*).

Considerable enhancement of bio-fuel cell performance has been achieved by utilizing electron mediators that promote transfer of electrons between immobilized-enzyme cofactors and the electrode surface. Of particular note are osmium complexes with bipyridyl (bpy) ligands (e.g. Os(4,4'-dimethyl-2.2'-bipyridyl)2L) (*10*) which enable lowering the electroactivation overvoltage for the electron transfer step. The use of CNT's as the electrocatalyst for enzyme nanowiring and to support mechanically the electrode assembly was the breakthrough achievement enabling miniaturization of biofuel-cells and considering them for implantation. The implanted biofuel-cells may serve with biosensors and other implanted microdevices as an integrated system.

A microfuel cell enhanced with AuNP has been investigated by Deng et al. (*113*). They used a 3D microporous gold electrode functionalized with AuNP and glucose dehydrogenase enzyme (GDH). These electrodes show high catalytic activity toward oxidation of glucose. The enhancement due to AuNP was 16-fold in comparison to a similar fuel-cell without AuNP. This effect is likely due to the nanoparticle catalyzed enhanced intermolecular electron-transfer, similar to that described by Carver et al. (*114*) for cytochrome c (Cyt c) and cobalt(III) phenanthroline complex $Co(phen)_3^{3+}$. The reaction of oxidation of the reduced Cyt c (Fe^{2+}) by $Co(phen)_3^{3+}$ has been found to be accelerated by a factor of up to 10^5 by AuNP@TX, where TX is an ethoxy thiol $HS-(C_2H_5)_5(OC_2H_5)_4X$ with an attached binding functionality X = [(C=O)(NH)(CY)COO⁻], where Y = CH_2COO^- or $CH_2C_6H_5$. According to Carver et al. (*114*), there is a direct electron transfer between Cyt c and $Co(phen)_3^{3+}$, while AuNP serves as a reversible binder of the donor and the acceptor, increasing their local concentrations and facilitating their close contact for efficient electron exchange. The electron mediation through AuNP@TX appears to be less likely due to the long pathway for electron tunneling.

Taniguchi and coworkers (*115*) have developed a D-fructose detection based on the direct heterogeneous electron transfer reaction of fructose dehydrogenase adsorbed onto multi-walled carbon nanotubes synthesized on platinum electrode. Multi-walled carbon nanotubes (MWCNTs) were synthesized directly on platinum plate electrodes by the chemical vapor deposition (CVD) method. The D-fructose detection system can be further developed to scalable biofuel cell utilizing D-fructose as a biofuel.

Other Functional Nanoparticles

Magnetic Nanoparticles

Magnetic nanoparticles have been widely utilized in biochemical studies and assays since they offer an easy way for separation and cleaning steps. Extensive reviews exist on this subject and we will not attempt to repeat them. It is however worth mentioning a new unexpected application of iron transport and storage proteins, ferritins, for well-controlled deposition of very small magnetic islands on a plane surface by thermal decomposition of ferritins.

Recently, Taniguchi and coworkers (*116*) have demonstrated the fabrication of size-controlled two-dimensional iron oxide nanodots derived from the heat treatment of ferritin molecules self-immobilized on modified silicon surfaces. The site-selected and size-controlled iron nanoparticles were prepared on coplanar surfaces via microcontact printing of SAM-modified Au/mica electrodes and controlled-potential electrolytic reactions (*117, 118*). By preparing ferritins with predefined number of iron ions stored inside the protein molecules, the nanodots with well-controlled size can be readily deposited to provide building blocks for unique structural designs.

Optical Sensing Methods

Optical biosensing methods based on unique optical properties of AuNPs and AgNPs have been widely investigated (*119–122*). Mirkin and coworkers (*123*) have demonstrated the immobilization of oligonucleotides on metal nanoparticles and their use for surface plasmon coupling-based sensitive colorimetric analyses in DNA-hybridization sensing, The electromagnetic coupling of surface plasmon (SP) in metal nanoparticle assemblies linked by complementary oligonucleotides has been analyzed by Schatz group (*124*).

New optical sensing methods have emerged with the synthesis of quantum dots (QD), small semiconductor nanoparticles exhibiting size-tunable fluorescence emission and single electron charging phenomena. Typical semiconductor nanoparticles experiencing quantum phenomena and stable under aqueous and biotic conditions are CdS, CdSe, CdTe, PbS, ZnS, and others. Dual semiconductor systems have also been investigated. These nanoparticles can be protected by a SAM shell, similar to the monolayer protected metal clusters (*125*). Size-tunable oligonucleotide-functionalized QDs have been developed by Maye and coworkers (*126–129*). Goldman et al. (*130*) have developed a bioanalytical system based on four different antibodies labeled with four-color QDs for determination of multiple analytes. Analytical applications of QDs for biosensing bioelectronic devices have been reviewed by Sheehan et al. (*131*). Clinical applications of fluorescent nanoparticles for targeted molecular imaging of tumors in living subjects have been discussed by Gao et al. (*132*).

Several optical modalities based on plasmonic properties of gold and silver nanoparticles have been employed for imaging cancer cells, including optical coherence tomography (*133*), reflectance confocal microscopy (*134*), acoustic spectroscopy (*135, 136*), and surface-enhanced resonance Raman spectroscopy (SERRS) (*137, 138*). Effects of nanoparticle size and shape and surface morphology on SERRS spectra have been described by Lin et al. (*139*).

Resonance Elastic Light Scattering (RELS)

RELS is a fast two-step process in which light is momentarily absorbed by a nanoparticle (e.g. AuNP, AgNP) followed by immediate coherent emission of absorbed photons in all directions without any energy loss. RELS occurs when polarizable particles are subjected to the oscillating electric field of a beam of light which induces oscillating dipoles in the particles and these dipoles radiate light to the surroundings. The RELS spectra of scattering intensity I_{sc} vs. emission wavelength λ_{em} are obtained at 90° angle from the incident (excitation) light beam, at constant excitation wavelength λ_{ex}, and show a narrow Gaussian peak centered at $\lambda_{em} = \lambda_{ex}$. The narrow linewidth (about $\Delta\lambda \approx 15$ nm) confirms that the effects due to radiation broadening, density fluctuation, fluorescence, and inelastic Raman scattering are negligible. The excitation beam monochromator is either scanned simultaneously with the detector beam monochromator ($\Delta\lambda = 0$) or set at a constant excitation wavelength λ_{ex}.

RELS is inherently sensitive to the interparticle distance, aggregation and the dielectric function of the medium surrounding the metal nanoparticles. The high sensitivity of RELS to metal nanoparticle assembly is due to the excitation of the surface plasmon (SP) in metal nanoparticles and coupling of local SP oscillations during the close approach of particles in the supramolecular structure formation.

The strong sixth-power dependence of elastic scattering intensity I_{sc} on the nanoparticle diameter a follows from the Rayleigh equation for light scattering from small particles:

$$I_{sc} = I_0 N \frac{\left(1 + \cos^2\theta\right)}{2R^2} \left(\frac{2\pi}{\lambda}\right)^4 \frac{\left([n_2 - n_1]^2 - 1\right)}{\left([n_2 - n_1]^2 + 2\right)} \left(\frac{a}{2}\right)^6$$

$$(8)$$

where n_1 and n_2 are the refractive indices for the solution and particles, respectively, λ is the wavelength of incident light beam, θ is the scattering angle, N is the number of particles, and I_0 is the constant.

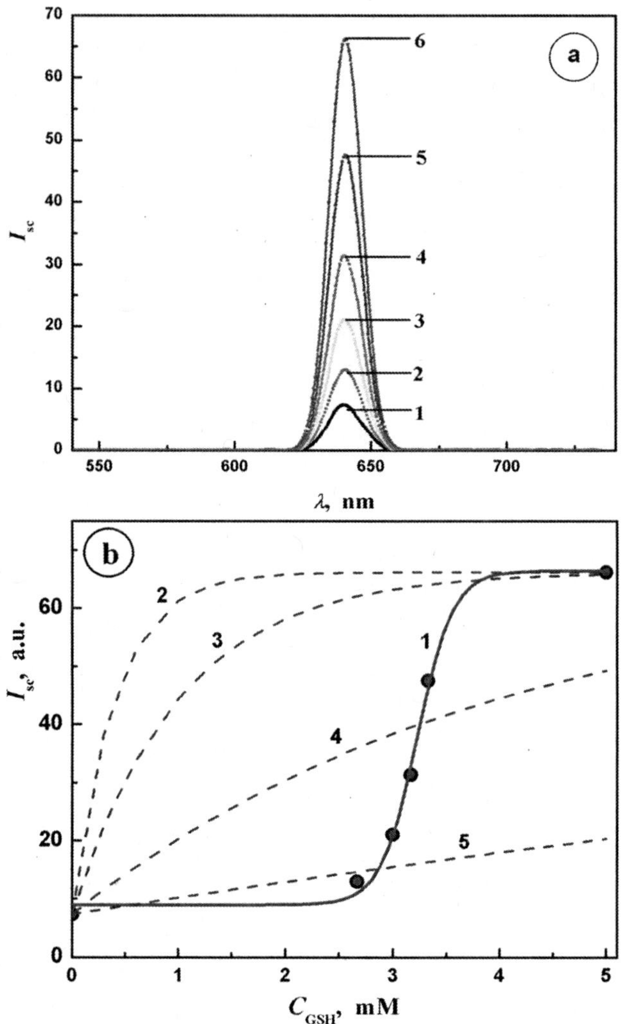

Figure 17. (a) Elastic light scattering spectra for 10.1 nM AuNP$_{5nm}$ for different concentrations of GSH, recorded within 1 min of GSH injection, C$_{GSH}$ [mM]: (1) 0, (2) 2.67, (3) 3.0, (4) 3.17, (5) 3.33, (6) 5; (b) Experimental dependence of I$_{sc}$ vs. C$_{GSH}$ (curve 1, points) fitted with Boltzmann threshold function (line) and (2-5): calculated curves for a hypothetical pseudo-first order Langmuirian kinetics showing the absence of threshold characteristics in Langmuirian model, k [M^{-1}s^{-1}]: (2) 40, (3) 17, (4) 4, (5) 0.8; τ = 60 s; incident beam wavelength: λ$_{ex}$ = 640 nm. (Reprinted with permission from Ref. (142). Copyright 2010 Elsevier.)

The RELS technique has been originally developed as a sensitive technique for the analytical determination of biomolecules (DNA, proteins) and organic complexes (*140, 141*). Recently, the RELS spectroscopy has been applied by Stobiecka and Hepel (*142–145*) to study gold and silver nanoparticle assembly.

The assembly of AuNP has been induced by injecting different biomarkers of oxidative stress (glutathione, homocysteine, cysteine). Typical light scattering spectrum for 5 nm diameter AuNP$_{5nm}$) in solution is presented in Figure 17a, curve 1, at constant excitation wavelengths, $\lambda_{ex} = 640$ nm, respectively. After the addition of glutathione to the gold nanoparticle solution, a strong enhancement of RELS intensity has been observed (Fig. 17a, curves 2–6). It is caused by the aggregation of the glutathione-capped gold nanoparticles due to the interparticle interactions of zwitterionic GSH molecules present in the shells of the AuNP@GSH (at pH 3.27, the GSH exists in the neutral form).

Sensor Array Fabrication

Microelectrode-based biosensors can be designed to form sensor arrays to enhance the device functionality. Performing simultaneous analysis for multiple analytes using multi-sensor arrays substantially reduces the analysis time and provides broader information about the sample components. Therefore, considerable efforts have been paid to develop new technologies for sensor array fabrication and to scaling down the larger biosensors already developed for laboratory use.

Novel Fabrication Methods

Single nanosensors and nanosensor arrays can be fabricated by an ultra-high resolution UV, X-ray, electron-beam, and focused ion-beam technologies, utilized in modern electronic industry for fine integrated circuit and advanced microprocessor manufacturing. New nanofabrication technologies based on scanning probe lithography (SPL) have recently emerged (*146–158*). Two main technologies of the SPL, dip-pen nanolithography (DPN) (Scheme 3), developed by Mirkin et al. (*146–150*) and a scanning multi-probe nanolithography (SMPN), proposed by Hong and Mirkin (*147*), will be discussed briefly.

SPL, which is based on well-established and commonly used advanced surface imaging techniques, atomic force microscopy (AFM) and scanning tunneling microscopy (STM), as well as the new near-field scanning optical microscopy (NSOM), is a high-resolution technique able to print very small features on smooth surfaces down to 10-50 nm size.

Another technique, ink-jet printing (IJP), appears as a competitive alternative to other printing techniques for low resolution microstructure designs. Evolved from personal printers for table microcomputers, IJP has achieved resolutions down to 1 μm size features. Importantly, IJP can print using metal or semiconductor nanoparticle-based inks and it is easily controlled from many graphics programs, including Adobe Photoshop and Illustrator, Microsoft Power Point, and Corel software.

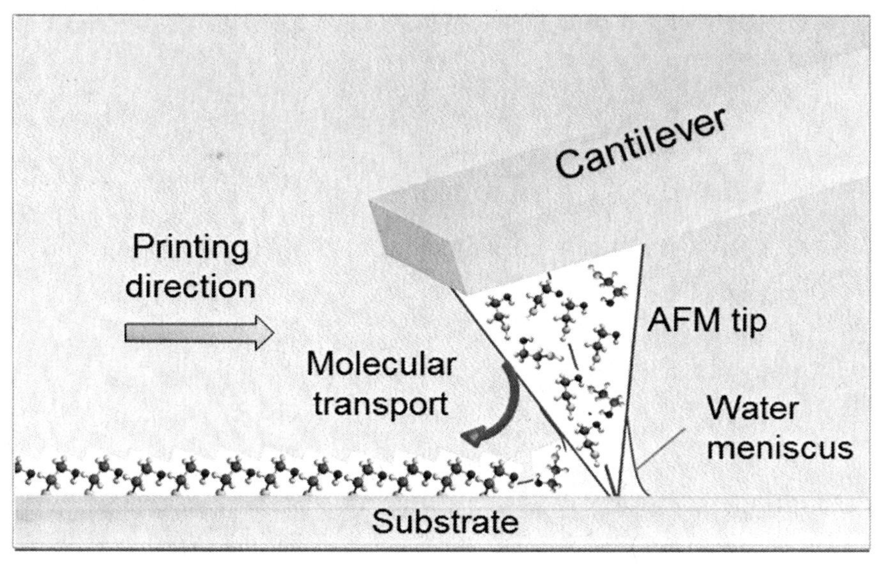

Scheme 3. Principles of dip-pen nanolithography

Ink-Jet Printing (IJP)

Ink-jet printing is based on throwing onto a target substrate picolitter droplets of an ink containing nanoparticles which could be formed into micrometer dots and lines (*159*). In this way, with a metal or semiconductor nanoparticle inks, it is possible to print electronic tracks and devices, such as the solar panels, resistors, diodes, etc. In this printing technology, the nanoparticle outer shell consists usually of a surfactant or ligand such as the alkyl chains (*160*) or other polymer chains (*161*). While the nanoparticle shells are necessary to maintain colloid stability and prevent nanoparticle aggregation, they are generally non-conductive which leads to the insulating character of a printed structure. Hence, thermal treatment (sintering) is required resulting in ligand decomposition and recrystallization of metal cores (*162*). It is promising to use silver nanoparticles (AgNPs) in electronic industry to fabricate printed circuits due to the excellent electric and thermal conductivity of silver. This can conveniently be performed by using the ink-jet printing technology which provides currently a micrometer resolution of the printed structures. The formation and properties of AgNPs have been studied extensively in recent years due to potential applications of this nanomaterial in electrooptic devices (*163, 164*), modern electronics (*165, 166*), nanomedicine (*167*), cell biology (*168*) and others (*169*).

So far, the IJP is the only microprinting method able to write directly core-shell nanoparticle-based patterns on any rigid or flexible substrates.

Silver Nanoparticles for Ink-Jet Printing

For ink-jet applications, AgNPs synthesized by thermal decomposition of silver alkanoate have been the most widely investigated nanomaterial so far (*170*). In this method, silver alkanoate is obtained by the reaction of silver nitrate and fatty acid dissolved in aqueous sodium hydroxide solution. The composition of the nanoparticle forming bath is very important: 5.2g of $AgNO_3$, 7.5g of myristic acid, 1.2g of NaOH per 100 mL of distilled water (*171*). The silver alkanoate is then thermally decomposed in an inert atmosphere (Fig. 1). According to Nasagawa et al. (*171*), the decomposition should be carried out in a nitrogen atmosphere at 250 °C, while Shim et. al. (*172*) suggested the decomposition reaction at 210°C with fatty acid dissolving in a non-polar, slowly evaporating solvent, such as 1-octadecene. Both processes result in silver powder composed of aggregated ligand-stabilized AgNPs with narrow size distribution.

The silver ink is prepared by redispersing the core-shell AgNP in a non-polar organic solvent, *e.g. n*-tetradecane. According to Puchalski et al. (*166*), tracks printed with this ink gain conductance after treatment at 240–250 °C for 30 min. For AgNP characterization, a colloid in cyclohexane has been advised (*166*).

Single-Probe SPL Nanoprinting

With various types of scanning probes and using AFM or STM, small features on surfaces can be formed by selective surface oxidation, etching, or deposition of atoms or molecules. Surface-protecting adsorbed molecules can also be removed from the surface in a well-defined pattern, for instance, molecules from a functional-SAM formed on a surface. As the SAM is designed for binding other molecules, e.g. oligonucleotides, proteins, or viruses, the patches without SAM will act as voids. In order to deposit nanoparticles at well-defined positions on a surface, a DPN with AFM tip has been used. The AFM tip is first saturated with a solution containing molecules or nanoparticles that are to be deposited. These molecules adsorb on the tip and are then transferred by diffusion along the surface of a water meniscus, formed between the tip and the substrate, onto the substrate surface during the printing process. Using this method, trenches 12-100 nm in size have been made on Ag, Au, and Pd.

The degree and rate of the ink-spreading on a substrate depends on the radius of curvature of the probe, the linear probe-velocity, humidity, and the reactivity of the ink with the substrate (*173*).

Parallel Nanoprinting using SPL

Since the serial printing with a single-probe is slow and time consuming, a parallel printing method utilizing scanning multi-probe nanolithography (SMPN) has been proposed (*147, 155*). The parallel, large-volume printing enables simultaneous writing of many patterns because the individual printing probes can be actuated independently by mechanically deflecting the corresponding AFM probes. The cantilever arrays can also heat locally a polymeric film to write a pattern of holes. A parallel printing of virus arrays with SPL has been described by Mirkin group (*174*). Massively parallel SPL with 55,000-probe writing head has recently been described (*156*). At the moment, the capability of SMPN achieved so far consists of writing arrays of identical patterns using passive nonactuated probes. Writing arrays of different patterns by thermal actuation of probes has also been reported.

The novel polymer-pen lithography (PPL) invented also by Mirkin group (*175, 176*) is promising as an inexpensive new means of printing large area patterns. One of the applications of PPL is printing protein microarray chips (*176*).

Conclusions

In summary, the emergence of functional nanoparticle nanotechnology has provided vast opportunities to improve the performance and enhance the sensitivity of bioelectronic devices by utilizing high electrocatalytic activity, high electric conductance, and unique optical and mechanical properties of nanoparticles. The utility of functional nanoparticles embedded in films of various biodevices, including implantable enzyme biosensors, immunosensors, molecularly templated biomimetic polymer films for biosensing applications, BioFETs, logic gates, and others have been assessed. The immense progress achieved in these devices still awaits on a translation to widespread applications, however, the perspectives for many of these technologies are very bright. The scaling down of biosensing devices provides a significant cost reduction of biologicals and the ease of use of dispensable biosensors provides hope for points of care in remote locations where the sophisticated instrumental techniques are not available. The cytotoxicity studies and sensor biofouling investigations are underway and should bring soon a general guidance concerning safe operation of bioelectronic devices based on functional nanoparticles.

References

1. Wang, J. *Analyst* **2005**, *130*, 421–425.
2. Wang, J. *Anal. Chim. Acta* **2003**, *500*, 247–257.
3. Merkoci, A.; Aldavert, M.; Marin, S.; Alegret, S. *Trends Anal. Chem.* **2005**, *24*, 341–349.
4. Guo, S.; Dong, S. *Trends Anal. Chem.* **2009**, *28*, 96–109.
5. Murphy, L. *Current Opinion Chem. Biol.* **2006**, *10*, 177–184.
6. Katz, E.; Willner, I.; Wang, J. *Electroanal.* **2004**, *16*, 19–44.
7. Zimmet, P. *J. Intern. Med.* **2000**, *247*, 301–310.
8. Wilson, R.; Turner, A. P. F. *Biosens. Bioelectron.* **1992**, *7*, 165–185.
9. Willner, B.; Katz, E.; Willner, I. *Curr. Opin. Biotechnol.* **2006**, *17*, 589–596.
10. Barriere, F.; Kavanagh, P.; Leech, D. *Electrochim. Acta* **2006**, *51*, 5187–5192.
11. Georganopoulou, D. G.; Carley, R.; Jones, D. A.; Boutelle, M. G. *Faraday Discuss.* **2000**, *116*, 291–303.

12. Xiao, Y.; Patolsky, F.; Katz, E.; Hainfield, J. F.; Willner, I. *Science* **2003**, *299*, 1877–1881.

13. Collins, P. G.; Bradley, K.; Ishigami, M.; Zettl, A. *Science* **2000**, *287*, 1801.

14. Guiseppi-Elie, A.; Lei, C. H.; Baughman, R. H. *Nanotechnology* **2002**, *13*, 559–564.

15. Wang, S. G.; Zhang, Q.; Wang, R.; Yoon, S. F.; Ahn, J.; Yang, D. J.; Tian, J. Z.; Li, J. Q.; Zhou, Q. *Electrochem. Commun.* **2003**, *5*, 800–803.

16. Guiseppi-Elie, A.; Rahman, A. R. A.; Shukla, N. K. *Electrochim. Acta* **2010**, *55*, 4247–4255.

17. Gao, M.; Dai, L.; Wallace, G. G. *Synth. Metals* **2003**, *137*, 1393–1394.

18. Wang, J.; Musameh, M. *Analyst (London)* **2004**, *129*, 1–2.

19. Munge, B.; Liu, G.; Collins, G.; Wang, J. *Anal. Chem.* **2005**, *77*, 4662.

20. Liu, G.; Lin, Y. *Electrochem. Commun.* **2006**, *8*, 251–256.

21. Liu, G.; Lin, Y. *Anal. Chem.* **2006**, *78*, 835–843.

22. Wang, J.; Zhang, X. J. *Anal. Chem.* **2001**, *73*, 844–847.

23. Guiseppi-Elie, A.; Choi, S. H.; Geckeler, K. E.; Sivaraman, B.; Latour, R. A. *NanoBiotechnol.* **2008**, *4*, 9–17.

24. Ahamad, A. J. S.; Lee, J. J.; Rahman, M. A. *Sensors* **2009**, *9*, 2289–2319.

25. Wang, J.; Li, M.; Shi, Z.; Li, N.; Gu, Z. *Microchem. J.* **2002**, *73*, 325–333.

26. Dale, N.; Hatz, S.; Tian, F.; Llaudet, E. *Trends Biotechnol.* **2005**, *23*, 420–428.

27. Vaddiraju, S.; Tomazos, I.; Burgess, D. J.; Jain, F. C.; Papadimitrakopoulos, F. *Biosens. Bioelectron.* **2010**, *25*, 1553–1565.

28. Vasylieva, N.; Barnych, B.; Meiller, A.; Maucler, C.; Pellegioni, L.; Lin, J. S.; Barbier, D.; Marinesco, S. *Biosens. Bioelectron.* **2011**, *26*, 3993–4000.

29. Heller, A.; Feldman, B. *Chem. Rev.* **2008**, *108*, 2482–2505.

30. Wang, J. *Chem. Rev.* **2008**, *108*, 814–825.

31. Murray, A. R.; Kisin, E.; Leonard, S. S.; Young, S. H.; Kommineni, C.; Kagan, V. E.; Kastranova, V.; Shvedova, A. A. *Toxicology* **2009**, *257*, 161–171.

32. Singh, S.; Solanki, P. R.; Pandey, M. K.; Malhotra, B. D. *Sens. Actuators, B* **2006**, *115*, 534–541.

33. Madaras, M. B.; Popescu, I. C.; Ufer, S.; Buck, R. P. *Anal. Chim. Acta* **1996**, *319*, 335–345.

34. Schneider, J.; Grundig, B.; Renneberg, R.; Cammann, K.; Madaras, M. B.; Buck, R. P.; Vorlop, K.-D. *Anal. Chim. Acta* **1996**, *325*, 161–167.

35. Berberich, J. A.; Yang, L. W.; Bahar, I.; Russell, A. J. *Acta Biomater.* **2005**, *1*, 183–191.

36. Graebner, H.; Georgi, U.; Huttl, R.; Wolf, G. *Thermochim. Acta* **1998**, *310*, 101–105.

37. Collins, A.; Nandakumar, M. P.; Csoregi, E.; Mattiasson, B. *Biosens Bioelectron* **2001**, *16*, 765–771.

38. Romero, M. R.; Garay, F.; Baruzzi, A. M. *Sens. Actuators, B* **2008**, *131*, 590–595.

39. Maines, A.; Ashworth, D.; Vadgama, P. *Food Technol. Biotechnol.* **1996**, *34*, 31–42.

40. Odaci, D.; Telefoncu, A.; Timur, S. *Bioelectrochemistry* **2010**, *79*, 108–113.

41. Akyilmaz, E.; Yorganci, E. *Biosens. Bioelectron.* **2008**, *23*, 1874–1877.

42. Soldatkin, O. O.; Peshkova, V. M.; Dzyadevych, S. V.; Soldatkin, A. P.; Jaffrezic-Renault, N.; El'skaya, A. V. *Mater. Sci. Eng. C* **2008**, *28*, 959–964.

43. Nipro Diagnostics, Inc., 2012; www.niprodiagnostics.com.

44. Rahman, A. R. A.; Justin, G.; Guiseppi-Elie, A. *Biomed Microdevices* **2009**, *11*, 75–85.

45. Guiseppi-Elie, A. *Anal. Bioanal. Chem.* **2011**, *399*, 403–419.

46. Laird, A. M.; Miller, P. R.; Kilgo, P. D.; Meredith, J.; Chang, M. C. *J. Trauma: Inj., Infect., Crit. Care* **2004**, *56*, 1058–1062.

47. Michaud, L. J.; Rivara, F. P.; Longstreth, W. T.; Grady, M. S. *J. Trauma* **1991**, *31*, 1356–1362.

48. Yendamuri, S.; Fulda, G. J.; Tinkoff, G. H. *J. Trauma: Inj., Infect., Crit. Care* **2003**, *55*, 33–38.

49. Frost, M. C.; Meyerhoff, M. E. *Curr. Opin. Chem. Biol.* **2002**, *6*, 633–641.

50. Wang, J. *Electroanalysis* **2001**, *13*, 983–988.

51. Wilson, G. S.; Gifford, R. *Biosens. Bioelectron.* **2005**, *20*, 2388–2403.

52. Wickramansighe, Y.; Yang, Y.; Spencer, S. A. *J. Fluoresc.* **2004**, *14*, 513–520.

53. Gerritsen, M.; Kros, A.; Lutterman, J. A.; Nolte, R. J. M.; Jansen, J. A. *J. Mater. Sci.: Mater. Med.* **2001**, *12*, 129–134.

54. Wisniewski, N.; Moussy, F.; Reichert, W. M. *Fresenius' J. Anal. Chem.* **2000**, *366*, 611–621.

55. Valdastri, P.; Menciassi, A.; Arena, A.; Caccamo, C.; Dario, P. *IEEE Trans. Inf. Technol. Biomed.* **2004**, *8*, 271–278.

56. Castillo, P. M.; Herrera, J. L.; Fernandez-Montesinos, R.; Caro, C.; Zaderenko, A. P.; Mejias, J. A.; Pozo, D. *Nanomedicine* **2008**, *3*, 627–635.

57. Shin, S. H.; Ye, M. K.; Kim, H. S.; Kang, H. S. *Int. Immunopharmacol.* **2007**, *7*, 1813–1818.

58. Yen, H. J.; Hsu, S. H.; Tsai, C. L. *Small* **2009**, *5*, 1553–1561.

59. Roe, D.; Karandikar, B.; Bonn-Savage, N.; Gibbins, B.; Roullet, J. B. *J. Antimicrob. Chemother.* **2008**, *61*, 869–876.

60. Jain, J.; Aurora, S.; Rajwade, J.; Khandelwal, S.; Paknikar, K. M. *Mol. Pharm.* **2009**, *6*, 1388–1401.

61. Amato, E.; Diaz-Fernandez, Y. A.; Taglietti, A.; Pallavicini, P.; Pasotti, L.; Cucca, L.; Milanese, C.; Grisoli, P.; Dacarro, C.; Fernandez-Hechavarria, J. M.; Necchi, V. *Langmuir* **2011**, *27*, 9165–9173.

62. Klippstein, R.; Fernandez-MontesinosR.; Castillo, P. M.; Zadarenko, A. P.; Pozo, D., Silver nanoparticles interactions with the immune system: implications for health and disease. In *Silver Nanoparticles*; Perez, D. P., Ed.; InTech: Vienna, 2010; pp 309–323.

63. Wilson, G. S.; Hu, Y. *Chem. Rev.* **2000**, *100*, 2693–2704.

64. Warsinke, A.; Benkert, A.; Scheller, F. W. *Fresenius' J. Anal. Chem.* **2000**, *366*, 622–634.

65. Halamek, J.; Hepel, M.; Skladal, P. *Biosens. Bioelectron.* **2001**, *16*, 253–260.

66. Pribyl, J.; Hepel, M.; Halamek, J.; Skladal, P. *Sens. Actuators, B* **2003**, *91*, 333–341.

67. Pribyl, J.; Hepel, M.; Skladal, P. *Sens. Actuators, B* **2006**, *113*, 900–910.

68. Bilitewski, U. *Anal. Chem.* **2000**, *72*, 692 A–701 A.

69. Stobiecka, M.; Hepel, M. *Biosens. Bioelectron.* **2011**, *26*, 3524–3530.

70. Tiefenauer, L. F.; Kossek, S.; Padeste, C.; Thiebaud, P. *Biosens. Bioelectron.* **1997**, *12*, 213–223.

71. Huang, H. Z.; Liu, Z. G.; Yang, X. R. *Anal. Biochem.* **2006**, *356*, 208–214.

72. Tang, H.; Chen, J.; Nie, L.; Kuang, Y.; Yao, S. *Biosens. Bioelectron.* **2007**, *22*, 1061–1067.

73. Zhang, S.; Huang, F.; Liu, B. H.; Ding, J. J.; Xu, X.; Kong, J. L. *Talanta* **2007**, *71*, 874–881.

74. Chen, X.; Wang, Y.; Zhou, J.; Yan, W.; Li, X.; Zhu, J. *Anal. Chem.* **2008**, *80*, 2133–2140.

75. Dequaire, M.; Degrand, C.; Limoges, B. *Anal. Chem.* **2000**, *72*, 5521–5528.

76. Chu, X.; Fu, X.; Chen, K.; Shen, G. I.; Yu, R. Q. *Biosens. Bioelectron.* **2005**, *20*, 1805–1812.

77. Liao, K. T.; Huang, H. J. *Anal. Chim. Acta* **2005**, *538*, 159–164.

78. Chen, J.; Tang, J.; Yan, F.; Ju, H. *Biomaterials* **2006**, *27*, 2313–2321.

79. Wu, L.; Chen, J.; Du, D.; Ju, H. *Electrochim. Acta* **2006**, *51*, 1208–1214.

80. Chen, J.; Yan, F.; Tan, F.; Ju, H. *Electroanalysis* **2006**, *18*, 1696–1702.

81. Bergveld, P. *IEEE Trans. Biomed. Eng. BME* **1970**, *17*, 70–71.

82. Janata, J. *Electroanalysis* **2004**, *16*, 1831–1835.

83. Spiegel, J. v. d.; Lauks, I.; Chan, P.; Babic, D. *Sens. Actuators, B* **1983**, *4*, 291–298.

84. Jin, Y.; Yao, X.; Liu, Q.; Li, J. *Biosens. Bioelectron.* **2007**, *22*, 1126–1130.

85. Xu, J. J.; Zhao, W.; Luo, X. L.; Chen, H. Y. *Chem. Commun.* **2005**, 792–794.

86. Jeong, S.; Shim, H. C.; Kim, S.; Han, C. S. *ACS Nano* **2010**, *4*, 324–330.

87. Oelssner, W.; Zosel, Z.; Guth, U.; Pechstein, T.; Babel, W.; Connery, J. G.; Demuth, C.; Gansey, M. G.; Verburg, J. B. *Sens. Actuators, B* **2005**, *105*, 104–117.

88. Milgrewa, M. J.; Riele, M. O.; Cumming, D. R. S. *Sens. Actuators, B* **2005**, *111-112*, 347–353.

89. Ariga, K.; Ji, Q.; Hill, J. P. *Adv. Polym. Sci.* **2010**, *229*, 51–87.

90. deSilva, A. P.; Gunaratne, H. Q. N.; McCoy, C. P. *Nature* **1993**, *364*, 42–44.

91. deSilva, A. P.; Uchiyama, S. *Nat. Nanotechnol.* **2007**, *2*, 399–410.

92. Margulies, D.; Melman, G.; Shanzer, A. *Nat. Mater.* **2005**, *4*, 768–771.

93. Ashkenasy, G.; Ghadiry, M. R. *J. Am. Chem. Soc.* **2004**, *126*, 11140–11141.

94. Baron, R.; Lioubashevski, O.; Katz, E.; Niazov, T.; Willner, I. *J. Phys. Chem. A* **2006**, *110*, 8548–8553.

95. Margulies, D.; Felder, C. E.; Melman, G.; Shanzer, A. *J. Am. Chem. Soc.* **2007**, *129*, 347–354.

96. Margulies, D.; Hamilton, A. D. *J. Am. Chem. Soc.* **2009**, *131*, 9142–9143.

97. Zhou, J.; Melman, G.; Pita, M.; Ornatska, M.; Wang, X. M.; Melman, A.; Katz, E. *ChemBioChem.* **2009**, *10*, 1084–1090.

98. Niazov, T.; Baron, R.; Katz, E.; Lioubashevski, O.; Willner, I. *Proc. Natl. Acad. Sci. U.S.A.* **2006**, *103*, 17160–17163.

99. deSilva, A. P.; James, M. R.; McKinney, B. O. F.; Pears, D. A.; Weir, S. M. *Nat. Mater.* **2006**, *5*, 787–790.

100. Gupta, T.; Boom, M. E. v. d. *Angew. Chem., Int. Ed.* **2008**, *47*, 2260–2262.

101. Gianneschi, N. C.; Ghadiri, M. R. *Angew. Chem., Int. Ed.* **2007**, *46*, 3955–3958.

102. Credi, A.; Balzani, V.; Langford, S. J.; Stoddart, J. F. *J. Am. Chem. Soc.* **1997**, *119*, 2679–2681.

103. Frezza, B. M.; Cockroft, S. L.; Ghadiri, M. R. *J. Am. Chem. Soc.* **2007**, *129*, 14875–14879.

104. Kikkeri, R.; Grunstein, D.; Seeberger, P. H. *J. Am. Chem. Soc.* **2010**, *132*, 10230–10232.

105. Ogawa, A.; Maeda, M. *Chem. Comm.* **2009**, 4666–4668.

106. Johnson, P. B.; Christy, R. W. *Phys. Rev. B* **1972**, *6*, 4370.

107. Kelly, K. L.; Coronado, E.; Zhao, L. L.; Schatz, G. C. *J. Phys. Chem. B* **2003**, *107*, 668.

108. Skládal, P. *J. Braz. Chem. Soc.* **2003**, *14*, 491–502.

109. Hepel, M. Electrode-Solution Interface Studied with Electrochemical Quartz Crystal Nanobalance. In *Interfacial Electrochemistry. Theory, Experiment and Applications*; Wieckowski, A., Ed.; Marcel Dekker, Inc.: New York, 1999; pp 599–630.

110. Hepel, M.; Cateforis, E. *Electrochim. Acta* **2001**, *46*, 3801–3815.

111. Stobiecka, M.; Deeb, J.; Hepel, M. *Electrochem. Soc. Trans.* **2009**, *19*, 15–32.

112. Srivastava, R.; Mani, P.; Hahn, N.; Strasser, P. *Angew. Chem., Int. Ed.* **2007**, *46*, 8988–8991.

113. Deng, L.; Wang, F.; Chen, H.; Shang, L.; Wang, L.; Wang, T.; Dong, S. *Biosens. Bioelectron.* **2008**, *24*, 329–333.

114. Carver, A. M.; De, M.; Bayraktar, H.; Rana, S.; Rotello, V. M.; Knapp, M. J. *J. Am. Chem. Soc.* **2009**, *131*, 3798–3799.

115. Tominaga, M.; Nomura, S.; Taniguchi, I. *Biosens Bioelectron.* **2009**, *24*, 1184–1188.

116. Tominaga, M.; Matsumoto, M.; Soejima, K.; Taniguchi, I. *J. Colloid Interface Sci.* **2006**, *299*, 761–765.

117. Tominaga, M.; Miyahara, K.; Soejima, K.; Nomura, S.; Matsumoto, M.; Taniguchi, I. *J. Colloid Interface Sci.* **2007**, *313*, 135–140.

118. Tominaga, M.; Ohira, A.; Kubo, A.; Taniguchi, I.; Kunitake, M. *Chem. Commun.* **2004**, *13*, 1518–1519.

119. Mauriz, E.; Calle, A.; Abad, A.; Montoya, A.; Hildebrandt, A.; Barcelo, D.; Lechuga, L. M. *Biosens. Bioelectron.* **2006**, *21*, 2129–2136.

120. Malinsky, M. D.; Kelly, K. L.; Schatz, G. C.; Duyne, R. P. V. *J. Am. Chem. Soc.* **2001**, *123*, 1471–1482.

121. Taton, T. A.; Lu, G.; Mirkin, C. A. *J. Am. Chem. Soc.* **2001**, *123*, 5164–5165.

122. Haes, A. J.; Duyne, R. P. V. *J. Am. Chem. Soc.* **2002**, *124*, 10596–10604.

123. Cao, Y.; Jin, R.; Mirkin, C. A. *J. Am. Chem. Soc.* **2001**, *123*, 7961–7962.

124. Lazarides, A. A.; Kelly, K. L.; Jensen, T. R.; Schatz, G. C. *J. Mol. Struct. Theochem.* **2000**, *529*, 59–63.

125. Lee, J.; Govorov, A. O.; Dulka, J.; Kotov, N. A. *Nano Lett.* **2004**, *4*, 2323–2330.

126. Han, H.; Francesco, G. D.; Maye, M. M. *J. Phys. Chem. C* **2010**, *114*, 19270.

127. Maye, M. M.; Gang, O.; Cotlet, M. *Chem. Comm.* **2010**, *33*, 6111.

128. Zylstra, J.; Amey, J.; Miska, N. J.; Pang, L.; Hine, C. R.; Langer, J.; Doyle, R. P.; Maye, M. M. *Langmuir* **2011**, *27*, 4371.

129. Han, H.; Zylstra, J.; Maye, M. M. *Chem. Mater.* **2011**, *23*, 4975.

130. Goldman, E. R.; Clapp, A. R.; Anderson, G. P.; Uyeda, H. T.; Mauro, J. M.; Medintz, I. L.; Mattoussi, H. *Anal. Chem.* **2005**, *76*, 684–688.

131. Sheehan, A. D.; Quinn, J.; Daly, S.; Dillon, P.; O'Kennedy, R. *Anal. Lett.* **2003**, *36*, 511–537.

132. Gao, J.; Chen, K.; Luong, R.; Bouley, D. M.; Mo, H.; Qiao, T.; Gambhir, S. S.; Cheng, Z. *Nano Lett.* **2012**, *12*, 281–286.

133. Gobin, A. M.; Lee, M. H.; Halas, N. J.; James, W. D.; Drezek, R. A.; West, J. L. *Nano Lett.* **2007**, *7*, 1929–1934.

134. Javier, D. J.; Nitin, N.; Levy, M.; Ellington, A.; Richards-Kortum, R. *Bioconjugate Chem.* **2008**, *19*, 1309–1312.

135. Pan, D.; Pramanik, M.; Senpan, A.; Allen, J. S.; Zhang, H.; Wickline, S. A.; Wang, L. V.; Lanza, G. M. *FASEB J.* **2011**, *25*, 875–882.

136. Zerda, A. d. l.; Kim, J. W.; Galanzha, E. I.; Gambhir, S. S.; Zharov, V. P. *Contrast Media Mol. Imaging* **2011**, *6*, 346–369.

137. Tam, N. C.; Scott, B. M.; Voicu, D.; Wilson, B. C.; G., G. Z. *Bioconjugate Chem.* **2010**, *21*, 2178–2182.

138. Kircher, M. F.; de la Zerda, A.; Jokerst, J. V.; Zavaleta, C. L.; Kempen, P. J.; Mittra, E.; Pitter, K.; Huang, R.; Campos, C.; Habte, F.; Sinclair, R.; Brennan, C. W.; Mellinghoff, I. K.; Holland, E. C.; Gambhir, S. S. *Nat. Med.* **2012** (advance online publication).

139. Lin, X. M.; Cui, Y.; Xu, Y. H.; Ren, B.; Tian, Z. Q. *Anal. Bioanal. Chem.* **2009**, *394*, 1729–1745.

140. Pasternack, R. F.; Bustamante, C.; Collongs, P. J.; Giannetto, A.; Gibbs, E. J. *J. Am. Chem. Soc.* **1993**, *115*, 5393–5399.

141. Pasternack, R. F. *Science* **1995**, *269*, 5226.

142. Stobiecka, M.; Coopersmith, K.; Hepel, M. *J. Colloid. Interf. Sci.* **2010**, *350*, 168–177.

143. Stobiecka, M.; Deeb, J.; Hepel, M. *Biophys. Chem.* **2010**, *146*, 98–107.

144. Stobiecka, M.; Hepel, M. *Sens. Actuators, B* **2010**, *149*, 373–380.

145. Stobiecka, M.; Hepel, M. *Biomaterials* **2011**, *32*, 3312–3321.

146. Ginger, D. S.; Zhang, H.; Mirkin, C. A. *Angew. Chem., Int. Ed.* **2004**, *43*, 30–45.

147. Hong, S.; Mirkin, C. A. *Science* **2000**, *288*, 1808–1811.

148. Lee, K.-B.; Lim, J.-H.; Mirkin, C. A. *J. Am. Chem. Soc.* **2003**, *125*, 5588–5589.

149. Piner, R. D.; Zhu, J.; Xu, F.; Hong, S.; Mirkin, C. A. *Science* **1999**, *283*, 661–663.

150. Smith, J. C.; Lee, K. B.; Wang, Q.; Finn, M. G.; Johnson, J. E.; Mrksich, M.; Mirkin, C. A. *Nano Lett.* **2003**, *3*, 883.

151. Kramer, S.; Fuierer, R. R.; Gorman, C. B. *Chem. Rev.* **2003**, *103*, 4367–4418.

152. Ryu, K. S.; Wang, X.; Shaikh, K.; Bullen, D.; Goluch, E.; Zou, J.; Liu, C.; Mirkin, C. A. *Appl. Phys. Lett.* **2004**, *85*, 136–138.

153. Salaita, K.; Wang, Y.; Mirkin, C. A. *Nat. Nanotechnol.* **2007**, *2*, 145–155.

154. Braunschweig, A. B.; Huo, F.; Mirkin, C. A. *Nature Chem.* **2009**, *1*, 353–358.

155. Lenhert, S.; Sun, P.; Wang, Y.; Fuchs, H.; Mirkin, C. A. *Small* **2007**, *3*, 71–75.

156. Salaita, K.; Wang, Y.; Fragala, J.; Vega, R. A.; Liu, C.; Mirkin, C. A. *Angew. Chem., Int. Ed.* **2006**, *45*, 7220–7223.

157. Li, K. B.; Park, S. J.; Mirkin, C. A.; Smith, J. C.; Mrksich, M. *Science* **2002**, *295*, 1702–1704.

158. Huo, F.; Zheng, G.; Liao, X.; Giam, L. R.; Chai, J.; Chen, X.; Shim, W.; Mirkin, C. A. *Nat. Nanotechnol.* **2010**, *5*, 637–640.

159. Fuller, S. B.; Wilhelm, E. J.; Jacobson, J. M. *J. Microelectromech. Sys.* **2002**, *11*, 54–60.

160. Jiang, P.; Xie, S. S.; Peng, S. J.; Gao, H. J. *Appl. Surf. Sci.* **2002**, *191*, 240–246.

161. Tsuji, T.; Thang, D. H.; Okazaki, Y.; Nakanishi, M.; Tsuboi, Y.; Tsuji, M. *Appl. Surf. Sci.* **2008**, *254*, 5224–5230.

162. Smith, P. J.; Shin, D. Y.; Stringer, J. E.; Derby, B.; Reis, N. *J. Mater. Sci.* **2006**, *41*, 4153–4158.

163. Farlandand, A. D. M.; Duyne, R. P. V. *Nano Lett.* **2003**, *3*, 1057–1062.

164. Ha, C. S.; Park, J. W.; Ullah, M. H. o-Phenylenediamine encapsulated silver nanoparticles and their applications for organic light-emitting devices. In *Silver Nanoparticles*; Perez, D. P., Ed.; InTech: Vienna, 2010; pp 153–160.

165. Xue, F.; Liu, Z.; Su, Y.; Varahramyan, K. *Microelectron. Eng.* **2006**, *83*, 298–302.

166. Puchalski, M.; Kowalczyk, P. J.; Klusek, Z.; Olejniczak, W., The applicability of global and surface sensitive techniques to characterization of silver nanoparticles for Ink-Jet printing technology. In *Silver Nanoparticles*; Perez, D. P., Ed.; InTech: Vienna, 2010; pp 63–78.

167. Shahverdi, A. R.; Fakhimi, A.; Shahverdi, H. R.; Minaian, S. *Nanomedicine* **2007**, *3*, 168–171.

168. Huangand, H.; Yang, X. *Carbohydrate Res.* **2004**, *339*, 2627–2631.

169. Klabunde, K. J. *Nanoscale Materials in Chemistry*; John Wiley and Sons: New York, 2001.

170. Khanna, P. K.; Kulkarni, D.; Beri, R. K. *J. Nanoparticles Res.* **2008**, *10*, 1059–1062.

171. Nagasawa, H.; Maruyama, M.; Komatsu, T.; Isoda, S.; Kobayashi, T. *Phys. Status Solidi A* **2002**, *191*, 67–76.

172. Shim, I. K.; Lee, Y.; Lee, K. J.; Joung, J. *Mater. Chem. Phys.* **2008**, *110*, 316–321.

173. Rozhok, S.; Sun, P.; Piner, R.; Lieberman, M.; Mirkin, C. A. *J. Phys. Chem. B* **2004**, *108*, 7814–7819.

174. Vega, R. A.; Shen, C. K. F.; Maspoch, D.; Robach, J. G.; Lamb, R. A.; Mirkin, C. A. *Small* **2007**, *3*, 1482–1485.

175. Huo, F.; Z, Z.; Zheng, G.; Giam, L. R.; Zhang, H.; Mirkin, C. A. *Science* **2008**, *321*, 1658–1660.

176. Zheng, Z.; Daniel, W. L.; Giam, K. R.; Huo, F.; Zheng, G.; Mirkin, C. A. *Angew. Chem., Int. Ed.* **2009**, *48*, 7626–7629.

Biological Applications of SERS Using Functional Nanoparticles

Yasutaka Kitahama,[1] Tamitake Itoh,[2] Prompong Pienpinijtham,[3] Sanong Ekgasit,[3] Xiao Xia Han,[1] and Yukihiro Ozaki[*,1]

[1]Department of Chemistry, School of Science and Technology, Kwansei Gakuin University, Sanda, Hyogo 669-1337, Japan
[2]Nano-bioanalysis Research Group, Health Research Institute, National Institute of Advanced Industrial Science and Technology (AIST), Takamatsu, Kagawa 761-0395, Japan
[3]Sensor Research Unit, Department of Chemistry, Faculty of Science, Chulalongkorn University, Bangkok 10330, Thailand
[*]E-mail: ozaki@kwansei.ac.jp

The purpose of this review is to outline the state-of-the art of biological applications of SERS using functional nanoparticles. The review consists of two parts. In the former part, mechanism of SERS is discussed in considerable detail with the particular emphasis on "twofold" EM enhancement mechanism. Quantitative evaluation of twofold EM enhancement is demonstrated. In the latter part, some examples of biological applications of SERS will be reported. First, highly sensitive detections of biologically important small molecules, e.g., iodide, thiacyanate, and estrogen are by SERS will be discussed. Second, we introduce the strategy for protein detections by SERS. Label-free detection and Raman dye labeled detection methods for them are compared. Finally, SERS detection of living cells, *Escherichia coli* and yeast, are described in detail.

© 2012 American Chemical Society

I. General Introduction

Surface-enhanced Raman scattering (SERS) is an extraordinary candidate for detecting and characterizing biological and biomedical molecules, because Raman cross-sections of these molecules attached to Au or Ag nanostructures can be enhanced by a factor of 10^{10}-10^{14} (*1–8*). In fact, SERS has actively been applied to the nondestructive and ultrasensitive characterization of biomolecules, and has thus attracted increasing interest in the field of life sciences, including in DNA-, protein-, cell-, and bacterial studies. The enhancement originates from the near electromagnetic (EM) field or charge transfer (CT) interactions between nanostructures and the attached molecules on them. Thanks to the enhancement, detection time of SERS considerably decreases compared to conventional Raman scattering. SERS for biological and biomedical molecules generally uses near-infrared (NIR) lasers, which can reduce the risk of damaging the molecules even by applying high power. The high specificity of vibrational spectra and the sensitivity to the aqueous environment increase the significance of SERS to study living biological systems. Also, SERS active nanoparticles (NPs) bring advantages of detecting or tracking different known biomolecules over fluorescent tags. The usage of such fluorescent tags suffer from confused overlapping fluorescence spectra broader than SERS spectra and non-uniform photobleaching rates, which brings us several potential complications.

II. Mechanism of Surface-Enhanced Raman Scattering

2.1. Introduction

On Ag or Au NPs, optical responses of adsorbed molecules are enhanced. It is widely known in the field of surface enhanced spectroscopy. One of the most common examples may be SERS for the researchers involved in analytical chemistry fields (*9–12*). The enhancement factors of SERS up to 10^{14} allow us to measure spectra of single molecules (SMs) (*13–17*). Raman spectra have distinct vibrational bands known as "molecular fingerprints", which enable us to attribute molecular species in details (*18–21*). Despite the significant impact of SERS in basic research, fundamental issues such as lack of conclusive evidence for validating the mechanism underlying SERS. We attempted here to resolve the issue by identifying correlations between SERS and its mechanism. There are mainly two mechanisms underlying SERS, which are based on two different mechanisms of enhancement: one is electromagnetic (EM) mechanism and another is chemical one (*10, 11*). EM mechanism is characterized by twofold EM enhancement of Raman process induced by plasmon resonance (*10, 15, 22–29*). Chemical enhancement is characterized by shifting of Raman scattering in non-resonance to that in resonance through the formation of charge transfer complexes between adsorbed molecules and metal surface (*11, 30–34*). Both mechanisms have been experimentally investigated in detail and found to be valid. Thus, in an effort to find out which mechanism is dominant, quantitative evaluation of SERS based on exclusive one mechanism is important. In the section, we focus ourselves on recent experimental investigations of EM mechanism using systems composed of single AgNP aggregates and rhodamine 6G (R6G) dye molecules. Thanks to the protection of functional groups of R6G, chemical interaction between Ag surfaces and π-electrons of R6G is weak, indicating we can expect exclusive contribution on EM mechanism to SERS in the system (*24*). Here, we do not differenciate SERS and SER(resonance)RS, because EM mechanism is common to both SERS and SERRS.

2.2. Evidence of "Twofold" EM Enhancement

EM mechanism is described as "twofold" EM enhancement of Raman process. The first enhancement is increasing in Raman excitation efficiency (a molecule temporarily absorbed a photon and goes into a virtual state) due to coupling between incident light and plasmon resonance, and the second one is increasing in Raman emission efficiency (the molecule emits a photon and goes back to different vibrational state in the ground state) due to coupling between Raman light and plasmon resonance (*22–29*). The twofold EM enhancement factor of SERS theoretically reaches up to 10^8 to 10^{14} at interparticle junctions in Ag and Au NP aggregates (*15, 34*). Imura et al. demonstrated that the positions are limited to crevasses between two NPs (*35*). The first and second enhancement has not been independently treated; therefore, to obtain conclusive evidence of EM mechanism the first and second enhancement should be independently evidenced.

The twofold EM enhancement provides us with a simple expression. It is a product of EM enhancement of the incident and Raman light (*22–24*). Thus, the total enhancement factor M_{total} of SERS is given by

$$M_{total} = \left| \frac{E^{Loc}(\lambda_L)}{E^I(\lambda_L)} \right|^2 \times \left| \frac{E^{Loc}(\lambda_L \pm \lambda_R)}{E^I(\lambda_L \pm \lambda_R)} \right|^2 = \left| M_1(\lambda_L) \right|^2 \times \left| M_2(\lambda_L \pm \lambda_R) \right|^2$$

(1)

where E^I and E^{Loc} are the amplitudes of the incident and local electric fields, respectively; λ_L is the excitation wavelength; $+\lambda_R$ and $-\lambda_R$ are the wavelengths of the anti-Stokes- and Stokes-shifted Raman scattering, respectively; and M_1 and M_2 are the first and the second enhancement factors, respectively. The spectrum of $M_2(\lambda)$ is similar to that of plasmon resonance, because $M_2(\lambda)$ was generated by plasmon involved in the enhancement (*24, 28, 36–43*). Thus, dependence of SERS spectra on plasmon resonance spectra is expected and can be a key for identifying "twofold" EM enhancement. However, plasmon resonance spectra of AgNP aggregates are usually complicated because of overlapping between dipolar and multipolar plasmon modes (*40*). Thus, observation of the dependence has been difficult for us. However, this difficulty can be resolved by selecting AgNP aggregates which dominantly exhibit dipolar plasmon, because we can exclude the complexity induced by mode overlapping. Note that we also check coupling of the dipolar plasmon with SERS. In other words, the dipolar plasmon should satisfy the following two criteria: polarization dependence of a plasmon resonance maximum follows a cosine-squared law, and SERS maxima and plasmon resonance maxima have the same polarization dependence to each other. As a consequence of applying these two criteria, coupling of dipolar plasmon resonance with SERS is identified (*24, 36–43*).

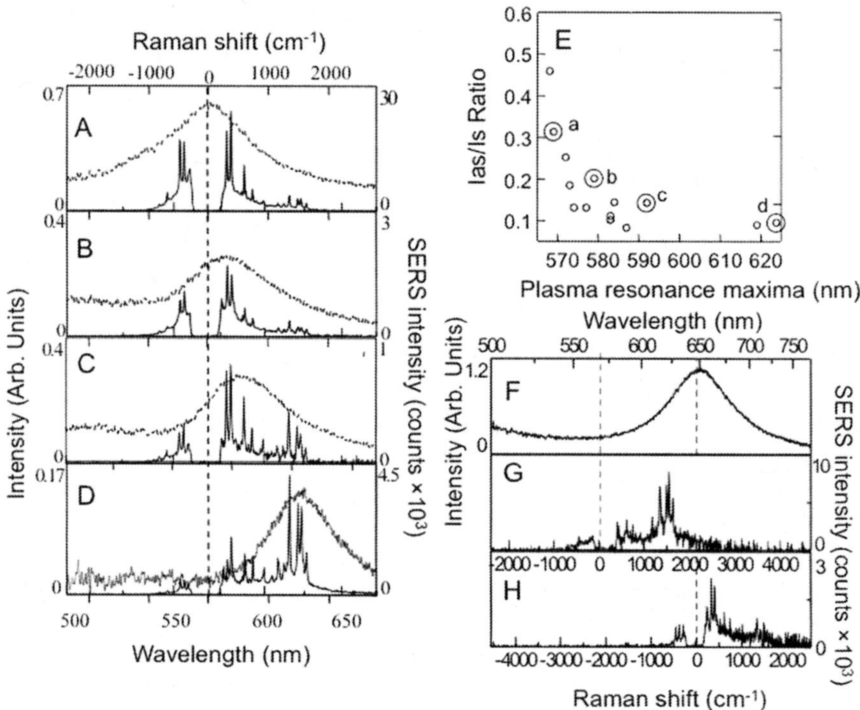

Figure 1. (A–D) Anti-Stokes and Stokes SERS spectra (solid lines) and plasmon resonance spectra (dotted lines) from four representative AgNP aggregates. The vertical dashed line indicates the excitation laser wavelength. Plasmon resonance maxima in (A–D) were observed at 568, 577, 588, and 623 nm. (E) Plasmon resonance maxima dependence of anti-Stokes to Stokes intensity ratios. Doubly-circled points indicated by (a–d) correspond to the data in Figs. 1(A-D). (F) Plasmon resonance spectra, (G) SERS spectra excited at 568 nm, and (H) SERS spectra excited at 647 nm. All spectra in (F)-(H) were obtained from the same AgNP aggregate. (Reproduced with permission from reference (41). Copyright 2007 American Physical Society.)

In the context of twofold EM enhancement, the dependence of SERS spectra on plasmon resonance spectra is observed as selective enhancement of SERS bands close to plasmon resonance maxima. Figure 1A-D show dipolar plasmon resonance and SERS spectra from five AgNP aggregates. SERS intensities close to plasmon resonance maxima look selectively enhanced. To confirm it, Figure 1E summarizes plasmon resonance maxima dependence of the anti-Stokes to Stokes intensity ratios for SERS maxima at 553 nm ($+634$ cm^{-1}) and 580 nm (-634 cm^{-1}). According to Eq. 1 this ratio is given by $M_1(\lambda_L)M_2(\lambda_L-\lambda_R)/M_1(\lambda_L)M_2(\lambda_L+\lambda_R)$. Thus, the first enhancement factor $M_1(\lambda_L)$ cancels. The reduction in the ratio $M_2(\lambda_L-\lambda_R)/M_2(\lambda_L+\lambda_R)$ in Figure 1E can be explained as a decrease in the coupling efficiency of anti-Stokes light and plasmon resonance due to separation of plasmon resonance maxima from anti-Stokes bands. This correlation supports the conjecture that the second enhancement is associated with the origin of variations in anti-Stokes and Stokes SERS spectra. Figure 1F-H show plasmon resonance and excitation wavelength dependence of SERS spectra for single AgNP aggregates for R6G. The intensities of SERS bands around 630 nm in Figure 1G, and 660 nm in Figure 1H are larger than those around 588 nm and 710 nm. It is apparent that the SERS bands close to the plasmon resonance maxima are selectively enhanced and that this effect is excitation wavelength dependent. These observations demonstrate that second enhancement is involved in the enhancement of SERS intensity.

By the above observation of selective enhancement of SERS bands, we established that variations of SERS spectra originate from second enhancement. However, this consideration intrinsically does not exclude the uncertainty due to inhomogeneity in the measurements of multiple NP aggregates. To exclude the uncertainty, we should demonstrate evidence of twofold EM enhancement with "one" NP aggregates. An optical property of plasmon resonance is useful for the demonstration. That is the refractive index dependence of plasmon resonance maximum (42, 44–46). It is well known that plasmon resonance changes its maximum wavelength according to surrounding media. Twofold EM enhancement factors depend on plasmon resonance spectra and plasmon resonance depends on the refractive index of surrounding media; thus, we can predict that SERS spectra are changed by increasing the refractive index of the surrounding medium. This experiment does not need measurements on multiple NP aggregates, because plasmon resonance can be changed by simply replacing the surrounding medium.

To investigate the dependence of SERS spectra on the refractive index of surrounding medium, we measured single AgNP aggregates without and with covering index matching oil layer (refractive index $n = 1.5$, thickness 20 μm). AgNP aggregates in such samples were located at the interface between air ($n = 1.0$) and glass ($n = 1.5$). The effective refractive index of the interface n_{eff} is ~1.3 without oil layer (45). AgNP aggregates were totally covered with oil for which the refractive index is the same as that of glass. Thus, n_{eff} is 1.5 with oil layer. Figure 2A-F shows plasmon resonance and SERS spectra from three representative single AgNP aggregates without and with an oil layer. Here, we observe red-shifts in the plasmon resonance maxima (Figure 2A-F). The red-shifts are induced by an increase in the refractive index (44–46). As shown in Figure 2A-F, large changes are commonly observed for SERS spectra and plasmon resonance maxima. This observation indicates that the spectral changes in SERS are caused by red-shifts in plasmon resonance maxima. An AgNP aggregate whose plasmon resonance maximum located around 700 nm in Figure 2E does not show SERS bands when covered with oil layer. The disappearance may be due to the lack of spectral overlap between SERS and the twofold EM enhancement spectra.

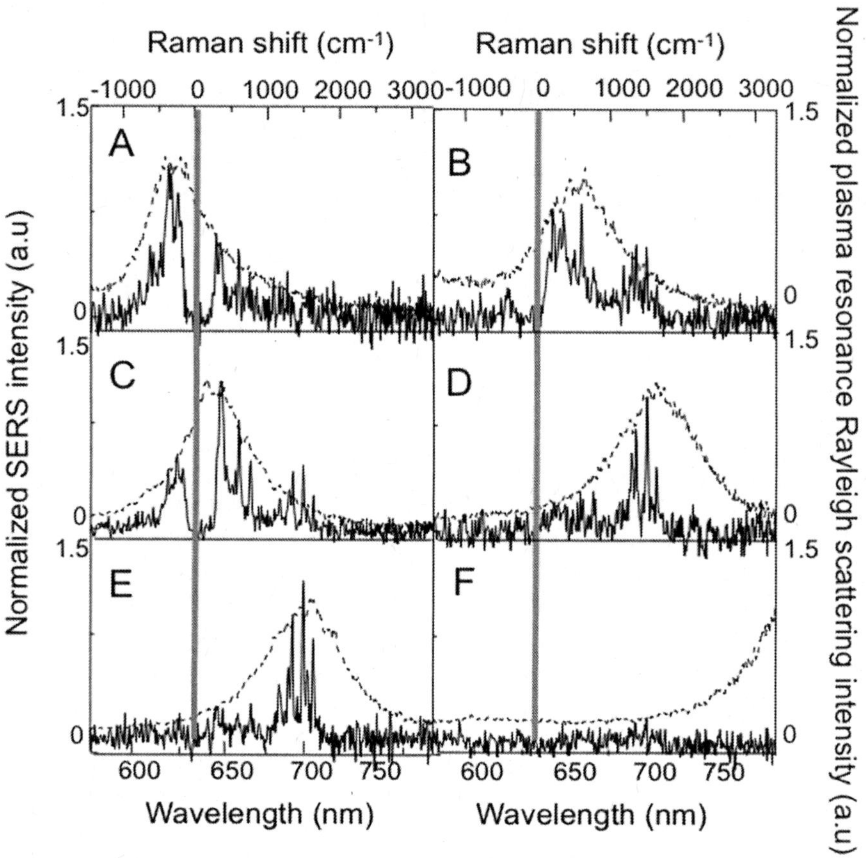

Figure 2. Plasmon resonance and SERS spectra of three representative single AgNP aggregates without (A, C, E) and with (B, D, F) covering oil. (Reproduced with permission from reference (42). Copyright 2009 American Institute of Physics.)

The changes in the SERS spectra induced by surrounding media are analyzed based on the twofold EM enhancement as indicated in Eq. 1. SERS spectra are calculated as the products of spectra of M_{total} and Raman scattering derived from ensemble AgNP aggregates (42). We calculated spectra of plasmon resonance and M_{total} for a dimer of AgNPs using n_{eff} = 1.3 and 1.5 by finite-difference time-domain (FDTD) method. Figures 3A and 3B show that an increase in n_{eff}

causes a red-shift in both spectral maximum of the plasmon resonance and M_{total}. Figures 3C and 3D show the calculated SERS spectra using $n_{eff} = 1.3$ and 1.5, respectively. The calculated changes in SERS spectra are consistent with the experimental ones in Figure 2A-D. Figure 3E shows the plot of the intensity ratios of anti-stokes (619 nm) to stokes SERS (707 nm) bands against plasmon resonance maxima λ_p. The inset of Figure 3E shows the calculated ratios against λ_p based on twofold EM enhancement. The calculated ratios using $n_{eff} = 1.3$ and 1.5 are well consistent with experimental ones. The consistency verifies EM mechanism of SERS excluding the uncertainty due to inhomogeneity in the measurements of multiple NP aggregates.

2.3. Quantitative Evaluation of Twofold EM Enhancement

Evaluation of EM mechanism in the previous section verified that twofold EM enhancement plays an important role in SERS. However, twofold EM enhancement factors are determined by plasmon resonance, and plasmon resonance are determined by the morphology of AgNP aggregates. Thus, on the basis of measurements of the morphology of AgNP aggregates, rigorous verification of EM mechanism is needed. In this section, plasmon resonance and SERS spectra from single isolated AgNP aggregates are combined with SEM images to accomplish one-to-one correspondence among optical properties of plasmon resonance, that of SERS, and the morphology of AgNP aggregates for the further evaluation of EM mechanism. The experimental observations are compared with FDTD calculations based on individual morphology of the AgNP aggregates. In this section, we use conventional Raman scattering from ensemble AgNP aggregates for the evaluation.

Here, we add the position coordinate of molecules r in $M_1(\lambda_L)M_2(\lambda)$ to Eq. 1. Thus, $M_1(\lambda_L,r)M_2(\lambda,r)$ is changed into

$$M_1(\lambda_L, r)M_2(\lambda, r) = \left|\frac{E^{loc}(\lambda_L, r)}{E^I(\lambda_L)}\right|^2 \times \left|\frac{E^{loc}(\lambda, r)}{E^I(\lambda)}\right|^2, \quad (2)$$

where r is an arbitrary position of molecules. To obtain twofold EM enhancement factor, $M_1(\lambda_L,r)M_2(\lambda,r)$ in Eq. 2 is summed over the location of dye molecules on the AgNP aggregate, and is thus rewritten as

$$\sum_{i=1}^{N} M_1^i(\lambda_L, r_i)M_2^i(\lambda, r_i). \quad (3)$$

M_1^i and M_2^i are enhancement factors of excitation and emission light at the center of gravity r_i of a molecule i, and N is the total number of molecules. Accordingly, the total cross-section spectra $\sigma_{total}(\lambda_L,\lambda)$ is given by

$$\sigma_{\text{total}}(\lambda_L, \lambda) = \sum_{i=1}^{N} [\sigma_{\text{RS}}(\lambda_L, \lambda) + q\sigma_{\text{FL}}(\lambda_L, \lambda)] M_1^i(\lambda_L, r_i) M_2^i(\lambda, r_i)$$

where $\sigma_{\text{RS}}(\lambda_L,\lambda)$ is conventional Raman cross-section, q is inverse of decay enhancement factor (43), and $\sigma_{\text{FL}}(\lambda_L,\lambda)$ is conventional fluorescence cross-section. For an AgNP dimer, the largest EM field is generated only at the crevice of AgNP dimer (14, 16, 20, 27), and this largest EM field at the crevice dominates $M_1(\lambda_L,r)M_2(\lambda,r)$. Thus, $\sum_{i=1}^{N} M_1^i(\lambda_L, r_i) M_2^i(\lambda, r_i)$ is approximately written as $M_1(\lambda_L,r_o)M_2(\lambda,r_o)$ for a single-molecule in the crevice even if some additional molecules exist at positions other than the crevice. Here, r_o denotes the position at the crevice.

We have identified that most of AgNP aggregates showing both dipolar plasmon resonance and SERS activity are dimers (24). Theoretical work has predicted that structural anisotropy of Ag dimers induces optical anisotropy in both plasmon resonance and SERS (23). To verify this prediction, we measured anisotropy of both plasmon resonance and SERS spectra, and then compared the measurements with FDTD calculations. Figure 4A-C shows a SEM image, polarization dependence of plasmon resonance, and that of SERS from one AgNP dimer. Figure 4F shows that calculated polarization dependence of plasmon resonance of the Ag dimer in Figure 4E well-reproduces the experimental polarization dependence in Figure 4B. Figures 4F and 4G show that structural anisotropy of the AgNP dimer induces splitting of plasmon resonance into the longitudinal mode at 620 nm and transverse mode < 400 nm, and coupling of the longitudinal plasmon resonance with light polarized along the long axis of the AgNP dimer generated the largest $M_1(\lambda_L,r)$ at the dimer crevice. The agreement between experimental and calculated polarization dependence suggests that the calculated spatial distribution of $M_1(\lambda_L,r)$ in Figure 4G reflects local EM fields around an AgNP dimer. Thus, SERS in Figure 4C is generated by EM field coupled with the longitudinal plasmon at crevices. Indeed, Figures 4D and 4H show that experimental anisotropy of plasmon resonance and $M_1(\lambda_L,r)$ well agreed with the calculated anisotropy. Note that r was selected at an arbitrary position in the crevice in Figure 4I. The agreement is consistent with the prediction that anisotropy of AgNP dimers causes anisotropy of SERS.

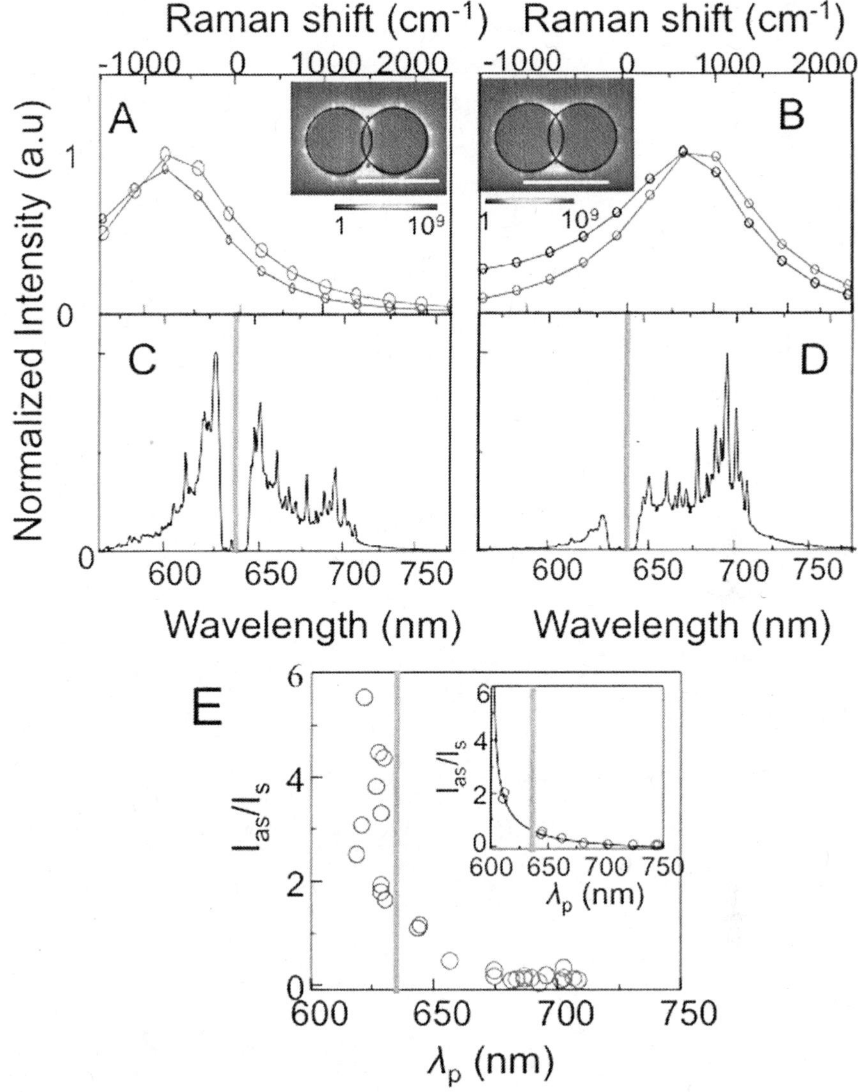

Figure 3. Calculated spectra of plasmon resonance (red lines) and SERS enhancement factors (blue lines) at (A) $n_{eff} = 1.3$ and (B) $n_{eff} = 1.5$. Insets: spatial distribution of SERS enhancement factors at plasmon resonance maxima. Scale bars are 50 nm. Calculations of SERS spectra at (C) $n_{eff} = 1.3$ and (D) $n_{eff} = 1.5$ following eq. 1. (E) Experiments of λ_p dependence of intensity ratios of anti-Stokes to Stokes SERS bands at 619 nm and 707 nm. Insets: calculations following Eq. 1. Blue and red circles show the ratios of anti-Stokes to Stokes SERS bands at 619 nm and 707 nm without and with covering oil, respectively. (Reproduced with permission from reference (42). Copyright 2009 American Institute of Physics.)

(A)
(E)
y
z
x
(I)
(B)
(F)
300
250
200
150
100
50
0
180
150
120
90
60
30
0
σp (×10-12 cm2)
500 600 700
Wavelength (nm)
(C)
(G)
35
28
21
14
7
0
σtotal (×10-15 cm2)
200
M1(λL, r)
1

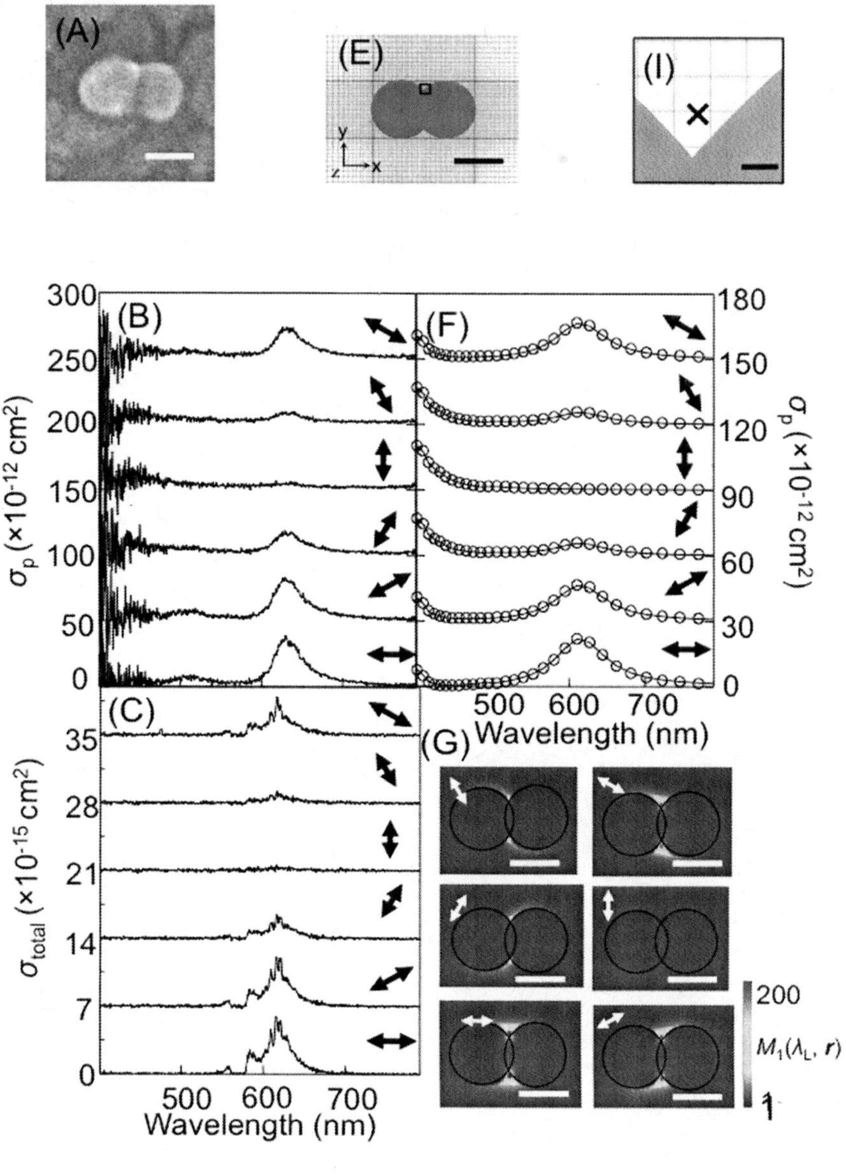

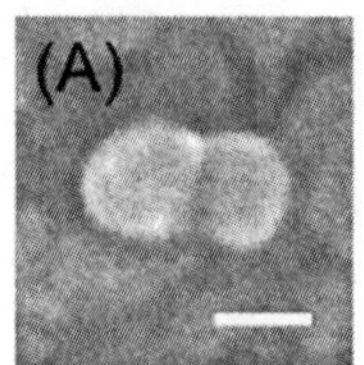

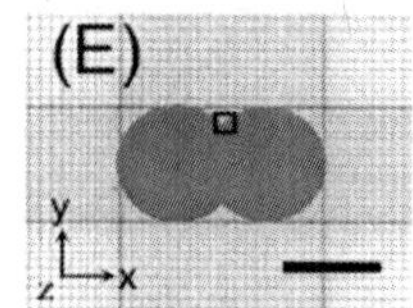

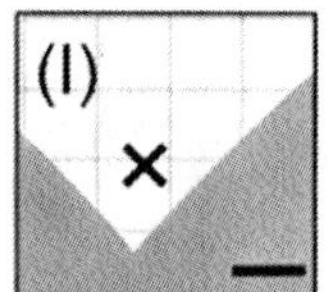

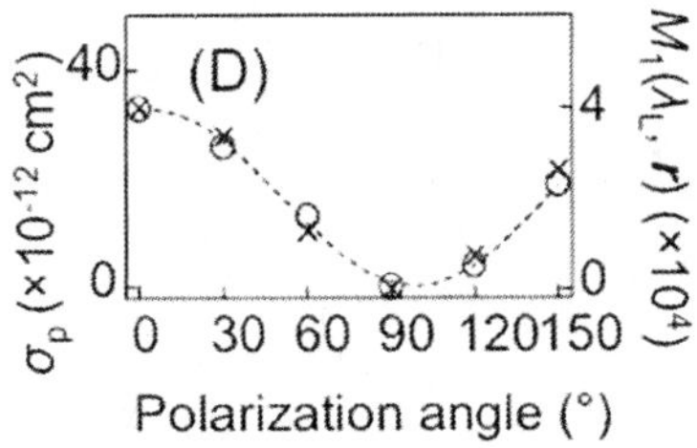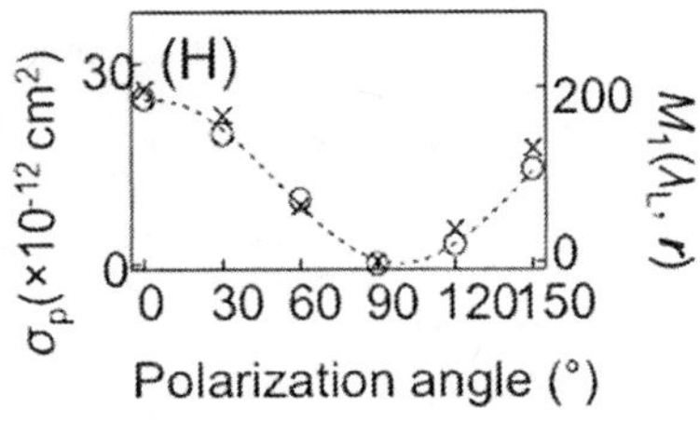

Figure 4. (A) SEM image, (B) polarization dependence of experimental plasmon resonance spectra, and (C) polarization dependence of experimental SERS spectra of a AgNP dimer. (D) Polarization angle dependence of cross-sections of experimental plasmon resonance (open circles with a dashed curve) and experimental EM enhancement factors (crosses with a dashed curve). (E) Modeled structure of an AgNP dimer for the FDTD calculation, (F) polarization dependence of a calculated plasmon resonance spectrum, and (G) polarization dependence of calculated spatial distribution of first EM enhancement factors. (H) Polarization angle dependence of cross-sections of calculated plasmon resonance (open circles with a dashed curve) and calculated EM factors (crosses with a dashed curve) as a function of polarization angles. (I) Magnified view in the vicinity of the crevice of the AgNP dimer in (E). Scale bars in (A, E, G, and I) are 50 nm, 50 nm, 50 nm and 1.5 nm, respectively. Mesh size 1.5 nm, refractive index 1.3 of the surrounding medium, and circularly-polarized incident light (532 nm) were selected in the FDTD calculation. (Reproduced with permission from reference (24). Copyright 2010 American Physical Society.)

To obtain spectral shapes of $M_1(\lambda_L,r)M_2(\lambda,r)$, not only $M_1(\lambda_L,r)$ but also $M_2(\lambda,r)$ should be evaluated. Figure 5A-C shows a SEM image of an Ag dimer, an experimental plasmon resonance spectrum, and an experimental SERS spectrum, respectively. Figure 5D-F shows the shape of an Ag dimer in calculation, a calculated plasmon resonance spectrum, and a spectrum of $M_1(\lambda_L,r_0)M_2(\lambda,r_0)$, respectively. The cross in Figure 5H indicates r_0 selected for the calculation for Figure 5F. Note that we selected r_0 at an arbitrary position in the crevice in Figure 5H simply because we do not consider rigorously the absolute value of $M_1(\lambda_L,r_0)M_2(\lambda,r_0)$ in the calculations in Figure 5E-5G. Figures 5B and 5E show that the experimental plasmon resonance spectrum well-agreed with the calculated one. The agreement indicates that the morphology in Figure 5E is valid for the FDTD calculation condition. Figures 5E and 5F show that the spectral shape of plasmon resonance was quite similar to that of $M_1(\lambda_L,r_0)M_2(\lambda,r_0)$. The similarity is reasonable because EM enhancement is induced by the plasmon resonance (*24*). Figure 5C shows that the SERS spectrum around 610 nm was selectively enhanced by the plasmon resonance in Figure 5B. We found that the selectively enhanced SERS spectra are common to all our experimental results (*24, 41–43*).

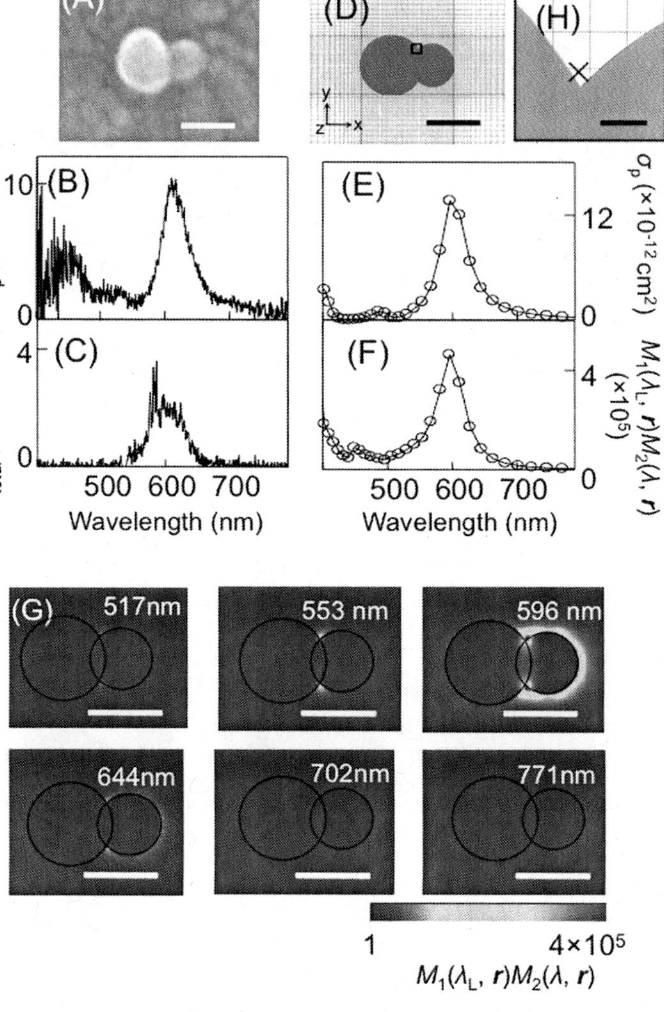

Figure 5. (A) SEM image, (B) experimental plasmon resonance spectrum, and (C) experimental SERS spectrum, (D) modeled structure in the FDTD calculation, (E) calculated plasmon resonance spectrum, and (F) calculated spectrum of EM enhancement factor of an Ag dimer. (G) Calculated spectrum of spatial distribution of twofold EM enhancement factors. (H) Magnified view in the vicinity of the crevice of the Ag dimer in (D). The SEM image, the plasmon resonance spectrum, and the SERS spectrum in (A)-(C) were observed from the identical AgNP. Spectrum of twofold EM enhancement factors in (F) was calculated at the position of cross in (H). Scale bars in (A, D, G, and H) are 50 nm, 50 nm and 1.5 nm, respectively. Mesh size 1.5 nm, refractive index 1.3 of the surrounding medium, and circularly-polarized incident light (532 nm) were selected in the FDTD calculation. (Reproduced with permission from reference (24). Copyright 2010 American Physical Society.)

To quantitatively evaluate twofold EM enhancement factors using FDTD calculation with Yee cells, in which every components of an electric and a magnetic field are set at the center of a side and of a plane in a cubic cell, we examined the accuracy in our calculations by comparing FDTD calculations with self-consistent ones under the identical condition. Note that self-consistent calculation offers an exact solution to symmetrical AgNP dimers (*23*). We further address an issue about the use of Yee cells in FDTD calculation (*47*). The intensity of EM field is maximized at a surface and exponentially decays outside with increasing distance from the surface. However, for a surface having a curvature comparable to mesh size, the calculated exponential decay and the peak intensity of EM field will be smoothed and smaller than the real intensity. This issue leads to underestimation of EM field intensity at the surface, and is not negligible for the surface having small curvature (*24*). Fortunately, a factor of underestimation can be reduced by the use of smaller meshes. We selected the morphology of an Ag dimer equivalent to that in Figure 2A of Ref. (*23*), and evaluated the factor of underestimation with mesh size of 0.98, 0.5, and 0.2 nm. Figure 6A shows the morphology of the Ag dimer. The x-axis is set to overlap with the long axis of the dimer. An EM field reaches the theoretical maximum at the local face-to-face surfaces along the x-axis, as shown in the square in Figure 6A. Note that we used electric flux density instead of an EM field to avoid discontinuity of the field across the surface. Figure 6B shows the magnified view of the square in Figure 6A. Figure 6C shows a spectrum of $M_1(\lambda,r)^2$ calculated by the use of mesh size 0.98 nm. The spectral maximum locates at ~ 418 nm, which agrees with Figure 2A in Ref. (*23*). This agreement provides evidence that the plasmon resonance for the FDTD calculation of $M_1(\lambda, r)^2$ is correct. Here, we explain the criterion for the accuracy in the calculation of $M_1(\lambda,r)^2$. The distance d outside from the surface, where the electric flux density is reduced by a factor of $|1/e|^2$, is defined as

$$d = \lambda/(2\pi\sqrt{-\varepsilon(\lambda)})$$

. Here, ε is a dielectric constant of Ag (or a vacuum) (*48, 49*). Along the long axis of a dimer, the ratio of the decay length of bulk Ag d_{Ag} to that of vacuum d_v remains constant ($d_v/d_{Ag} \sim 2.36$) (*48, 49*). Thus, we can evaluate the accuracy in FDTD calculation from the deviation of the ratio from ~ 2.36. Figure 6D shows the decay curves of the electric flux density at 418 nm with several mesh sizes. The ratios evaluated are 0.18, 0.4, and 2.0 with mesh size 0.98, 0.5, and 0.2 nm, respectively. The value of 2.0 is satisfactory considering that FDTD calculation using 0.2 nm mesh requires a computational time longer than 20 days. The calculated electric flux density at the center of the gap is well consistent with self-consistent calculation of Figure 2B in Ref. (*23*) as shown by the cross in Figure 6D.

Finally, we calculate SERS cross-section spectra with the mesh size of 0.2 nm to quantitatively evaluate twofold enhancement factors. In Figure 7, we show a set of SERS cross-section spectra acquired from the same dimer with three different incident wavelengths (Figure 7C-E). SERS cross-sections were calculated by multiplying $M_1(\lambda_L, r_o)M_2(\lambda, r_o)$ with $\sigma_{RS}(\lambda_L,\lambda) + q\sigma_{FL}(\lambda_L,\lambda)$ (Figure 7H-J), where r_o (the cross in Figure 7K) is the location of the center of a R6G molecule. The center locates at ~ 0.8 nm from the Ag surface (*50–52*). This distance is sufficiently large for blocking charge transfer interaction between R6G molecules and Ag surfaces. The shape of the SERS spectra is correctly reproduced in the calculations, and moreover, the experimental cross-sections are consistent with the calculations within a factor of ~2, which is acceptable considering the huge enhancement factor of 10^9 involved in the SERS process. This quantitative consistency of the experimental SERS spectra with the calculations provided us with convincing evidence for EM mechanism.

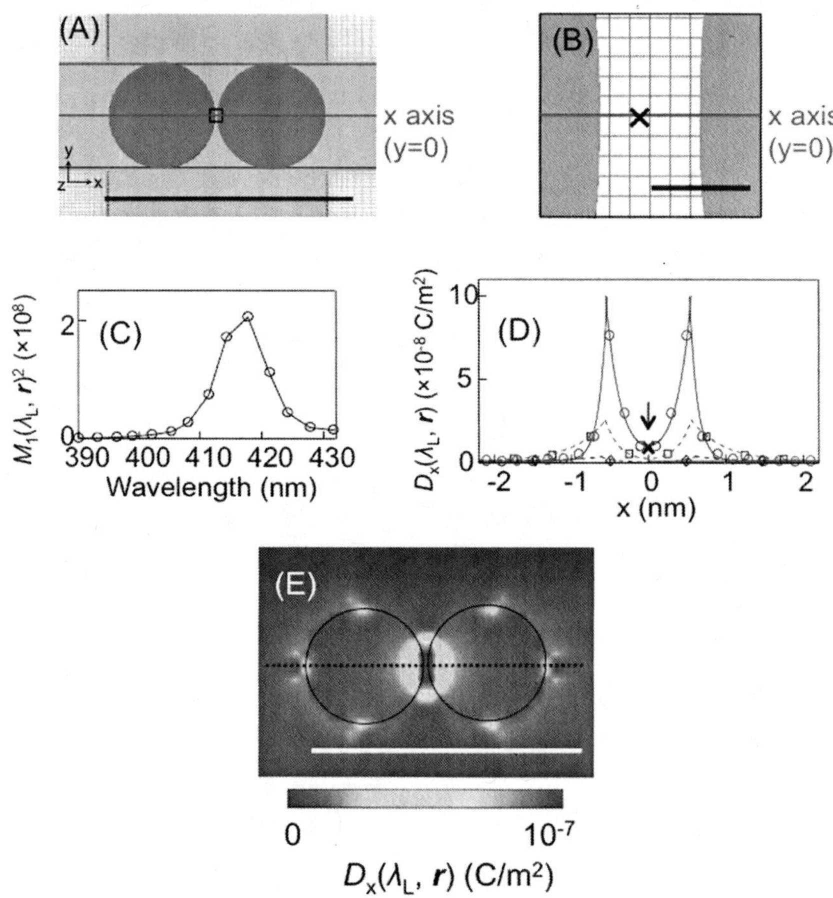

Figure 6. (A) Modeled structure of an Ag dimer in the FDTD calculation. (B) Magnified view of the vicinity of the gap in the modeled structure. (C) Calculation spectrum of the twofold EM enhancement factor at the cross in (B). (D) Calculated cross-section of spatial distribution of the electric flux density (black broken line, red broken line and blue solid line; mesh size is 1.5 nm, 0.5 nm and 0.2 nm respectively). (E) Spatial distribution of the electric flux density D_x. The cross-pointed by the arrow in (D) indicates the calculated enhancement factor in Figure 2(A) of Ref. (17). Note that the cross-section in (D) indicates the electric flux density along the dashed line in (E). Scale bars in (A, B, and E) is 50 nm, 1.0 nm, and 50 nm, respectively. (Reproduced with permission from reference (24). Copyright 2010 American Physical Society.)

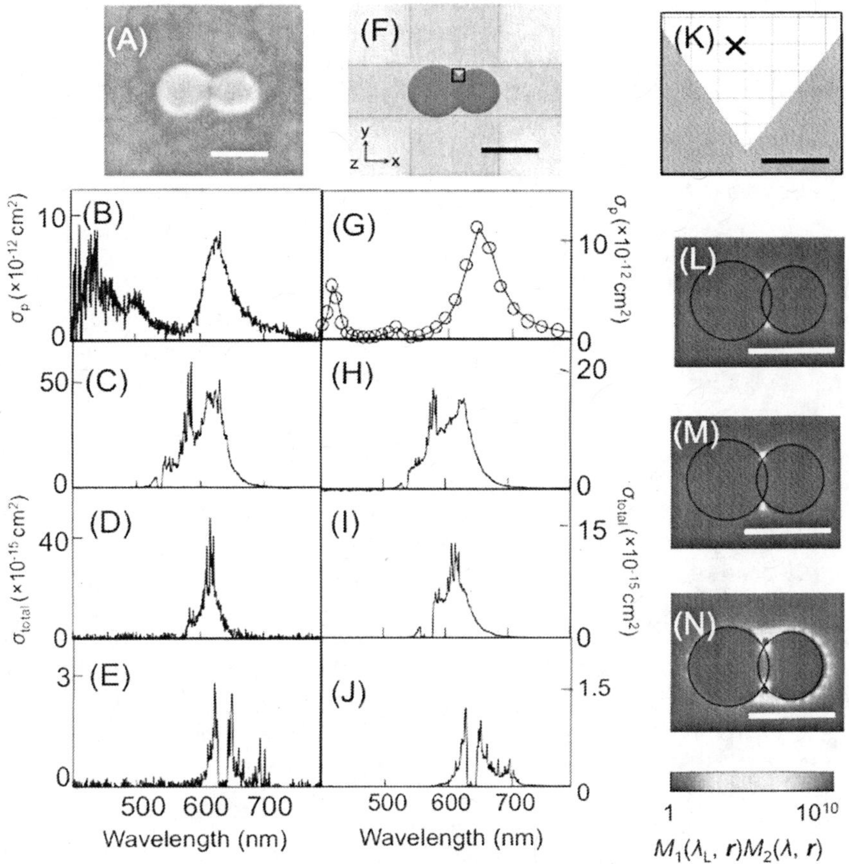

Figure 7. (A) SEM image of an Ag dimer, (B) experimental plasmon resonance spectrum, and experimental SERS spectra excited at (C) 532, (D) 561, and (E) 633 nm. (F) Modeled structure of an Ag dimer in the FDTD calculation, (G) calculated plasmon resonance spectrum and calculated SERS spectra excited at (H) 532, (I) 561, and (J) 633 nm. (K) Magnified view in the vicinity of the crevice of the model structure in (F). Spatial distribution of calculated twofold EM enhancement factor excited at (L) 532, (M) 561, and (J) 633 nm. Experimental plasmon resonance maximum in (B) was ~ 620 nm, its FWHM was ~60 nm, and its cross-section was ~1 × 10^{-11} cm². Circularly-polarized incident light (532, 561, and 633 nm), the mesh size 0.2 nm, and the refractive index 1.3 of the surrounding medium were selected in the FDTD calculation. Scale bars in (A, F, L, and M) are 50 nm. Scale bar in (K) is 0.5 nm. (Reproduced with permission from reference (24). Copyright 2010 American Physical Society.)

III. Biological Application

3.1. SERS Analysis of Small Biologically Important Molecules

3.1.1. Introduction

Iodide (I⁻) and thiocyanate (SCN⁻) ions play an important role in the human body and health science. In the thyroid gland, thyroid hormones, namely triiodothyronine (T3) and thyroxine (T4), which control many metabolic activities in human body are produced from I^- (*53*). The deficiency or excess of iodide is a cause of many diseases such as goiter, hypothyroidism, and hyperthyroidism (Grave's disease) (*53, 54*). In human body fluids (*e.g.*, serum, saliva, and urine), a small amount of SCN^-, which is produced by the digestion of some vegetables or thiocyanate-containing foods (*e.g.*, milk and cheese), can be found (*55, 56*). However, a higher level of SCN^- is usually found when tobacco smoke is inhaled. The presence of SCN^- in body fluids, especially saliva, is a crucial evidence to indicate cyanide exposure and is determined as a biomarker to discriminate nonsmokers and smokers (*55–57*). Here, we show that starch-reduced AuNPs have successfully been used to sensitively and selectively determine both I^- and SCN^- concentrations by SERS technique.

For some SERS-inactive biomolecules, e.g., phenolic estrogens, changing them to be SERS-active is a new idea we proposed for their detection. It is an indirect method, differing from the Raman-dye labeled strategy because the target molecules are changed to azo dyes and the surface-enhanced resonance Raman scattering (SERRS) spectra correspond to the target molecules.

3.1.2. Starch-Reduced Au Nanoparticles and Their Application

3.1.2.1. Starch-Reduced Au Nanoparticles

Starch is a carbohydrate consisting of glucose monomers connected via glycosidic bonds (*58*). Their molecular structures are shown in Figure 8. Starch is not commonly employed as a strong reducing agent to generate electrons for metal-ion reduction because of a small number of reducing ends in starch molecules. However, starch can be employed as a reducing agent to produce spherical AuNPs via alkaline degradation, as reported in our previous work (*59*).

The potential of starch to act as a reducing agent for AuNP synthesis is illustrated in Figure 9. The reducing efficiency of starch can be increased by raising the pH of starch solution. At conditions of high acidity, a strong absorption band at 315 nm, which attributes to the charge transfer from Cl^- to Au^{3+} center of $AuCl_4^-$ complex (*60*), is observed. This implies that no Au^{3+} reduction occurs.

The systems of higher pH display systematic changes of optical appearances and broad bands over 500 – 800 nm in localized surface plasmon resonance (LSPR) extinction spectra, which indicate an occurrence of large AuNPs (*61*). Under alkaline conditions, starch is degraded to shorter chains (*62–66*) while the final alkaline degradation products of starch are well-known to be carboxylic acids (*e.g.*, formic acid, acetic acid, lactic acid, glycolic acid), which do not have the reducing property (*62*). Based on the "Nef-Isbell mechanism" (*63, 65*), alkali degrades starch molecule via β-elimination and concomitantly generates the intermediates with aldehyde or α-hydroxy ketone moieties, which should be reducing species in this synthesis, as illustrated in Figure 10. The shortened starch chains can be continuously β-eliminated, and produce more reducing species. Thus, if there are Au^{3+} ions available in the solution, those intermediates will reduce Au^{3+} to AuNPs. Furthermore, the steric power of starch chain to stabilize AuNPs is reduced due to the degradation. As a result, the system with extensive degradation (1.0 M NaOH) could not stabilized AuNPs, which causes aggregation and precipitation, as shown in Figure 9.

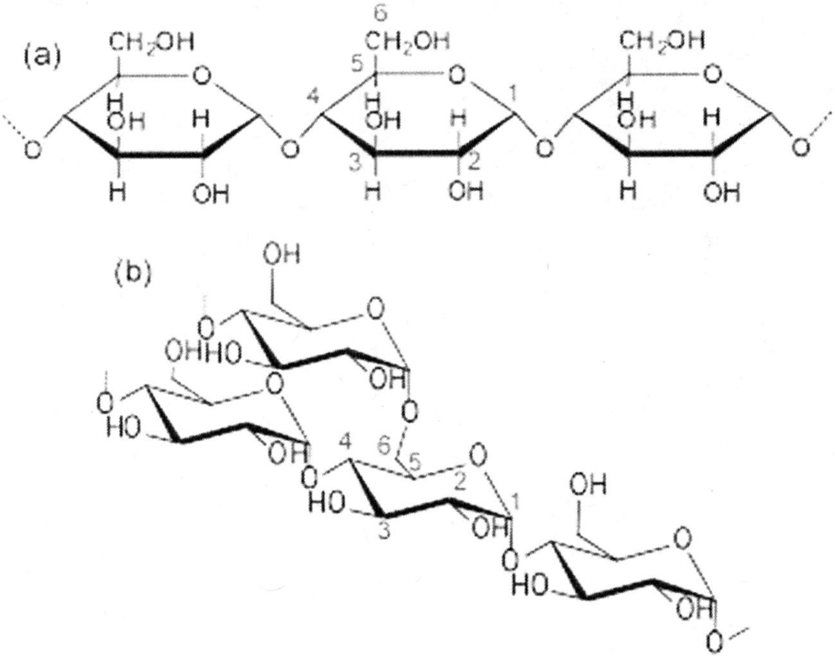

Figure 8. Structures of (a) amylose and (b) amylopectin.

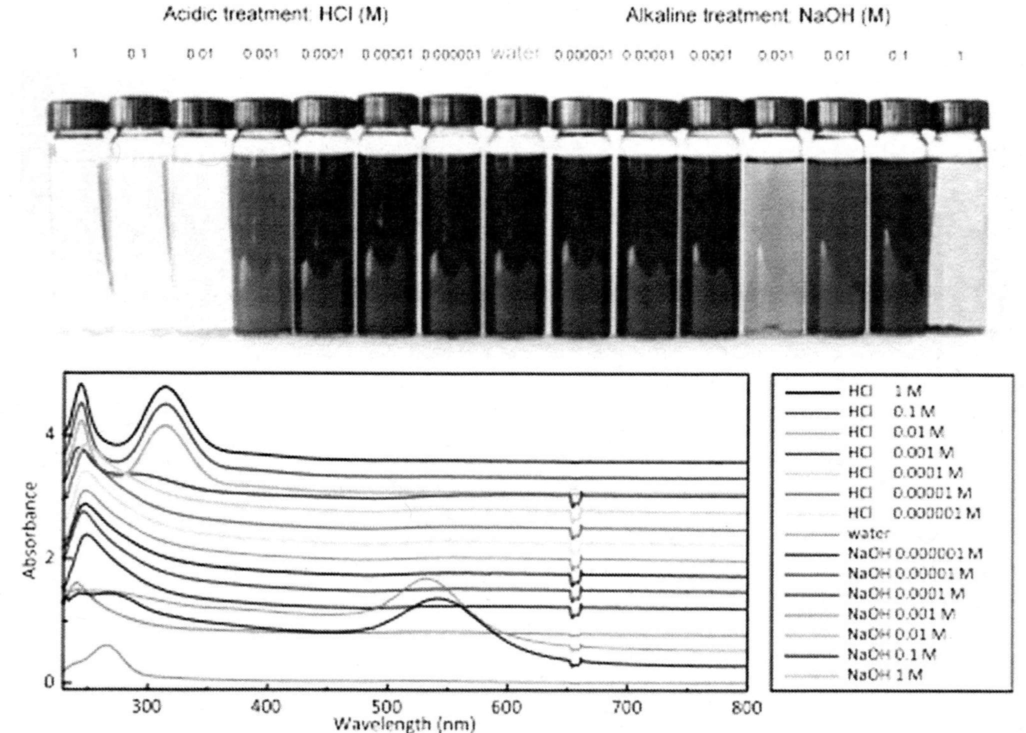

Figure 9. Photograph and corresponding UV – visible spectra of Au colloids after 5-day reaction. The 2 % starch and 400 ppm Au^{3+} solutions (pH 7) were separately incubated with HCl (0.000001 – 1 M), NaOH (0.000001 – 1 M), or water before mixing at room temperature. (Reproduced with permission from reference (59). Copyright 2012 Springer.)

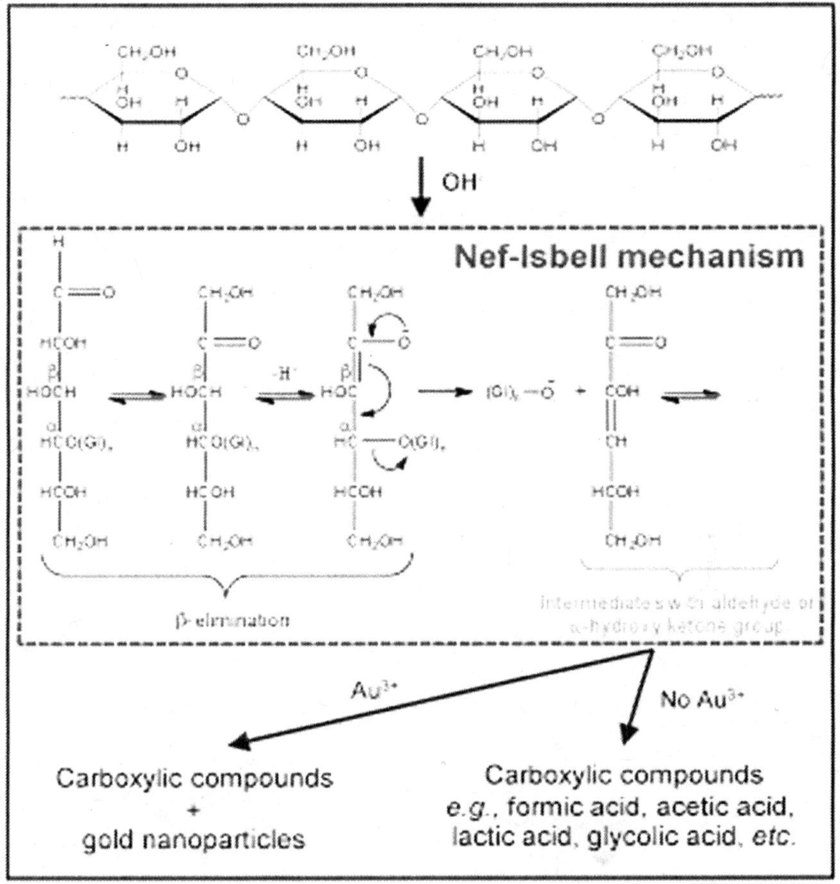

Figure 10. A proposed mechanism for the formation of Au NPs via alkaline degradation of starch. (Reproduced with permission from reference (59). Copyright 2012 Springer.)

Additionally, the size of spherical AuNPs can be selectively controlled from nanometer-sized up to submicrometer-sized regimes by adjusting pH of the Au^{3+} ion solution. The aqueous Au^{3+} ion exhibits speciation characteristics of various forms ($AuCl_4^-$, $AuCl_3(OH)^-$, $AuCl_2(OH)_2^-$, $AuCl(OH)_3^-$, and $Au(OH)_4^-$) depending on pH. Each complex possesses different reduction potentials. Tetrachloroaurate complex, $AuCl_4^-$, has the highest reduction potential while tetrahydroxyaurate complex, $Au(OH)_4^-$, has the lowest (*67, 68*). At higher pH of gold ion solution, it is more difficult to reduce gold ion, and then larger AuNPs are produced, as shown in Figure 11. However, large Au particles are also produced at a very low NaOH concentration (0.01 M NaOH) because the species with aldehyde or α-hydroxy ketone moieties have no good reducing efficiency under poor alkaline conditions.

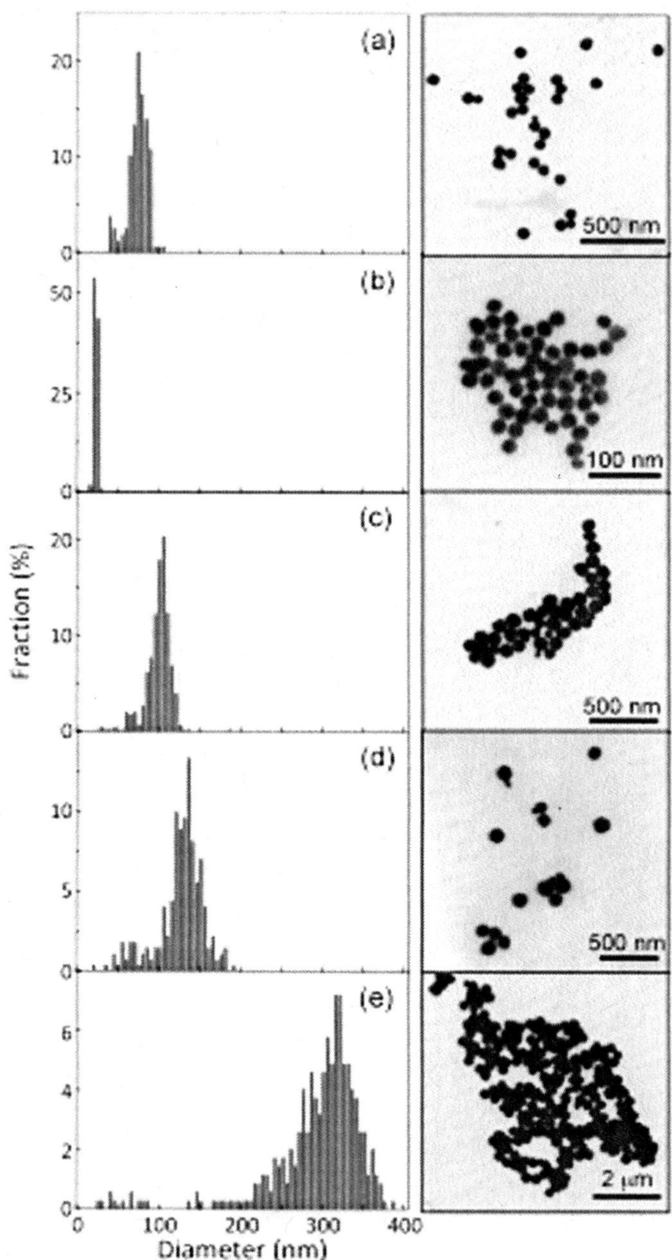

Figure 11. TEM images and the corresponding particle size distributions of AuNPs synthesized at 80 °C. A 2% starch solution with 0.02 M NaOH was employed as a reducing agent. The concentrations of NaOH in Au³⁺ solution were (a) 0.01, (b) 0.02, (c) 0.04, (d) 0.06, and (e) 0.2 M. (Reproduced with permission from reference (59). Copyright 2012 Springer.)

This method offers several advantages such as simplicity, environmentally friendly raw materials and products, possibility for a large-scale production, non-complicated route, and inexpensive instruments. Moreover, due to the biocompatibility of starch, starch-reduced AuNPs could also be applied in applications that require biocompatibility.

3.1.2.2. Determination of Iodide and Thiocyanate Concentrations Using SERS of Starch-Reduced Au Nanoparticles

Fortunately, starch-reduced AuNPs produced via the alkaline degradation process of starch show a sharp peak at 2125 cm^{-1} due to the $-C\equiv C-$ stretching mode of by-products from the alkaline degradation of starch (*69*), which stabilize AuNPs. An explanation of the origin of the $-C\equiv C-$ functional group may lie in the process of alkaline degradation of starch–"the Nef-Isbell mechanism". After β-elimination, the generated 6-carbon species consists of enol ($-CH=COH-$) groups (see Figure 10), which is the most possible position to be dehydrated and form a $-C\equiv C-$ functional group. This peak is very useful because its position is far from the fingerprint region and the positions of other Raman bands. Naturally, I$^-$ strongly adsorbs on a Au surface, and the SERS intensity at 2125 cm^{-1} decreases upon I$^-$ adsorption because of the replacement of $-C\equiv C-$ species by I$^-$ on the Au surface. Also, SCN- adsorbs very well on a Au surface, resulting in the appearance of a strong peak at around 2100 cm^{-1} due to the $-C\equiv N$ stretching vibration (*69*). Therefore, these two peaks can be used as a probe for I$^-$/SCN$^-$ detections (*70*), as shown in Figure 12.

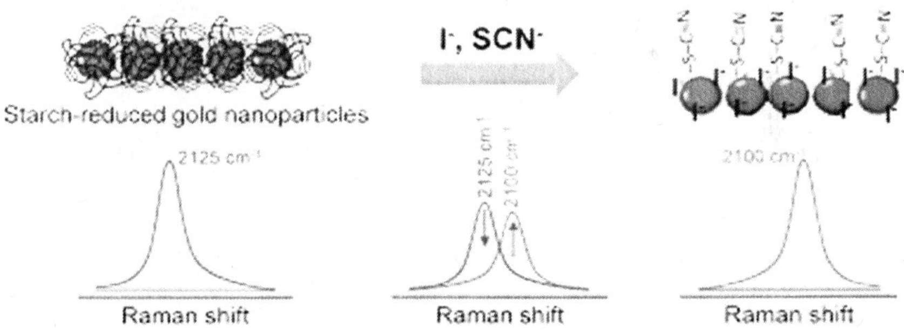

Figure 12. A scheme of the proposed method for I⁻ and SCN⁻ detection. (Reproduced with permission from reference (70). Copyright 2011 American Chemical Society.)

The detection limit of I$^-$ anion for this technique was 0.01 μM, and the variation of the intensity of 2125 cm^{-1} band was nearly linear as a function of the I$^-$ concentration in the range of 0.2 – 2.0 μM, as shown in Figure 13. Moreover, at I$^-$ concentration lower than 0.2 μM, the decrease in intensity can be explained by Langmuir adsorption isotherm. Since the peak height at 2125 cm^{-1} is exploited to estimate the I$^-$ content, this peak conversely limits the range of measurement. The total range of I$^-$ determination is 0.01 – 2.0 μM. For the measurements of SCN$^-$, the peak around 2100 cm^{-1} was fitted by a Voigt function to reveal two composite peaks at high and low wavenumbers attributed to the –C≡C– and –C≡N stretching modes, respectively, as shown in Figure 14a. The composite peak at the higher wavenumber shows a maximum at 2125 cm^{-1}, and the peak intensity (I_{high}) decreases with an increase in KSCN concentration. For the composite peak at the lower wavenumber, the intensity increases as a function of KSCN concentration, and the peak shifts to higher wavenumbers, as shown in Figure 14b. The peak finally shifts to 2100 cm^{-1}, which is characteristic of the –C≡N stretching mode of thiocyanate adsorbed on a Au surface (*71*), because of changes in the molecular structure from S=C=N$^-$ to –S–C≡N. The intensity of the composite peak at low wavenumber (I_{low}) does not show linearity, and the Langmuir equation cannot be used to describe because the increase in intensity of this peak is from a combination of adsorption and changes in molecular structure. However, the concentration range of the SCN$^-$ measurement is 0.05 – 50 μM, which is wider than that of the I$^-$ measurement.

Any anions cannot interfere with the measurement of SCN$^-$ because of the direct measurement via the peak of –C≡N stretching mode from SCN$^-$, which is very strong and far from fingerprint region. In order to test interferences of I$^-$ measurement, SERS spectra of drop-dried Au films soaked in 1 mM solutions containing various kinds of anions were measured. There is no significant change in the SERS intensity at 2125 cm^{-1} (I_{2125}/I_{520}), as shown in Figure 15, because the interactions between other anions and Au are not strong enough to overcome the adsorption of –C≡C– species. Anions with a sulfur atom, such as S^{2-}, may obstruct I$^-$ determination because of a higher adsorptivity of sulfur on Au surface. However, the interference will occur at a very high concentration of interferences. The peak at 2125 cm^{-1} will completely disappear, and I$^-$ concentration cannot be determined when the concentration of S^{2-} is higher than 10 μM.

The capability of the proposed SERS-based I$^-$/SCN$^-$ determination in a real complex matrix was carried out by using human serum as a media. The SERS spectrum of the Au film soaked in the human serum does not show an obvious change compared to that soaked in water because the biomolecules and other interferences in the human serum cannot overwhelm the steric hindrance of starch by-products that stabilize the Au surface, and thus they cannot strongly adsorb on the Au surface. Even the peak intensity is deviated from those in the water system, the trend of intensity change still looks similar and can be used to determine the I$^-$ and SCN$^-$ concentrations.

Using this method and starch-reduced AuNPs, the I$^-$/SCN$^-$ measurements can be performed directly without separating the anions by a chromatographic process; only a Raman microspectrometer is needed. It is very easy to fabricate the SERS substrate for the measurement. Starch-reduced AuNPs can be utilized immediately after synthesis without any surface modification, treatment, or purification. Thus, this method is easy, quick, and inexpensive.

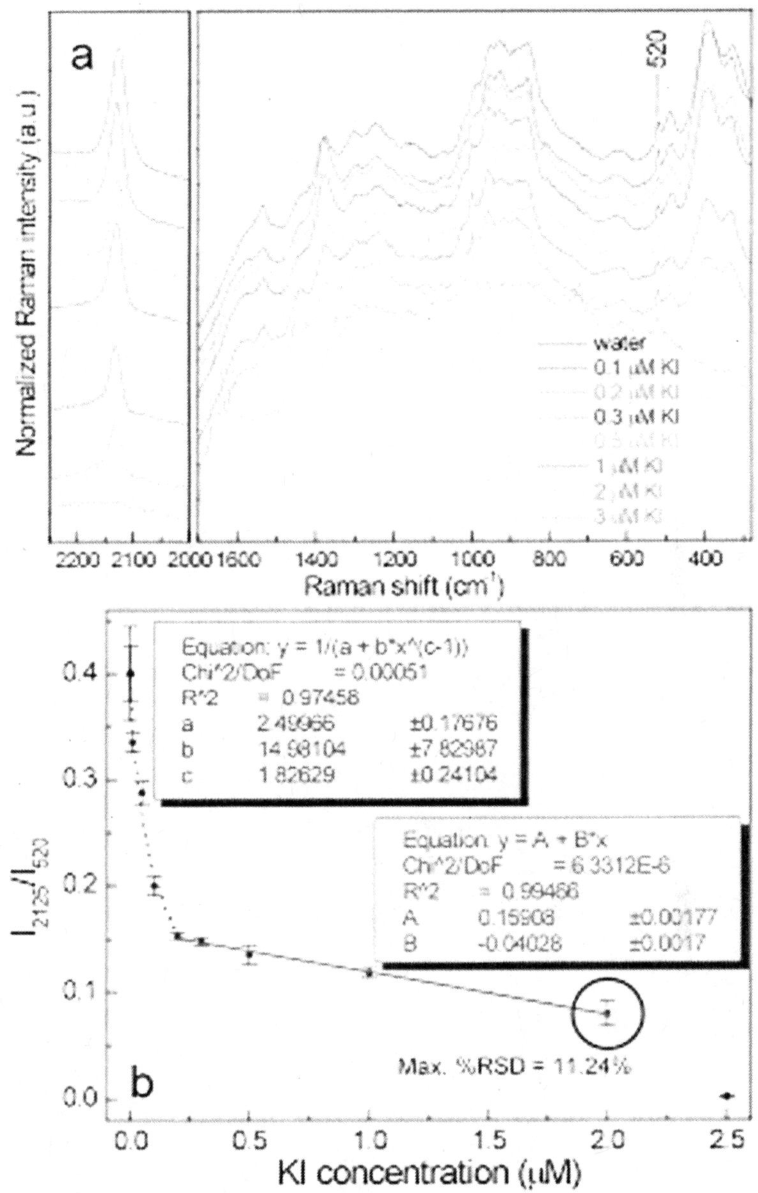

Figure 13. (a) SERS spectra of starch-reduced AuNPs soaked in different concentrations of KI solutions. (b) Plot of the intensity ratio between 2125 and 520 cm⁻¹ (I_{2125}/I_{520}) versus the KI concentration. The red solid line is fitted by a linear function, and the dotted blue line is fitted by a Langmuir adsorption isotherm. (Reproduced with permission from reference (70). Copyright 2011 American Chemical Society.)

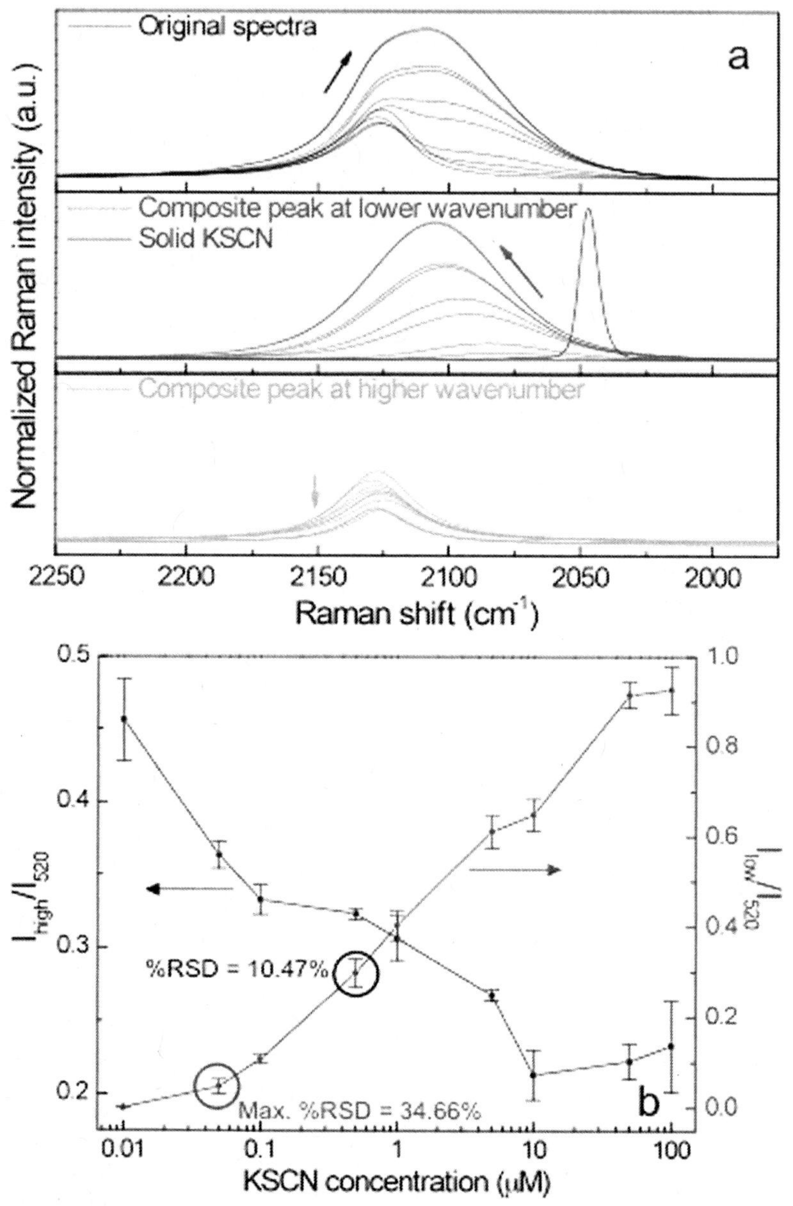

Figure 14. (a) SERS spectra of starch-reduced AuNPs soaked in different concentrations of KSCN solutions. The spectra were fitted by the Voigt function composing two peaks. The arrows show the trends of spectral changes with increasing KSCN concentration. (b) Plots of I$_{high}$/I$_{520}$ and I$_{low}$/I$_{520}$ versus the KSCN concentration. (Reproduced with permission from reference (70). Copyright 2011 American Chemical Society.)

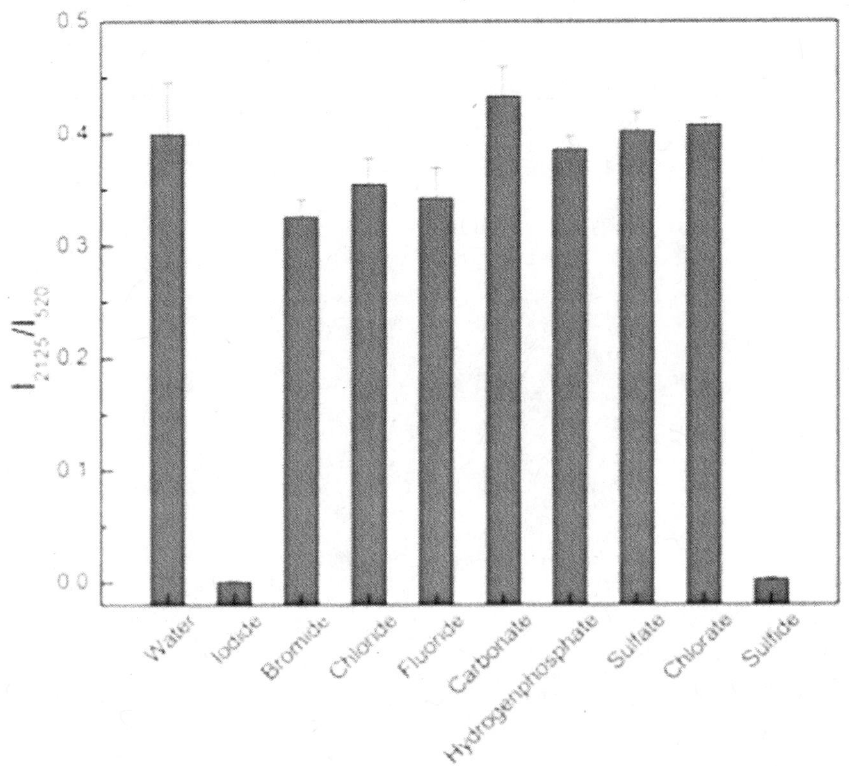

Figure 15. Comparison of I_{2125}/I_{520} of starch-reduced Au NPs soaked in 1 mM solutions containing different anions. (Reproduced with permission from reference (70). Copyright 2011 American Chemical Society.)

3.1.3. *Azo Coupling Reaction-Based Ultrasensitive Detection of Phenolic Estrogens Using SERRS*

Studies have shown that many adverse health effects are associated with human exposure to dietary or environmental estrogens. Development of rapid and highly sensitive detection methods for estrogens is therefore very important and necessary to maintain hormonal concentration below the safety limit. Herein, we demonstrate a simple and rapid approach to detect trace amounts of phenolic estrogen based on SERRS (*72*). Due to an azo coupling reaction between diazonium ions and the phenolic estrogens, azo compounds are formed with strong SERRS activity, which allows phenolic estrogen determination at sub-nanomolar levels in solution (Figure 16).

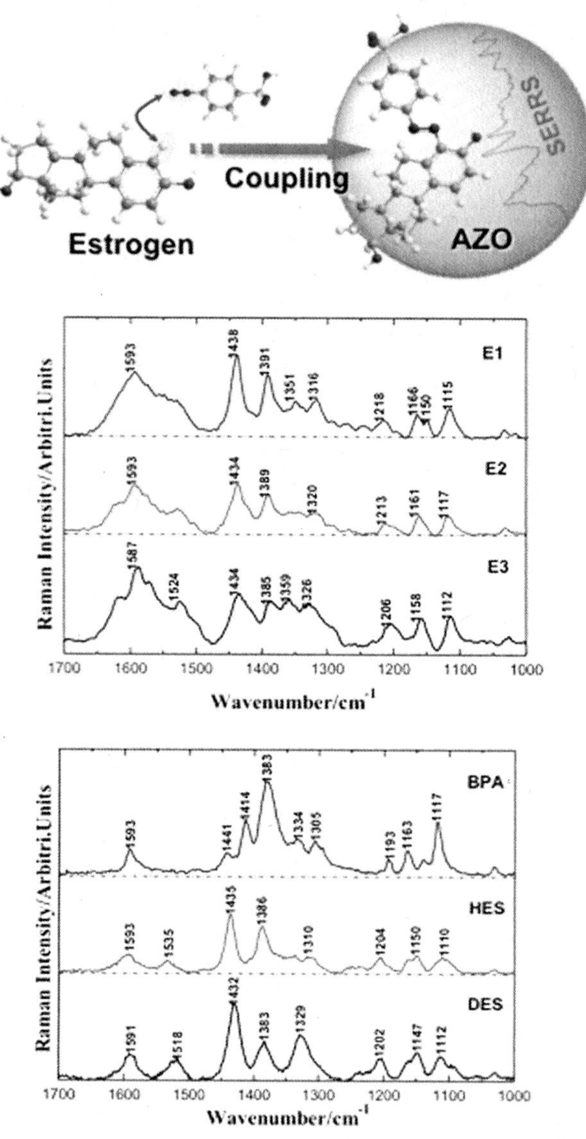

Figure 16. Coupling reaction-based estrogen detection and SERRS spectra of E1 (estrone), E2 (estrodiol), E3 (estriol), BPA (bisphenol A), HES (hexestrol), DES (diethylstilbestrol)-derived azo dyes. (Reproduced with permission from reference (72). Copyright 2011 American Chemical Society.)

After the coupling reaction, most of the phenolic estrogen-derived azo compounds are red or dark red with a maximum absorption at ~500 nm. Furthermore, we found that the coupling reaction was completed very rapidly (within 1 minute), and that its maximum absorption was almost stable within one hour. The limit of detection of the proposed method for target estrogen detection in solution was around 0.1 ppb (0.1 µg/kg), which is comparable to the lowest value reported using mass spectrometry and surface plasmon resonance-based methods (*72*).

The proposed protocol has multiplexing capability because each SERRS fingerprint of the azo dyes specifically corresponds to the related estrogen. Because of their azo structure, the comparable affinity of the azo dyes to the Ag NPs enabled parallel adsorption followed by overlap of the SERRS bands, confirming the application potential of the proposed approach for the detection of estrogen mixtures. It is universal and highly selective not only for phenolic estrogens but also for other phenolic molecules, even in complex systems. Moreover, because diazonium ions can react with other benzene derivatives (e.g., aminobenzene), this proposed protocol is also useful for their identification.

3.2. SERS Analysis of Proteins

3.2.1. Introduction

Generally, there are two kinds of strategies for SERS-based detection of biological molecules. As can be seen in Figure 17, one is label-free detection and the other is Raman dye-labeled detection (*6, 73, 74*). Label-free detection is a method of acquiring intrinsic SERS spectra of target biomolecules. It is a simple, direct and reliable approach, but its sensitivity is not high enough in some cases especially for those biomoleculs with small Raman-cross sections. In the case of Raman-dye labeled detection, SERS spectra of extrinsic labels are used as representatives of masked target biomolecules. Recent years, lots of Raman-dye labeled NPs probes have been developed for biomolecule detection based on biomolecule-ligand interactions. It is an indirect approach with high sensitivity; however, it is not always reliable because nonspecific bindings of the ligands may cause false positive results. In this chapter, we describe our several important strategies for the detection of proteins by SERS.

3.2.2. Label-Free Detection

3.2.2.1. Analytical Technique for Multiprotein Detection Based on Western Blot and SERS

We developed a new analytical procedure in combination of Western blotting and SERS for label-free protein detection, designated "Western—SERS", consisting of protein electrophoresis, Western blotting, colloidal Ag staining and SERS detection (Figure 18A). A novel method of Ag staining for Western blot that uses Ag NPs, an excellent SERS-active substrate, was proposed. The most important point of this method is that it allows *in situ* multiple protein detection on one nitrocellulose (NC) membrane.

The present method offers dual advantages of simplicity and high sensitivity. Comparing to mass spectrometry, we can detect label-free proteins directly on a NC membrane without time-consuming procedures of stripping and digestion. Moreover, the detection limit of the "Western—SERS" is almost consistent with the detection limit of colloidal Ag staining (2 ng/ band), and SERS signals do not self-quench unlike fluorescence. Thus, the new method has great potential for identifying proteomic components or proteins of differential expression in certain proteome (*75*).

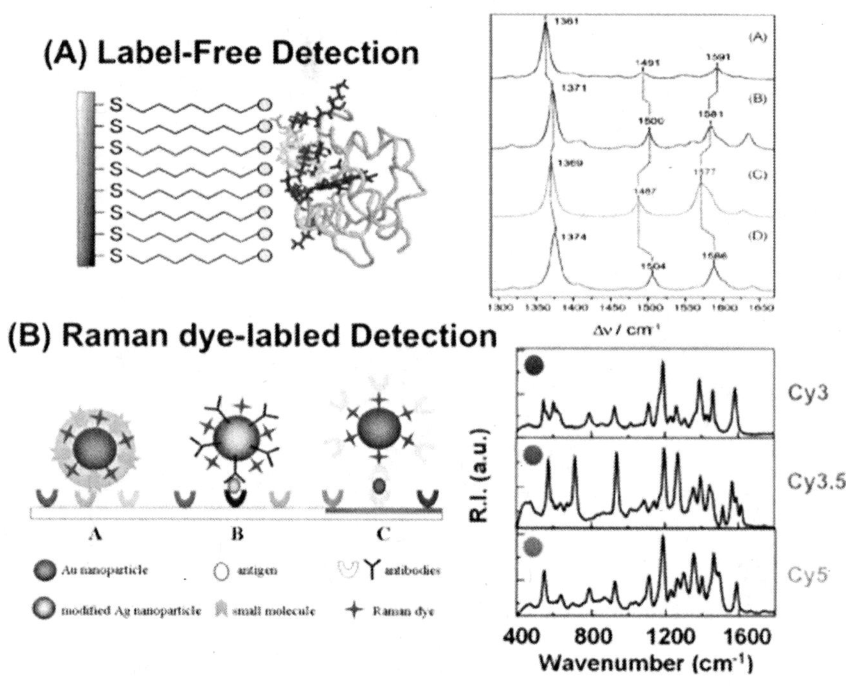

Figure 17. SERS-based label-free detection (A) and (B) Raman dye-labeled protein detection. (Reproduced with permission from reference (73), (74) and (6). Copyright, 2004, 2003, and 2009 American Chemical Society and Springer, respectively).

3.2.2.2. Detection of Proteins in Solution

It is difficult for roughened metal surfaces and dried colloids to provide reproducible SERS spectra especially for simple proteins without chromophores because of denaturation and different orientation of analytes on metal surface. On the other hand, no previous SERS-based study for proteins allows routine detection of label-free proteins with relatively high sensitivity in aqueous solution because halide ions, which are commonly used aggregation reagent, can form a strongly bonded surface layer that repels the adsorption of target proteins.

In our study, we use an acidified sulfate, which weakly binds to Ag surface, instead of haloid as an aggregation reagent. In this way, we have observed concentration-dependent SERS spectra of not only some hemoproteins but also some simple proteins in aqueous solution with highly spectral reproducibility and relatively high sensitivity (*76*).

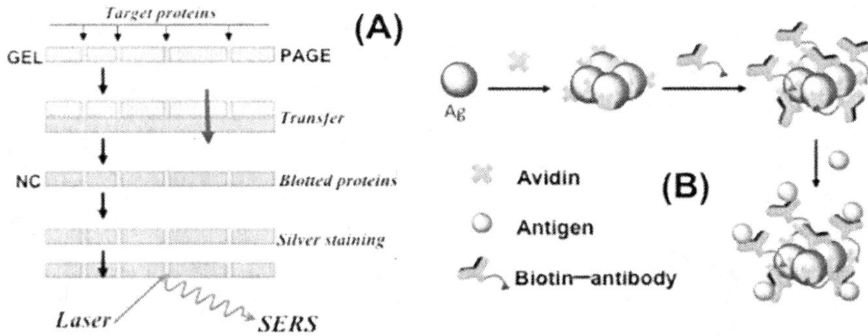

Figure 18. Procedure of Western—SERS (A) and (B) label-free indirect immunoassay based on avidin-induced Ag aggregates. (Reproduced with permission from reference (75) and (77). Copyright 2008 and 2011 American Chemical Society and John Wiley & Sons, respectively).

3.2.2.3. Indirect Immunoassay Using an Avidin-Induced SERS Substrate

Maintaining the bioactivity of the antibody or antigen after attaching them to metal NPs is crucial for immunoassays, and on the other hand, immunocomplexes normally have small Raman cross sections. These two important points should be considered before performing SERS-based label-free immunoassays. Development of a label-free immunoassay using SERS remains a challenge, because the SERS-active substrates currently used failed to combine biocompatibility and high sensitivity.

As is shown in Figure 18B, we developed a simple and effective protocol for a SERS-based label-free immunoassay using an avidin-aggregated Ag colloid. The Ag NPs are induced to aggregate by strong electrostatic interactions with avidin, and biotin-conjugated proteins are subsequently attached to these Ag aggregates by the biotin–avidin recognition, and finally immunoreaction would occur on these Ag aggregates. Since SERS is particularly sensitive to the first layer of adsorbates and their orientations, the differences in the SERS spectra before and after antigen involvement could be used to probe protein–protein binding events. The most important aspects of this protocol include the aggregation of the Ag NPs, and the mode of attachment of the antibodies to the Ag surface, which allow highly sensitive detection of antigens and more biocompatible protein adsorption over previous relevant methods (*77*).

3.2.3. *Raman Dye Labeled Detection*

3.2.3.1. *Fluorescein Isothiocyanate Linked Immunoabsorbent Assay Based on SERRS*

By using fluorescein isothiocyanate (FITC) as a Raman probe, we have developed a simple and sensitive method for an immunoassay based on SERRS. A SERRS-based immunoassay in a microtiter plate is reported. We have employed the main pretreatment method of enzyme-linked immunoabsorbent assay (ELISA) to the present study, and taken the advantages of good solid supports of ELISA and high sensitivity of SERRS together.

The proposed method has several advantages for immunoassay. First, we can determine the concentration of antigens via the intensity of a SERRS signal of FITC molecules that are attached to antibodies, and without an enzyme reaction the process it is simple and time saving. Second, one can conveniently obtain SERRS spectra of FITC directly from Ag aggregates on the bottom of a microtiter plate. Third, by using SERRS of FITC, the approach is sensitive enough to detect antigens at the concentration of 0.2 ng/mL, which is comparable to ELISA. Results are presented to demonstrate that the proposed SERRS-based approach may have great potential in high-sensitivity immunoassays (*78*).

3.2.3.2. *Simplified Protocol for Detection of Protein-Ligand Interactions via SERRS and Surface-Enhanced Fluorescence*

Protein microarray with its versatile applications plays an important role in high throughput proteomic studies. A protein microarray and SERS based protocol for detection of protein—ligand recognitions has been developed. After interactions between proteins and their corresponding ligands, we employed colloidal Ag staining for producing active substrates for SERS and surface-enhanced fluorescence (SEF). The same instrument is used for observing both images and spectra of fluorescence emission and Raman scattering in our studies.

This protocol exploits several advantages of simplicity over other SERS and SEF-based related methods because of the protein staining-based strategy for Ag NP assembling, high sensitivity from SERRS and SEF, and high stability in photostability compared to fluorescence-based protein detection methods. Therefore, it has great potential in highly sensitive and throughput chip-based protein function determination (*79*).

3.2.3.3. Protein-Mediated Sandwich Strategy for SERS: Application to Versatile Protein Detection

SERS spectra of Raman labels from Ag substrates are susceptible to change because of different orientations on metal surfaces or conformational changes due to intermolecular interactions as described in 3.2.2.2. For reproducible SERS-based protein identification, immunoassay and drug screening, metal sandwich substrates bridged by proteins have been created (*80*).

The sandwich architectures are fabricated based on a layer-by-layer (LbL) technique. The first Au monolayer is prepared by self-assembling of Au NPs on a poly (diallyldimethylammonium chloride) (PDDA)-coated glass slide. The second Au or Ag layer is produced by the interactions between proteins in the middle layer of the sandwich architecture and the metal NPs (Figure 19).

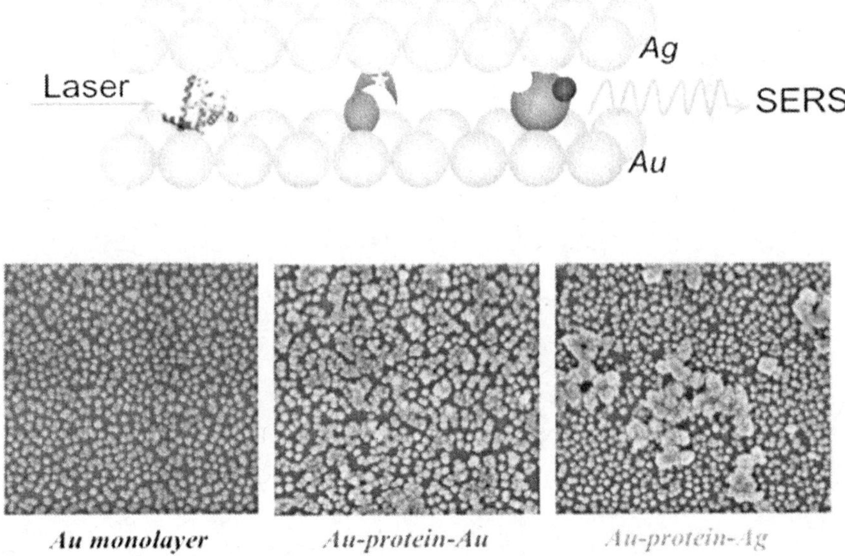

Figure 19. Metal sandwich SERS-active substrates for versatile protein detection. (Reproduced with permission from reference (6) and (80). Copyright 2009 Springer and American Chemical Society, respectively.)

Many SERS studies have revealed that SERS spectra from Au substrates are much better in stability than those from Ag substrates, while the SERS enhancement ability of Au substrates is much weaker than that of Ag. In our studies, owing to the stable adsorption on the first Au monolayer, the target molecules between two metal layers show much better SERS/SERRS spectra in reproducibility than those from Ag aggregates.

The most significant feature of this method compared with other SERS-based methods for protein detection is that a self-assembling Au NP monolayer is used for both capturing proteins and producing the SERS-active substrate with the second Ag layer, making the proposed sandwich substrate more accessible and sensitive than other Au based SERS methods, and more reproducible than other Ag-based SERS measurements.

Enormous SERS "hot spots" may emerge among metal aggregates formed on the edges between the two metal layers and between the junctions of metal NPs in the second metal layer. SERRS/SERS spectra with both high sensitivity and reproducibility, which are crucial for an analytical method, were obtained. By using these metal sandwich substrates, SERS-based versatile protein detections (i.e., identification, immunoassay, and drug screening) (Figure 19) have been effectively carried out. The reproducibility of target molecules from Au—protein—Ag sandwiches is comparable with those from Au—protein—Au sandwiches.

Moreover, SERRS intensities about seven times stronger can be obtained from the Au—protein—Ag sandwiches compared with those from the Au—protein—Au sandwiches. All the results presented in this study indicate that the proposed sandwich strategy holds great promise in SERS-based structural and functional proteomic studies (*80*).

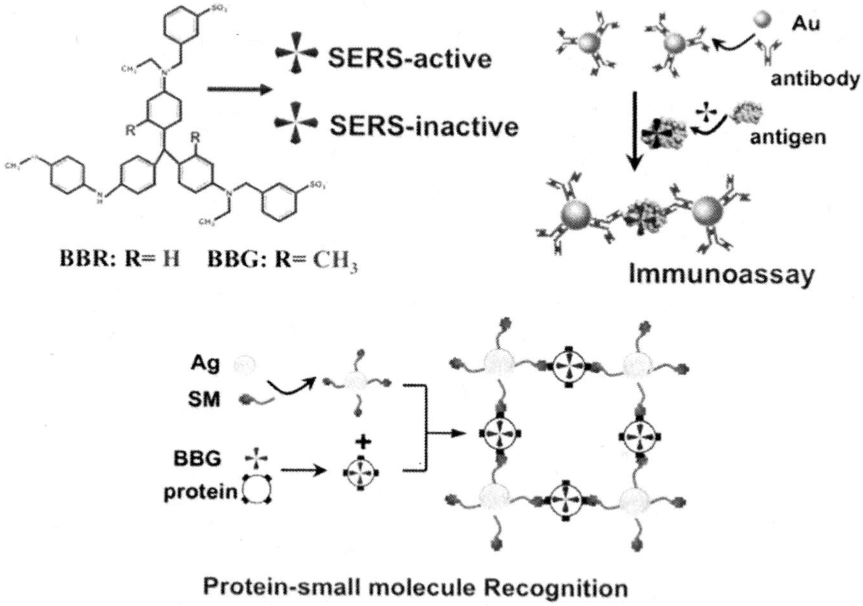

Figure 20. Coomassie brilliant blue as Raman probes for protein-ligand recognition. (Reproduced with permission from reference (81). Copyright 2010 American Chemical Society.)

Coomassie brilliant dyes have high affinity to proteins and high Raman activity, based on which we have employed brilliant blue R-250 (BBR) and brilliant blue G-250 (BBG) as SERS labels to probe protein—ligand recognitions (Figure 20). This method differs from previously proposed methods in that target proteins are labeled rapidly before biological recognitions without procedures of separation and purification, rather than attaching Raman labels to metal NPs, which significantly simplifies the Raman dye labeling procedure.

In typical assays, ligand-functionalized metal NPs assemble by target protein-specific bindings and this assembly sequentially turns on electromagnetic enhancement of Raman scattering of the proposed labels. The method with its advantages of rapidness, high sensitivity and spectral multiplexing has great potential in probing protein—molecule recognitions not only in solution but also on flexible solid substrates (*81*).

Another application of these dyes is probing protein concentration. Protein concentration determination, known as protein assay, is a basic biochemical method and often necessary before processing protein samples. Currently, the measurement used most commonly in protein assays is absorbance of light, and there are several colorimetric assays for proteins including Lowry, Biuret, Bradford and bicinchoninic acid assays. The Bradford assay with its advantages of rapidness, convenience and relative sensitivity has been used extensively. This protein assay is based on an absorbance shift from 465 to 595 nm of a band arising from BBG when binding to proteins. The disadvantage of the Bradford assay is, however, its narrower linear concentration range, which presents a severe problem when the concentration of a target protein is outside the range.

In a protein—BBG liquid mixture, we found that the amount of unbound BBG molecules remarkably decreases with the increase of the protein amount due to high affinity of BBG to proteins, which results in the decrease of SERS intensity of unbound BBG after the addition of Ag NPs.

We found that there was a direct relationship between the SERS intensity of unbound BBG molecules and the target protein concentration, and accordingly, the target protein concentrations can be determined by the SERS intensity of unbound BBG molecules in a protein—BBG mixture. Silicon, with a Raman peak at 520 cm^{-1}, was used as an internal standard to eliminate instrumental variables (e.g., laser power) and differences in focus. Because of high sensitivity and selectivity of SERS, this new SERS-based method allows detecting proteins over a much wider concentration range with lower limit of detection than Bradford assay and other protein assays commonly used in biochemical laboratories (*82*).

3.3. SERS Analysis of Living Cell

3.3.1. SERS Analysis of Escherichia coli

3.3.1.1. Introduction

It is important to rapidly detect and identify bacteria, especially harmful pathogens, which cause food poisoning, water contamination, and bioterrorism. One of notorious bacteria is *Escherichia coli*, which sometimes causes serious food poisoning (for example, *E. coli* O157:H7) and urinary tract infection. *E. coli* is a kind of gram-negative bacteria. Their cell wall consists of an inner membrane, a thin peptidoglycan, and an outer membrane (*83*), whereas gram-positive cell membrane such as yeast is covered with only a thick peptidoglycan (*84*). The outer membrane contains a lipopolysaccharide, which causes a fever in a human body and called endotoxin (*85*). Bacteria have been hitherto classified by morphological characteristics, biochemical tests, and genomic analysis. However, these are time-consuming, because cell culturing to grow colonies large enough for classification requires an overnight in a suitable medium. Recently, spectroscopy has been applied to detection of bacteria. SERS is particularly suitable to detect bacteria rapidly because of spectral multiplexing by the sharp vibrational modes and the high-sensitivity. SERS spectra of bacteria such as *E. coli* have been obtained mainly by three methods after that bacteria are centrifuged, rinsed in water, and then re-centrifuged: [i] mixing with noble metal colloidal solution (*86–90*), [ii] reduction of noble metal ion to the NPs on the cell (*91–95*), and [iii] using SERS-active substrate (*96–100*).

3.3.1.2. Mixing with Ag Colloidal Solution

For the method [i], bacteria are suspended in distilled water. The suspension of bacteria is mixed with Ag colloidal solution. This is mainly prepared by Lee-Meisel procedure, in which trisodium citrate is added to sliver nitrate aqueous solution with heating and stirring (*101*). Thus the bacteria are coated with AgNPs as shown in Figure 21a. Figure 21b shows SERS spectra of the Ag-coated *E. coli* MC4100 at various concentrations of the bacteria, 10^3–10^5 cfu/mL (*86*), where cfu is an abbreviation for colony forming units. The SERS spectrum at 10^5 cfu/mL is almost the same as those obtained by other groups (*90, 92*).

It is noted that bacteria are characterized to genus level (*89, 96, 97*), and moreover *E. coli* is discriminated to strain level by cluster analysis (*89*). Briefly, principal components analysis was used to reduce the dimensionality of the SERS data. Then discriminant function analysis (DFA) discriminated between groups on the basis of these retained principal components (PCs). Figure 22a depicts a composite dendrogram generated from the hierarchical cluster analysis using PC-DFA models produced by the differences in the similarities between the SERS spectra at 400–980 cm^{-1}, which is the most informative area for adenosine, inosine, adenine, and hypoxanthine, of various bacteria as shown in Figure 22b. Also in the case of seven *E. coli*, a composite dendrogram was generated to investigate the potential of SERS to discriminate to the strain level as depicted in Figure 22c.

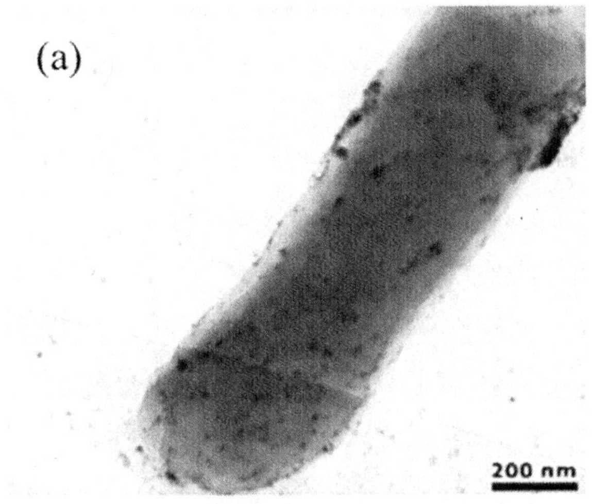

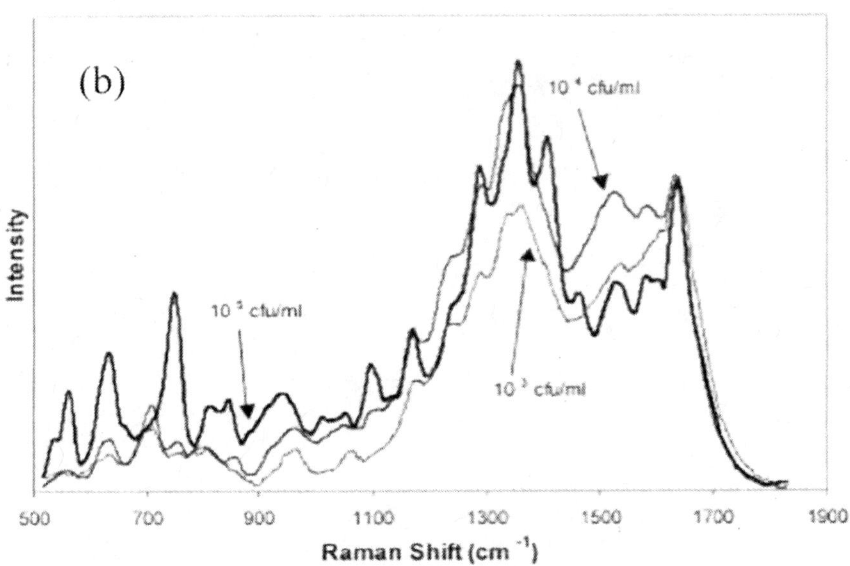

Figure 21. (a) Transmission electron microscope (TEM) image of Ag-coated E. coli MC4100. (b) SERS spectra of the Ag-coated E. coli at various concentrations of the bacteria. (Reproduced with permission from reference (86). Copyright 2006 Springer.)

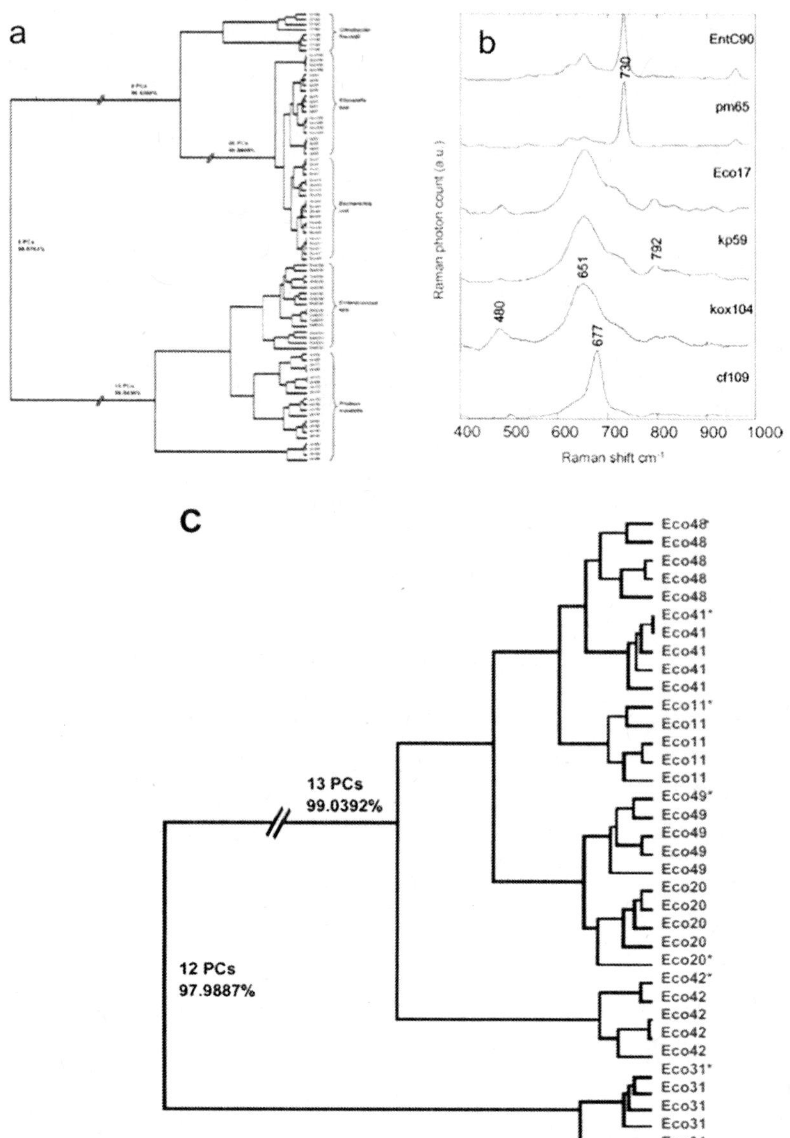

Figure 22. (a) A composite dendrogram generated from the hierarchical cluster analysis using PC-DFA models for various bacteria, Citrobacter freundii including cf109, Klebsiella spp. including kp59 and kox108, E. coli including Eco17, Enterococcus spp. including EntC90, and Proteus mirabilis including pm65 (from the top to the bottom). (b) Processed SERS spectra of EntC90, pm65, Eco17, kp59, kox108, and cf109. (c) A composite dendrogram generated from the hierarchical cluster analysis using PC-DFA models for seven E. coli. (Reproduced with permission from reference (89). Copyright 2004 American Chemical Society.)

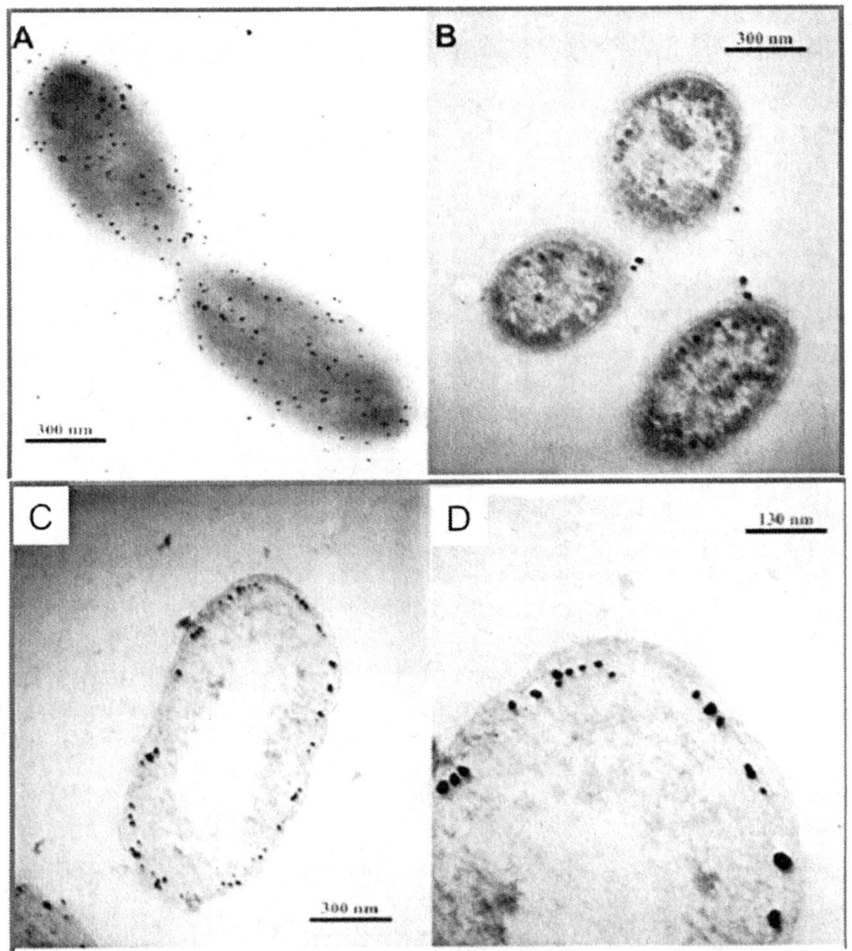

Figure 23. TEM images of G. sulfurreducens coated with AgNPs prepared by that the cells were exposed to Ag nitrate in buffer solution, and then acetate was added for reduction, (a) whole cells; (b) cross section through cells. (c, d) TEM cross section images through the cells infused with Au NPs prepared by that the cells were exposed to chloroauric acid in buffer solution, and then hydrogen was bubbled through the solution for reduction. (Reproduced with permission from reference (91). Copyright 2008 American Chemical Society.)

3.3.1.3. Reduction of Ag Ion on the Cell

For the method [ii], bacteria are suspended in sodium borohydride aqueous solution as electron donor, rinsed in water, and then re-suspended in Ag nitrate aqueous solution. Silver ion is chemically reduced to the NPs on the cell surface.

On the other hand, AgNPs are formed inside the cell by reverse order of addition of the reagents, in which bacteria are suspended in Ag nitrate aqueous solution, rinsed in water, and then re-suspended in sodium borohydride aqueous solution (*92*). Just for reference, transmission electron microscopy (TEM) images of Ag and Au NPs on and inside *Geobacter sulfurreducens*, respectively (*91*), are shown in Figure 23. Figure 24a and 24b describe that SERS spectra of *E. coli* coated with AgNPs, *E. coli* wall fraction, and flavin adenine dinucleotide (FAD) are almost the same (*92*). FAD is located in the cell wall and plays an important role in respiratory processes in a living cell (*92, 93*). On the other hand, Figure 24c represents that SERS spectrum of *E. coli* infused with AgNPs is different from the extracellular SERS spectrum in Figure 24b. In the intracellular SERS spectrum, the strong peaks at 930 and 1400 cm^{-1} are usually assigned to carboxylate stretching. Figure 24d shows SERS spectrum of *E. coli* protoplasmic fraction. However, this is similar to neither the extracellular nor intracellular SERS spectra. The spectrum, especially the peaks at 735 and 1330 cm^{-1}, is due to adenine-containing moieties mostly from denatured DNA (*92*).

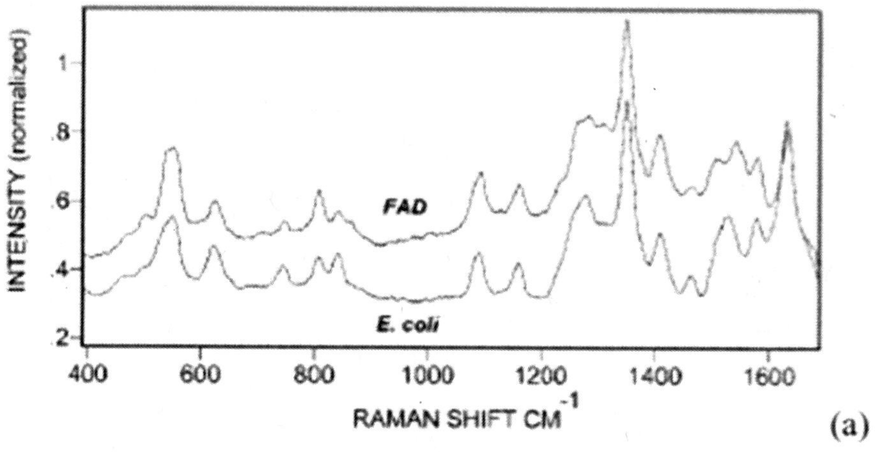

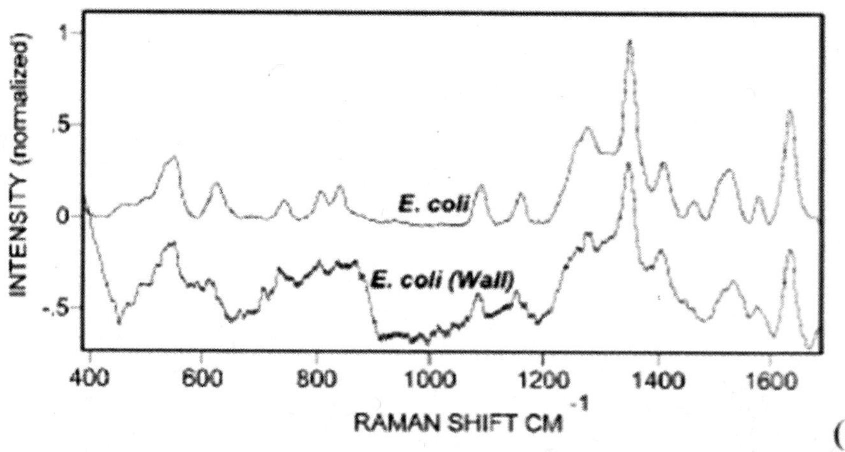

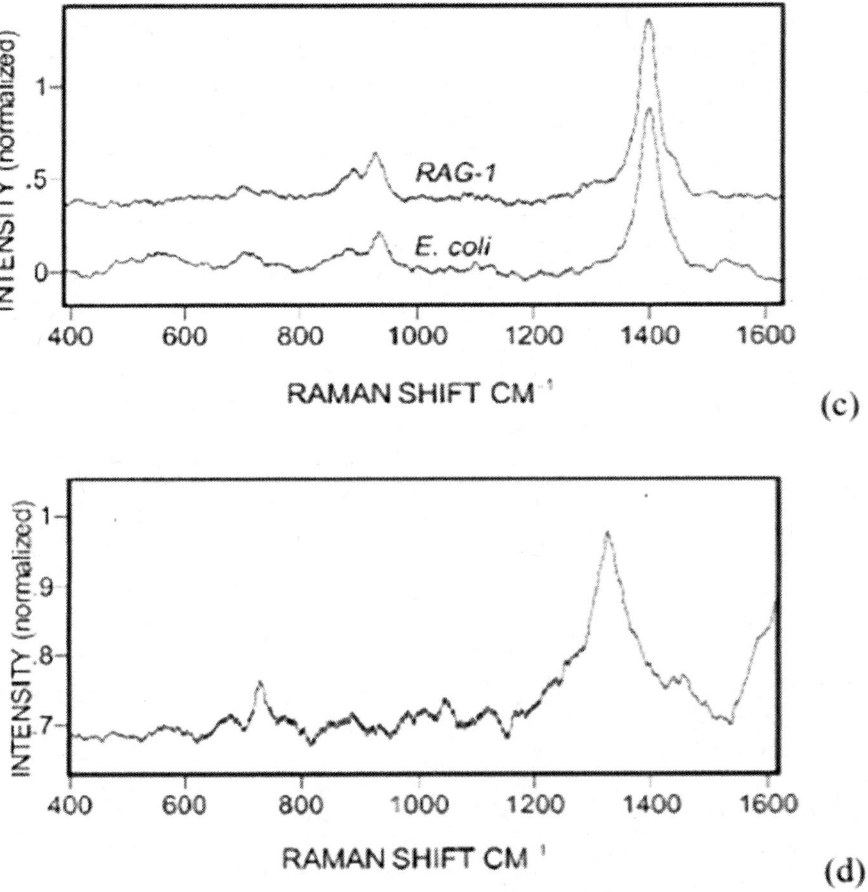

Figure 24. (a) SERS spectra of flavin adenine dinucleotide (FAD) and E. coli coated with AgNPs. (b) SERS spectra of E. coli coated with AgNPs and E. coli wall fraction. (c) SERS spectra of gram-negative strain Acinetobacter calcoaceticus RAG-1 and E. coli infused with AgNPs. (d) SERS spectrum of E. coli protoplasmic fraction. (Reproduced with permission from reference (92). Copyright 2004 Society for Applied Spectroscopy.)

Photo-reduction at a laser focal point on a glass surface in an Ag nitrate aqueous solution has already been reported as on-demand synthesis of SERS-active AgNPs (*102, 103*). The photo-reduction was combined with optical trapping of a single bacterium, which can avoid undesirable perturbation due to immobilization of a cell on a glass surface; namely, SERS from a photo-reduced Ag nanoaggregate on a single optically-trapped bacterium was measured by alternatively irradiating coaxially-focused green and near-infrared (NIR) lasers at the same position (*95*). By using a dark field illumination, the optical microscopic images of *E. coli* before and after optical trapping and photo-reduction, which are schematically depicted in Figure 25a–25c, were observed. Figure 25d shows that *E. coli* swayed in the solution before optical trapping. Figure 25e shows that an optically-trapped single *E. coli* was aligned vertically, namely along the laser beam direction. After trapping a single *E. coli*, the green laser beam was focused on the optically-trapped *E. coli*. Figure 25f shows an orange spot at the focal point after a few seconds of green laser irradiation. Note that the NIR laser was turned off just before focusing the green laser beam. After the NIR laser was turned off, the *E. coli* with Ag nanoaggregates was still trapped by the focused green laser and again drifted away after the green laser was also turned off. Simultaneous focusing of both NIR and green laser beams into the sample suspension generated a bubble at the focal point. The bubble disturbed the optical trapping and SERS measurement of *E. coli*. The bubble is likely induced by temperature elevation of the photo-reduced Ag nanoaggregate whose plasmon resonated with the highly-intense NIR laser light. Indeed, the plasma resonance spectrum of an ensemble AgNPs (like Figure 25h) has a broad line width (*42*). Thus, the plasma resonance of the photo-reduced Ag nanoaggregate can be resonated with the NIR laser light. *E. coli* without Ag nanoaggregates can be stably trapped by simultaneously focusing the laser beams in water. By using a scanning electron microscope (SEM), an *E. coli* before and after the appearance of the orange spot (Figure 25d) was observed. Figure 25g shows that an *E. coli* was not adsorbed by Ag nanoaggregates before photo-reduction. Figure 25h shows that an Ag nanoaggregate that consists of 10 or more AgNPs was formed. The Ag nanoaggregate, which looks like bright spot, broke outer membrane of the cell wall and sunk in the cell surface.

SERS spectrum was measured from an orange spot (Figure 25f) that may be from an Ag nanoaggregate on a single *E. coli* (Figure 25h). Note that photo-reduced Ag nanoaggregates covering all area of the focal point enable us efficient SERS detection. The required accumulation time for SERS detection was only 5 and 15 seconds at the longest, in contrast to 30–120 seconds for ensemble SERS measurements (*86, 90, 92*), in which only parts of the illumination area are covered with Ag nanoaggregates. Figure 26a shows that two broad peaks were always located at ~1550 and ~1300 cm^{-1}, and a sharp peak was often observed at ~750 cm^{-1}. The broad peaks are similar to averaged SERS spectra of amorphous carbon formed by photo- and thermo-degradation of carbon monoxide (*104*). In the each SERS spectrum of amorphous carbon before averaging, however, sharp peaks are observed at random positions unlike the present SERS spectrum (*104*). The present spectrum is similar to a SERS spectrum from *E. coli* in an Ag colloidal suspension at a low concentration of *E. coli* (*86*), 10^3 cfu/mL in Figure 21b. Time-resolved SERS spectra were measured from the photo-reduced Ag nanoaggregate on the single *E. coli*. During the measurement, the *E. coli* was continuously illuminated with the green laser. Figure 26b shows that the sharp peak at ~750 cm^{-1} became prominent after reduction of the broad peaks at ~1300 and ~1550 cm^{-1}. This may indicate detection of two kinds of molecules. One candidate which exhibits SERS peaks at ~1300 and ~1550 cm^{-1} is the molecules existing on the cell surface, namely FAD. Then the other candidate that exhibits SERS peaks at ~750 cm^{-1} is the molecules inside the cell because of penetration of a photo-reduced Ag nanoaggregate inside the cell by an increase in its size as indicated in Figure 25h. In the SERS spectrum from an ensemble of Ag-coated *E. coli*, the peak at ~750 cm^{-1} has been attributed to tyrosine or an adenine moiety (*91, 92*). To confirm the attribution, we observed conventional SERS spectrum of cytoplasmic fraction of *E. coli* from a mixture of a disrupted *E. coli* suspension and a concentrated Ag colloidal solution by a centrifuge with 0.5 M NaCl, which facilitates the molecule to adsorb on the Ag surfaces. Figure 26c shows sharp peaks at ~740, ~1340, and ~1470 cm^{-1} in the spectrum. These peaks are attributed to adenine, in particular the prominent peak at ~733 cm^{-1} is due to the ring-breathing mode. Appearance of three peaks in Figure 26c may indicate that dense NaCl solution changes an upright or at least tilted orientation of adsorbed molecules (*105*). Therefore, the changes in the time-resolved SERS spectra are likely attributed to amorphous carbon or FAD on *E. coli* and an adenine moiety inside *E. coli*.

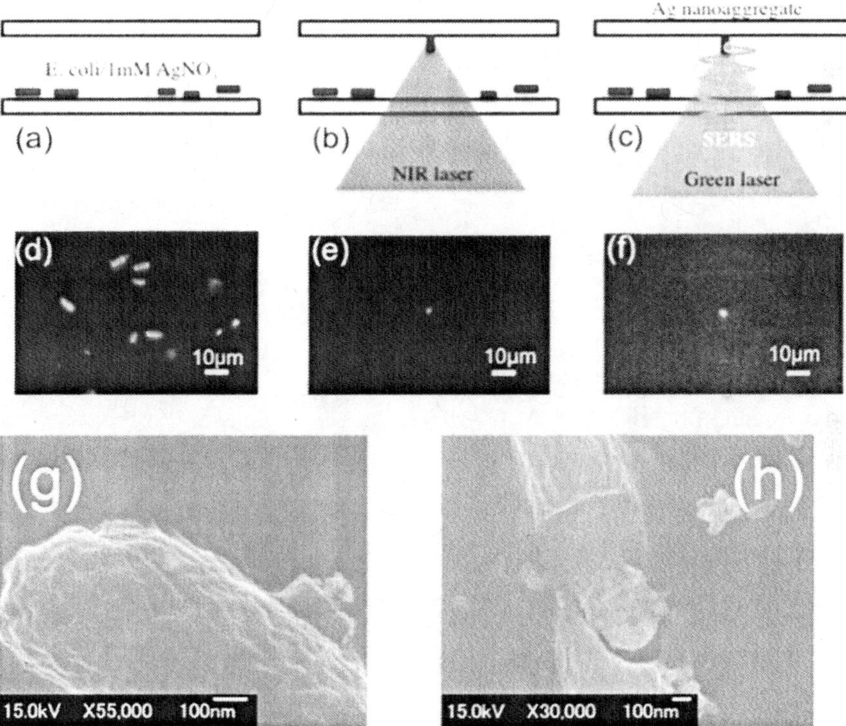

Figure 25. (a)–(c) Schematic experimental procedure for SERS measurement from a Ag nanoaggregate on a single bacterium by optical trapping using near-infrared (NIR) laser and photo-reduction using green laser. (d)–(f) The corresponding dark-field microscope images of E. coli JM109. SEM images of E. coli JM109 (g) before and (h) after the photo-reduction. (Reproduced with permission from reference (95). Copyright 2011 The Chemical Society of Japan.)

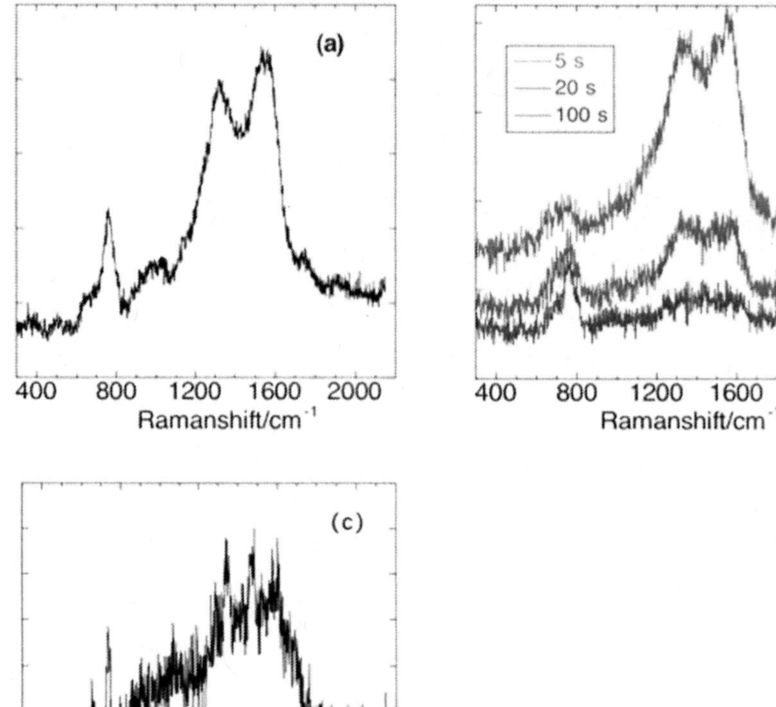

Figure 26. (a) SERS spectrum and (b) time-resolved SERS spectra from a Ag nanoaggregate on a single optically-trapped E. coli JM109. The accumulation times per a spectrum were 15 s in (a) and 5 s in (b). (c) SERS spectrum of cytoplasmic fraction of E. coli JM109 accumulated for 240 s. (Reproduced with permission from reference (95). Copyright 2011 The Chemical Society of Japan.)

3.3.1.4. SERS-Active Substrate

For the method [iii], SERS-active substrate are formed by an in situ Au ion doped silica sol-gel procedure (*96, 97*), an immobilization of metal colloid NPs on silanized glass surface (*98, 99*), a sputter-coating thin film of Ag on a glass cover slip (*100*), and so on. Figure 27 shows an example of SERS-active substrate (*96*). It is noted that single cell has been measured by using optical trapping and SERS-active substrate. Single cell measurements using SERS will be able to be applied to the *in situ* discrimination between pathogenic and non-pathogenic bacteria without cultivation and screening, in which there are possibilities of infection and contamination, whereas ensemble measurements simply provide SERS spectra averaged over various bacteria, and thus discrimination of bacteria without cultivation and screening is difficult. An optically-trapped cell is vertically translated towards a glass slide coated with Au colloid NPs (*98*). NIR light transmitted through and around the cell subsequently excites the AuNPs. The enhanced electromagnetic field extends a short distance, hundreds of nanometers, away from the substrate surface and influence constituents of the cell and not the entire cell; namely, this can be used to selectively probe the cell. In another case (*99*), a metal-coated silica bead, which is sufficiently transparent while the presence of metal islands, is used as optically trappable probe for SERS spectroscopy. A laser beam at 785 nm was used both for the optical trapping and for the excitation. The beads can be placed and scanned with nanometric accuracy near cell.

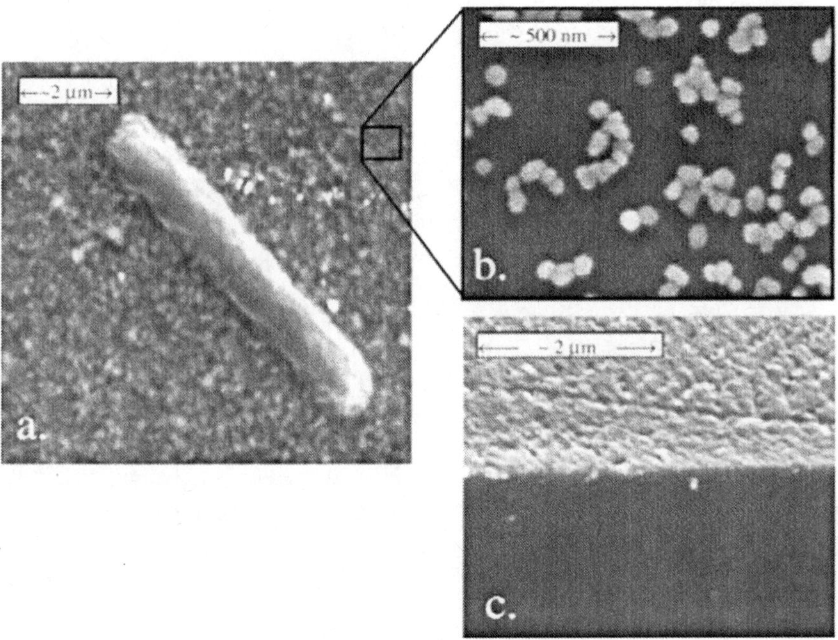

Figure 27. SEM images of (a) Bacillus anthracis on a Au nanoparitcles which cover silica substrate, (b) the SERS substrate at ~10-fold magnification, and (c) the SERS substrate looking toward the cleaved edge, in which Au and silica appear the bright and dark regions, respectively. (Reproduced with permission from reference (96). Copyright 2005 American Chemical Society.)

In the above studies, *E. coli* has been discriminated to strain level by hierarchical cluster analysis of the spectra from the ensemble, although various groups have obtained almost the same spectra. In both the ensemble and single cell measurements, the extracellular, intracellular, and plasmic SERS spectra show different peaks, which are attributed to the constituents on and inside the cell of *E. coli*. The SERS spectrum of single *E. coli* can be measured by using optical trapping. Next, it is introduced that variations in SERS spectra of single yeast using AgNP dimers have been discussed in terms of inhomogeneous protein distribution on the cell wall.

3.3.2. SERS Analysis of Yeast Cell

3.3.2.1. Introduction

In the work, yeast cells are selected as the target of SERS analysis, because yeast is one of the most extensive model eukaryotic systems for various basic and applied fields of life science, medicine and biotechnology (*84*). Among yeast cell organelles, cell walls attract much attention because it is the first line of defense of the host to a fungal infection from the standpoint of pathology, and in chemotherapy it is the first target of an antifungal agent (*84*). Among the cell wall components (glucan, mannan, chitin and proteins), proteins in the outermost layer plays an important role in cellular functions because of their sensitivity towards different biological functions (*106*). To understand surface and interfacial chemistry and membrane protein dynamics in single living yeast cell, we need a non-destructible spectroscopic technique with high sensitivity. SERS has been tremendously used for single molecule detections of cellular proteins due to the high affinity of Au and AgNPs towards different protein molecules (*107, 108*). There are excellent studies for the extraction and characterization of cell membrane proteins under the area of biotechnology (*84, 109–111*). These techniques are much complicated and needs high expertise. Furthermore, most of them are destructible methods. Recent SERS analysis on the cell wall of single living yeast reveals that nanoscale detections of membrane protein signals are possible using SERS (*112, 113*). The objective of the present work is to perform SERS mapping of living yeast (*Sachharomycea cerevicea*) cell walls for the characterization of outermost cell wall proteins. These types of investigations can leads to the complete understanding of membrane protein dynamics in living cells.

3.3.2.2. SERS Measurement and Analysis of Single Living Yeast Cells

Figure 28a and b shows the dark-field images of yeast cells without and with AgNPs, respectively. Yeast cells automatically adsorbed AgNPs without any treatment. The colored spots in Figure 28b distributed heterogeneousely on the cell surface correspond to isolated or aggregated AgNPs. It is interesting that the color of spots in Figure 28b largely varies from one spot to another. The color changes can be explained by the size and shape of AgNPs. Figure 29a and b shows the AFM images of the yeast cell without and with AgNPs, respectively. The presence of AgNPs on the yeast cell wall is further confirmed by a height trace analysis in our earlier studies (*112*). AFM measurements revealed that about 20% of adsorbed AgNPs form dimers. Figure 30a and b shows the dark-field and the corresponding SERS image of AgNPs adsorbed on the cell wall, respectively. Bright spots in Figure 30b correspond to SERS active AgNPs. It can clearly be seen in red, green, and yellow colors. Red or yellow colored spots showed SERS activity, and the percentage of SERS light spots was around 20% of all light spots of AgNPs under the experimental conditions. The distribution of dimers in Figure 29b is consistent with that of SERS light spots in Figure 30b. SERS intensity change with respect to time is known as 'blinking'. Figures 31a to 31e show the blinking of SERS light spots within the time period of 5.0 seconds. Suggested explanations of SERS blinking include the behavior of thermal diffusion of single molecules in and out of localized EM fields (*114*), binding and orientational changes of single molecules and structural changes of NPs (*115*) etc. All these effects are reported from single or a few molecules. Blinking can also be due to the photodecomposition effect. In such a case, there will be 2 broad bands at about 1350 and 1580 cm^{-1}. These bands are not observed in the present SERS spectra. So in the current yeast cell surfaces, SERS blinking can be considered to be molecular fluctuations at spatially confined nanometer scale interstitial sites at AgNP dimers.

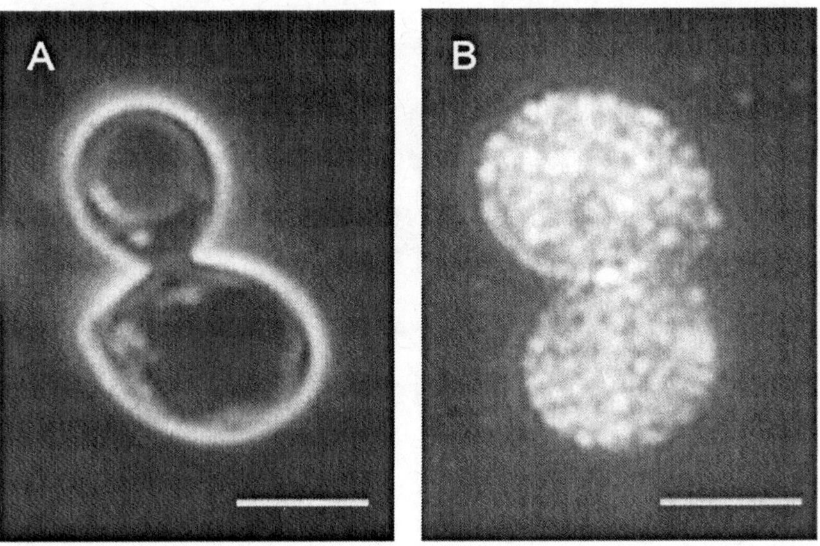

Figure 28. (A) Dark field image of yeast cells without AgNPs on cell walls. (B) Dark field image of yeast cells with AgNPs on cell walls. (× 100) (scale 10 μm). (Reproduced with permission from reference (113). Copyright 2009 Springer.)

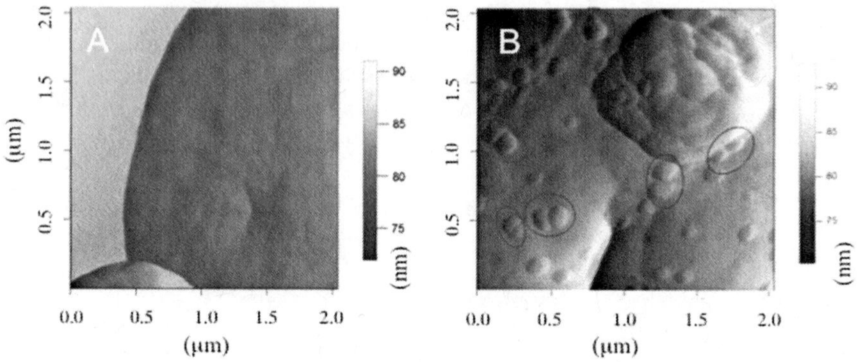

Figure 29. (A) AFM images of yeast cells without AgNPs on cell walls. (B) AFM images of yeast cells with AgNPs on cell walls. (Reproduced with permission from reference (112). Copyright 2008 American Institute of Physics.)

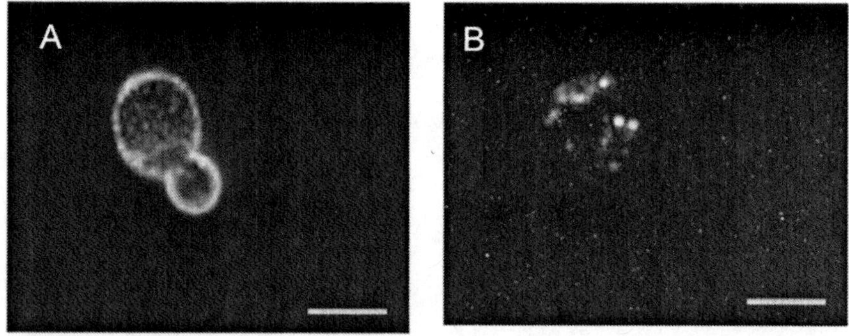

Figure 30. (A) Dark field image of yeast cells with AgNPs on cell walls. (B) The corresponding SERS image of yeast cells with AgNPs on cell walls. (× 60) (scale 10 μm). (Reproduced with permission from reference (113). Copyright 2009 Springer.)

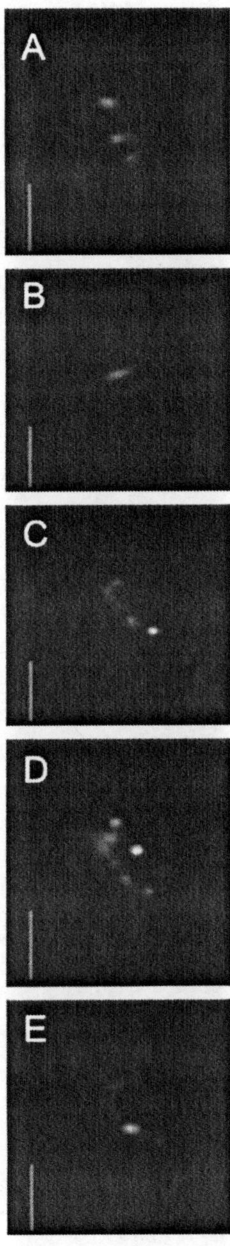

Figure 31. Time series of blinking of SERS active AgNPs on a single yeast cell wall at (A) 0 s, at (B) 5 s, at (C) 10 s, at (D) 15 s, and at (E) 20 s. (× 60) (scale 10 μm). (Reproduced with permission from reference (112). Copyright 2008 American Institute of Physics.)

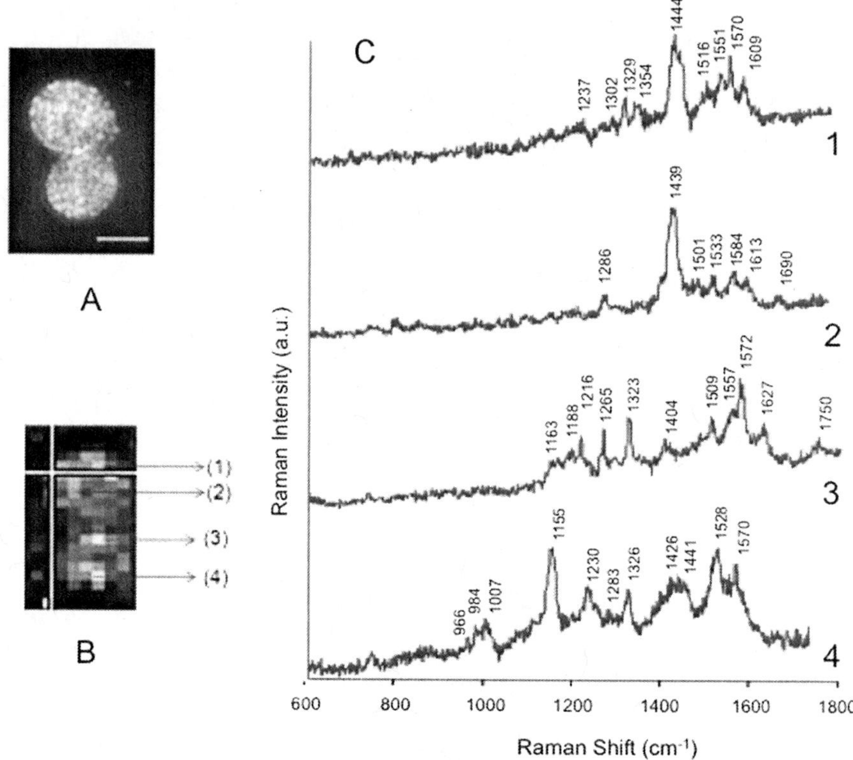

Figure 32. (A) Dark field image of yeast cells with AgNPs on cell walls. (B) The corresponding SERS mapping image of yeast cells with AgNPs on cell walls. (C) The spectra (1) to (4) has been measured from the positions indicated in (B). (Reproduced with permission from reference (113). Copyright 2009 Springer.)

Figure 32 shows SERS mapping results of the single yeast cell wall. SERS spectra measured from different points are also different to each other. The deference may be related to the inhomogeneity of yeast cell surfaces (*112*). By the measurement of the SERS spectra of glucan, mannan and chitin, which are the cell wall components of yeast, we found that the origin of the cell wall SERS spectra is proteins attached to mannan, namely mannoproteins (*112*). In order to get a clearer picture, mannoprotein extracted from the yeast cell walls has been analyzed by SERS spectroscopy. Figure 33 shows three different SERS spectra obtained from different AgNP aggregates. Main the characteristic Raman peaks in mannoprotein have been attributed to $C-NH_2$ stretch (1329 cm-1), $C-NH_2$ stretch (1354 cm-1), δ NH + ν CN (1516 cm-1) and δ NH + ν CN (1570 cm-1) (*91, 108, 116*). Here, also we can notice that all the important Raman peaks are in between 1200 cm-1 and 1700 cm-1 like those of yeast cell walls. Also, there are many SERS peaks common to yeast cell wall SERS peaks, indicating that SERS peaks of yeast cell wall are from mannoprotein in cell walls. We consider that binding affinity between AgNPs and mannoprotein results in the exclusive appearance of SERS spectra of mannoprotein. We still do not have conclusive reason of the exclusive appearance. The binding affinity of molecules to the AgNPs strongly depends on molecular surface charge, and is much stronger for cationic groups than for anionic ones in the case of Lee-Meisel colloidal AgNPs used in the present work (*101*). For this reason, cationic groups usually dominate SERS spectra and anionic ones are almost invisible. Thus, surface charge distribution of mannoproteins needs to be studied in more detail by controlling surface charges of AgNPs.

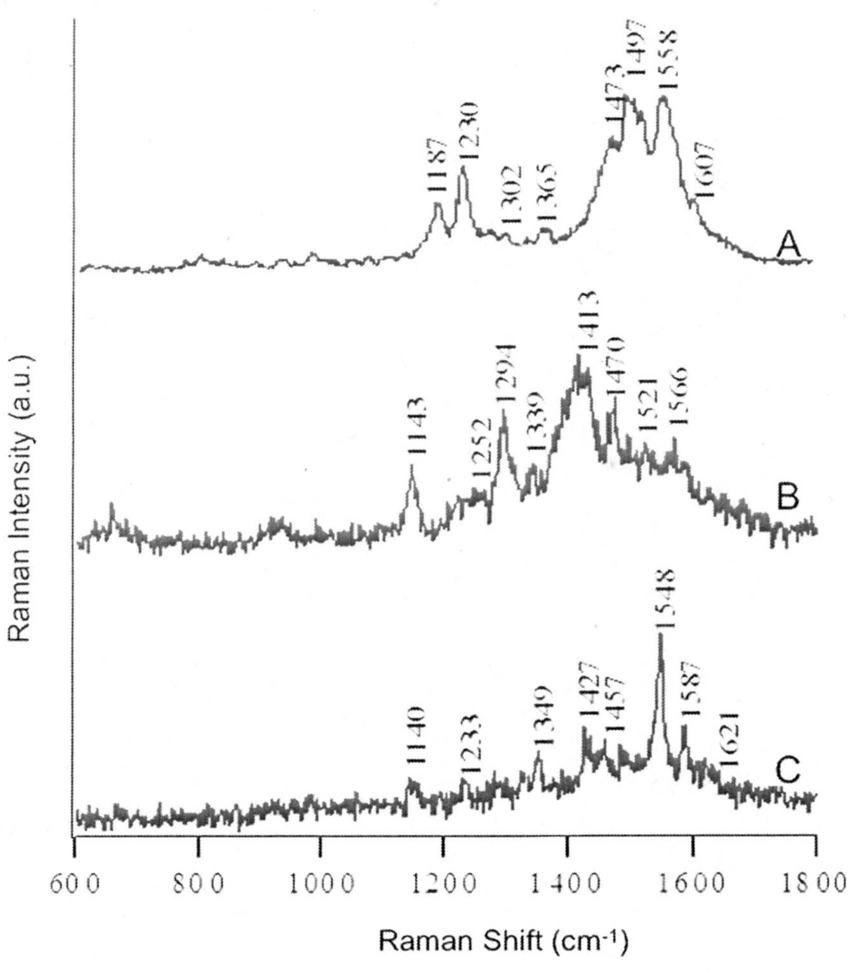

Figure 33. (A—C) SERS spectra of extracted mannoprotein from different AgNPs. (Reproduced with permission from reference (113). Copyright 2009 Springer.)

Finally, we point out the possibility and difficulty to obtain reproducible SERS spectra from cell wall surfaces. We have demonstrated that SERS enables us to detect single biomolecules, but the detected SERS spectra have always time and position dependence owing to dynamic interactions between NPs surface and molecules. The proteins are intrinsically dynamic systems with various factors that affect the unique interaction with AgNPs, i.e., flexibility/mobility, accessibility, polarity, exposed surface and turns. To obtain better reproducible SERS spectra, we need to largely improve the spectral, temporal, and spatial resolution of high speed and inexpensive spectroscopic systems.

References

1. Aroca, R. *Surface-Enhanced Vibrational Spectroscopy*; John Wiley & Sons Ltd.: Chichester, UK, 2006.
2. Kneipp, K.; Moskovits, M.; Kneipp, H. *Surface-Enhanced Raman Scattering: Physics and Applications*; Springer: Berlin, 2006.
3. Le Ru, E. C.; Etchegoin P. G. *Principles of surface-enhanced Raman spectroscopy and related plasmonic effects*; Elsevier: Amsterdam, Netherlands, 2009.
4. Itoh, T.; Sujith, A.; Ozaki, Y. In *Frontiers of Molecular Spectroscopy*; Laane, J., Ed.; Elsevier: Amsterdam, Netherlands, 2009; pp 289–320.
5. Stiles, P. L.; Dieringer, J. A.; Shah, N. C.; Van Duyne, R. P. *Annu. Rev. Anal. Chem.* **2008**, *1*, 601.
6. Han, X. X.; Zhao, B.; Ozaki, Y. *Anal. Bioanal. Chem.* **2009**, *394*, 1719.
7. Han, X. X.; Zhao, B.; Ozaki, Y. *Trends in Anal. Chem.*, in press.
8. Qian, X. M.; Nie, S. M. *Chem. Soc. Rev.* **2008**, *37*, 912.
9. Fleischmann, M.; Hendra, P. J.; McQuillan, A. J. *Chem. Phys. Lett.* **1974**, *26*, 163.
10. Jeanmaire, D. L.; Van Duyne, R. P. *J. Electroanal. Chem.* **1977**, *84*, 1.
11. Albrecht, M. G; Creighton, J. A. *J. Am. Chem. Soc.* **1977**, *99*, 5215.
12. Murphy, D. V.; Von Raben, K. U.; Chang, R. K.; Dorain, P. B. *Chem. Phys. Lett.* **1982**, *85*, 43.
13. Kneipp, K.; Wang, Y.; Kneipp, H.; Perelman, L.; Itzkan, I.; Dasari, R. R.; Feld, M. *Phys. Rev. Lett.* **1997**, *78*, 1667.
14. Nie, S.; Emory, S. *Science* **1997**, *275*, 1102.
15. Xu, H.; Bjerneld, E.; Käll, M.; Borjesson, L. *Phys. Rev. Lett.* **1999**, *83*, 4357.
16. Micheals, A.; Nirmal, M.; Brus, L. *J. Am. Chem. Soc.* **1999**, *121*, 9932.
17. Dieringer, J. A.; Lettan, R. B., II; Scheidt, K. A.; Van Duyne, R. P. *J. Am. Chem. Soc.* **2007**, *129*, 16249.
18. Cao, Y. C.; Jin, R.; Mirkin, C. A. *Science* **2002**, *297*, 1536.
19. Qian, X.; Peng, X. H.; Ansari, D. O.; Goen, Q. Y.; Chen, G. Z.; Shin, D. M.; Yang, L.; Young, A. N.; Wang, M. D.; Nie, S. M. *Nat. Biotechnol.* **2007**, *26*, 83.
20. Anker, J. N.; Hall, W. P.; Lyandres, O.; Shah, N. C.; Zhao, J.; Van Duyne, R. P. *Nat. Mater.* **2008**, *7*, 442.

21. Li, J. F.; Huang, Y. F.; Ding, Y.; Yang, Z. L.; Li, S. B.; Zhou, X. S.; Fan, F. R.; Zhang, W.; Zhou, Z. Y.; Wu, D. Y.; Ren, B.; Wang, Z. L.; Tian, Z. Q. *Nature* **2010**, *464*, 392.

22. Inoue, M.; Ohtaka, K. *J. Phys. Soc. Jpn* **1983**, *52*, 3853.

23. Xu, H.; Aizpurua, J.; Käll, M.; Apell, P. *Phys. Rev. E* **2000**, *62*, 4318.

24. Yoshida, K.; Itoh, T.; Tamaru, H.; Biju, V.; Ishikawa, M.; Ozaki, Y. *Phys. Rev. B* **2010**, *81*, 115406.

25. Wang, D.; Kerker, M. *Phys. Rev. B* **1981**, *24*, 1777.

26. Moskovits, M. *Rev. Mod. Phys.* **1985**, *57*, 783.

27. Pettinger, B. *J. Chem. Phys.* **1986**, *85*, 7442.

28. Xu, H.; Wang, X.; Persson, M. P.; Xu, H. Q.; Käll, M.; Johansson, P. *Phys. Rev. Lett.* **2004**, *93*, 243002.

29. Le Ru, E.; Etchegoin, P. *Chem. Phys. Lett.* **2006**, *423*, 63.

30. Lombardi, J. R.; Birke, R. L.; Lu, T.; Xu, J. *J. Chem. Phys.* **1986**, *84*, 4174.

31. Otto, A.; Mrozek, I.; Grabhorn, H.; Akemann, W. *J. Phys.: Condens Mater.* **1992**, *4*, 1143.

32. Campion, A.; Kambhampati, P. *Chem. Soc. Rev.* **1998**, *27*, 241.

33. Birke, R. L.; Znamenskiy, V.; Lombardi, J. R. *J. Chem. Phys.* **2010**, *132*, 214707.

34. Wu, D.; Li, J.; Ren, B.; Tian, Z. *Chem. Soc. Rev.* **2008**, *37*, 1025.

35. Imura, K.; Okamoto, H.; Hossain, M.; Kitajima, M. *Nano Lett.* **2006**, *6*, 2173.

36. Itoh, T.; Hashimoto, K.; Ozaki, Y. *Appl. Phys. Lett.* **2003**, *83*, 2274.

37. Itoh, T.; Hashimoto, K.; Ikehata, A.; Ozaki, Y. *Appl. Phys. Lett.* **2003**, *83*, 5557.

38. Itoh, T.; Hashimoto, K.; Ikehata, A.; Ozaki, Y. *Chem. Phys. Lett.* **2004**, *389*, 225.

39. Itoh, T.; Kikkawa, Y.; Biju, V.; Ishikawa, M.; Ikehata, A.; Ozaki, Y. *J. Phys. Chem. B* **2006**, *110*, 21536.

40. Itoh, T.; Biju, V.; Ishikawa, M.; Kikkawa, Y.; Hashimoto, K.; Ikehata, A.; Ozaki, Y. *J. Chem. Phys.* **2006**, *124*, 134708.

41. Itoh, T.; Yoshida, K.; Biju, V.; Kikkawa, Y.; Ishikawa, M.; Ozaki, Y. *Phys. Rev. B* **2007**, *76*, 085405.

42. Yoshida, K.; Itoh, T.; Biju, V.; Ishikawa, M.; Ozaki, Y. *Appl. Phys. Lett.* **2009**, *95*, 263104.

43. Yoshida, K.; Itoh, T.; Biju, V.; Ishikawa, M.; Ozaki, Y. *Phys. Rev. B* **2009**, *79*, 085419.

44. Itoh, T.; Asahi, T.; Masuhara, H. *Jpn. J. Appl. Phys.* **2002**, *41*, L76.

45. Itoh, T.; Uwada, T.; Asahi, T.; Ozaki, Y.; Masuhara, H. *Can. J. Anal. Sci. Spectrosc.* **2007**, *52*, 130.

46. Sönnichsen, C.; Geier, S.; Hecker, N.; Plessen, G.; Feldmann, J.; Ditlbacher, H.; Lamprecht, B.; Krenn, J.; Aussenegg, F.; Chan, V.; Spatz, J.; Möller, M. *Appl. Phys. Lett.* **2000**, *77*, 2949.

47. Taflove, A.; Hagness, S. *Computational electrodynamics*, 3rd ed.; Artech House: Norwood, 2005.

48. Ong, T.; Celli, V.; Marvin, A. *J. Opt. Soc. Am. A* **1994**, *11*, 759.

49. See, for example, Reather, H. *Surface plasmons on smooth and rough surfaces and on gratings*; Springer-Verlag: New York, 1988; Chapter 2.

50. Sen, T.; Sadhu, S.; Patra, A. *Appl. Phys. Lett.* **2007**, *91*, 043104.

51. Sen, T.; Patra, A. *J. Phys. Chem. C* **2008**, *112*, 3216.

52. Saini, S.; Srinivas, G.; Bagchi, B. *J. Phys. Chem. B* **2009**, *113*, 1817.

53. Preedy, V. R.; Burrow, G. N.; Watson, R. R. *Comprehensive Handbook of Iodine: Nutritional, Biochemical, Pathological, and Therapeutic Aspects*; Academic Press: New York, 2009.

54. *WHO/FAO Human Vitamin and Mineral Requirements: Report of a joint FAO/WHO expert consultation Bangkok, Thailand*; WHO/FAO: Rome, 2002.

55. Tamošiūnas, V.; Padarauskas, A.; Pranaitytė, B. *Chemija* **2006**, *17*, 21–24.

56. Galanti, L. M. *Clin. Chem.* **1997**, *43*, 184–185.

57. Toraño, J. S.; van Kan, H. J. M. *Analyst* **2003**, *128*, 838–843.

58. Murphy, P. In *Handbook of hydrocolloids*; Phillips, G. O., Williams, P. A., Eds.; Woodhead Publishing Ltd.: Cambridge, U.K., 2000; pp 41–65.

59. Pienpinijtham, P.; Thammacharoen, C.; Ekgasit, S. *Macromol. Res.* DOI: 10.1007/s13233-012-0162-7.

60. Usher, A.; McPhail, D. C.; Brugger, J. *Geochim. Cosmochim. Acta* **2009**, *73*, 3359–3380.

61. Link, S.; El-sayed, M. A. *J. Phys. Chem. B* **1999**, *103*, 4212–4217.

62. Krochta, J. M.; Hudson, J. S.; Tillin, S. *J. Am. Chem. Soc. Div. Fuel Chem.* **1987**, *32*, 148–156.

63. Golova, O. P.; Nosova, N. I. *Russ. Chem. Rev.* **1973**, *42*, 327–338.

64. Jackson, D. S.; Choto-Owen, C.; Waniska, R. D.; Rooney, L. W. *Cereal Chem.* **1988**, *65*, 493–496.

65. Knill, C. J.; Kennedy, J. F. *Carbohydr. Polym.* **2003**, *51*, 281–300.

66. Han, J.-A.; Lim, S.-T. *Carbohydr. Polym.* **2004**, *55*, 193–199.

67. Ji, X.; Song, X.; Li, J.; Bai, Y.; Yang, W.; Peng, X. *J. Am. Chem. Soc.* **2007**, *129*, 13939–13948.

68. Zhang, H.; Xu, J. J.; Chen, H. Y. *J. Phys. Chem. C* **2008**, *112*, 13886–13892.

69. Socrates, G. *Infrared and Raman Characteristic Group Frequencies: Tables and Charts*, 3rd ed.; John Wiley & Sons Ltd: Chichester, U.K., 2001.

70. Pienpinijtham, P.; Han, X. X.; Ekgasit, S.; Ozaki, Y. *Anal. Chem.* **2011**, *83*, 3655–3662.

71. Bron, M.; Holze, R. *Electrochim. Acta* **1999**, *45*, 1121–1126.

72. Han, X. X; Pienpinijtham, P.; Zhao, B.; Ozaki, Y. *Anal. Chem.* **2011**, *83*, 8582–8588.

73. Murgida, D. H.; Hildebrandt, P. *Acc. Chem. Res.* **2004**, *37*, 854–861.

74. Cao, Y. C.; Jin, R.; Nam, J.-M.; Thaxton, C. S.; Mirkin, C. A. *J. Am. Chem. Soc.* **2003**, *125*, 14676–14677.

75. Han, X. X.; Jia, H. Y.; Wang, Y. F.; Lu, Z. C.; Wang, C. X.; Xu, W. Q.; Zhao, B.; Ozaki, Y. *Anal. Chem.* **2008**, *80*, 2799–2804.

76. Han, X. X.; Huang, G. G.; Zhao, B.; Ozaki, Y. *Anal. Chem.* **2009**, *81*, 3329–3333.

77. Han, X. X.; Chen, L.; Ji, W.; Xie, Y. F.; Zhao, B.; Ozaki, Y. *Small* **2011**, *7*, 316–320.

78. Han, X. X.; Cai, L. J.; Guo, J.; Wang, C. X.; Ruan, W. D.; Han, W. Y.; Xu, W. Q.; Zhao, B.; Ozaki, Y. *Anal. Chem.* **2008**, *80*, 3020–3024.

79. Han, X. X.; Kitahama, Y.; Tanaka, Y.; Guo, J.; Xu, W. Q.; Zhao, B.; Ozaki, Y. *Anal. Chem.* **2008**, *80*, 6567–6572.

80. Han, X. X.; Kitahama, Y.; Itoh, T.; Wang, C. X.; Zhao, B.; Ozaki, Y. *Anal. Chem.* **2009**, *81*, 3350–3355.

81. Han, X. X.; Chen, L.; Guo, J.; Zhao, B.; Ozaki, Y. *Anal. Chem.* **2010**, *82*, 4102–4106.

82. Han, X. X.; Xie, Y.; Zhao, B.; Ozaki, Y. *Anal. Chem.* **2010**, *82*, 4325–4328.

83. Beveridge, T. J. *J. Bacteriol.* **1999**, *181*, 4725–4733.

84. Osumi, M. *Micron* **1998**, *29*, 207–233.

85. Rietschel, E. T.; Brade, H.; Holst, O.; Brade, L.; Muller-Loennies, S.; Mamat, U.; Zahringer, U.; Beckmann, F.; Seydel, U.; Brandenburg, K.; Ulmer, A. J.; Mattern, T.; Heine, H.; Schletter, J.; Loppnow, H.; Schonbeck, U.; Flad, H. D.; Hauschildt, S.; Schade, U. F.; Di Padova, F.; Kusumoto, S.; Schumann, R. R. *Curr. Top. Microbiol. Immunol.* **1996**, *216*, 39–81.

86. Sengupta, A.; Mujacic, M.; Davis, E. J. *Anal. Bioanal. Chem.* **2006**, *386*, 1379–1386.

87. Sengupta, A.; Laucks, M. L.; Davis, E. J. *Appl. Spectrosc.* **2005**, *59*, 1016–1023.

88. Liu, Y.; Chao, K.; Nou, X.; Chen, Y.-R. *Sens. Instrum. Food Qual.* **2009**, *3*, 100–107.

89. Jarvis, R. M.; Goodacre, R. *Anal. Chem.* **2004**, *76*, 40–47.

90. Jarvis, R. M.; Goodacre, R. *Chem. Soc. Rev.* **2008**, *37*, 931–936.

91. Jarvis, R. M.; Law, N.; Shadi, I. T.; O'Brien, P.; Lloyd, J. R.; Goodacre, R. *Anal. Chem.* **2008**, *80*, 6741–6746.

92. Zeiri, L.; Bronk, B. V.; Shabtai, Y.; Eichler, J.; Efrima, S. *Appl. Spectrosc.* **2004**, *58*, 33–40.

93. Zeiri, L.; Bronk, B. V.; Shabtai, Y.; Czégé, J.; Efrima, S. *Colloids Surf., A* **2002**, *208*, 357–362.

94. Efrima, S.; Bronk, B. V. *J. Phys. Chem. B* **1998**, *102*, 5947–5950.

95. Kitahama, Y.; Itoh, T.; Ishido, T.; Hirano, K.; Ishikawa, M. *Bull. Chem. Soc. Jpn.* **2011**, *84*, 976–978.

96. Premasiri, W. R.; Moir, D. T.; Klempner, M. S.; Krieger, N.; Jones, G., II; Ziegler, L. D. *J. Phys. Chem. B* **2005**, *109*, 312–320.

97. Premasiri, W. R.; Gebregziabher, Y.; Ziegler, L. D. *Appl. Spectrosc.* **2011**, *65*, 493–499.

98. Alexander, T. A.; Pellegrino, P. M.; Gillespie, J. B. *Appl. Spectrosc.* **2003**, *57*, 1340–1345.

99. Bálint, Š.; Kreuzer, M. P.; Rao, S.; Badenes, G.; Miškovský, P.; Petrov, D. *J. Phys. Chem. C* **2009**, *113*, 17724–17729.

100. Biju, V.; Pan, D.; Gorby, Y. A.; Fredrickson, J.; McLean, J.; Saffarini, D.; Lu, H. P. *Langmuir* **2007**, *23*, 1333–1338.

101. Lee, P. C.; Meisel, D. J. *J. Phys. Chem.* **1982**, *86*, 3391–3395.

102. Bjerneld, E. J.; Svedberg, F.; Käll, M. *Nano Lett.* **2003**, *3*, 593–596.

103. Itoh, T.; Biju, V.; Ishikawa, M.; Ito, S.; Masuhara, H. *Appl. Phys. Lett.* **2009**, *94*, 144105.

104. Kudelski, A.; Pettinger, B. *Chem. Phys. Lett.* **2000**, *321*, 356–362.

105. Giese, B.; McNaughton, C. *J. Phys. Chem. B* **2002**, *106*, 101–112.

106. Kapteyn, J. C.; Ende, H. V. D.; Klis, F. M. *Biochim. Biophys. Acta* **1999**, *1426*, 373.

107. Kneipp, J.; Kneipp, H.; McLaughlin, M.; Brown, D.; Kneipp, K. *Nano Lett.* **2006**, *6*, 2225.

108. Podstawka, E.; Ozaki, Y.; Proniewicz, M. *Appl. Spectrosc.* **2004**, *58*, 570.

109. Chaffin, W. L.; Pez-Ribot, J. L.; Casanova, M.; Gozalbo, D.; Marti´nez, J. P. *Microbiol. Mol. Biol. Rev.* **1998**, *62*, 130.

110. Klis, F. M.; Boorsma, A.; De Groot, P. W. A. *Yeast* **2006**, *23*, 185.

111. Sumita, T.; Yoko-o, T.; Shimma, Y.; Jigami, Y. *Eukaryotic Cell* **2005**, *4*, 1872.

112. Sujith, A.; Itoh, T.; Abe, H.; Anas, A. A.; Yoshida, K.; Biju, V.; Ishikawa, M. *Appl. Phys. Lett.* **2008**, *92*, 103901.

113. Sujith, A.; Itoh, T.; Abe, H.; Yoshida, K. I.; Kiran, M. S.; Biju, V.; Ishikawa, M. *Anal. Bioanal. Chem.* **2009**, *394*, 1803–1809.

114. Futamata, M.; Maruyama, Y.; Ishikawa, M. *J. Phys. Chem. B* **2003**, *107*, 7607.

115. Kudelski, A.; Pettinger, B. *Chem. Phys. Lett.* **2004**, *383*, 76.

116. Podstawka, E.; Ozaki, Y.; Proniewicz, L. M. *Appl. Spectrosc.* **2004**, *58*, 581.

Chapter 10

Current Progress on Surface-Enhanced Raman Scattering Chemical/Biological Sensing

Justin L. Abell,[*,a] Jeremy D. Driskell,[b] Ralph A. Tripp,[c] and Yiping Zhao[d]

[a]Department of Biological and Agricultural Engineering, University of Georgia, Athens, GA 30602
[b]Department of Chemistry, Illinois State University, Normal, IL 61790
[c]Department of Infectious Diseases, University of Georgia, Athens, GA 30602
[d]Department of Physics and Astronomy, University of Georgia, Athens, GA 30602
[*]E-mail: jabell@uga.edu

For several decades, surface-enhanced Raman scattering (SERS) has proven to be an extremely sensitivity detection technique with superior chemical specificity. Continuing advancements in nanostructure fabrication have dramatically advanced SERS capabilities as a powerful chemical and biological sensing platform. Despite this potential, to date SERS has yet to become a mainstream, commercially viable detection technology. This chapter highlights advancements in SERS-based biosensing techniques using silver nanorod (AgNR) substrates fabricated by oblique-angle deposition (OAD). Specifically, the superior features of highly sensitive and reproducible OAD-AgNR substrates for SERS sensing are discussed with an emphasis on biodetection applications. The chapter also covers the progress made towards realizing OAD as a versatile technique capable of future SERS-device development. In addition, the potential for OAD integration with traditional nanofabrication techniques and manufacturing processes to ultimately fulfill the need for a mainstream SERS technology is also discussed.

© 2012 American Chemical Society

Introduction

Raman spectroscopy (RS) can be used to probe the chemical composition of a sample by measuring the inelastic scattering of photons incident upon molecular bonds (*1*). This interaction manifests as a shift in the energy of the scattered photons, of which a highly resolved spectrum containing a wealth of chemical information can be obtained; the high spectral resolution produces a unique signature for each type of compound analyzed. In addition to providing a direct measure of the chemical composition, information regarding the molecular orientation, configuration, and bond strength can also be ascertained. These characteristics thus provide RS with a distinct advantage over other analytical techniques such as absorbance or fluorescence spectroscopy, which suffer from poor specificity as only few, large broad spectral features are achievable. Furthermore, unlike infra-red (IR) spectroscopy, RS is for the most part insensitive to the presence of water for most applications, opening the possibility for analysis of biological samples. Finally, RS is possible with simple, portable equipment, and is therefore not hindered by large, expensive instrumentation like mass spectrometry. Despite these attractive characteristics, however, RS suffers from very poor sensitivity; alas, such a shortcoming disqualifies traditional RS from any sort of trace chemical detection and analysis. Fortunately, this stifling drawback can be overcome by using plasmonics, ultimately allowing RS to be employed as one of the most sensitive analytical techniques currently known.

When molecules are adsorbed on nanostructured noble metal surfaces, e.g. silver and gold, the excitation beam used to generate the Raman scattering can also induce intense localized electromagnetic (EM) fields at the surface of the metal nanostructures. These EM-fields can greatly enhance the Raman process, and such an effect is known as surface-enhanced Raman scattering (SERS) and provides an enormous increase in the sensitivity of RS. The magnitude of this enhancement is characterized by the enhancement factor (EF) and is routinely on the order of 10^6-10^8 (*1*). Under special conditions, extreme enhancement allows for single molecule detection (*2, 3*).

Since its discovery in the mid-70's (*4*), SERS has attracted the interest of researchers from many different fields of expertise. The high sensitivity and superior chemical specificity, in conjunction with the other aforementioned analytical benefits of RS, has driven researchers to evaluate SERS as a platform for trace chemical and biological detection. The research community has recently focused on developing a more complete understanding of the SERS mechanism by which to better engineer surfaces for maximization of the enhanced signals. Researchers have also been evaluating the capabilities of SERS as a chemical and bioanalytical detection platform for various translational applications (*5*). Despite this increased level of research, SERS has yet to significantly penetrate into the commercial and industrial sectors, and its use has been reserved almost exclusively to research laboratories (*6*). This chapter examines a novel fabrication technique known as oblique-angle deposition (OAD) and how it can be used to generate silver nanorod (AgNR) SERS substrates with the requisite analytical qualities for a wide-range of bioanalytical applications. This chapter also discusses recent advancements regarding AgNR-SERS biodetection, highlights the development and application of OAD-AgNR substrates for integration with large-scale manufacturing processes, and demonstrates the versatility necessary for SERS device development.

SERS Substrate Fabrication

Conventional Nanofabrication of SERS Substrates

To achieve the surface enhancement effect, the essential requirements are that the surface be composed of a particular metal (i.e. gold, silver, or copper) and have appropriate nanoscale features/topologies to induce the necessary EM enhancement during optical excitation (*1, 7, 8*). A multitude of different types of SERS substrates have been developed over the years to meet these needs; however, each type of substrate has a unique set of advantages and disadvantages that are intimately correlated with the method of fabrication (Table 1). Indeed, one of the primary barriers to stifle the development of SERS as a mainstream technology entails the issue of SERS substrate fabrication (*6*).

Table 1. Comparison of different SERS substrate fabrication processes

Substrate	*Advantages*	*Disadvantages*
ORC electrode	Surface can be regenerated *in situ*	Poor measurement reproducibility and surface uniformity
EBL	Precise control of nanostructure positions and morphology; excellent reproducibility and uniformity	Time consuming; small sensing area
Suspended colloids	Easy fabrication	Low sensitivity
Immobilized colloids	Higher sensitivity than suspended colloids	Special techniques needed to improve uniformity and reproducibility
PVD template methods	Allows for precise nanoscale morphology; can be very uniform with large sensing area	Limited nanostructure design (typically restricted to rudimentary structures)

Oxidation-reduction cycling (ORC) of a metal electrode surface was the first type of SERS substrate used (*4*), and is still widely employed for fundamental SERS studies, but precise control of the nanoscale surface morphology is unfeasible to obtain on a routine basis. Although such a technique allows for the surface to be regenerated *in situ*, the difficulty to achieve uniform, reproducible nanoscale surface morphology with ORC translates to poor SERS measurement reproducibility. In contrast, electron beam lithography (EBL) allows for extremely precise, controllable, and reproducible arrays of nanostructures; however, each individual nanostructure must be fabricated one at a time in a serial manner (9). Considering the sub-micron feature size requirement of SERS-active nanostructures, only a relatively small sensing area (typically only tens of microns across) can be fabricated in a reasonable amount of time. With current manufacturing technology, such a structure-by-structure fabrication method is completely unsuited for the batch-style fabrication necessary for large-scale production.

Metal colloids are another popular SERS substrate due to their ease of fabrication. In this case, a metal precursor solution such as $AgNO_3$ or $HAuCl_4$ is mixed with a reducing (e.g. $NaBH_4$) and capping agents (e.g. citrate) to generate dispersed colloids (10). The SERS enhancement of the dispersed colloid can be greatly improved by adding an aggregating agent such as a salt. However, these liquid phase substrates demonstrate relatively low sensitivity as the SERS-active colloids are suspended throughout a 3-dimensional volume, rather than being constrained to the focal plane of the excitation and collection optics of the measurement setup. Immobilizing the colloid to the surface of a solid support can circumvent the low-sensitivity problem, but controlling the uniform dispersion of the nanoparticles onto the fixed support over a larger area is challenging.

Another popular means to fabricate SERS-substrates entails physical vapor deposition (PVD). Traditionally used by the semiconductor and optical coating industries to uniformly deposit smooth, thin films over a large area, PVD has been successfully used to produce SERS-active nano-structured metal surfaces as well. For example, depositing ultra-thin films of Ag or Au produces nano-sized metal "island" structures on the substrate surface (11). Alternatively, SERS-active surfaces can be fabricated using PVD to deposit thin metal films onto a pre-patterned template. Deep-UV lithography (12), chemical etching (13), or casting non-metal (e.g. SiO_2) nanospheres onto a surface (14) are typical methods to pattern a surface prior to metal deposition. These PVD methods have been heavily employed throughout the SERS community, but generally do not provide much control over the resulting nanostructure, i.e. only relatively simple shapes (spheres, squares, or lines) can be used as the template. In addition, they may suffer from many of the same disadvantages discussed with the other types of substrates, namely poor surface uniformity, small sensing area, and/or low sensitivity. To address these issues, we will turn our attention to SERS substrate fabrication using a different PVD method.

Oblique-Angle Deposition (OAD)

During PVD, vapor originating from the source material arrives at a substrate surface in a unidirectional, line-of-sight manner. If a flat surface is oriented perpendicular to the incident vapor, the vapor material will condense onto the substrate and form a smooth, continuous film. However, the substrate surface may also be rotated to a grazing angle with respect to the incident vapor. In this latter case, known as oblique angle deposition (OAD), the resulting film structure becomes highly sensitive to the surface roughness (15). During the initial stages of OAD, the depositing material forms small islands that begin to block, or shadow, vapor from depositing on the surface immediately behind them.

This results in preferential deposition of the vapor material onto these nucleation centers. With continued deposition, these nuclei begin to acquire an anisotropic structure, ultimately growing into rod-shaped structures that uniformly tilt towards the vapor incident direction and align parallel to each other. The resulting tilt angle, rod length, rod width, and rod separation can be controlled by tuning the vapor deposition rate, time, and angle. Furthermore, changing the deposition angle (either continuously or incrementally) during the deposition allows the formation of bends or joints that can yield multi-layer structures, such as zigzags or helices. In addition, most materials compatible with traditional PVD can also be implemented with OAD to produce nanorod films using many different materials. Previous studies have found that Ag nanorod arrays produced by OAD can act as excellent SERS-active substrates.

OAD-Generated Silver Nanorod Arrays for SERS

Performing OAD with a SERS-active material, e.g. Ag, produces uniform films of aligned and tilted silver nanorods (AgNR) arrays that have superior SERS characteristics (*16–19*). These substrates possess many of the requisite qualities of an ideal SERS substrate: 1) The AgNR substrates demonstrate very high sensitivity and good point-to-point and substrate-to-substrate reproducibility; 2) The PVD nature of the AgNR fabrication means that uniform, large area substrates can be reproducibly fabricated in a batch-style approach, thus making this method attractive for large-scale manufacturing; 3) Assuming that the substrate surface is sufficiently smooth and flat, the AgNR films can be readily deposited onto substrates of any size or shape; 4) The low-temperature nature of OAD means that it is compatible with other fabrication and manufacturing processes, and can therefore be used in conjunction with temperature-sensitive materials such as plastics; and 5) The anisotropic nature OAD fabrication allows for the development of a diverse set of SERS-active nanostructures that can be customized for specific SERS applications.

To fabricate the AgNR substrates, an electron-beam deposition system can be used to deposit silver onto cleaned glass slides, typically 1×1 cm^2 in size, which has been discussed in detail in previous publications (*17, 18*). Briefly, after loading the substrates into the evaporation chamber, the chamber pressure is evacuated to $\leq 10^{-6}$ Torr before starting the deposition. The deposition rate and thickness is monitored by a quartz crystal microbalance (QCM). The substrate surface normal is initially kept parallel with the incident vapor direction; thus, smooth thin films are generated in this configuration. Prior to depositing any silver, a Ti film 20 nm thick is typically deposited to act as an adhesive interface between the glass and the subsequent Ag film High purity silver (99.999%) is vaporized at a nominal deposition rate of 0.2-0.4 nm/s. A 100-500 nm Ag thin film may be deposited to act as a base layer for the subsequent AgNR deposition. To deposit the AgNR array, the substrate is rotated 86° relative to the vapor direction allowing for the formation of the AgNRs. The length of the nanorods is controlled via the deposition time and/or deposition rate. The schematic diagram in Figure 1 illustrates the OAD fabrication process.

The morphological and SERS characteristics of the OAD-AgNR SERS substrates have been previously reported (*17, 18, 20*). Scanning electron microscopy (SEM) was used to observe the morphology of the resulting AgNR film after fabrication via OAD. The nanorods show a very uniform tilting direction, with a tilting angle of 71° with respect to the surface normal. The average nanorod length is ~900 ± 100 nm with a diameter of ~100 ± 30 nm and an average inter-rod gap of 100-200 nm. The density of the nanorod array on the surface is ~13 ± 0.5 nanorods/μm^2. Figure 2 shows a representative SEM image of the AgNRs. To characterize the SERS response, a 2 μl droplet of the Raman reporter trans-1,2-bis(4-pyridyl)-ethene (BPE) diluted in methanol to a concentration of 10^{-7} M was applied to the substrate. The droplet spread out uniformly and completely covered the 1× 1 cm^2 AgNR sample and the methanol quickly dried. The SERS spectrum was acquired by probing the AgNRs with a confocal micro-Raman system equipped with a 20× objective and a 785 nm excitation laser; the laser power was set to ~15 mW and the acquisition time was 1s.

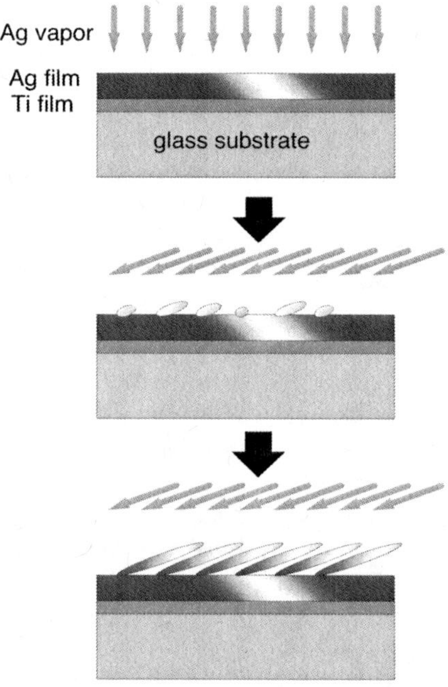

Figure 1. Schematic diagram of the OAD process to fabricate AgNRs. From top to bottom: Ti and Ag films are deposited onto a clean glass substrate, then the substrate is rotated to a grazing angle (e.g. 86°) relative to the Ag vapor at which point the depositing Ag forms nucleation centers, the Ag deposition continues whereby the nucleation centers elongate to rod structures.

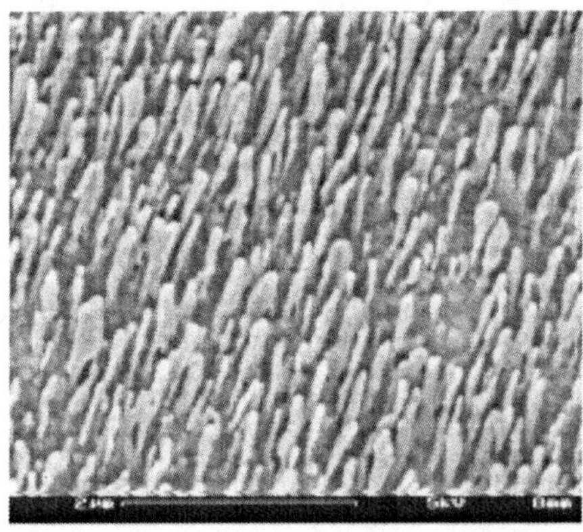

Figure 2. SEM image of AgNRs (QCM = 2000 nm) deposited on Ag film. (Reproduced with permission from Ref (18). Copyright 2008 American Chemical Society.)

Based on the area and the concentration of the analyte, the resulting surface coverage of the BPE is estimated to be only ~ 0.01%, yet a high quality BPE spectrum with a high signal-to-noise ratio is readily achieved. To determine how the nanorod length effects the resulting SERS enhancement, AgNR substrates with different average nanorod lengths were deposited and the enhancement factor was estimated to be at a maximum (~5 × 10⁸) when the nanorod length is ~ 850 nm. The reproducibility of the SERS measurements was performed by probing multiple points on a single substrate as well as comparing the SERS signal intensity from different substrates. Point-to-point measurements on a single substrate, as well as substrate-to-substrate measurements yield a relative standard deviation of signal intensity of < 15%. In addition, employing a dynamic measurement technique to translate, or raster, the detection spot continuously across the surface during a measurement can effectively average out subtle variations resulting from spatially-dependent variations in the AgNR morphology or analyte distribution (21). The point-to-point relative standard deviation can be further reduced to ~3%, which indicates that the AgNR films can overcome the poor reproducibility that plagues many SERS studies.

Other experimental factors affecting the SERS response of the AgNR substrate have also been investigated. The polarization of the excitation beam has a noticeable effect on the measured SERS intensity (22). This polarization-dependent behavior is due to the anisotropic nature of the tilted nanorod arrays. The measured SERS intensity is shown to be approximately 25% greater when the E-field of the excited light is perpendicular to the AgNR growth direction. Moreover, the angle of the excitation beam relative to the AgNR tilting direction has also been shown to have significant effect (23). For BPE adsorbed onto the AgNR surface, the measured SERS intensity shows a steady 8-fold increase as the angle of excitation beam is increased to 45° with respect to the surface normal, and then decreases sharply as the angle is increased further. This observation has been explained with a classical electrodynamic dipole radiation model assuming the BPE is adsorbed to the AgNRs in a vertical manner. The values predicted by this theory very closely match those of the empirical data (*24*). The effect of the underlying surface beneath the AgNR film has been shown to greatly affect the sensitivity of the measurements (*18, 25*). Different materials have been evaluated for the underlying film such as Si, Ti, Al, and Ag. By increasing the reflectivity of this surface, the SERS signal can be increased. In fact, by depositing a thin film of silver (typically 100-500 nm in thickness) prior to depositing the nanorod film, the SERS intensity can be improved by approximately one order of magnitude.

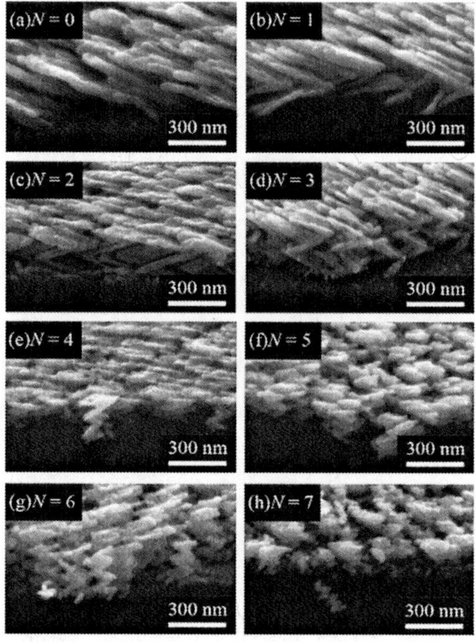

Figure 3. Bent AgNRs deposited with increasing number of bends N by rotating the substrate 180° for each arm. The total nanostructure length is kept constant at ~650 nm. (Reproduced with permission from reference (26). Copyright 2012 American Institute of Physics.)

These tilted, straight nanorods provide a suitable substrate for highly sensitive, reproducible SERS measurements. However, the OAD technique offers the necessary versatility to fabricate more intricate SERS-active nanostructured substrates. By changing the azimuthal angle (i.e. the substrate is rotated within the substrate plane) during the deposition, the nanorods will continue to grow towards the vapor source, but now at a different angle relative to the original nanorod growth direction. For example, if the azimuthal angle is rotated by 180°, the nanorod growth direction will also change by 180°, resulting in a 'bent', or chevron-shaped, nanorods (*15*). The length of each arm between bends can be readily adjusted by controlling the time that the substrate is held at each orientation during the deposition. Using the aforementioned deposition scheme, and repetitively alternating the deposition angle by 180°, yields zigzag AgNRs (Figure 3) (*26*). Figure 3 shows 8 different zigzag AgNR substrates, each varying the number of bends N, but keeping the total length L of the nanostructure the same at 655 ± 50 nm; thus, for a given deposition, the length of each arm of the zigzag structure is $L/(N+1)$. The SERS signal of the Raman reporter R6G, shows a 3-fold intensity increase as N increases from 0 to 4. We believe that regions of high enhancement, i.e. hot spots, are located at the bends between the arms, and by increasing the number of hot spots the signal intensity should increase. However, when $N > 4$ the number of hot spots still increases, but a significant decrease in the measured intensity is observed. To explain this, we point out that the aspect ratio γ of the arms decreases as N increases, and when $4 \leq N \leq 7$, this means $3 \leq \gamma \leq 4$. This change in the aspect ratio could be causing the localized surface plasmon resonance (LSPR) frequency to be moving away from that of the 785 nm excitation source, ultimately yielding less efficient enhancement.

Moreover, using different azimuthal angles can form more intricate nanorod structures. For example, changing the azimuthal angle by only 90° after each nanorod arm deposition, and repeating this procedure for $N \geq 3$ arms (here, N refers to number of arms, not bends) will yield a square-helix structure. We performed this procedure to form helical AgNRs with $N = 3$ to 8 arms, but in contrast to the zigzag nanorods, the length of each arm was fixed at ~ 800 nm regardless of N (Figure 4A) (*27*). Using BPE as a Raman probe, and dispersing an equal number of molecules uniformly onto to each substrate, we find that the SERS intensity increase by approximately 5-fold when the number of arms increases from $N = 3$ to $N = 5$; when $N > 5$, the SERS intensity levels off and does not show any further significant change (Figure 4B). Just as with the zigzag AgNRs, we believe that the number of SERS 'hot spots' is proportional to the number of bends in the helical structure, which are localized at the joints of the helix. On the other hand, because the excitation laser intensity is attenuated as it penetrates the AgNR film, the BPE molecules adsorbed deeper inside the film will be only weakly excited, and have less contribution to the overall signal. The resulting SERS signal of these BPE molecules will likewise be attenuated as it propagates back through the film towards the detector. Thus, a complete loss of signal could occur for BPE molecules adsorbed below the first 5 arms of the AgNR helical film. Another explanation for the lack of change in SERS intensity beyond $N = 5$ is that the BPE molecules may be completely adsorbed to the top 5 layers when the dilute (10^{-5} M) BPE solution is applied, i.e. the BPE analyte is being filtered out of the sample solution by the top several layers of the AgNRs before a significant number of molecules can penetrate deeper into the film.

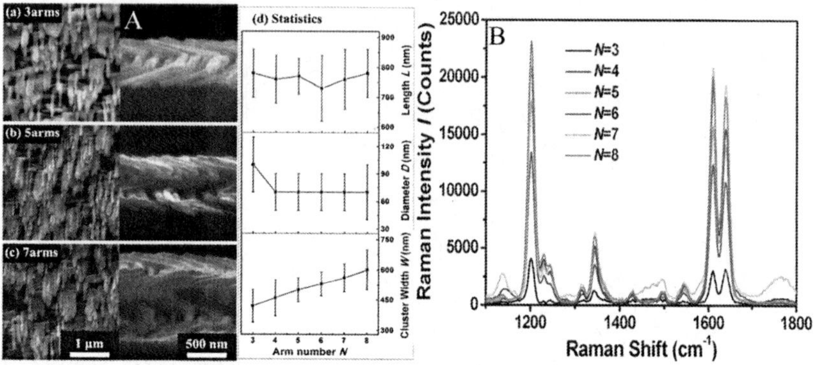

Figure 4. A) SEM images of helical AgNRs (right column = top view, left column = cross section) with number of arms N = a) 3, b) 5, and c) 7; d) corresponding morphological statistics for arm length, arm diameter, and cluster width. B) The SERS spectra of BPE adsorbed to helical AgNRs with N = 3 to 8. (Reproduced with permission from Ref (27). Copyright 2011 The Royal Society of Chemistry.)

AgNRs for Biodetection

SERS provides remarkable sensitivity and high chemical specificity; therefore, a large interest has emerged to develop SERS as a biodetection platform. Likewise, we have shared this same motivation, and have actively investigated the use of highly sensitive and reproducible OAD-generated AgNR substrates to detect and analyze various biologically relevant samples such as virus, bacteria, as well as biomarkers such as microRNA (miRNA). The following sections highlight the progress that has been made towards realizing the use of AgNR substrates as practical SERS biosensors.

SERS Virus Detection

Rapid, reliable viral detection is a non-trivial necessity for the medical and public health sectors. Faster and more accurate detection of viral infection is critical for proper clinical treatment and epidemic prevention. Traditionally, this detection necessity has relied upon antibody-based techniques such as enzyme-linked immunosorbant assays (ELISA) or DNA amplification methods such as polymerase chain reaction (PCR). These techniques require an extrinsic label (i.e. a fluorophore or radiolabel) to indirectly detect pathogens. This labeling approach, however, increases the cost, complexity, and processing time of the detection assays. The following sections highlight how AgNR SERS substrates offer a potential alternative for simple, rapid, and direct label-free viral detection.

AgNRs have been evaluated for label-free SERS detection of respiratory syncytial virus (RSV), i.e. without the use of an extrinsic label or capture molecule (*28*). In this study, the specificity of AgNRs SERS substrates to differentiate between different types of virus, including adenovirus, rhinovirus, HIV, RSV, and influenza was evaluated, as were the intra-species samples, *i.e.*, different strains, of the RSV and influenza. This study did not use surface-immobilized probe molecules such as antibodies to capture the viral antigens at the surface of AgNRs. SERS detection of virus was therefore evaluated in the presence of complex biological backgrounds.

All of the aforementioned virus samples except influenza were propagated and harvested in the Vero cell lysate (VCL) and centrifuged to remove cellular debris. The virus titers were determined to range from 5×10^6 to 1×10^7 PFU/ml. A similar process was used for uninfected Vero cells to generate the negative control (VCL without virus). Influenza virus was propagated in embryonated chicken eggs and the virus titers were determined to be 10^7-10^8 EID$_{50}$. The corresponding negative control was naïve allantoic fluid. Approximately 1 μl of these viral samples were applied to AgNR substrates and allowed to dry.

The resulting spectra of many of the different virus species show significant similarities yet notable differences as well. These differences suggest that different types of virus can be discriminated solely based on the intrinsic SERS signal without the use of a capture probe or labeling molecules. However, because the virus is suspended in a complex biological matrix background (VCL) one can assume that the viral component is co-adsorbing to the AgNR surface with many other matrix components. To help differentiate the background from the viral signal, the spectra of the RSV+VCL sample was compared to that of the negative control (VCL only) as well as purified RSV (Figure 5A). The RSV purification was accomplished via sucrose gradient centrifugation, followed by dialysis against PBS. The spectra from these samples show that the purified RSV and the VCL+RSV share peaks at 1066 cm^{-1} (assigned as a C-N stretching mode), 835 cm^{-1} (assigned to tyrosine), and 545/523 cm^{-1} (doublet peak, assigned to the disulfide stretch), which are either very weak or absent in the VCL spectra (highlighted with grey bands in Figure 3A). As might be expected, the VCL and RSV+VCL spectra also demonstrate some spectral features that are not found in the purified RSV, such as a large double peak between 1600 and 1650 cm^{-1} and a peak at 650 cm^{-1}, which are presumably originating from components in the cellular debris of the VCL. It is important to note that some spectral features are exclusive to the RSV spectra, e.g. peaks arising from the C-N stretching vibration, tyrosine, or disulfide bridge; however, these same components are common chemical constituents of proteins and may therefore be expected to detected from ubiquitous components undoubtedly present in many biological samples including the VCL. Admittedly, the exact reason as to why these spectral bands are not detectable in the presence of protein-containing samples such as VCL is elusive. but this may be due to the preferential binding of viral components that contain more of these particular chemical constituents.

The specificity of the AgNR SERS substrates was further investigated by measuring the SERS signal of several different strains of a single virus species. Four different strains of RSV (A2, A/Long, B1, and ΔG) were applied to AgNRs and the spectra collected in a manner similar to that previously described. As expected, the four different strains produce very similar spectra, and demonstrate subtle differences in the relative intensity or positions of the spectral peaks (Figure 5B). Some spectral variability exists between different points measured within the same sample, but averaging these replicate measurements and normalizing to the highest peak intensity yields spectra that can be used to visually determine spectral differences between the highly similar strains. For example, the main peak for A2, B1, and ΔG occurs at 1042-1045 cm^{-1} (C-N stretching), while this same band is blue shifted slightly to 1055 cm^{-1}. Moreover, the A/Long also demonstrates a few unique peaks at 877 and 633 cm^{-1} not observed with the other three strains.

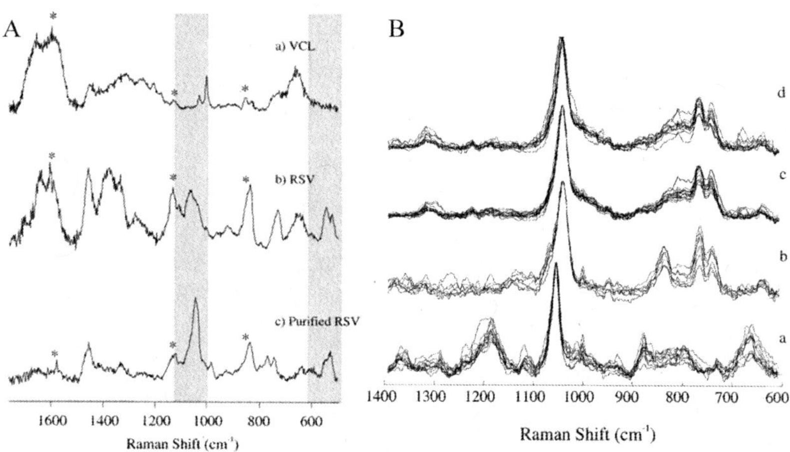

Figure 5. A) Representative SERS spectra of a) uninfected VCL, b) RSV-infected cell lysate, and c) purified RSV. The grey bands highlight the spectral regions with features present in the samples containing RSV, but not those containing only VCL. The asterisks identify spectral features that appear to result from the background. B) SERS spectra of different RSV strains: a) A/Long, b) B1, c) ΔG, and d) A2. Each plot shows multiple points on replicate substrates measured for each strain; each spectrum was normalized to the peak intensity of the most intense band at ~1045 cm^{-1}. (Reproduced with permission from Ref (28). Copyright 2006 The American Chemical Society.)

To test the sensitivity of virus detection by the AgNRs, RSV ΔG sample was applied to AgNR substrates at various concentrations. After the samples dried, the viral signature could be detected when the initial sample concentration was 100 PFU/ml. Considering the volume of virus sample (0.3 μl), the sample area covered on the AgNRs after the sample dried (~1 mm^2), the virus concentration

(100 PFU/mL), and the area of the sample spot probed during measurement, *i.e.*, the laser spot size (<100 μm diameter), this corresponds to the detection of well below "1 PFU" per measurement area. Indeed, "1 PFU" likely represents many virus particles, however, it is not unlikely that the observed signal is not originating from intact virus, but rather viral components from virus that have disintegrated during the experiment. Because the components of the virus (e.g. proteins, nucleic acids, etc.) are present as numerous subunits, the effective concentration of these components is significantly higher that of the intact virus. These components are also likely to be more evenly distributed across the surface of the AgNRs, rather than being contained to discrete points (intact virus) on the surface. Moreover, the observed 'virus' signal may indeed be in part originating from host factors, i.e. biochemical components generated by the cells in response to infection, which will likely adsorb to the AgNR surface alongside the viral components. Further investigation of these possibilities will provide further insight into SERS virus detection.

Using SERS for Label-Free Strain Discrimination of Rotavirus

Rotavirus (RV) strains are categorized according to three components, i.e. and inner core, inner capsid, and outer capsid. Variations of the two outer capsid proteins VP7 and VP4, are used to classify RV as G and P genotypes, respectively. The most severe infections in humans are caused by five G variants (G1-G4 and G9) and three P variants (P1A, P1B, and P2A). Traditional immunochromatographic assays rely on the conservative inner capsid VP6, but do not provide information regarding the G and P genotypes, which is critical for monitoring and controlling disease. To elucidate the genotype, ELISA and RT-PCR methods are typically employed, but these methods are expensive, labor intensive, and require specific antibodies or primers. Thus, our investigation focused on label-free and antibody-free AgNR-SERS to quantitatively detect and discriminate different strains of RV in a biological matrix (*29*).

MA104 cells were used to propagate eight RV isolates representative of the most commonly identified G and P genotypes of RV (YO, Wa, ST-3, S2, RV5, RV4, RV3, and F45). The RV cell lysates were diluted to a rotavirus titer of 10^5 fluorescent focus forming units (ffu)/mL before applying to the AgNRs. MA104 cell lysate without RV was used as a negative control. Measured spectra were baseline-corrected and normalized to the 633 cm^{-1} peak for easier visual comparison.

From a visual perspective, the spectra from the eight different strains of RV show a large degree of similarity in the overall spectral profile with the most prominent bands located at 1003, 1030, 1045, and 1592 cm^{-1}, however variations, namely in the intensity of these bands, are visually evident (Figure 6). In contrast, the spectra from the eight different RV strains are markedly different from the MA106 negative control. Although some of the differences between the RV-positive and RV-negative samples are obvious, and could indeed be due to direct detection of the RV virus, at this point we cannot rule out the possibility that we are indeed detecting the virus-induced host factors.

To classify the visually similar spectra of the eight different RV strains, we used partial least squares discriminate analysis (PLS-DA). This is a multivariate calibration technique that utilizes the entire spectral region (400-1800 cm^{-1}) to determine the best-fit relationship between a matrix containing the sample spectra (i.e. descriptor matrix) and a matrix containing the sample identities (i.e. class matrix). PLS-DA minimizes the contribution of spectral features that vary *within* a particular sample type, while maximizing the contribution of spectral features that vary *between* sample types. Cross validation (using Venetian blinds) was used to validate the model whereby 90% of the total spectral measurements were used to build the model, and the remaining 10% are treated as unknowns to be assigned to a class. This was iterated 10 times until every spectrum is withheld as an unknown and classified.

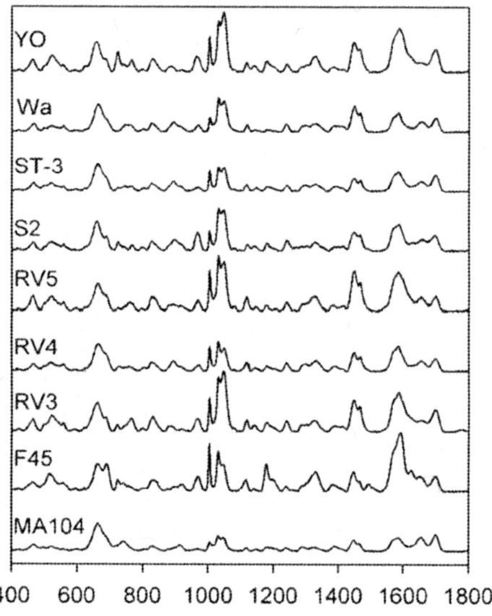

Figure 6. Average SERS spectra for eight strains of rotavirus and the negative control (MA104 cell lysate). Spectra are baseline-corrected, normalized to the 633 cm^{-1} peak intensity, and offset for visualization. (Reproduced with permission from Ref (29). Copyright 2010 Plos One.)

Using a model with only two classes, i.e. RV-positive and RV-negative, PLS-DA is able to discriminate the RV samples from the negative controls with 100% accuracy. The ability to discriminate the RV samples according to either their P-type or G-type was also evaluated, where four and five classes, respectively, were used to build the model. For the P-typing, only a single spectrum from the Wa (P[8]) strain is incorrectly classified with the negative control class. This yields a specificity of 100% and a sensitivity of >98% (here specificity and sensitivity are measures of false positives and false negatives, respectively). A similar analysis performed for the G-type classification yields a specificity of 99% and a sensitivity > 96%. Finally, PLS-DA was used to classify the SERS spectra at the strain level using nine classes (one for each strain; Figure 7). The specificity and sensitivity are found to be > 99% and 100%, respectively, which demonstrates satisfactory results for the PLS-DA classification of rotavirus SERS spectra obtained with AgNR substrates. Moreover, the fact that one can classify the individual strains using the AgNR-SERS with PLS-DA, suggests that SERS can indeed detect the virus rather than host factors.

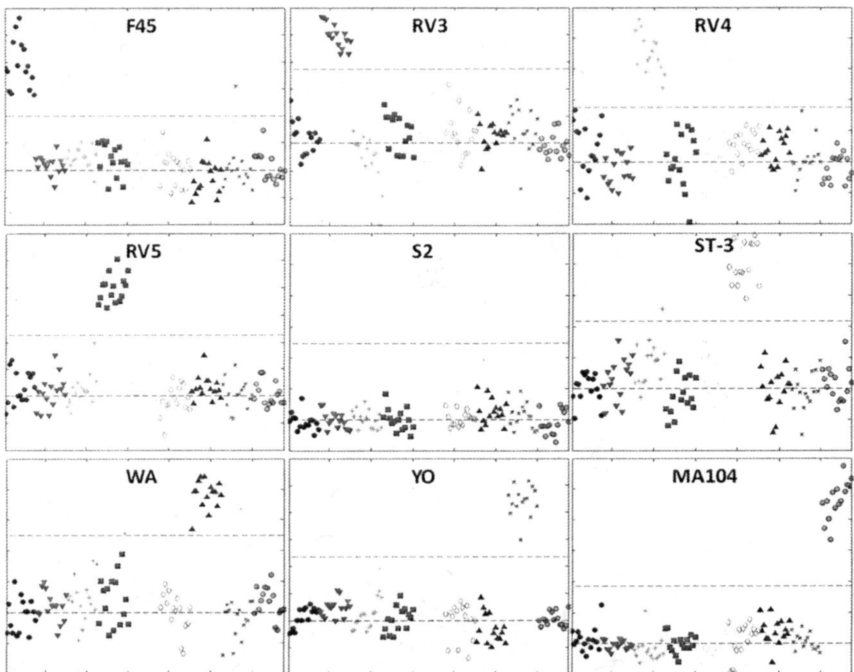

Figure 7. Cross-validation results for strain classification of RV-infected samples and negative control (MA104 cell lysate). The x-axis represents the sample number; samples that lie above the red dashed threshold are identified as positive for a particular strain of RV (or negative control), and those that fall below the line are classified as negative for a particular strain. The black dashed line is a guide for the eye representing a predicted Y value of zero. Values from x- and y-axis have been omitted. (Adapted with permission from Ref (29). Copyright 2010 Plos One.)

Partial least squares (PLS) in conjunction with AgNR SERS measurement for quantitative analysis was also investigated. The RV samples were diluted to known concentrations using MA104 cell lysate as a diluent. PLS is able to accurately predict the RV concentrations between 10^5-10^6 ffu/mL. Below ~10^5 ffu/mL, however, PLS predicts concentrations higher than the true values. This observation could be due to the fact that spectra become dominated by the complex background matrix of the cell lysate when the viral components are present at lower concentrations.

SERS Detection of Bacteria

Although detection of virus is critical, improved detection of other pathogens such as bacteria is likewise eagerly sought. Beyond the obvious medical ramifications, bacteria can have large economic consequences in industrial sectors, namely food processing and packaging, where *Salmonella* and *E.coli* are common culprits leading to costly recalls of food products. The current mainstay of bacteria detection is performed using cell cultures, ELISAs, or PCR, which can be very time consuming, expensive, and/or prone to poor specificity, namely false-positives. Thus, we have investigated the potential utility of AgNR-SERS to fulfill this niche as a bacterial biosensor.

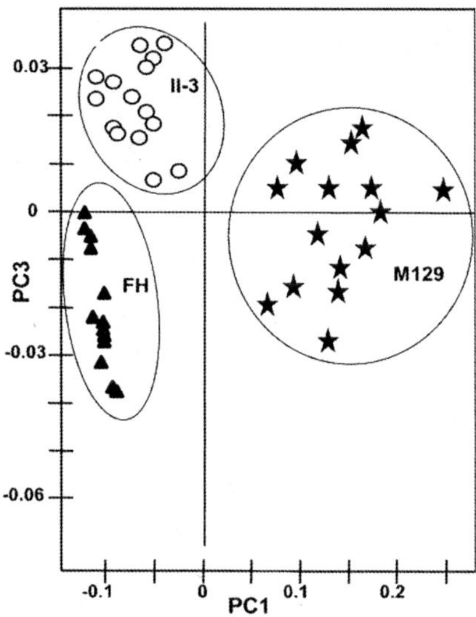

Figure 8. PC3 vs PC1 scores for n = 15 spectra of three different M. pneumoniae strains FH, M129 and II-3. (Reproduced with permission from Ref (30). Copyright 2010 Plos One.)

SERS Detection of Mycoplasma Pneumoniae in Clinical Samples

Mycoplasma pneumoniae clinical throat swab samples were used in addition to lab-cultured M129, FH, and II-3 strains (*30*). The lab cultures were centrifuged and washed 3× before finally fixing the cells with formalin and serially diluting in water. The clinical samples were diluted 1:100 in water. As expected, the three different strains of lab-cultured *M. pneumonia* show a high degree of spectral similarity. We employed a multivariate method known as principle component analysis (PCA), which is an unsupervised method that can classify spectra based on their covariance. PCA uses the covariance of a dataset to reduce the dimensionality of the spectra, i.e. from several thousand dimensions (one dimension for each value of the spectrum's x-axis) to a few dimensions (i.e. the principle components,

PCs). The data can therefore be classified or grouped according to the PC scores. Figure 8 shows that the spectra (*n* = 15 for each strain) of the three strains of *M. pneumoniae* can be grouped according to their PC1 and PC3 values, where each data point represents one spectrum.

The feasibility of detecting the *M. pneumoniae* in clinical samples was tested by first using simulated clinical samples in which the M129 cultured strain was spiked into uninfected throat swap samples at concentrations between 8.2×10^4 to 8.2×10^1 culture forming unit (CFU)/µl samples. Unspiked throat swab samples were used as a negative control. PLS-DA was employed using two classes, i.e. *M. pneumoniae*-positive or -negative, (using the same cross validation process described in the previous section). The M129 *M. pneumoniae* spectra could be accurately classified as *M. pneumoniae*-positive at concentrations of ≥ 82 CFU/µl, which is comparable to the previously reported PCR limit. Finally, PLS-DA was performed on the AgNR-SERS spectra of the true clinical samples previously shown to be positive for *M. pneumoniae* with PCR and culture. The spectra for all 10 specimens were classified with > 97% accuracy (Figure 9).

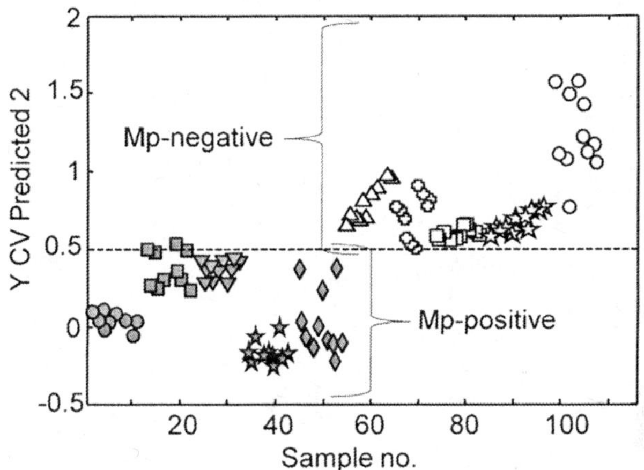

Figure 9. PLS-DA predictions of true clinical samples. Grey symbols and open symbols represent samples previously shown to be M. pneumoniae-negative or –positive, respectively, by culture and real-time PCR; each symbol type represents a sample collected from a different person. (Reproduced with permission from Ref (30). Copyright 2010 Plos One.)

SERS Discrimination of Different Types of Bacteria Samples

Six types of bacteria were analyzed in this study (*31*): Generic *Esherichia coli* (EC), *Esherichia coli* O157:H7 (H7), *Esherichia coli* DH 5α (DH), *Staphylococcus aureus* (SA), *Staphlylococcus epidermidis* (SE), and *Salmonella typhimuium* 1925-1 poultry isolate (ST). The cells were grown in trypticase soy broth and stocks yielded ~10^9 CFU/mL. Excess growth media was removed by centrifuging and washing with de-ionized (DI) water 3× before preparing final dilutions in DI water. 2.0 μL of bacteria suspensions were applied to the AgNRs, which spread out to cover an area of approximately 2 mm in diameter before drying. Measurements were acquired with a portable fiber optic probe system equipped with a beam spot diameter of ~100 μm. Therefore, a sample concentration of 10^8 CFU/ml, yields approximately 500 cells within the beam spot during measurement.

The SERS spectra of the four different bacteria show a significant similarity. For example, they all show many of the same predominant spectra bands, such as ~730 cm^{-1} and ~1325 cm^{-1} which are attributed to the ring-breathing and ring-stretching mode, respectively, of the adenine base group. In addition they share a peak at ~1450 cm^{-1} which is attributed to the CH_2 deformation mode of proteins. However, the relative peak intensities of these bands vary between the different species. In addition, EC and ST share a strong band at ~550 cm^{-1}, which is assigned to carbohydrate. We point out that because SERS can only detect molecules within very close proximity to the metal surface, one may reasonably conclude that the composition of the cell wall will strongly affect the SERS signal. However, in this study, bacteria that have disparate cell wall compositions, i.e. Gram-positive (SA and SE) and Gram-negative (EC and ST) did not show the expected spectral disparity. The fact that we see a strong nucleotide signal (730 and 1325 cm^{-1}) suggests that the AgNRs are in more intimate contact with the internal components than expected. One possibility is that the components from inside the cells are 'leaking' from the cell during the drying process, where they may ultimately adsorb to the AgNR surface. Figure 10 shows an SEM image of the EC cells after drying onto the AgNR substrate, where we can see the cells readily splay across the nanorods.

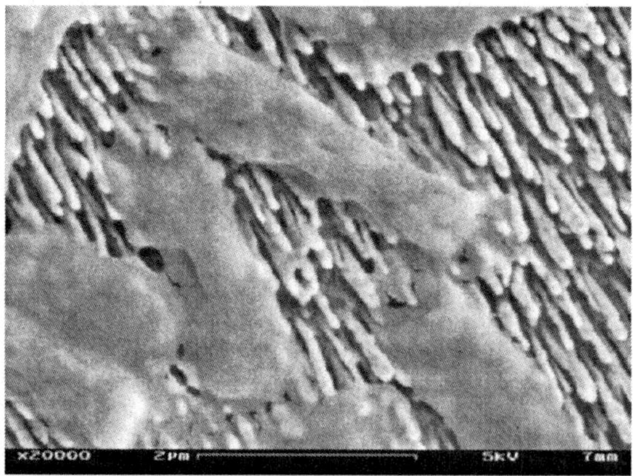

Figure 10. SEM image of E.coli on AgNR substrates. (Reproduced with permission from Ref (31). Copyright 2008 Society of Applied Spectroscopy.)

Similar to *M. pneumonia* studies, PCA was used to classify the highly similar SERS spectra acquired for the four different species of bacteria, where four different groupings are readily resolved when plotting the PC1 vs. PC2 values of the spectra. Also analyzed were two-component mixture samples (EC+ST and EC+SA), where bacteria are mixed in a 1:1 ratio prior to applying to the AgNRs. After plotting the PC1 vs. PC2 values of the mixture and pure bacteria spectra, three distinct groups were obvious corresponding to the pure EC, the mixture, and ST/SA. In both of the EC+ET and EC+SA samples, the average PC1 value was in between that of the EC and the ET/SA. In addition, PCA was able to produce three distinct groups for the three different *E. coli* strains EC, H7 and DH.

Because SERS directly detects the chemical constituents of a sample, a living cell and a dead cell could contain many of the same components, therefore SERS detection may fall short in the ability to discriminate between living and dead bacteria. To evaluate this, *E.coli* H7 and DH cells were first boiled in a water bath at 100° C for 10 minutes prior to applying to the AgNRs. The resulting spectra show a very noticeable difference between the dead and alive cells for both strains of *E.coli* (Figure 11). For example, the dead cells show a significantly reduced intensity of the predominant spectral bands at ~550, 735, 1330, and 1450 cm^{-1}. This is evidence that AgNR-SERS analysis has the potential to differentiate between viable and non-viable cells, but the exact cause for the observed spectral changes after cell death is elusive.

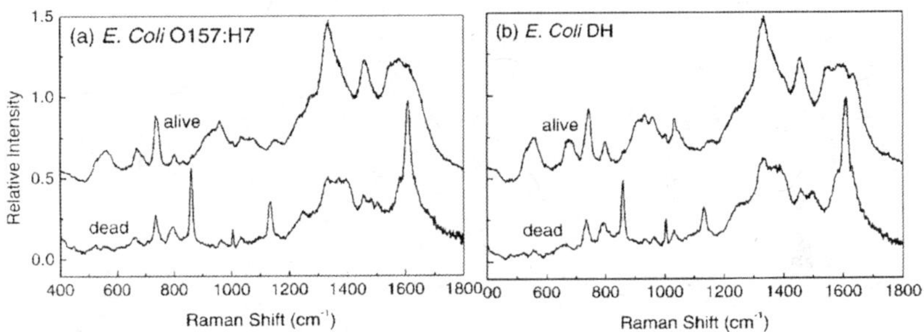

Figure 11. SERS spectra of dead and alive E.coli cells for a) H7 strain and b) DH strain. (Reproduced with permission from Ref (31). Copyright 2008 Society of Applied Spectroscopy.)

The Detection of microRNA Biomarkers

The ability of the AgNR SERS substrates to directly detect 'whole' pathogens with sufficient specificity to differentiate between different species as well as different strain variants without the use of an external label has been demonstrated. However, rather than directly detecting pathogens, many biosensors detect a broad class of molecules referred to as biomarkers that can range from nucleotides to proteins to oligopeptides to small organic metabolites. These biomarkers do not necessarily play a direct role in the disease mechanism, but may simply be byproducts of the diseased state. One particular class of biomarkers that has gained considerable attention recently is microRNA (miRNA) (*32, 33*). These are small sequences of non-coding regulatory RNA, typically 19-25 nucleotides long, have demonstrated significant physiological importance, such as regulation of cell development, division, and apoptosis, and may serve as biomarkers for certain types of cancer.

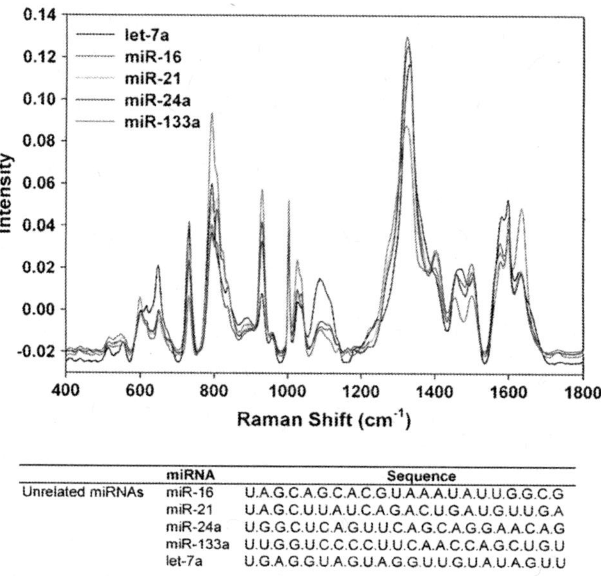

	miRNA	Sequence
Unrelated miRNAs	miR-16	U.A.G.C.A.G.C.A.C.G.U.A.A.A.U.A.U.U.G.G.C.G
	miR-21	U.A.G.C.U.U.A.U.C.A.G.A.C.U.G.A.U.G.U.U.G.A
	miR-24a	U.G.G.C.U.C.A.G.U.U.C.A.G.C.A.G.G.A.A.C.A.G
	miR-133a	U.U.G.G.U.C.C.C.C.U.U.C.A.A.C.C.A.G.C.U.G.U
	let-7a	U.G.A.G.G.U.A.G.U.A.G.G.U.U.G.U.A.U.A.G.U.U

Figure 12. Baseline-corrected and unit-vector normalized SERS spectra for miRNA sequences. Each spectrum is an average of n = 18 measurements. The table shows the sequences for each miRNA represented in the plot. (Reproduced with permission from Ref (34). Copyright 2008 Elsevier.)

The feasibility of the AgNRs SERS analysis to detect and differentiate between different sequences of the let-7 family of miRNAs was evaluated (*34*). After applying 1μg of pure miRNA sequence (suspended in 1 μL of water) to the AgNR substrates and allowing the sample to dry, the SERS spectra was measured. To allow for easier visual and quantitative comparison, the recorded spectra were unit-vector normalized and baseline corrected. Each miRNA sequence yields very uniform spectra for multiple measurements obtained from multiple substrates. Figure 12 shows the average spectra and corresponding nucleotide sequences for the let-7a and four different unrelated sequences from the miR family, where we can see noticeable differences in the relative peak intensities for different sequences. However, the peak positions are very similar for all the sequences, and this is expected as these different sequences share many of the same RNA components, *i.e.* A, G, and U nucleotide, albeit at different ratios. We also point out that the miR sequences also contains C nucleotide, which is absent in let-7a. These different ratios of nucleotides are therefore expected to be the physical source of the spectral variations for different sequences.

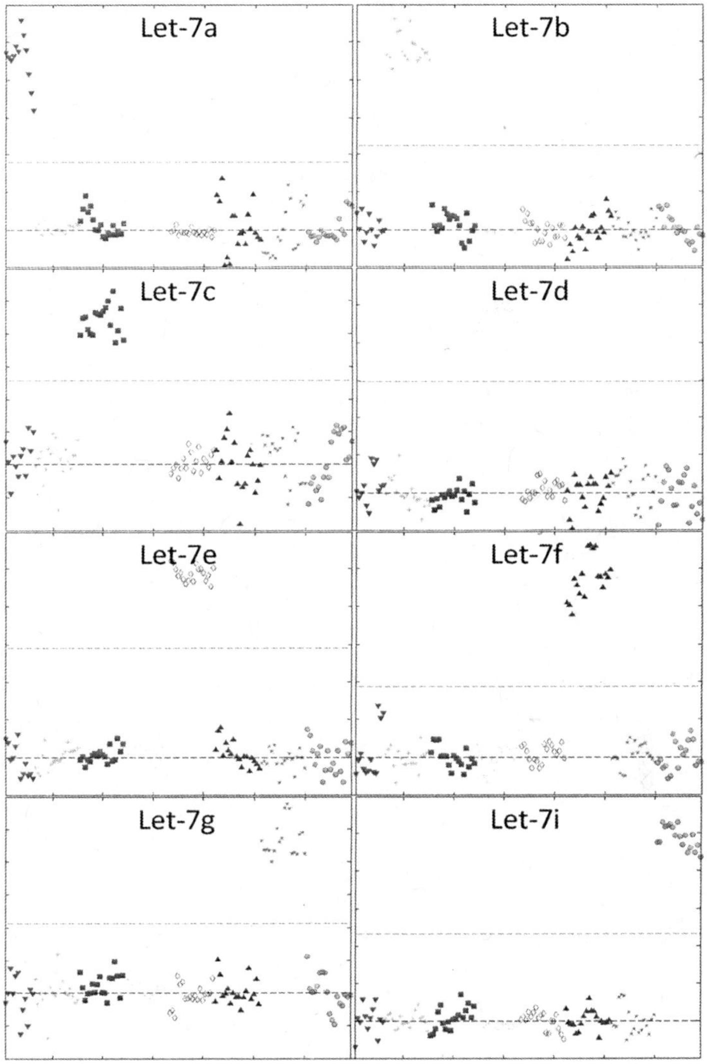

Figure 13. PLS-DA results after cross validation of SERS spectra of eight different miRNA sequences from the let-7 family. The x-axis represents the sample number. Samples above the red dashed line are classified as belonging to the designated class for that plot, while those below are predicted to not belong to that class. The black dashed line is a guide for the eye representing a predicted Y value of zero. The x- and y-axis values have been omitted. (Adapted with permission from Ref (34). Copyright 2008 Elsevier.)

To differentiate and classify these similar spectra in a more quantitative and objective manner, PLS-DA was used and the entire spectral region 400-1800 cm^{-1} of all the spectra was used to build the model. To validate the model, cross-validation (Venetian blinds) was used where 8/9th of 90 collected miRNA spectra (5 classes × 18 replicates) were used to construct the model, while the remaining 1/9th are classified using this model. The resulting sensitivity and specificity are both found to be 100%.

To further evaluate the discriminatory power of AgNR SERS measurements and PLS-DA with regards to miRNA sequence detection, we applied this same approach to eight different sequences within the let-7 family, which have 71-95% similarity, with some sequences differing by only a single nucleotide. The PLS-DA results are shown in Figure 13. Remarkably, after cross-validation 100% sensitivity and 99-100% specificity were achieved, despite the large degree of spectral similarity between the different sequences.

This demonstrates the ability of the AgNR SERS analysis, in conjunction with statistical approaches such as PLS-DA, to discriminate highly similar miRNA spectra, but only pure samples were evaluated. In many biodetection scenarios, however, it is expected that a mixture of miRNAs will be present within a sample. With this motivation, the ability of the AgNRs to quantitatively detect miRNAs in a mixture was evaluated (*35*). miR-133a and let-7a were mixed together in various ratios with a total miRNA concentration of 1.0 µg/µL, applied to the AgNRs, and allowed to dry. Similar to the aforementioned report, the pure miR-133a and let-7a spectra are similar but with different relative band intensities. Concordantly, a mixture containing 0.6 µg of let-7a and 0.4 µg of miR-133a yields a spectrum with many of the peaks demonstrated band intensities intermediate to those observed in either of the pure samples.

PLS was used to quantify these differences, and correlate them to the concentrations of the miR-133a and let-7a miRNA present in the sample. Plotting the miRNA concentrations predicted by external PLS validation with the true concentrations yields an R^2 value of 0.99 for both the let-7a and miR-133a. This approach was further extrapolated to predict the let-7a concentration in a mixture containing up to 5 different miRNA sequences (1 µg/µL total miRNA concentration). Regardless of this rather complex mixture, the externally validated PLS regression analysis yields an R_2 value of 0.975 and a root mean square of error of prediction (RMSEP) of 10.1 µM which indicates a good agreement between the true and predicted concentrations.

AgNR SERS for Other Environmental and Bioagent Detection

The aforementioned studies show a broad range of biological detection capabilities, but are not an exhaustive account of the applicability of AgNR SERS analysis, as other biological and environmental applications have likewise been investigated. For example, DNA aptamers have been used to successfully capture viral proteins at the surface of AgNRs, improving the SERS detection specificity for these bioagents (*36*). Food adulterants such as melamine have been shown to be detectable with AgNRs (*37*). Environmentally hazardous agents such uranium compounds (*38*) and pentachlorinated biphenyls (PCBs) (*39*) have also been analyzed. AgNR-SERS detection of highly carcinogenic food toxins such as aflaotoxins, which reside in common products such as peanuts have also been investigated (*40*).

Integrating OAD–AgNR SERS Substrates with Other Technology

In the previous sections, AgNR substrates were shown to demonstrate the necessary SERS sensitivity, specificity, and reproducibility to be used as a practical bioanalytical SERS platform for the label-free detection of pathogens and biomarkers. However, before SERS can be integrated as a functional and widely used mainstream biodetection technology other challenges still need to be addressed; the AgNR-OAD fabrication must show enough versatility to be integrated with other fabrication techniques, and provide a means for further development into realistic sensing devices.

OAD Substrate Fabrication Versatility

In order for SERS to be implemented on a large scale, SERS-active surfaces should be capable of integration with other technology and fabrication processes for improved functionality and device development. Unfortunately, many of the previously discussed fabrication methods fall short in this regard. However, OAD offers a number of advantages; in particular, OAD-generated nanorod films can be uniformly deposited onto any surface that is sufficiently flat and smooth. This lack of constraint means that a substrate can be of any size or shape, from whole wafer to $< 1mm^2$. In fact, the SERS-active AgNR films can be deposited onto the tips of ~500 μm diameter optical fibers (Figure 14) (*41*). This is accomplished by first polishing the tip of the fiber to a smooth, flat surface, and then positioning the fiber in the deposition chamber so that the Ag vapor deposits on tip at the requisite 4°. To test the SERS response of the AgNRs, 2 μL of BPE was applied to the AgNR-coated fiber tip and allowed to dry. Other fibers were immersed in adenine solution for 60 s, removed and allowed to dry. Both BPE and adenine could be detected when the sample concentrations were as low as 100 nM.

As noted, a surface only needs sufficient smoothness for uniform AgNR deposition. In addition, OAD is considered to be a low temperature fabrication method, as the substrate temperature is typically less than 100°C during the Ag deposition. These two conditions imply that a variety of materials can be used to support the AgNR film, including silicon, glass, metal, plastics, composites, etc. Although SERS surfaces, including AgNRs, are typically fabricated onto rigid silicon or glass substrates, we have investigated the use of flexible materials such as polyethylene terephthalate (PET) sheets or polydimethylsiloxane (PDMS) films as the substrate (*42*). Not only do we find that AgNRs deposited onto these substrates provide SERS characteristics comparable to those deposited onto the more conventional glass or silicon, but also that a significant amount of the SERS enhancement is maintained after subjecting these substrates to bending or stretching. A picture of a flexible AgNR-PET substrate is shown in Figure

15A. For example, using BPE as a Raman probe we show that the SERS signal intensity of the AgNR-PDMS sheets does not demonstrate an appreciable change when the film is stretched to 35% beyond its original length. In addition, the signal intensity does not decrease below the initial un-stretched value even after 100 cycles of 10% stretching, and interestingly, the signal intensity actually shows a noticeably increase after approximately 25 cycles (Figure 15B). The exact cause of this increased SERS sensitivity is unclear, but is likely results from microscopic changes the AgNR film morphology such as buckling which is clearly observed with the SEM after the stretching tests.

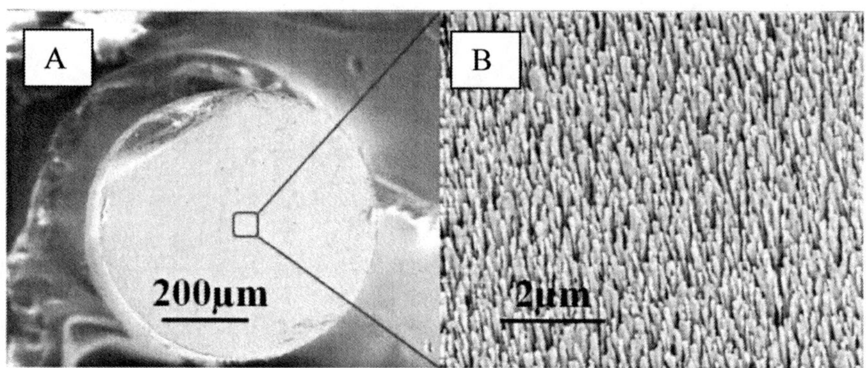

Figure 14. SEM image of the AgNR-coated optical fiber tip A) low magnification showing entire cross section of fiber, and B) high magnification showing uniform AgNR film. (Reproduced with permission from Ref (41). Copyright 2011 Elsevier.)

Although the shadowing effect resulting from nanoscale topology on the substrate is responsible for the formation of the individual nanorods, inducing the shadowing effect on a larger, micron-scale allows for the selective deposition of AgNRs onto pre-patterned surfaces. Such a technique may be useful for the development of MEMS-based SERS devices. To demonstrate this, the 'tracks' of a standard compact disk (CD) can be used as a micro-patterned surface after the protective polycarbonate film is etched with acid. The tracks of the CD form a series of grooves with a period of approximately 1.5 µm with sufficient height to act as shadowing centers for OAD. Using the standard deposition conditions outlined previously, AgNRs can be selective deposited onto the ridges of a CD. Figure 16 shows that the AgNRs can be grown parallel or perpendicular to the groove direction with great precision. More intricate surface patterns could also be employed. For merely illustrative purposes, Figure 17A shows an SEM image of a microchip that has used as a supporting substrate, where we can see the AgNR conforms precisely to the underlying architecture. Pre-patterning a silicon wafer with anisotropic dry etching produces pits that have side walls that form 4-7° angle with the wafer surface normal. Depositing Ag vapor at an angle directly perpendicular to the surface of the wafer therefore yields the appropriate OAD angles for deposition of SERS-active AgNRs onto the sidewalls of the pits (Figure 17B) (*43*).

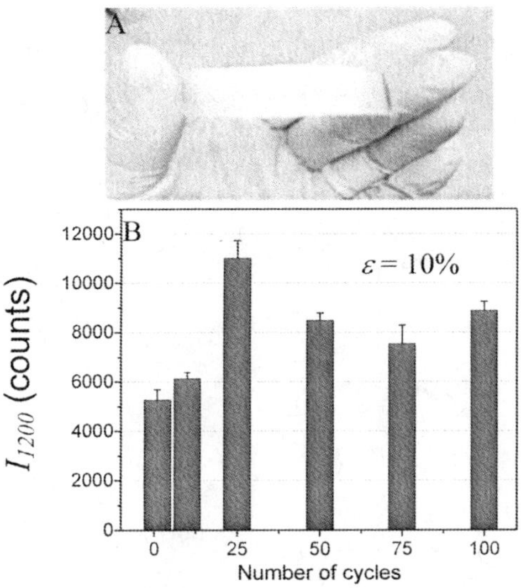

Figure 15. A) Picture of AgNRs on flexible PET substrate. B) 1200 cm^{-1} peak intensity (I_{1200}) of BPE on AgNRs/PDMS substrate as a function of stretching cycle with 10% strain. (Reproduced with permission from Ref (42). Copyright 2012 The Royal Society of Chemistry.)

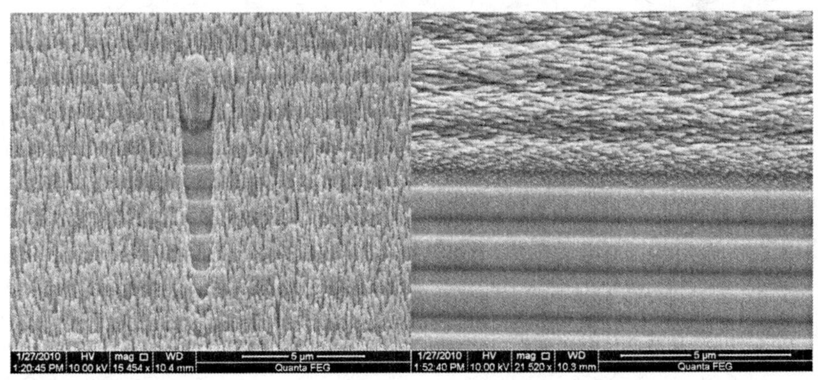

Figure 16. AgNRs deposited onto a CD with nanorod growth direction perpendicular and parallel to the CD groove direction. The image on the left shows a defect so that the underlying groove pattern can be seen.

Figure 17. SEM images of AgNRs deposited onto micro-patterned surfaces: A) selective and precise AgNR growth on surface architecture of microchip. B) OAD deposition of AgNRs deposited onto walls of micropits etched into Si wafer. (Part B reproduced with permission from Ref (43). Copyright 2011 Institute of Physics.)

Further Engineering the AgNR Surface

The AgNR substrates are chemically active and may degrade with prolonged exposure to the atmosphere or biological samples. To prolong the shelving life and broaden the potential for more realistic SERS bio-applications the AgNRs must have improved stability in environments that would normally degrade the Ag surface, which may be accomplished by thin, conformal coatings of more stable materials. For example, biological samples typically contain a relatively high concentration of Cl⁻ (e.g. ~150 mM), which is detrimental to the Ag surface as Cl⁻ will react with Ag to form AgCl, ultimately etching the AgNR structure and mitigate its high SERS sensitivity. In contrast, gold is very stable in such environments, but unfortunately fabrication of Au nanorods via OAD is very expensive and not currently a feasible alternative. Wet-chemical synthesis techniques, on the other hand, are a much more practical means of fabricating Au nanostructures and surfaces. With this motivation in mind, we have investigated the use of a galvanic replacement reaction (GRR) to deposit Au onto the AgNR surface (*44*). As the GRR name implies, Au⁺ ions in solution galvanically displace Ag atoms from the nanorod, leading to the eventual formation of an Au shell on the AgNR surface. The study shows that initially a porous Au/Ag alloy forms on the AgNRs and becomes more compact and uniform as the reaction proceeds. Longer reaction times ultimately lead to a complete removal of the Ag core and de-alloying yielding a pure Au shell. Figure 18 shows SEM images of the AgNRs treated to progressively GRR times ($t = 0$ to 50 min); the nanorod structure becomes increasingly thicker with prolonged reaction time. Although the SERS intensity of the Au/AgNR film decreases to about 30% of that of the uncoated AgNRs after 20 minutes of the GRR treatment, the stability is vastly improved for extended (18 hr) exposure to 100 mM Cl⁻ compared to the bare AgNRs, which shows a drastic decrease in intensity after a similar treatment (Figure 19). This demonstrates that the Au/AgNR substrates are more suitable for many biological applications. The observed drop in SERS intensity after GRR treatment is observed when the excitation wavelength is 785 nm; however, because the Au-coating induces a red shift in the LSPR of the nanorods, we believe that the SERS sensitivity could be recovered if the excitation wavelength could be tuned to allow for more efficient coupling with the LSPR. Regardless, the large enhancement factor of the AgNRs ($\sim5\times10^8$) suggests that a significant portion of the enhancement can be sacrificed without mitigating the high sensitivity.

Multi-well AgNR SERS Substrates for High-Throughput Screening

Because the AgNRs can be uniformly deposited over a large area, a relatively large wafer, such as a microscope slide, can be used as a substrate. After the AgNR are uniformly deposited over the surface, a molding technique can be employed to pattern an array of 40 uniform wells using PDMS (*45*). The side walls of the μL-sized wells are composed of PDMS and the bottom surface contains the SERS-active AgNR film. This has the advantage of minimizing substrate consumption, while providing a uniform array of substrates for parallel, multi-sample SERS screening applications. In addition, the confining properties of the wells help confine samples to a regular, pre-defined area of the AgNRs, which can yield improved reproducibility for measurements of biological samples. This type of "SERS chip" has been widely used by our lab as well as many of our collaborators for various bioanalytical SERS applications, including toxins, DNA, bacteria, and protein, many of which have been discussed in the previous section. Figure 20 shows a picture of the AgNR SERS chip.

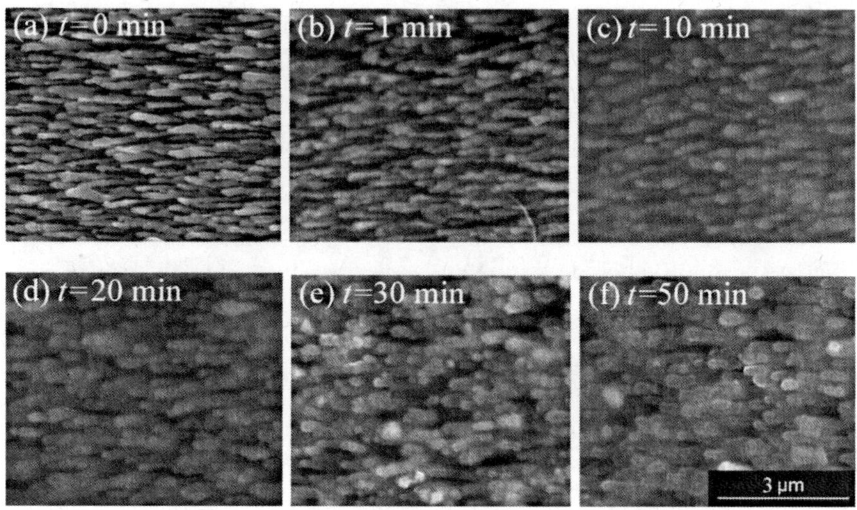

Figure 18. SEM images of the AgNRs treated to progressively longer GRR times. The scale bar is the same for all images. (Reproduced with permission from Ref (44). Copyright 2012 The Royal Society of Chemistry.)

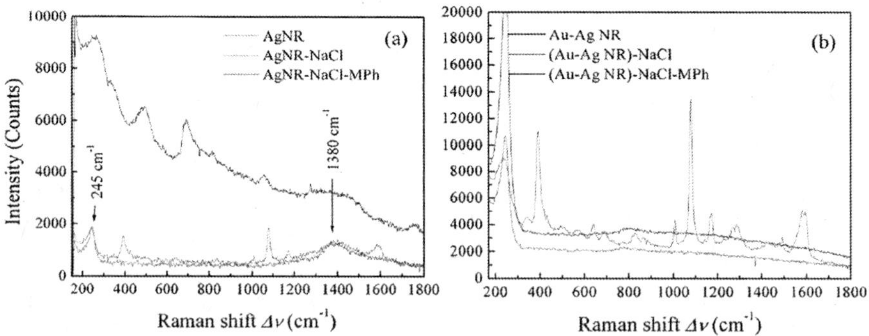

Figure 19. SERS spectra before and after treatment with Cl⁻ for 18 hours, followed by adding Raman reporter molecule MPh for a) bare AgNRs and b) AgNRs pre-treated with GRR for 20 minutes. (Reproduced with permission from Ref (44). Copyright 2012 The Royal Society of Chemistry.)

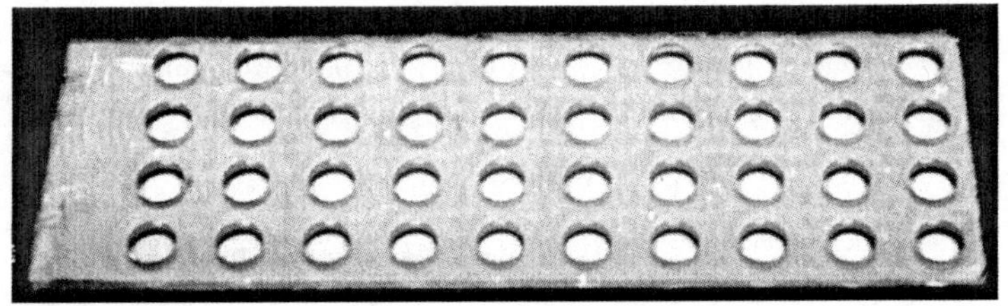

Figure 20. Picture of a 40-well array SERS chip patterned with PDMS.

Ultrathin Layer Chromatography SERS for Sample Mixture Separation and SERS Detection

SERS analysis requires the analyte to be within close proximity (preferably adsorbed) to the SERS-active surface before a SERS signal can be measured. The adsorption step, which we refer to as 'sampling', is traditionally accomplished by applying a small droplet of a relatively pure sample solution to the AgNR substrate, where the analyte is concentrated on the AgNR surface as the sample solvent dries. Another common method entails immersing the substrate into a volume of the sample solution and allowing the analyte to passively adsorb. Although these sampling techniques seem straightforward enough, a number of complexities can arise. For example, if a particular analyte does not have strong affinity to adsorb to the metal surface, or if other components in the sample are present at a higher concentration and/or have a stronger affinity to adsorb to the metal surface, these extraneous components will out-compete the analyte of interest for the limited number of adsorption sites. In these cases the analyte will not be detected. Thus, mixtures have proven to be especially troublesome for SERS analysis, but because most applications will involve a mixture of some sort (e.g. food, blood, etc), special attention needs be paid to the sampling procedure before SERS detection can be adequately employed.

Separation and analysis of mixtures has been previously established for many non-SERS applications. One mainstay technique is ultra-thin layer chromatography (UTLC), which has been well established as an efficient and effective way to separate compounds in a mixture. The UTLC principle is based on the varying affinities of different compounds to dissolve in a mobile liquid phase (e.g. solvent) and adsorb to a thin-layer solid phase (i.e. a chromatography plate). Traditionally, the mobile phase consists of solvents such as methanol and/or acetonitrile, while the plate is a porous film typically composed of SiO_2 or Al_2O_3. The sample is initially spotted onto the plate and one edge of the plate is immersed in a reservoir of the mobile phase. Capillary action then causes the mobile phase to migrate along the length of the UTLC plate. The migrating mobile phase propagates through the sample spot, which induces migration of the sample components. The rate of migration of the components is dependent on their specific affinity for the plate surface and solubility in the mobile phase. Thus, the different components initially localized to a single spot on the plate become dispersed as a series of spatially resolved bands. The compound isolated within a band can then be visually identified by referencing a standard, or scraped off and further identified via chemical analysis (e.g. absorbance spectroscopy). We have therefore integrated these working principles, along with the highly specific and sensitive chemical detection capability of SERS to produce a simple yet more selective detection platform for chemical analysis of mixtures.

The AgNRs demonstrate significantly different affinities to adsorb different analytes. In addition, the AgNR film is very porous and significant capillary activity can be observed with various solvents. The ability of AgNR as a UTLC plate has been accomplished by the aforementioned UTLC procedure (46). We have analyzed several different 2- or 4- component mixtures containing different analytes such as methyl orange (MO), cresol red (CR), methylene violet 2B (MV), BPE, rhodamine 6G (R6G), and/or melamine. 0.1 µL samples are first spotted onto the AgNR 'plate', and allowed to dry. After placing the edge of the plate into a reservoir containing the mobile phase, capillary action propagates the mobile phase along the plate and through the sample spots. After the solvent front equilibrates and stops migrating, the AgNR substrate is removed from the mobile phase reservoir and allowed to dry. A spatial SERS mapping is accomplished by measuring the SERS response at incremental points starting from initial sample spot and measuring every 0.5 mm until the entire length of the migration distance is measured.

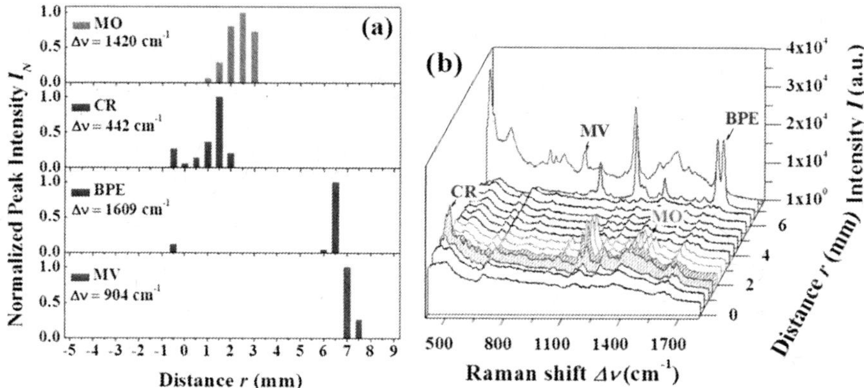

Figure 21. UTLC-SERS separation of MO, CR, BPE, MV using methanol as a mobile phase: a) the normalized peak intensity of the signature peaks used for each of the four analytes as a function of position UTLC-AgNR substrate, and b) the recorded spectra as a function of position. The initial mixture sample spot is at r = 0. (Reproduced with permission from Ref (46). Copyright 2012 The Royal Society of Chemistry.)

Prior to UTLC, the initial sample spots were measured with SERS and show spectral features of all the components of the mixture convoluted as a single spectrum. After inducing the mobile phase migration, the SERS-mapping demonstrates spatially-distinct spectral response. For example, the melamine + R6G mixture shows that the melamine band has been constrained to the initial sample spot due melamine's high adsorption affinity for the silver surface, whereas the R6G, which has a weak affinity to bind to the silver surface, has been carried with the solvent front during migration. Indeed, the melamine and R6G show very contrasting behavior, and their physical separation is readily achieved on the AgNR substrates during UTLC treatment. However, a mixture is likely to contain more than two components and the compounds may not demonstrate such contrasting behavior during UTLC separation. This situation can be tested when a mixture MO, CR, BPE, and MV are spotted onto the AgNR substrate, and using methanol as the mobile phase. Figure 21 shows the corresponding spectra and signature peak intensity as a function of position relative to the original sample spot after UTLC. The different spectral profiles of each of the four analytes can be resolved at various regions within the mobile phase migration zone. The chromatogram quantitatively demonstrates the spectral intensity of each analyte at different locations. We can see that the MO and CR bands show significant overlap at $r = 1.5$ mm. Although this overlap would be troublesome for traditional UTLC that use low specificity detection (i.e. absorbance), the high chemical specificity inherent to the SERS measurements allows for the chemical signature to still be identified (and quantified) despite the presence of spectral interference from the other components.

SERS Flow Cell

Whereas the UTLC-SERS is a valuable tool for spatial separation of analytes, some applications may require temporal detection of analytes temporally separated within a sample. Hence, many detection platforms have been integrated with flow injection analysis (FIA) or microfluidic (MF) devices to continuously monitor the time-dependent composition of a sample, such as the eluent from liquid chromatography separation. For SERS, most of these devices have employed suspended colloids continuously injected and mixed (on chip) with a sample solution containing the analyte of interest. SERS measurements are then acquired 'downstream' after the analyte has had sufficient time to adsorb to the suspended colloids. On the other hand, AgNRs have been evaluated for use within an FIA or MF system using a fixed substrate design.

AgNRs can be deposited onto a 5 mm diameter 200 μm-thick glass wafer. This substrate can then be placed within a reservoir of a PDMS micro-channel, and a glass coverslip (with a photoresist-patterned rectangular channel) can be attached to form the top and side surfaces of the sensing channel (Figure 22A). Such a device has been used to test the stability of AgNRs for *in situ* detection under continuous flow conditions; after prolonged flow times of > 10 minutes, pre-adsorbed BPE maintains a near continuous signal suggesting that the AgNR structures do not degrade at flow rates of at least 200 μl/min. We also see that it is possible to obtain Langmuir isotherms for increasing analyte concentration injected over the substrate (Figure 22B). These results demonstrate the feasibility of using AgNRs for real-time *in situ* SERS detection in FIA or MF devices.

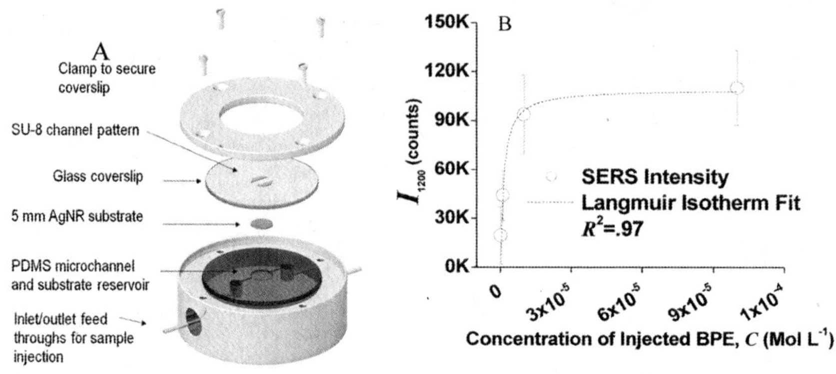

Figure 22. A) Disassembled AgNR flow cell for real-time in situ SERS analysis. B) Steady-state concentration-dependent I_{1200} for BPE solution injection through flow cell with Langmuir isotherm fitting.

Moving OAD-AgNRs Towards Commercialization

Before a technology can be commercialized, it must be capable of reproducible, high-throughput, and cost-effective fabrication. Traditional PVD has demonstrated these characteristics for decades, and has become an established and widely-used fabrication process in the semiconductor and optical coating industries. Considering that OAD is a particular type of PVD process, the same characteristics should apply. In our lab, a 5×6 in^2 metal plate, on which the glass or silicon substrates are fixed, is centered over the Ag source material during OAD in a $2 \times 2 \times 3$ ft^3 chamber. In theory, however, because the Ag vapor deposits on every surface in the chamber having a line of sight with the source material, many more substrate holders could be employed. However, considering the vapor flux (i.e. deposition rate) is dependent on distance and angle relative to the source material, we have constructed an umbrella-like substrate holder that attaches to the top of the chamber, and holds ten 1×3 in^2 wafers equidistant and centered over the source material (*47*). In addition, this holder can adjust the polar angles of each substrate simultaneously during deposition with a pneumatic mechanism so that a titanium and a silver thin film (the underlayer) can first be deposited at a vapor incident angle of 0° (i.e. the substrate surface perpendicular to the incident vapor direction), followed by rotation to 86° so that nanorods can be deposited. The entire holder can also be rotated continuously about the axis of vapor plume so that any remaining variation in position-dependent deposition can be mitigated, helping to assure more substrate-to-substrate uniformity. A picture of the custom substrate holder is presented in Figure 23.

As we have already point out, the deposition angle is one possible parameter to control the nanorod morphology, however, additional manufacturing control of the nanorod growth can be induced by changing other parameters such as the surface temperature of the substrate. Surface temperature is a critical factor for

PVD, and especially for OAD nanorod growth mechanism. This is because 'hot' vapor atoms condensing onto a 'cold' substrate surface are in a dynamic state, and can diffuse across the surface until they finally bind in low-energy configuration. Increasing the temperature of the surface allows the adatoms to seek more thermodynamically favorable states before binding, whereas decreasing the temperature will reduce the surface diffusion time of adatoms forcing them to adopt kinetically favorable states. Under the latter conditions, more anisotropic nanorod growth is favored. The fact that adatoms surface diffusion biases isotropic growth in the lateral direction, reducing this effect via low temperature (LT) deposition results in nanorod growth in the more anisotropic direction (*i.e.* rod-lengthening direction). Therefore, the effect of LT AgNR deposition using a liquid N_2-cooled substrate holder was investigated. Liquid N_2 is pumped through a copper block substrate holder which is able to achieve a temperature of -130° C. Glass substrates are attached to the block with conductive copper tape, and the deposition is carried out in a manner identical to that described for the standard AgNR substrates. First, 20 nm Ti and 200 nm Ag thin films are deposited while the substrate is at room temperature and directly facing the vapor source. Afterwards, the substrate is rotated 86° and the liquid nitrogen is introduced into the substrate holder.

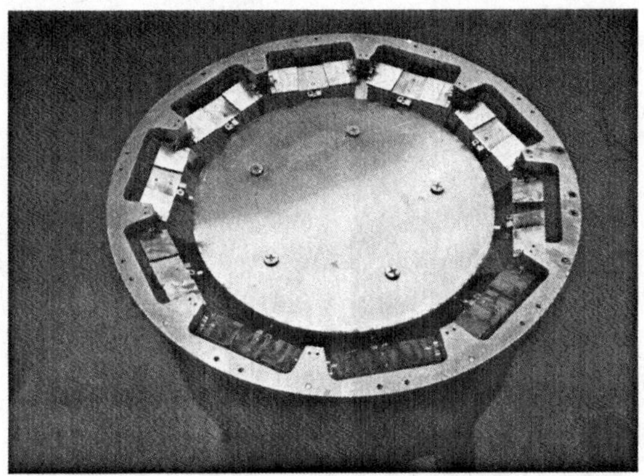

Figure 23. A rotating substrate holder designed for uniform large-scale OAD. (Reproduced with permission from Ref (47). Copyright 2012 The International Society for Optics and Photonics.)

The resulting AgNR morphology is noticeably different from the nanorods deposited at room temperature (RT). The nanorods are thinner with more of a 'blade' than 'rod' appearance. In addition, the LT-OAD nanorods demonstrate a longer overall length compared to the RT-OAD nanorods for given amount of deposited material. For example, when the total deposition thickness (*i.e.* the thickness reported by the QCM) is fixed at 2000 nm for a RT deposition, the resulting nanorods have a length of ~1200 nm. However, for a LT deposition, the same nanorod length can be achieved for only ~550 nm QCM thickness. Thus, the LT deposition is a significantly more efficient means of fabricating SERS-active Ag nanostructures as significantly less material is required, clearly an attractive manufacturing aspect. In addition, for a fixed nanorod length, the AgNRs deposited at LT show improved SERS intensity (unpublished data).

Although the substrate fabrication issues are of critical importance for implementation of SERS as a mainstream technology, instrumentation issues must also be addressed. SERS research is in a large part motivated by the fact that SERS is a highly sensitive technique. However, the fact that it also benefits from rapid measurements, in addition to simple, portable instrumentation, further motivates the desire to develop SERS into a cost-effective, yet highly deployable technology. Traditional biodetection techniques, on the other hand, such as immunoassays, PCR, and cell culturing typically require a fully equipped, sterile laboratory setting staffed with trained technicians. Thus, samples must be transported from the field to a laboratory, adding significant time before critical results are obtained. Likewise, chemical detection techniques such as mass spectroscopy or gas chromatography require heavy, bulky, yet delicate equipment that must be operated with trained personnel. In contrast, the relatively simple characteristics of a standard Raman instrumentation permits the development of handheld Raman devices. In this case, the handheld instrument is often integrated with spectral recognition software that provides an untrained user with a direct chemical readout.

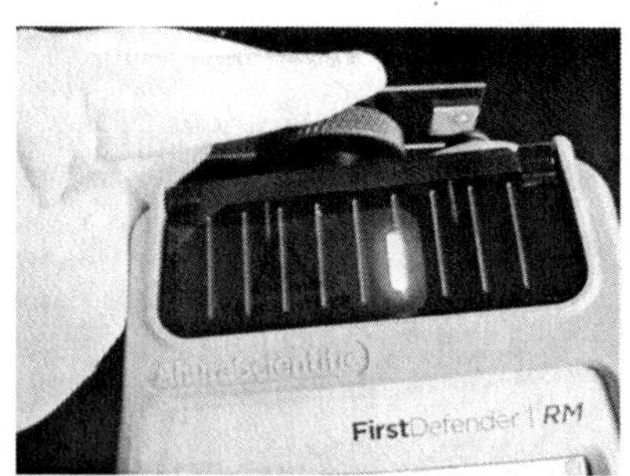

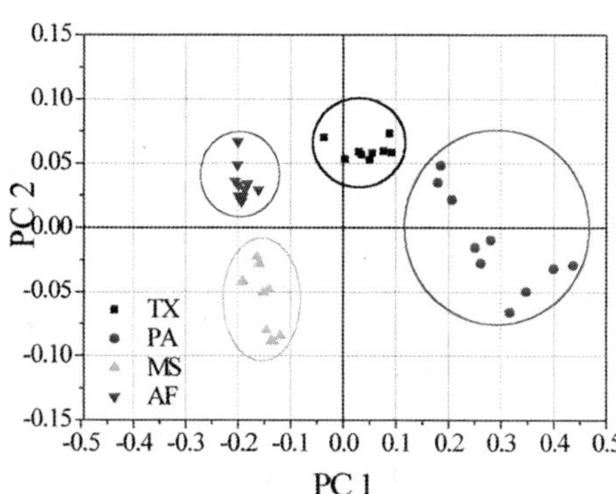

Figure 24. A) Picture of handheld First Defender Raman system measuring the SERS response of 1 × 1 cm² AgNR substrate; B) PC2 vs PC1 plot for of the viral samples (TX, PA, and MS) and the negative control (AF) measured with the handheld Raman device. (Reproduced with permission from Ref (48). Copyright 2012 The International Society for Optics and Photonics.)

Such a handheld Raman system has been used with AgNRs to detect virus within a biological sample matrix (*48*). Three types of avian influenza virus A/Mute Swan/MI/06/451072-2/2006 (MS; H5N1), A/chicken/Pennsylvania/13609/1993 (PA; H5N2), and A/chicken/TX/167280-4/02 (TX; H5N3) were propagated in embryonated chicken eggs. Allantoic fluid (AF) from mock-infected eggs were used as negative controls. After harvesting, the samples were diluted 100-fold and applied to the AgNR substrates. A First Defender RM handheld Raman system (Thermo Fisher Scientific Inc.) was used to acquire SERS spectra from the samples applied to the AgNRs. Although the handheld system lacks the high sensitivity of larger, more conventional microscope-based Raman systems, sufficient spectral quality is achieved so that PLS-DA accurately differentiates between virus-positive and –negative samples. Using PCA the spectra could be accurately grouped according to strain. Figure 24 shows a picture of the handheld device with AgNR substrate, and the corresponding PCA plots for the 3 different viruses. In conjunction with aforementioned discussion about AgNR substrate fabrication, these results suggest that AgNR-SERS platforms are indeed well poised to become a viable commercial SERS substrate.

Conclusions and Future Prospective

To conclude, we turn our attention to some AgNR-SERS developments that are possible to realize in the near future. For one, label-free DNA microarrays could be fabricated in a manner similar to the traditional microarrays, i.e. spotting probe samples to spatially confined points in an array format on a support substrate. The only difference would be the replacement of the glass microscope slide with the AgNR slide, and changing the corresponding silane chemistry to that of thiols. Our lab is currently working with collaborators to use AgNRs for label-free detection of nucleic acid hybridization via immobilized DNA probes. Extrapolation of our proof-of-principle technique to that of an array format could pave the way for label-free DNA SERS-chips. Secondly, although this chapter discusses biodetection from liquid samples, vapor sampling is also a highly sought after capability. One of the primary problems with vapor sampling is that the agents of interest are present at relatively low concentrations. Nevertheless, air can indeed act as a medium for pathogens and environmental pollutants to cause harm. Directing environmental vapor onto the surface of the AgNRs could be accomplished using traditional means, i.e. a small fan, but more intricate ways, such as electrostatic sampling, i.e. creating convective flow via electric fields, that allow for more selective and precise sampling. Development of such a sampling device using AgNRs is currently underway. Finally, SERS analysis of true biological samples (e.g. blood, urine, saliva) will further benefit by improving surface selectivity for specific biochemical capture, or improving signal analysis for complicated mixtures. As previously pointed out, many complex samples yield a SERS signal that may be a rather complicated convolution of many component signals, ultimately making interpretation a daunting or even impossible task. Indeed, physical separation techniques such as the aforementioned UTLC-SERS may not always be feasible; in this case, statistical techniques such as PCA or

PLS-DA definitely provide much needed assistance for data interpretation, but mainly for classification purposes. Understanding the physical source of the spectral features, however, may at times be critical, especially if an unknown component is introduced (intentionally or accidentally) in the sample milieu. For this reason, reliable SERS databases containing vast libraries of spectra for different compounds will be invaluable to the future SERS user. We point out that for a given compound, such libraries will ultimately need to contain spectra collected under different conditions known to affect the SERS spectra for a particular compound, i.e. pH, solvent effects, *etc.* need to be accounted for.

Throughout this chapter, we have highlighted the many advantages afforded by SERS for biodetection applications. These have included high sensitivity, superior specificity, and rapid yet simple analysis. Perhaps more importantly, we have discussed many of the shortcomings currently stifling SERS implementation as a mainstream chemical and biological sensing technology. The OAD-AgNR substrates have been thoroughly evaluated for many biosensing applications such as virus, bacteria, and miRNA detection, and have shown promising results. Meanwhile, the utility of the AgNRs demonstrates improved detection capabilities when integrated with other established methodologies such as UTLC or statistical data analysis techniques such as PCA or PLS-DA. The fabrication and technological versatility of the OAD technique has also been reviewed in context of AgNR-SERS. These aspects indicate that the current research has only begun to scratch the surface of much potential advancement towards realizing highly anticipated SERS devices in the future.

Acknowledgments

The research described in this chapter has been funded by ARL (W911NF-07-2-0065, W911NF-07-R-0001-04), USDA (USDA CSREES grant number 2009-35603-05001), and NSF (ECCS-0304340, ECCS-0701787, ECCS-1029609, CBET-1064228). We would like to thank the other lab members and numerous collaborators that have contributed invaluable effort and insight into this work. Special thanks to: Professors Yao-wen Huang, Richard Dluhy, Duncan Krauss, and Jitenra Pratap Singh; collaborating scientists Dr. Simona Murph; and former and current graduate students Jing Chen, Yu Zhu, Chunyuan Song, Qin Zhou, Pierre Negri, Hsiao Chu, Steve Chaney, Yongjun Liu, Xiaomeng Wu, and Saratchandra Shanmukh.

References

1. Le Ru, E. C. *Principles of surface-enhanced Raman spectroscopy : and related plasmonic effects*, 1st ed.; Amsterdam: Elsevier, 2009.
2. Kneipp, K.; Wang, Y.; Kneipp, H.; Perelman, L. T.; Itzkan, I.; Dasari, R.; Feld, M. S. *Phys. Rev. Lett.* **1997,** *78*, 1667–1670.
3. Nie, S. M.; Emery, S. R. *Science* **1997,** *275*, 1102–1106.
4. Fleischmann, M.; Hendra, P. J.; McQuilla, A. J. *Chem. Phys. Lett.* **1974,** *26*, 163–166.

5.   Hudson, S. D.; Chumanov, G. *Anal. Bioanal. Chem.* **2009**, *394*, 679–686.

6.   Natan, M. J. *Faraday Discuss.* **2006**, *132*, 321–328.

7.   Banholzer, M. J.; Millstone, J. E.; Qin, L. D.; Mirkin, C. A. *Chem. Soc. Rev.* **2008**, *37*, 885–897.

8.   Vo-Dinh, T. *TrAC, Trends Anal. Chem.* **1998**, *17*, 557–582.

9.   Kahl, M.; Voges, E.; Kostrewa, S.; Viets, C.; Hill, W. *Sens. Actuators, B* **1998**, *51*, 285–291.

10.   Lee, P. C.; Meisel, D. *J. Phys. Chem.* **1982**, *86*, 3391–3395.

11.   Seki, H.; Philpott, M. R. *J. Chem. Phys.* **1980**, *73*, 5376–5379.

12.   Vo-Dinh, T.; Dhawan, A.; Norton, S. J.; Khoury, C. G.; Wang, H. N.; Misra, V.; Gerhold, M. D. *J. Phys. Chem. C* **2010**, *114*, 7480–7488.

13.   Perney, N. M. B.; Baumberg, J. J.; Zoorob, M. E.; Charlton, M. D. B.; Mahnkopf, S.; Netti, C. M. *Opt. Express* **2006**, *14*, 847–857.

14.   Van Duyne, R. P.; Hulteen, J. C.; Treichel, D. A. *J. Chem. Phys.* **1993**, *99*, 2101–2115.

15.   Hawkeye, M. M.; Brett, M. J. *J. Vac. Sci. Technol., A* **2007**, *25*, 1317–1335.

16.   Wachter, E. A.; Moore, A. K.; Haas, J. W. *Vib. Spectrosc.* **1992**, *3*, 73–78.

17.   Chaney, S. B.; Shanmukh, S.; Dluhy, R. A.; Zhao, Y. P. *Appl. Phys. Lett.* **2005**, *87*.

18.   Driskell, J. D.; Shanmukh, S.; Liu, Y.; Chaney, S. B.; Tang, X. J.; Zhao, Y. P.; Dluhy, R. A. *J. Phys. Chem. C* **2008**, *112*, 895–901.

19.   Zhou, Q.; Li, Z. C.; Yang, Y.; Zhang, Z. J. *J. Phys. D: Appl. Phys.* **2008**, *41*.

20.   Liu, Y. J.; Chu, H. Y.; Zhao, Y. P. *J. Phys. Chem. C* **2010**, *114*, 8176–8183.

21.   Abell, J. L.; Garren, J. M.; Zhao, Y. P. *Appl. Spectrosc.* **2011**, *65*, 734–740.

22.   Zhao, Y. P.; Chaney, S. B.; Shanmukh, S.; Dluhy, R. A. *J. Phys. Chem. B* **2006**, *110*, 3153–3157.

23.   Liu, Y. J.; Fan, J. G.; Zhao, Y. P.; Shanmukh, S.; Dluhy, R. A. *Appl. Phys. Lett.* **2006**, *89*, 173174.

24.   Liu, Y. J.; Zhao, Y. P. *Phys. Rev. B* **2008**, *78*, 075436.

25.   Zhou, Q.; Liu, Y. J.; He, Y. P.; Zhang, Z. J.; Zhao, Y. P. *Appl. Phys. Lett.* **2010**, *97*, 121902.

26.   Zhou, Q.; Zhang, X.; Huang, Y.; Li, Z. C.; Zhao, Y. P.; Zhang, Z. J. *Appl. Phys. Lett.* **2012**, *100*, 113101.

27.   Zhou, Q.; He, Y. P.; Abell, J.; Zhang, Z. J.; Zhao, Y. *Chem. Commun.* **2011**, *47*, 4466–4468.

28.   Shanmukh, S.; Jones, L.; Driskell, J.; Zhao, Y. P.; Dluhy, R.; Tripp, R. A. *Nano Lett.* **2006**, *6*, 2630–2636.

29.   Driskell, J. D.; Zhu, Y.; Kirkwood, C. D.; Zhao, Y. P.; Dluhy, R. A.; Tripp, R. A. *PLoS One* **2010**, *5*, e10222.

30.   Hennigan, S. L.; Driskell, J. D.; Dluhy, R. A.; Zhao, Y. P.; Tripp, R. A.; Waites, K. B.; Krause, D. C. *PLoS One* **2010**, *5*, e13633.

31.   Chu, H. Y.; Huang, Y. W.; Zhao, Y. P. *Appl. Spectrosc.* **2008**, *62*, 922–931.

32.   Stefani, G.; Slack, F. J. *Nat. Rev. Mol. Cell Biol.* **2008**, *9*, 219–230.

33.   Roush, S.; Slack, F. J. *Trends Cell Biol.* **2008**, *18*, 505–516.

34.   Driskell, J. D.; Seto, A. G.; Jones, L. P.; Jokela, S.; Dluhy, R. A.; Zhao, Y. P.; Tripp, R. A. *Biosens. Bioelectron.* **2008**, *24*, 917–922.

35. Driskell, J. D.; Primera-Pedrozo, O. M.; Dluhy, R. A.; Zhao, Y. P.; Tripp, R. A. *Appl. Spectrosc.* **2009**, *63*, 1107–1114.

36. Negri, P.; Kage, A.; Nitsche, A.; Naumann, D.; Dluhy, R. A. *Chem. Commun.* **2011**, *47*, 8635–8637.

37. Du, X. B.; Chu, H. Y.; Huang, Y. W.; Zhao, Y. P. *Appl. Spectrosc.* **2010**, *64*, 781–785.

38. Leverette, C. L.; Villa-Aleman, E.; Jokela, S.; Zhang, Z. Y.; Liu, Y. J.; Zhao, Y. P.; Smith, S. A. *Vib. Spectrosc.* **2009**, *50*, 143–151.

39. Zhou, Q.; Yang, Y.; Ni, J.; Li, Z. C.; Zhang, Z. J. *Physica E* **2010**, *42*, 1717–1720.

40. Wu, X.; Gao, S.; Wang, J.-S.; Wang, H.; Huang, Y.-W.; Zhao, Y. *Analyst* **2012**, *137*, 4226–4234.

41. Zhu, Y.; Dluhy, R. A.; Zhao, Y. P. *Sens. Actuators, B* **2011**, *157*, 42–50.

42. Singh, J. P.; Chu, H.; Abell, J.; Tripp, R. A.; Zhao, Y. *Nanoscale* **2012**, *4*, 3410.

43. Fu, J.; Cao, Z.; Yobas, L. *Nanotechnology* **2011**, *22*, 505302.

44. Song, C. Y.; Abell, J. L.; He, Y. P.; Murph, S. H.; Cui, Y. P.; Zhao, Y. P. *J. Mater. Chem.* **2012**, *22*, 1150–1159.

45. Abell, J. L.; Driskell, J. D.; Dluhy, R. A.; Tripp, R. A.; Zhao, Y. P. *Biosens. Bioelectron.* **2009**, *24*, 3663–3670.

46. Chen, J.; Abell, J.; Huang, Y.-w.; Zhao, Y. P. *Lab Chip* **2012**, *12*, 3096–3102.

47. Zhao, Y.; Harold, S. *SPIE* **2012**, *8401*, 84010O.

48. Chunyuan, S.; Jeremy, D. D.; Ralph, A. T.; Yiping, C.; Yiping, Z.; Augustus, W. F., III *SPIE* **2012**, *8358*, 83580I.

Chapter 11

Molecular Sensing Based on Surface-Enhanced Raman Scattering and Optical Fibers

Xuan Yang,[1,3] Damon A. Wheeler,[2] Claire Gu,[1] and Jin Z. Zhang[*,2]

[1]Department of Electrical Engineering, University of California
at Santa Cruz, Santa Cruz, California 95064, U.S.A.
[2]Department of Chemistry and Biochemistry, University of California
at Santa Cruz, Santa Cruz, California 95064, U.S.A.
[3]Lawrence Livermore National Laboratory,
Livermore, California 94550, U.S.A.
[*]E-mail: zhang@ucsc.edu

Molecular sensors based on surface-enhanced Raman scattering (SERS) and optical fibers have been widely used in chemical and biological detections due to their unique advantages: molecular specificity, high sensitivity, flexibility, and *in-situ* remote sensing capability. In this chapter, we review the development of fiber biosensors based on SERS and highlight several important milestones. Various fiber SERS configurations will be discussed with an emphasis on our recent techniques known as the double substrate "sandwich" structure and the liquid core photonic crystal fiber. They are based on a stronger electromagnetic field enhancement and/or increased SERS interaction volume. These fiber sensors were tested with dye molecules, proteins, and bacteria and showed extremely high sensitivity.

Introduction

The general demand for sensors for the detection of chemical and biological agents is much greater than before, with applications spanning environmental, food safety, medical, military, and security concerns. Most of the current sensing techniques tend to be either non-molecular specific, bulky, expensive, relatively inaccurate, or unable to provide real-time data. Clearly, alternative

© 2012 American Chemical Society

technologies are urgently needed for chemical and biological detection. An ideal sensing technology should meet the following requirements: highly sensitive, molecule-specific, reliable, easy to fabricate, low cost, label-free, real-time, compact and applicable to a large number of molecular species.

Among all sensors, optical sensors are desirable because they are usually very sensitive, often noninvasive, in some cases molecule-specific, and relatively inexpensive. The most common sensing technique underlying optical sensors is based on the fluorescence of a labeling dye whose fluorescence properties change upon interaction with the analyte (*1, 2*). Fluorescence-based sensors are highly sensitive, inexpensive, and easy to fabricate; however, these types of sensors are generally not molecule-specific due to the broad fluorescence spectrum.

Compared with fluorescence spectroscopy, Raman spectroscopy offers the advantage of molecular specificity and provides "fingerprint" information about specific molecules since each molecular has its own unique set of vibrational modes (*3*). However, the Raman signal is usually very weak due to the small scattering cross-section, and it is therefore difficult to detect the analyte at low concentrations. One way to enhance the Raman signal is to exploit resonance Raman spectroscopy which provides an enhancement factor of about $10^2 - 10^3$ (*4*), however the signal is still quite small even with resonance enhancement. Furthermore, resonance enhancement requires excitation light at shorter wavelengths than that typically used for normal Raman.

Alternatively, the promising method to enhance the Raman signal is surface-enhanced Raman spectroscopy (SERS). SERS is a process whereby the Raman scattering signal can be amplified by orders of magnitude due to the strong enhancement of the electromagnetic field by the surface plasmon resonance of the metallic nanostructures and surface chemical enhancement (*5, 6*). Since its discovery, SERS has attracted extensive attention and been well-employed in detection of a large number of chemicals and biological molecules (*7–9*).

While SERS provides the molecular specificity and high sensitivity, optical fibers have been used as SERS probes because of their low cost, flexibility, compactness, and remote sensing capability (*10–16*). Conventionally, optical fibers have been used to transmit the laser light and the Raman or SERS signal with low propagation loss; however, they can provide more functions than a simple transmission channel. In fact, optical fibers have attracted extensive attention as SERS probes because they can not only further increase the SERS signal but also facilitate the integration of a compact sensor system for in-situ remote sensing.

In this chapter, we will review the development of fiber SERS sensors and highlight several important milestones. Various fiber SERS configurations will be discussed with two main categories: the conventional singlemode/multimode optical fiber and the novel photonic crystal fiber. These fiber sensors were tested with small dye molecules, medium-sized proteins, and large and complex systems such as bacteria. All the results demonstrate optical fibers as a promising and versatile platform for highly sensitive and label-free SERS detections and characterizations of chemical and biological species.

Fiber Sensors Based on Conventional and D-Shaped Optical Fibers

The single multimode fiber SERS probe was first demonstrated by Mullen *et al.* in 1991 (*10*). In the following years, studies involving different shapes of fiber tips were tested, such as flat, angled and tapered fibers (*14–16*). Recently, various nanofabrication technologies have also been employed to pattern SERS substrates on the planar fiber tip (*17–20*). For example, Zhu *et al.* used oblique angle deposition to deposit silver nanorod array on the fiber end for SERS detection in a forward scattering configuration (*17*); Kostovski *et al.* demonstrated a nanoscale biotemplating approach to fabricate fiber SERS sensors, using nanoimprint lithography to replicate cicada wing antireflective nanostructures (*18, 19*); Smythe *et al.* developed a "decal transfer" technique that allows e-beam lithography-defined SERS substrates to be transferred to the facet of an optical fiber (*20*). End-tip fiber optic SERS probes have been shown to produce excellent results with high stability and portability, however, the small SERS-active region on the fiber tip limits their sensitivity. It is highly desirable to improve the detection sensitivity of SERS probes based on these conventional optical fibers.

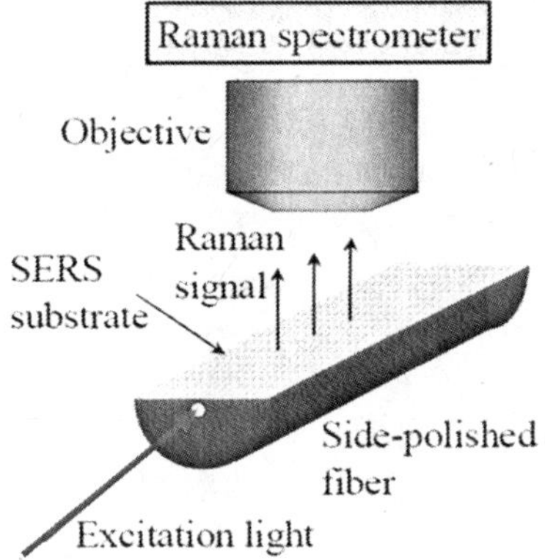

Figure 1. Schematic of the Raman probe with a D-shaped fiber coated with SERS substrate on the flat surface. Adapted with permission from ref. (21). Copyright 2005 American Institute of Physics.

To address the aforementioned issue, one method is to increase the SERS active surface area; in our study, we chose to use a D-shaped fiber configuration, in which the cross-sectional D shape was formed by side polishing the fiber, as shown in Figure 1 (*21*). Light can be coupled out of the polished fiber into silver or gold nanostructures coated on the polished surface, which can potentially increase the SERS-active region by several orders of magnitude. In this configuration, as much as 70% of light that coupled into the fiber may be absorbed into the SERS active surface across much of an 80,000 μm^2 surface as compared to ~50 μm^2 of an end-polished fiber of the same kind. This results in as much as three orders of magnitude increase in Raman scattered photons compared to end-tip fiber probes. Figure 2 shows the excellent and consistent test using Rhodamine 6G

(R6G) with light directly illuminating the fiber surface and coupled through the fiber. However, one major disadvantage of this scheme is that the Raman signal has to be collected from the sideway by the objective lens.

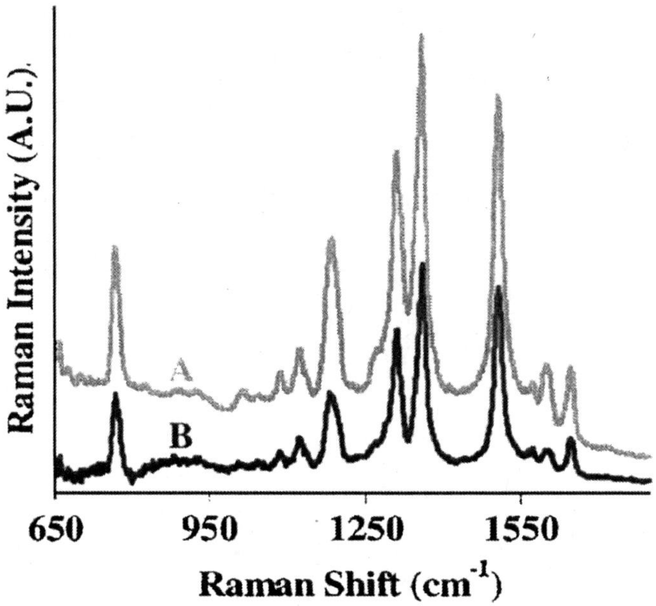

Figure 2. Representative SERS spectra of R6G on silver nanoparticles dried on a D-shaped fiber collected with excitation laser incident to fiber surface (A) and coupled into the fiber (B). Adapted with permission from ref. (21). Copyright 2005 American Institute of Physics.

Other than increasing the surface area of the SERS active region, another straightforward strategy for improving the SERS sensitivity is to increase the electromagnetic field. Towards this goal, we developed a configuration based on a tip coated multimode fiber (TCMMF) with a double-substrate "sandwich" structure using two substrates simultaneously (*22–24*). In this configuration, one type of silver nanoparticles (SNPs) is coated on the tip of a multimode fiber while the other type of SNPs is mixed with the target analyte molecules inside the solution. After dipping the coated fiber probe into the solution, these two types of SERS substrates will sandwich the analyte molecules where the electromagnetic field is stronger and therefore leads to an increased SERS signal.

In our original design, 5 nm-sized hexanethiolate-protected SNPs were used to coat the fiber tip and 25 nm-sized citrate-reduced SNPs were inside the solution. Figure 3 shows the performance of the original TCMMF compared to that by the bulk detection in which the laser light is directly focused onto the solution without using any optical fibers. It can be seen that the original TCMMF SERS probe can provides 2-3× sensitivity than the bulk detection.

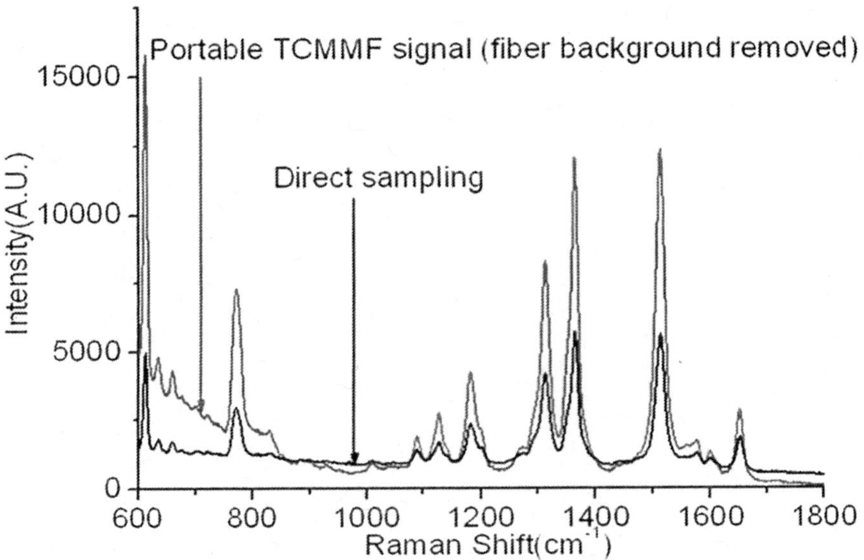

Figure 3. SERS spectra obtained by the TCMMF probe and the bulk detection using R6G as a test molecule. Adapted with permission from ref. (23). Copyright 2010 American Institute of Physics.

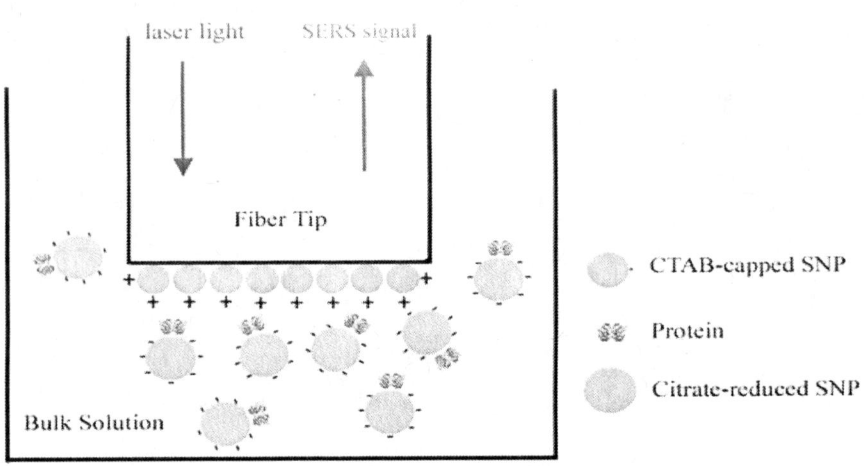

Figure 4. Schematic of the TCMMF SERS probe in aqueous protein detection. Adapted with permission from ref. (24). Copyright 2011 American Chemical Society.

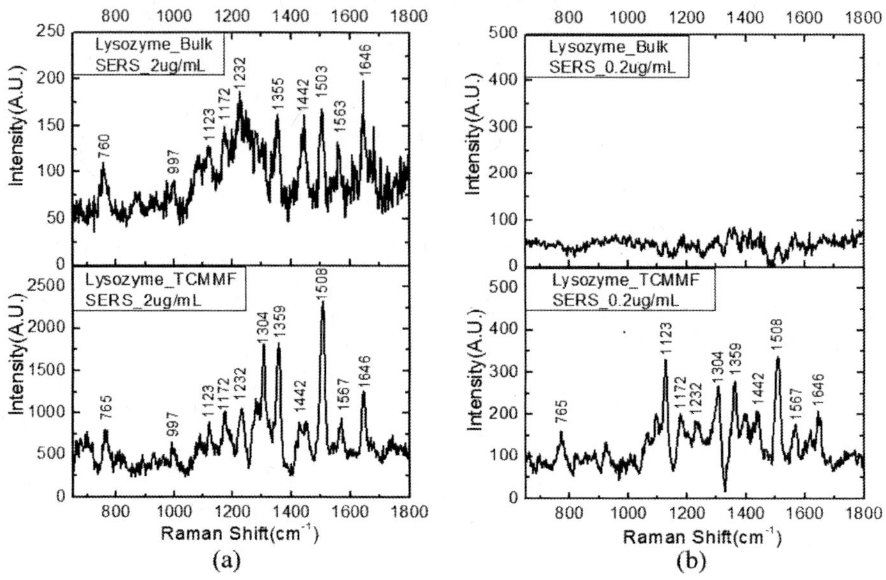

Figure 5. SERS spectra of lysozyme detected by bulk solution and TCMMF probe at various concentrations: (a) 2 µg/mL; (b) 0.2 µg/mL. Adapted with permission from ref. (24). Copyright 2011 American Chemical Society.

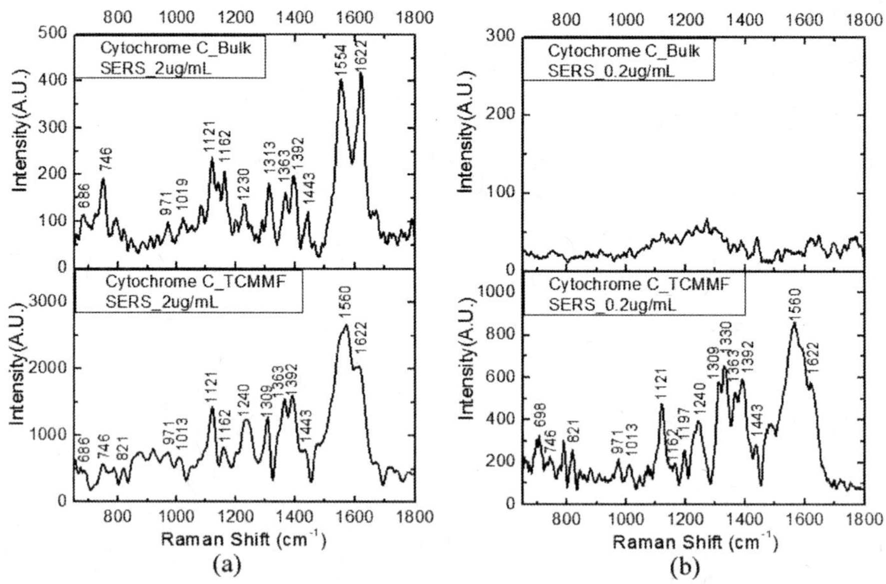

Figure 6. SERS spectra of cytochrome c detected by bulk solution and TCMMF probe at various concentrations: (a) 2 µg/mL; (b) 0.2 µg/mL. Adapted with permission from ref. (24). Copyright 2011 American Chemical Society.

However, the "sandwich" effect in our original design was relatively weak due to the small size of the SNPs coated on the fiber tip and the same type of charges (negatively-charged) carried by these two types of SNPs. In order to further improve the sensitivity, a new TCMMF probe with cetyltrimethylammonium bromide (CTAB)-capped positively-charged SNPs coated on the fiber tip was fabricated (*24*). Figure 4 shows the schematic of the new TCMMF probe and its application in SERS detections of proteins. Compared to the previous method, CTAB-capped SNPs have a larger size (25 nm vs. 5 nm) and an opposite charge with respect to the citrate-reduced SNPs in the bulk solution. With the oppositely-charged surfaces, the electrostatic force can decrease the gap distance between these two types of SNPs, facilitate the formation of the "sandwich" structure, lead to a stronger EM field and therefore increase the SERS signal. With the enhanced SERS signal by the double substrate "sandwich" structure, the TCMMF SERS probe can detect a lower concentration in aqueous protein solution.

As shown in Figure 5 and Figure 6 respectively, while the detection limit of lysozyme and cytochrome c in bulk detection is 2 µg/mL, the TCMMF can obtain a detection limit of 0.2 µg/mL, which is 1 order of magnitude lower than that by the bulk detection. This detection limit is comparable to and even slightly lower than that by the dried protein-silver film strategy by Culha *et al.* (*25*). Other than the high sensitivity, our TCMMF probe in aqueous solution can provide several unique advantages: high reproducibility, flexibility, in-situ remote sensing capability and also the integration capability with the portable Raman system (*23*).

Fiber Sensors Based on Novel Photonic Crystal Fiber

Since its discovery in the 1990s (*26, 27*), photonic crystal fiber (PCF) has emerged as a promising and powerful platform for a multitude of fascinating applications, including absorption (*28*), fluorescence (*29*), Raman (*30, 31*), and SERS (*32–40*). The unique microstructures of axially aligned air channels in PCF not only provide the photonic bandgap for the confinement of light inside the fiber core, but also offer opportunities for a strong interaction between the analyte and the light. In particular, PCF SERS sensors are becoming an increasingly attractive technique for various chemical, biological, medical, and environmental detections.

Generally, there are two types of PCF sensors based on SERS: one is the solid-core PCF (SCPCF) and the other is the hollow-core PCF (HCPCF). For SCPCF, the cladding holes are typically decorated with the SERS substrate (*e.g.*, SNPs) and filled with the analyte solution, and the SERS signal is generated by the interaction between the evanescent wave and the analyte. For example, Yan *et al.* utilized a SCPCF with four large air holes surrounding the solid silica core and the surface was coated with gold nanoparticles for SERS detection (*32*); Amezcua-Correa *et al.* explored the SERS characterization with SNPs coated on the inner wall of PCF using high-pressure chemical deposition technique (*33*); and Du's group reported a forward-propagating full-length SERS-active SCPCF platform with immobilized SNPs (*34, 35*).

However, the relatively weak evanescent wave in SCPCFs may not generate the SERS signal efficiently while most of the light confined in the central solid silica core leads to a large Raman background. In contrast, HCPCF usually has an air core in the center and a periodic structure in the cladding. The light can be well confined in the air core of HCPCF with the photonic bandgap guiding. With the sample solution inside the central core where the electromagnetic field is the strongest, most of the light is utilized for generation of the SERS signal. Moreover, as the light is confined inside the liquid core, the interaction between the light and silica wall can be very small, which leads to a greatly reduced fiber background.

The first attempt to combine HCPCF and SERS was performed by Yan *et al.* in 2006 (*36*). In this first study, we presented a HCPCF SERS probe coated with a layer of gold nanoparticles (GNPs) on the inner wall of the air holes. The analyte solution entered the air channels due to the capillary effect and was subsequently dried using a heating procedure. The laser light was coupled into one end, propagated along the HCPCF, and interacted with the SERS substrate and the molecules at the other end. The SERS signal was then transmitted back to the measuring tip and coupled out of the fiber into the Raman spectrometer. Figure 7 shows the cross-section of the HCPCF and the first proof-of-principle demonstration with the SERS spectrum of Rhodamine B (RhB) detected by the HCPCF (*36, 40*).

The original HCPCF probe can only work well when the sample solution is dried in the fiber, which highly limits the applications for *in vivo* sensing. In addition, the scheme of coating nanoparticles on the inner walls only utilized the light energy near the walls and has not fully taken advantage of the power inside the fiber for SERS detection. In addition, the quality of HCPCF needs improvement for fewer defects, lower insertion loss, and better modal confinement.

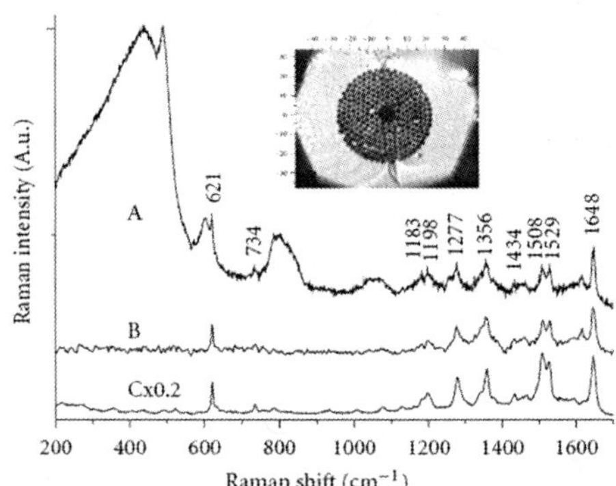

Figure 7. SERS spectra of RhB. Curve A: obtained from the measuring tip of the HCPCF probe (excitation power of ~4.7 mW, scan time of 20 s, and accumulation time of 5); Curve B: after fiber background subtraction; Curve C: obtained using a SERS substrate coated on a Si wafer (excitation power of ~0.47 mW, scan time of 20s, and accumulation time of 1). Inset: micrograph of the cross section of HCPCF. Adapted with permission from ref. (36). Copyright 2006 American Institute of Physics.

To overcome the problems mentioned above, a liquid core PCF (LCPCF) probe was proposed and demonstrated (*37*). The LCPCF probe was fabricated by sealing the cladding holes of the HCPCF and leaving the central hollow core open. The SERS substrate, in this case SNPs, was mixed with the analyte inside the solution rather than being coated on the walls. After dipping the sealed end of the HCPCF into the solution of SNPs/analyte, only the central core was filled due to the capillary force. In this experiment, a 10 cm-long segment of HCPCF was prepared with both ends carefully cleaved, and the cladding holes were sealed by inserting 2-3 mm of one tip of the HCPCF into a high-temperature flare (~1000 °C) for 3-5 s.

Figure 8 shows the schematic of LCPCF and the micrograph of the cross-section of the HCPCF after the sealing process. Various sample solutions including Rhodamine 6G (R6G), human insulin, and trypotophan were tested for the LCPCF probe's sensitivity in this study. We demonstrated that the LCPCF sensor is more sensitive than the original HCPCF probe, and this is attributed to the better confinement of both light and sample in the central core of the LCPCF and thereby increased interaction volume (*37*).

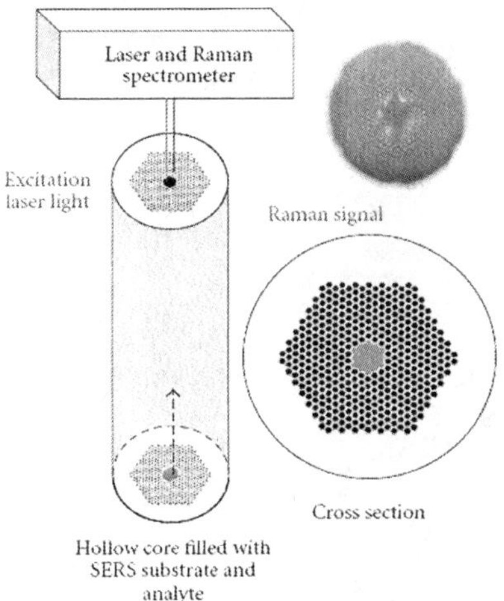

Figure 8. Schematic of the LCPCF SERS sensor and its cross-sectional view. Adapted with permission from ref. (37). Copyright 2007 American Institute of Physics.

To further improve the sensitivity, a better process control was developed in our following study (*38, 39*). A fusion splicer was used to seal the cladding holes instead of the high-temperature flare. By changing the arc power, arc duration, and positions of the fibers, the fusion splicer can generate heat more uniformly and seal the cladding holes more exactly than the flame method. In addition, we chose to place the dipped end under the microscope for better excitation and collection efficiency. After optimizing both the fabrication process and the detection scheme, we achieved a sensitivity enhancement of 100× by using the LCPCF compared to the bulk detection, in which the laser light was focused onto the sample solution without using any optical fibers (*38, 39*). With the sensitivity enhancement, we obtained a detection limit of 0.1 nM using R6G as a test molecule, as shown in Figure 9. This is the lowest detection limit ever reported by using HCPCF as a SERS probe.

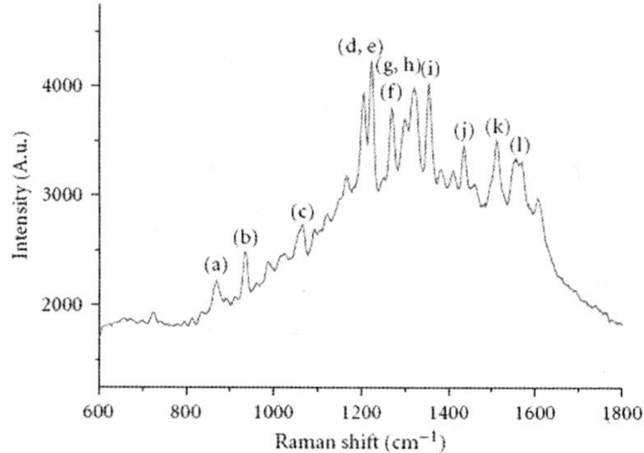

Figure 9. SERS signal from the detection of 0.1 nM R6G using a LCPCF probe. Adapted with permission from ref. (39). Copyright 2010 Optical Society of America.

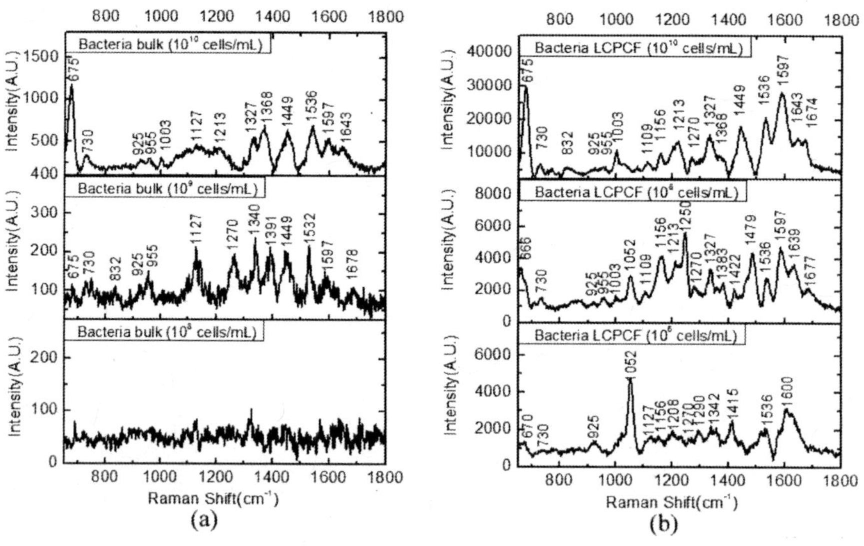

Figure 10. SERS spectra of MR-1 cells detected by (a) bulk detection (10^{10}, 10^9, 10^8 cells/mL) and (b) using the LCPCF probe (10^{10}, 10^8, 10^6 cells/mL), respectively. Adapted with permission from ref. (24). Copyright 2011 American Chemical Society.

In order to study LCPCF's application in the detection of interesting biological macromolecules, we employed it for the SERS detection of the live bacterial cells of *Shewanella oneidensis* MR-1 in our recent study (*24*). Figure 10 shows the SERS spectra of MR-1 cells collected by the bulk detection and using the LCPCF probe at various concentrations. While the detection limit in the bulk detection is 10^9 cells/mL, the LCPCF probe can obtain a detection limit down to 10^6 cells/mL.

In summary, three types of fiber SERS sensing schemes including D-shaped fiber, double-substrate "sandwich" structure, and PCF are discussed. For D-shaped fiber, similar to other conventional fiber SERS probes, it is very flexible and easy to integrate with the detection system although it does require collecting the Raman signal from the sideway for higher sensitivity performance. For double-substrate "sandwich" structure, it has an optrode configuration and is usually more sensitive than the D-shaped fiber; however, it needs to mix the metal nanoparticles with the analyte inside the solution, which might not be desirable for certain in-situ remote sensing applications. And for the PCF, it is highly sensitive due to the light confinement but tends to be more challenging for integration of the portable sensing systems (*23*).

Conclusion

The above results demonstrate that optical fibers provide a convenient platform for molecular sensors based on SERS, which are molecular specific, highly sensitive, flexible, and compact. These fiber SERS probes can be implemented in various interesting chemical and biological detections, including small organic dye molecules, medium-sized proteins, and large systems such as bacteria. Such optical fiber sensors are expected to have significant impact in chemical, biological, environmental, national security, and other applications.

Acknowledgments

The work presented here is funded by the National Science Foundation. The following collaborators, researchers, and graduate students have been contributing significantly to this project: Dr. Yi Zhang, Dr. Chao Shi, Dr. Rebecca Newhouse, Dr. Adam Schwartzberg, Prof. Shaowei Chen, Kazuki Tanaka, Qiao Xu, Alissa Y. Zhang, Dr. Fang Qian, Prof. Yat Li, Prof. Bin Chen, Prof. Changxi Yang, Prof. Guofan Jin, and Dr. Tiziana C. Bond.

References

1. Weiss, S. *Science* **1999**, *283*, 1676–1683.
2. Moerner, W. E.; Fromm, D. P. *Rev. Sci. Instrum.* **2003**, *74*, 3597–3619.
3. Kneipp, K.; Kneipp, H.; Itzkan, I.; Dasari, R. R.; Feld, M. S. *Chem. Rev.* **1999**, *99*, 2957.
4. Asher, S. A. *Anal. Chem.* **1993**, *65*, 201A–210A.
5. Campion, A.; Kambhampati, P. *Chem. Soc. Rev.* **1998**, *27*, 241–250.

6. Kneipp, K.; Kneipp, H.; Itzkan, I.; Dasari, R. R.; Feld, M. S. *J. Phys.: Condens. Matter* **2002**, *14*, R597–R624.

7. Carron, K. T.; Kennedy, B. J. *Anal. Chem.* **1995**, *67*, 3353–3356.

8. Tao, A.; Kim, F.; Hess, C.; Goldberger, J.; He, R.; Sun, Y.; Xia, Y.; Yang, P. *Nano Lett.* **2003**, *3*, 1229–1233.

9. Shanmukh, S.; Jones, L.; Driskell, J.; Zhao, Y.; Dluhy, R.; Tripp, R. A. *Nano Lett.* **2006**, *6*, 2630–2636.

10. Mullen, K. I.; Carron, K. T. *Anal. Chem.* **1991**, *63*, 2196–2199.

11. Gu, C.; Zhang, Y.; Schwartzberg, A. M.; Zhang, J. Z. *Proc. SPIE* **2005**, *5911*, 591108.

12. Volkan, M.; Stokes, D. L.; Vo-Dinh, T. *Appl. Spectrosc.* **2000**, *54*, 1842.

13. Stokes, D. L.; Vo-Dinh, T. *Sens. Actuators B* **2000**, *69*, 28.

14. Viets, C.; Hill, W. *J. Raman Spectrosc.* **2000**, *31*, 625.

15. Viets, C.; Hill, W. *J. Phys. Chem. B* **2001**, *105*, 6330.

16. Viets, C.; Hill, W. *Sens. Actuators, B* **1998**, *51*, 92.

17. Zhu, Y.; Dluhy, R. A.; Zhao, Y. *Sens. Actuators, B* **2011**, *157*, 42–50.

18. Kostovski, G.; White, D. J.; Mitchell, A.; Austin, M. W.; Stoddart, P. R. *Biosens. Bioelectron.* **2009**, *24*, 1531–1535.

19. Kostovski, G.; Chinnasamy, U.; Jayawardhana, S.; Stoddart, P. R.; Mitchell, A. *Adv. Mat.* **2011**, *23*, 531–535.

20. Smythe, E. J.; Dickey, M. D.; Bao, J.; Whitesides, G. M.; Capasso, F. *Nano Lett.* **2009**, *9*, 1132–1138.

21. Zhang, Y.; Gu, C.; Schwartzberg, A. M.; Zhang, J. Z. *Appl. Phys. Lett.* **2005**, *87*, 123105.

22. Shi, C.; Yan, H.; Gu, C.; Ghosh, D.; Seballos, L.; Chen, S.; Zhang, J. Z.; Chen, B. *Appl. Phys. Lett.* **2008**, *92*, 103107.

23. Yang, X.; Tanaka, Z.; Newhouse, R.; Xu, Q.; Chen, B.; Chen, S.; Zhang, J. Z.; Gu, C. *Rev. Sci. Instrum.* **2010**, *81*, 123103.

24. Yang, X.; Gu, C.; Qian, F.; Li, Y.; Zhang, J. Z. *Anal. Chem.* **2011**, *83*, 5888–5894.

25. Kahraman, M.; Sur, I.; Culha, M. *Anal. Chem.* **2010**, *82*, 7596–7602.

26. Birks, T. A.; Roberts, P. J.; Russell, P. S. J.; Atkin, D. M.; Shepherd, T. J. *Electron. Lett.* **1995**, *31*, 1941–1943.

27. Cregan, R. F.; Mangan, B. J.; Knight, J. C.; Birks, T. A.; Russell, P. S. J.; Roberts, P. J.; Allan, D. C. *Science* **1999**, *285*, 1537–1539.

28. Ritari, T.; Tuominen, J.; Ludvigsen, H.; Petersen, J. C.; Sorensen, T.; Hansen, T. P.; Simonsen, H. R. *Opt. Express* **2004**, *12*, 4080–4087.

29. Smolka, S.; Barth, M.; Benson, O. *Appl. Phys. Lett.* **2007**, *90*, 111101.

30. Pristinski, D.; Du, H. *Opt. Lett.* **2006**, *31*, 3246–3248.

31. Buric, M. P.; Chen, K. P.; Falk, J.; Woodruff, S. D. *Appl. Opt.* **2008**, *47*, 4255–4261.

32. Yan, H.; Liu, J.; Yang, C.; Jin, G.; Gu, C.; Hou, L. *Opt. Express* **2008**, *16*, 8300–8305.

33. Amezcua-Correa, A.; Yang, J.; Finlayson, C. E.; Peacock, A. C.; Hayes, J. R.; Sazio, P. J. A.; Baumberg, J. J.; Howdle, S. M. *Adv. Funct. Mater.* **2007**, *17*, 2024–2030.

34. Oo, M. K. K.; Han, Y.; Martini, R.; Sukhishvili, S.; Du, H. *Opt. Lett.* **2009**, *34*, 968–970.

35. Han, Y.; Tan, S.; Oo, M. K. K.; Pristinski, D.; Sukhishvili, S.; Du, H. *Adv. Funct. Mater.* **2010**, *22*, 2647–2651.

36. Yan, H.; Gu, C.; Yang, C.; Liu, J.; Jin, G.; Zhang, J.; Hou, L.; Yao, Y. *Appl. Phys. Lett.* **2006**, *89*, 204101.

37. Zhang, Y.; Shi, C.; Gu, C.; Seballos, L.; Zhang, J. Z. *Appl. Phys. Lett.* **2007**, *90*, 193504.

38. Shi, C.; Lu, C.; Gu, C.; Tian, L.; Newhouse, R.; Chen, S.; Zhang, J. Z. *Appl. Phys. Lett.* **2008**, *93*, 153101.

39. Yang, X.; Shi, C.; Wheeler, D.; Newhouse, R.; Chen, B.; Zhang, J. Z.; Gu, C. *J. Opt. Soc. Am. A* **2010**, *27*, 977–984.

40. Yang, X.; Shi, C.; Newhouse, R.; Zhang, J. Z.; Gu, C. *Int. J. Opt.* **2011**, *2011*, 754610.

Chapter 12

DNA Functional Gold and Silver Nanomaterials for Bioanalysis

Wei-Yu Chen, Yen-Chun Shiang, Chi-Lin Li, Arun Prakash Periasamy, and Huan-Tsung Chang*

Department of Chemistry, National Taiwan University, 1, Section 4, Roosevelt Road, Taipei 106, Taiwan
E-mail: changht@ntu.edu.tw

Deoxyribonucleic acids (DNA) exhibit many predominant capabilities such as specific binding affinities, catalytic activities, and/or chemical stability. On the other hand, gold and silver nanomaterials (NMs) possess unique size- and shape-dependence optical properties, large surface area, biocompatibility, and high stability. These properties have enabled the extensive use of DNA with gold or silver NMs in optical biosensors. DNA-conjugated gold nanoparticles have become most popular optical probes for various targets, with advantages of high sensitivity and selectivity. Fluorescent DNA-templated silver nanoclusters (DNA–Ag NCs) have features of molecule-like optical properties, easy preparation, and good biocompatibility. In this chapter, we highlight the synthesis of water-soluble DNA-functionalized gold and silver NMs, and their optical properties and applications in bioanalysis and cell imaging.

Introduction

Nanomaterials (NMs) have sizes in the range of 1–100 nm, which have become one of the most fascinating materials in the past ten years (*1–3*). They possess strong size- and shape-dependence chemical and physical properties, which are quite different from those of their corresponding bulk materials, mainly because of quantum effect (*4*). The past decade have witnessed progressive advances in the synthesis, characterization, and application of a variety of NMs,

© 2012 American Chemical Society

including gold nanoparticles (Au NPs) (*5*), gold nanodots (Au NDs) (*6*), quantum dots (QDs) such as CdSe and CdTe (*7, 8*), magnetic nanoparticles (MNPs) (*9, 10*), titanium dioxide nanoparticles (TiO_2 NPs) (*11*), silica nanoparticles (Si NPs) (*12*), carbon quantum dots (C–dots) (*13, 14*), graphene (*15, 16*), metal nanoclusters (NCs) (*17*), and so on. The as-synthesized NMs have been widely employed in many fields because of their unique size- and shape-dependence optical (*e.g.*, surface plasmon resonance (SPR), surface enhanced Raman scattering (SERS), and fluorescence), electronic, magnetic, and catalytic properties, which make them ideal candidates as signaling elements for being sensitive biosensors (*18–20*). In general, most NMs are prepared through bottom-up or top-down approaches; "bottom-up" approaches involve the self-assembly of small sized structures into larger structures and "top-down" approaches are production of nanoscale structures from large materials (*21*).

Although some of the NMs such as Au NMs have shown their strong interactions with analytes possessing thiol and amino residues, most of them do not provide high selectivity for specific analytes from complicated biological samples (*22*). In order to prepare functional Au NMs having high affinity towards analytes of interest, recognition elements such as small organic ligands, peptides, proteins, and deoxyribonucleic acid (DNA) have been conjugated to their surfaces through simple adsorption and covalent bonding (*23–25*). Relative to proteins, DNA is much more stable. More interestingly, some DNA molecules have been found selective for analytes, depending on their sequence and conformation (*26, 27*). These particular DNA molecules are short sequences of single-stranded DNA (ssDNA), which are called aptamers. Aptamers are commonly identified *in vitro* from vast combinatorial libraries through a process known as systematic evolution of ligands by exponential enrichments (SELEX) (*28–30*). Some aptamers even possess enzyme activity, including ribozymes and deoxyribozymes (DNAzymes) that are often coordinated to metal ions such as Mg^{2+} or Pb^{2+} as cofactors (*31, 32*). DNAzymes with the assistance of metal ions providing activity enhancement over 100-fold have been found, revealing their great potential in sensing of metal ions (*33, 34*). Relative to antibodies, aptamers are more stable and easier to synthesize, and are available at a lower cost, but have fewer applications because not many aptamers are available. Aptamers can be synthesized automatically and easily modified with a mercapto or amino group in the 5′- or 3′-end of the ssDNA. The thiol-aptamers are assembled onto the surface of Au NMs through Au–S bonding (*35, 36*). Amino group modified aptamers can be easily assembled onto the surface of bare or carboxyl groups modified Au NMs (*36*).

When compared to free aptamers, aptamers that are conjugated with NMs provide the advantages of greater resistance to nuclease digestion and higher affinity toward targets (*37, 38*). Equilibrium dissociation constants (K_d) of aptamers with their targets such as metal ions, small organic molecules, peptides, proteins, nucleic acids, carbohydrates, or even whole cells are usually in the range of picomolar (pM) to micromolar (μM), close to those of antibodies for antigens (*39*). Aptamer conjugated NMs are also renowned for biomedical applications (*40, 41*). For example, TBAs–hTBA$_{29}$ (a 29-base sequence providing TBA$_{29}$ functionality, a T$_3$ linker, and a 15-base sequence for hybridization) and hTBA$_{15}$ (a 15-base sequence providing TBA$_{15}$ functionality, a T$_3$ linker, and a 15-base sequence for hybridization) conjugated with Au NPs that are modified with captured DNA provide enzymatic inhibition of thrombin (*42*).

Among the NMs, Au NPs are most commonly used for the preparation of aptamers functionalized optical sensors, mainly because of their strong SPR absorption with extremely high absorption (extinction) coefficients (10^8–10^{10} M^{-1} cm^{-1}) in the visible region, along with simple preparation, high stability, and biocompatibility. In addition, the surface chemistry of Au NPs is versatile, allowing conjugation with various functional elements (*e.g.*, organic acids, aminothiols, amphiphilic polymers, antibodies, nucleic acids, and proteins) through strong Au–S or Au–N covalent bonding (Figure 1A) or through physical adsorption (Figure 1B) (*24, 25*). Loss of their recognition to analytes due to a change in conformation and desorption sometimes occur when the aptamers are adsorbed on the surfaces of Au NMs. In order to provide high affinity, covalent bonding of aptamers to Au NMs that had been modified with thiol compounds having a suitable spacer and functional groups is suggested (Figure 1C) (*43*). However, it is not an easy task and loss of hydrophilicity of Au NMs occurs sometimes.

Integrating nanotechnology with functional nucleic acids have opened up new strategies for biomolecules detection (*44–47*). Most popular sensing mechanism is based on hybridization of DNA bound to the Au NPs with complementary ssDNAs to form double helix strands, leading to crosslinking aggregation and thus the shift in SPR peaks from 520 nm to 650 nm (*48, 49*). In addition, "non-crosslinking" assays are applied for the detection of analytes based on their induced changes in the surface charge density of Au NPs. Relative to citrate-reduced Au NPs, ssDNA-modified Au NPs are more stable and remain red in color in solution containing NaCl at the concentration above 0.5 M (*50*). When the ssDNA on Au NPs surface is hybridized with its complementary DNA sequence, aggregation occur through the formation of double helix as a result of reduces in their surface negatively charged density (*50*). A decrease in surfaces negatively charged density and a change in the aptamer conformation (from random coil to secondary) also occurs in the aptamer conjugated Au NPs (Apt–Au NPs) in the presence of metal ions. Fan and co-workers demonstrated that potassium binding aptamer-modified Au NPs aggregated immediately upon the addition of K^+, through K^+-induced formation of four-stranded tetraplex (G-quartet) structure (*51*). In addition to colorimetric assays, Au NPs have been widely used in different kinds of analytical techniques, including fluorescence (*52, 53*), SERS (*54, 55*), SPR (*56, 57*) and mass spectrometry (MS) (*58–60*).

In contrast to Au NPs that absorb light, metal nanoclusters (NCs) that consist of a few atoms present a new type of fluorescent NMs for sensing of analytes of interest (*41, 61–65*). These metal NCs have molecule-like properties and no longer exhibit plasmonic properties (*17, 66, 67*). As their size is comparable to the de Broglie wavelength at the Fermi level, NCs display dramatically different optical and chemical properties from NPs with diameters greater than 2 nm (*17, 66, 67*). The energy levels tend to become closer upon increasing the number of atoms in NCs, resulting in size-dependent optical properties (*17, 66, 67*). In addition, the electronic close shell and odd-even effects play increasing roles in determining the dissociation energies of NCs; for example the dissociation energies of Ag NCs increase from 1 to 3 eV, when the number of Ag atoms increases from 2 to 25 (*68–70*). Among various metallic NCs, Au and Ag NCs are the most popular, mainly because of their fluorescence properties, biocompatibility, large Stokes shift, and high emission rates (*17, 41, 61, 65*). Au and Ag NCs that are prepared in the presence of ligands, including phosphine, thiolates, polymers, proteins, and DNA possess fluorescence at different wavelengths with significantly different quantum yields (*17, 61, 62, 65, 67*). Proteins such as bovine serum albumin (BSA) are commonly used as templates for the preparation of water-soluble, stable, and biocompatible Au NCs (*71–75*). On the other hand, DNA as templates are used for the preparation of Ag NCs, mainly because mismatched cytosine–cytosine pairs in DNA duplexes are stabilized by Ag^+ ions (*76–79*).

In the past decade, we have witnessed numerous DNA-functionalized Au and Ag NMs based sensing systems, including absorption, fluorescence, SERS, and amperometry. This chapter covers the synthesis and optical properties of DNA functional Au NPs and Ag NCs and their applications based on absorption and fluorescence detection modes.

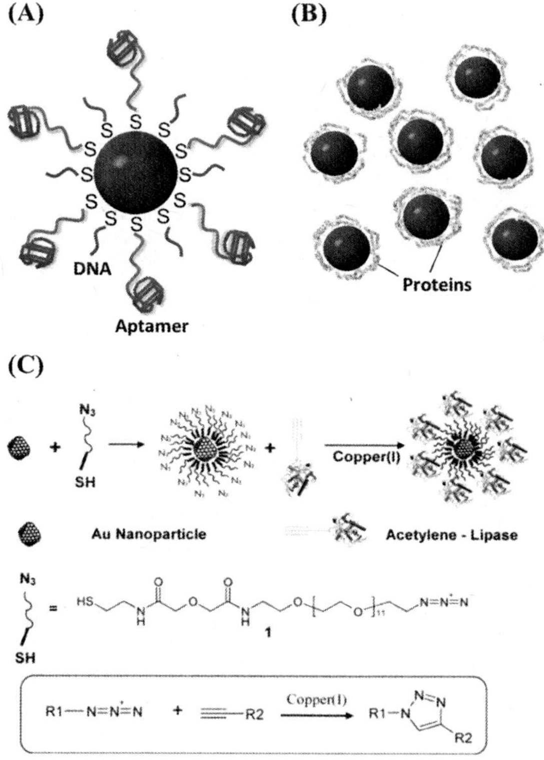

Figure 1. Functionalization of Au NPs through (A) covalent, (B) non-covalent (physical adsorption), and (C) covalent conjugation with a suitable spacer. Figure C is reprinted with permission from reference (43). Copyright 2006 American Chemical Society.

DNA functional Au NPs (DNA–Au NPs)

Preparation and Optical Properties of DNA Functional Au NPs

Preparation

Au NPs are usually prepared from Au^{3+} ions ($HAuCl_4$) with reducing agents such as citrate (*80, 81*). It is well known that the strength and concentration of a reducing agent are important factors affecting the seeding and growth rates of Au NPs (*80, 81*). A strong and/or high concentration of reducing agent is used for the preparation of small sizes of Au NPs. For example, $NaBH_4$ and citrate are separately used to prepare Au NPs having sizes smaller and larger than 5 nm, respectively. By controlling citrate concentration, different sizes of Au NPs are prepared. For example, 32- and 56-nm Au NPs are easily prepared by adding 1% trisodium citrate solutions (0.5 and 0.3 mL, respectively) rapidly to 0.01% $HAuCl_4$ solutions (50 mL) under reflux, in which the mixtures react for 8 min. Although citrate ions stabilize the Au NPs, they are readily replaced by thiol derivated DNA through Au–S bonding for further applications. Generally, preparation of DNA–Au NPs is carried out by simply mixing citrate-Au NPs and thiol-modified oligonucleotides in aqueous solution. Relative to citrate-Au NPs, DNA–Au NPs are more stable in solution containing high salt and/or at high temperature. The stability and hybridization efficiency of DNA–Au NPs are highly dependent on the surface density of DNA. At low surface density, DNA tend to form flat structures on the surfaces of Au NPs, leading to low stability and weak recognition ability. Salt aging of DNA–Au NPs in 100 mM NaCl overnight is effective to prepare stable DNA–Au NPs having greater surface density of DNA (*49*). However steric effects occur when the density of the DNA on the Au NPs surfaces is too high, leading to losses in the sensitivity and reproducibility. In one of our previous studies, we found that the surface density of DNA molecules on Au NPs affect the sensitivity for Hg^{2+} ions significantly (*82*). Thus, it is important to select suitable concentrations of DNA and Au NPs to prepare DNA–Au NPs. In order to minimize surface effect and thus to provide high hybridization efficiency, DNA containing a spacer sequence (*e.g.,* 15 bases of thymine) and a recognition sequence is recommended (*83*). The main advantages of preparation of DNA–Au NPs through covalent bonding include easy control of the density of DNA and stable DNA molecules on the surfaces of Au NPs. Although Au NPs can be simply conjugated with DNA through electrostatic attractions, they are not as stable as covalently bonded DNA–Au NPs in high-ionic-strength solution. In addition, control of the conformation of DNA on the Au NPs surface is difficult.

Optical Properties

The optical properties of Au NPs are mainly dominated by SPR, involving the collective oscillation of electrons at their surfaces, in resonance with the incident electromagnetic radiation (*84, 85*). The SPR absorption wavelengths of Au NPs are dependent on their size, shape, and refractive index; therefore, any changes in surface structure, aggregation, or medium's compositions may induce colorimetric changes in their dispersions. The SPR band observed at 520–530 nm for the Au NPs reveal the sizes ranging from 5–20 nm in diameter (*5*). Moreover, the SPR band is red shifted upon increasing the size of Au NPs. Unlike spherical Au NPs, Au nanorods and Au–Ag nanorods exhibit two SPR bands; for example, Au–Ag nanorods have a transverse band at 508–532 nm and longitudinal band at 634–743 nm, depending on their aspect ratio (length/width) (*86–88*). On the contrary to large Au NPs, small Au NPs (<2 nm) lack an apparent SPR band, but fluoresce more strongly. Such fluorescent Au NPs are often known as gold nanodots (Au NDs). Alkanethiol-protected Au NDs ranging in sizes from several atoms to small particles (<2 nm) fluoresce (quantum yields from 10^{-5} to 10^{-1}) in the region from blue light to the near-infrared (near-IR) (*6*). The Au NDs usually possess a biexponential emission decay, which is possibly attributed to the differential distribution of complicated luminescent pathways of polynuclear gold(I)–thiolate complexes, ligand-to-metal charge transfer (LMCT), and/or LMCT/metal centered triplets (*89, 90*).

Analytical Application of DNA–Au NPs

Detection of Metal Ions

Some trace amounts of metal ions are essential; for example, Cu^{2+} plays a vital part in the development and performance of the human nervous and cardiovascular systems (*91*). However, short or long term exposure to high levels of Cu^{2+} ions can lead to gastrointestinal disturbances or can cause serious damage to the liver and kidneys (*92*). Relative to Cu^{2+} ions, Cd^{2+}, Pb^{2+}, and Hg^{2+} are more toxic. Among these, Hg^{2+} is a most highly toxic and widespread pollutant ion, causing damages to the brain, nervous system, and the kidney even at very low concentrations (*93*). The long-term exposure of Pb^{2+} ions by infants and children delays their physical or mental development, whereas in adults it causes kidney problems and high blood pressure (*94*). On the other hand, the long-term exposure to Cd^{2+} ions causes kidney damage. The environmental protection agency (EPA) of the United States has set the safe limited concentrations of Cu^{2+}, Cd^{2+}, Pb^{2+}, and Hg^{2+} ions in drinking water as 20, 0.005, 0.015, and 0.002 ppm, respectively (*95*). Thus, specific and sensitive detection systems for monitoring these metal ions in ecosystems are highly demanded.

1. Mercury (Hg^{2+}) Ions

Polythymine (T_n) and 13-nm-diameter Au NPs were used to develop a selective and sensitive assay for the detection of Hg^{2+} ions, avoiding the need for conducting tedious processes and the use of expensive thiol-functionalized DNA molecules (*26*). The sensing mechanism involves the formation of DNA–Hg^{2+} complexes through T–Hg^{2+}–T coordination, varying the negative charge density and structures of the DNA strand on the Au NPs surfaces (Figure 2). The reduced degree of electrostatic repulsion among Au NPs leads to their aggregation, which is evidenced by the obvious change in color of the solution from red to purple.

Although qualitative detection of the existence of Hg^{2+} ions at the concentration of 0.25 μM is possible by the naked eye, absorption detection is required for more sensitive quantification. In order to obtain greater linearity, ratios of the absorbance (extinction) values at 650 nm over 520 nm against Hg^{2+} concentration are commonly plotted. The absorbance values at 650 and 520 nm are for the aggregated and dispersed Au NPs. This approach provides a limit of detection (LOD) at a signal-to-noise ratio (S/N) 3 of 250 nM for Hg^{2+} ions, with a great selectivity toward Hg^{2+} ions over other metal ions, including K^+, Mg^{2+}, Ca^{2+}, Cu^{2+}, Ni^{2+}, Co^{2+}, Zn^{2+}, Pb^{2+}, Cd^{2+}, and Fe^{2+}. Similarly, various T-rich DNA–Au NPs have been demonstrated for the sensitive (nM level) detection of Hg^{2+} ions based on Hg^{2+} ion-induced aggregation of Au NPs (*96–98*). These studies have shown the important roles of DNA sequence, DNA density, concentration of Au NPs, as well as solution pH, ionic strength, and compositions for determining the sensitivity and selectivity for Hg^{2+} ions.

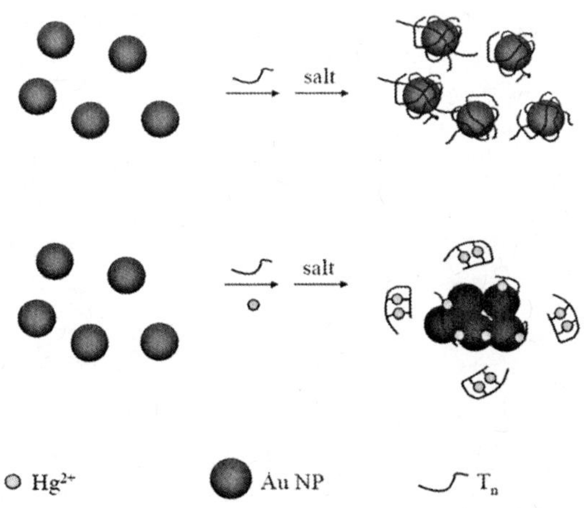

Figure 2. Schematic representation of a Hg^{2+} nanosensor based on high affinity between polythymine (T_n) and 13-nm-diameter Au NPs. Reprinted with permission from reference (26). Copyright 2008 Royal Society of Chemistry.

2. Lead (Pb^{2+}) Ions

Lu and co-workers demonstrated a system consisting of 5′-thiol-modified 12-mer DNA attached to 13-nm Au NPs, a DNAzyme (17E), and its substrate (Sub) for the detection of Pb^{2+} ions, in which the Sub hybridizes specifically to a DNA on each end, while the 17E acts as a recognition portion (*99*). These hybridizations cause aggregation of Au NPs, however, in the presence of Pb^{2+} ions, the 17E catalyzes hydrolytic cleavage of Sub and thus prevents the formation of Au NPs aggregates (no color change). This sensor system provides an LOD of sub-micromolar level for Pb^{2+} ions. Similarly, Lu and co-workers developed AuNP–DNAzyme conjugates for the detection of Pb^{2+} ions, with improved sensitivity (*100*). They chose a 8–17 DNAzyme and 13-nm Au NPs to construct an easy-to-use dipstick tests for Pb^{2+}, demonstrating a very high activity with a fast cleavage rate (estimated k_{obs} ~50 min^{-1} at pH 7.0) (*101*). In the absence of Pb^{2+}, Au NPs-uncleaved substrate is captured at the control zone *via* streptavidin-biotin interaction. In the presence of Pb^{2+}, the enzyme strand (17E) catalyzes the cleavage of the substrate (17S) at the single ribo-adenosine base and Au NPs-cleaved product migrates beyond the control zone, which is captured at the test zone by hybridization to complementary DNA. Unlike aptamer-based colorimetric tests, the presence of target does not cause immediate disassembly of the aggregates in the DNAzyme-based colorimetric tests. Pb^{2+} ions is visualized at a concentration down to ~0.5 μM. Lu and co-workers also constructed a label-free colorimetric sensor for on-site and real-time Pb^{2+} ions detection based on 8–17 DNAzyme (Figure 3) (*102*). Upon adding trishydroxymethylaminomethane (Tris) and NaCl to adjust ionic strength, followed by addition of Au NPs, the released ssDNA is adsorbed onto Au NPs, which prevents the individual red Au NPs to form blue aggregates under high-ionic-strength conditions (100 mM NaCl). In the absence of Pb^{2+} or in the presence of other metal ions however, no cleavage reaction occurs, and therefore the enzyme (17E(8))–substrate ((8)17S) complex is not able to stabilize the individual red Au NPs, resulting in purple–blue Au NP aggregates. The dynamic range of the sensor can be tuned simply by adjusting the pH.

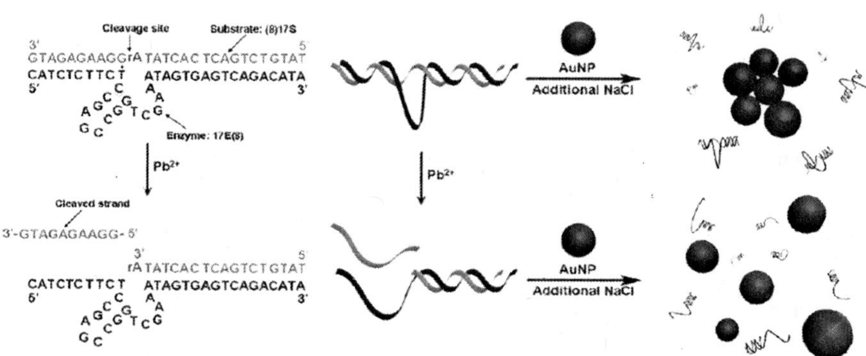

Figure 3. Left: Secondary structure of the DNAzyme complex, which consists of an enzyme strand (17E(8)) and a substrate strand ((8)17S). After lead-induced cleavage, 10-mer ssDNA is released which can adsorb onto a Au NP surface. Right: Schematic representation of the label-free colorimetric sensor. The lead-treated/-untreated complexes and NaCl are mixed with Au NPs. The Au NPs aggregate in the absence of lead but remain dispersed in the presence of lead. Reprinted with permission from reference (102). Copyright 2008 John Wiley & Sons.

Although these DNA–Au NPs are highly selective and sensitive for the detection of Pb^{2+}, use of high-cost DNA is a concern. Au NPs without conjugation with DNA have been used for the detection of Pb^{2+} ions. A colorimetric, label-free, non-aggregation-based Au NP probe was developed for the detection of Pb^{2+} ions in aqueous solution through acceleration of the leaching of Au NPs by thiosulfate ($S_2O_3^{2-}$) ions and 2-ME (Figure 4) (*103*). In the presence of Pb^{2+} ions and 2-ME, Pb–Au alloys are formed on the surfaces of the Au NPs. The formation of Pb–Au alloys accelerates the dissolution of Au NPs into solution, leading to dramatic decreases in the SPR absorption at 520 nm. The 2-ME/$S_2O_3^{2-}$–Au NP probe is highly sensitive and selective toward Pb^{2+} ions, with a linear detection range (2.5 nM−10 μM).

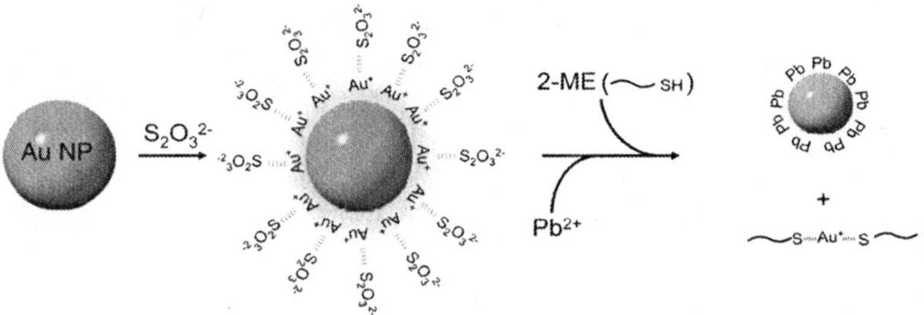

Figure 4. Cartoon representation of the sensing mechanism of the 2-ME/$S_2O_3^{2-}$–Au NP probe for the detection of Pb^{2+} ions. Reprinted with permission from reference (103). Copyright 2009 American Chemical Society.

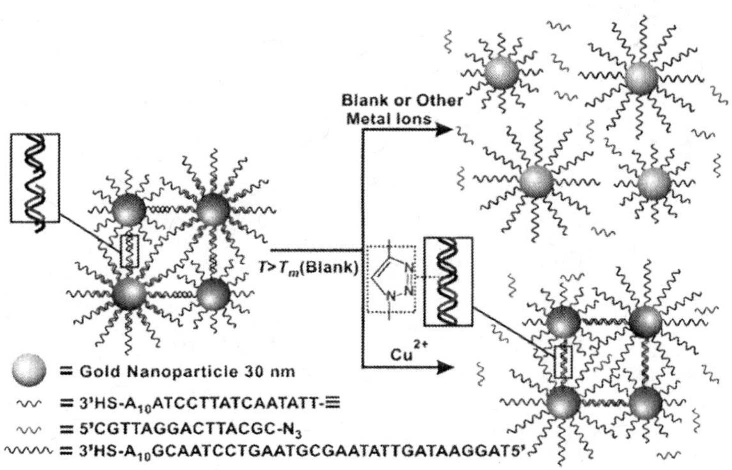

Figure 5. Schematic illustration representing the mechanism involved in the aggregation and dissociation of the Au NP probes used in the colorimetric detection of copper ions. Reprinted with permission from reference (104). Copyright 2010 John Wiley & Sons.

3. Copper (Cu^{2+}) Ions

A colorimetric method for the detection of Cu^{2+} ions (at concentrations down to 20 μM) based on densely functionalized DNA–Au NP conjugates using click chemistry was demonstrated (*104*). This approach relies on the ligation of two 15-bp oligonucleotides strands within polymeric aggregation of the DNA–Au NPs. Two sets of DNA–Au NP conjugates are prepared by functionalizing two batches of 30-nm Au NPs with different azide strands. The Au NPs template is modified with 3′-propylthiol-terminated 40-mer oligonucleotides, while the alkyne Au NPs are functionalized with 3′-propylthiolated and 5′-alkylated 25-mer oligonucleotides that are complementary to half of the DNA on the Au NPs template. Three oligonucleotides and the azide strand are complementary to the other half of the template strand. In the presence of Cu^{2+} ions, these oligonucleotides form three-stranded aggregates, leading to a color change from red to purple (Figure 5) and an increase in the melting temperature from 50.4 to 62.6°C.

Detection of Biomolecules and Cells

1. Small Organic Molecules

Small organic molecules play important roles in regulating cellular metabolism pathways to maintain biological balance. For example, adenosine triphosphate (ATP) transports chemical energy within cells for biosynthetic reactions, motility, and cell division. The aptamer specifically binding to ATP is listed in Table I, which was selected through SELEX (*105*). Also listed in Table I are the aptamers for heme (an iron-containing porphyrin) that shows extreme catalytic properties (*106–108*), cocaine, and kanamycin (*109–111*). Aptamers conjugated Au NPs are sensitive and selective for the detections of ATP, cocaine and kanamycin, mainly based on the analytes induced changes in the conformation of the aptamers, leading to aggregation of the Au NPs (*112, 113*).

Adenosine is an index for the regulation of renal hemodynamics, tubular reabsorption, and release of rennin, and thus its determination is essential. Conjugation of ATP-binding aptamer (ABA) onto Au NPs allows the development of a sensing strategy with good sensitivity (LOD = 10 nM) (*114*). In the absence of ATP, the color of ABA–Au NPs changes from red to purple, resulting from salt-induced aggregation. Through the interaction of ATP with ABA sequence, a unique G-quartet structure is formed and the surface charge density significantly increases. As a result, the ATP/ABA–Au NPs conjugates remain well dispersed at high-salt (2.5 M NaCl) conditions. The aptamer density on Au NPs surface is a decisive factor for stabilizing the Au NPs in the solutions. The optimized density of aptamer (about 35 aptamers per Au NPs) leads to the most sensitive ATP detection, with an LOD of 10 nM. Alternatively, a fluorescent method using Au NPs, two aptamers, and Oligreen was developed (*115*). Oligreen is an ssDNA binding dye, which fluoresces at 520 nm when excited at 480 nm. ABA and platelet-derived growth factor (PDGF)-binding aptamer (PBA) are used for specific interaction with ATP and for a signal reporter. Although Oligreen can bind to ATP–ABA complexes and PBA on the Au NP surfaces, its fluorescence is weak due to electron transfer from ATP–ABA–Oligreen complexes and PBA–Oligreen to the Au NPs. When more adenosine molecules are bounded to ABA sequence, more G-quartet structures are formed, which makes it difficult for PBA–Oligreen complex to access the surfaces of ABA–Au NPs (Figure 6) (*115*). As a result, more PBA–Oligreen complexes remain in the bulk solution, leading to an increase in fluorescence. In other words, the fluorescence of the solutions increases upon increasing adenosine concentration. This approach provides an LOD of 5.5 nM for adenosine.

Table I. Selected DNA aptamers for small organic molecules

Targets	Sequence	K_d (μM)	References
Adenosine/ATP	5′-ACC TGG GGG AGT ATT GCG GAG GAA GGT-3′	6.0	(*105*)
Cocaine	5′-GGG AGA CAA GGA TAA ATC CTT CAA TGA AGT GGG TCT CCC-3′	0.4–10	(*109*)
Hemin	5′-GTG GGT AGG GCG GGT TGG-3′	1.4	(*106*)
Kanamycin	5′-TGG GGG TTG AGG CTA AGC CGA-3′	0.079	(*114*)

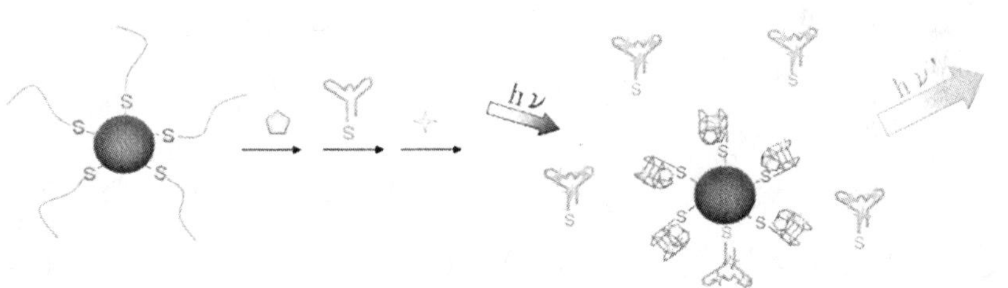

Figure 6. Schematic representation of the sensing assay using ABA–Au NPs and PBA to improve the sensitivity. Reprinted with permission from reference (115). Copyright 2010 Elsevier.

Adenosine-binding aptamers, DNAzymes, and Au NPs were used altogether to construct a DNAzyme-directed assembly of Au NPs in alumina thin layer chromatography plates, which were used as a colorimetric adenosine sensor for the detection of adenosine (*116*). Aptamer and DNAzyme are conjugated separately with different Au NPs through Au–S bonding. In the presence of a substrate, which is partially complemented with the aptamer and DNAzyme, assembly of Au NPs is formed, leading to the color change from red to blue on the alumina plate. Adenosine molecule activates the DNAzyme and thus cleaves the substrate oligonucleotides between Au NPs, resulting in the separation of Au NPs, as evidenced by the appearance of red color. This approach provides an LOD of 0.2 mM for adenosine. The sensing system is selective to adenosine, with a support of no interference from 5 mM guanosine, cytidine, and uridine.

2. Proteins

PDGF is a growth factor protein found in human platelet, with the growth-promoting activity toward fibroblasts, glial cells, and smooth muscle cells. PDGF is a ubiquitous mitogen and chemotactic factor for many connective tissue cells and an important biomarker for epithelial cancers since paracrine PDGF signal triggers stromal recruitment and epithelial-mesenchymal transition that affects tumor growth, angiogenesis, invasion, and metastasis (*117*). Its concentration in blood or urine in healthy people is low (< 1 nM), while concentration higher than 1 nM has been found in cancer patients. Because of low concentration and high concentrations of interference proteins, small molecules and salts, the detection of trace PDGF from complicated biological samples is not an easy task.

Various strategies using aptamers conjugated Au NPs have been demonstrated for the detection of PDGF. When PDGF molecules interact with the aptamer (5′-CAGGCTACGGCACGTAGAGCATCACCATGATCCTG-3′)-bounded Au NPs (PBA–Au NPs) through 1:2 fashion, Au NPs become aggregated, leading to decrease and increases in the absorbance (extinction) at 530 and 650 nm, respectively (*118*). Different concentrations of PDGF lead to various degrees of crosslinking of PBA–Au NPs. Upon increasing the PDGF concentration from 0–75 nM, aggregation of the PBA–Au NPs (8.4 nM) increases. As the concentration of PDGF is increased over 400 nM, the aggregated Au NPs reverse back to the dispersed form, mainly resulting from the saturation of PBA–Au NPs surfaces with PDGF. By constructing a plot of ratios of the absorbance values of PBA–Au NPs at 650 to 530 nm against the concentrations of three isomers of PDGF (PDGF-AA, PDGF-AB and PDGF-BB), this sensing system provides linear ranges within 25 and 150 nM. The sensing system is also sensitive for the detection of PDGF receptor-β. When PDGF receptor-β is introduced into PDGF-BB/PBA–Au NPs solution, PBA and PDGF receptor-β are competitively bound to PDGF-BB, reducing the degree of aggregation of PBA–Au NPs. Using this competitive binding assay, the LOD of this approach for PDGF receptor-β is 3.2 nM.

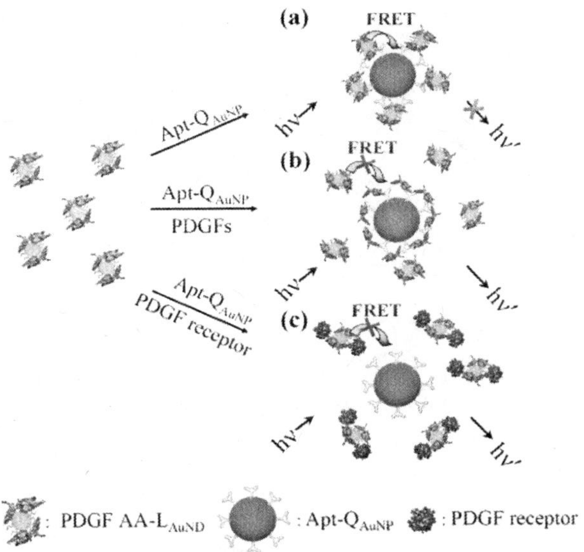

a h, Planck's constant; *ν*, frequency of light.

Figure 7. Schematic representation of PDGF and PDGF receptor based nanosensors that operate on the modulation of the photoluminescence quenching of PDGF-AA–Au NDs and PBA–AuNPs. Reprinted with permission from reference (121). Copyright 2008 American Chemical Society.

To improve the sensitivity for PDGF, a fluorescence quenching approach using PBA–Au NPs has been demonstrated by taking the advantage of high extinction coefficients of Au NPs (*119*).   N,N-dimethyl-2,7-diazapyrenium dication (DMDAP) is intercalated strongly within polynucleotide duplexes (*120*). When DMDAP is bound to PBA sequence on the Au NPs, the fluorescence of DMDAP is highly quenched by PBA–Au NPs. The fluorescence signals increase upon increasing PDGF concentration, resulting from the release of DMDAP from the surfaces of Au NPs. This approach provides high sensitivity (LOD = 8.0 pM) and selectivity for PDGF. Alternatively, PBA–Au NPs (13 nm in diameter) and PDGF bound photoluminescent Au NDs were used to develop a sensitive and selective assay for PDGF, in which the former acts as an acceptor (wavelength of maximum absorption were 520 nm), and the latter acts as a donor (wavelength of maximum excitation and emission are 375 and 520 nm, respectively) (*121*). The photoluminescent Au NDs having a diameter of 2.0 (± 0.1) nm were synthesized through reduction of hydrogen tetrachloroaurate (HAuCl$_4$) with tetrakis(hydroxymethyl)phosphonium chloride (THPC) and then through etching with 11-mercaptoundecanoic acid (11-MUA) under alkaline conditions. Through electrostatic and hydrophobic interactions of PDGF-AA with the 11-MUA–Au

NDs, PDGF-AA–Au NDs were prepared (*121*). Photoluminescence quenching occurs through a specific interaction of PDGF-AA with PBA on the surfaces of the PDGF-AA–Au NDs and PBA–Au NPs, respectively (Figure 7) (*121*). In the presence of PDGF in sample solution, the interaction of PDGF-AA–Au NDs with PBA–Au NPs is suppressed, leading to increase in photoluminescence intensity. In order to further improve the sensitivity and minimize the matrix interference from biological samples, a NP-assisted protein enrichment method was conducted in conjunction with the probe. This approach allows the detection of PDGF-AA at the concentration down to 10 pM and determination of the concentrations of PDGFs in the collected HTB-26 cell media and urine.

Since the PDGF-binding aptamer was used to prepare aptamer functional Au NPs (Apt–Au NPs) for the detection of PDGF (*118*), various Apt–Au NPs have been prepared for the detection of proteins, including human immunoglobulin E (IgE) and lysozyme as listed in Table II (*122–125*). SPR and electrochemical sensing strategies have provided good sensitivity for protein detection using Apt–Au NPs. A novel sandwich immunoassay for the detection of IgE using a goat anti-human IgE on SPR gold film as a capture and Apt–Au NPs as an amplification reagent provides an LOD of 1 ng/ml (*123*).

Table II. Selected DNA aptamers for proteins

Targets	Sequences	K_d (μM)	References
PDGF	5′-CAG GCT ACG GCA CGT- AGA GCA TCA CCA TGA TCC TG-3′	6.0	(*118*)
Thrombin	5′-GGT TGG TGT GGT TGG-3′	0.4–10	(*126*)
	5′-GTC CGT GGT AGG GCA- GGT TGG GGT GAC-3′		(*127*)
IgE	5′-ATC TAC GGG GCA CGT- TTA TCC GTC CCT CCT AGT- GGC GTG CCC C-3′	0.02	(*122*)
Lysozyme	5′-GAT GAA TTC GTA GAT-3′	0.03	(*124*)

Thrombin converts soluble fibrinogen into insoluble strands of fibrin in coagulation cascade and is considered as an important target when screening for new anticoagulants to interfere with blood coagulation. Thus, techniques that allow sensitive and selective detection of trace amounts of thrombin in biological samples is extremely important. There are two extended surfaces in human α-thrombin that are mainly composed of positively charged residues specifically bound to thrombin-binding aptamer (TBA, 5′-GGTTGGTGTGGTTGG-3′) (*126,*

127). Willner and co-workers demonstrated a thrombin assay based on thrombin induced aggregation of Au NPs, which further formed Au nanorods in aqueous solution containing Au^{3+} ions and cetyl trimethylammonium bromide (CTAB) (*128*). Addition of various concentrations of thrombin leads to formation of different levels of TBA-conjugated Au NPs aggregates as a result of the two aptamer binding sites of thrombin linked to more than one TBA-conjugated Au NPs. The absorbance of the enlarged Au NPs aggregates increases upon increasing the concentration of thrombin, leading to increases in the absorbance at 650 nm. This approach provides LODs of 20 and 2 nM for thrombin in liquid and solid phases, respectively. A DNA detection colorimetric probe was demonstrated, based on the modulation of the activity of thrombin by fibrinogen-modified Au NPs (Fib–Au NPs) in the presence of bivalent TBAs (*129*). The bivalent TBAs interact specifically with thrombin, suppressing its activity toward Fib–Au NPs. The inhibitory effect of designed TBAs with targeted complementary sequence toward thrombin is highly dependent on the concentration of target DNA.

3. Cancer Cells

Lack of understanding of chronic diseases such as cancers at the molecular level has made the pre-stage diagnosis extremely challenging. There are only few biomarkers available for precise identification of small amounts of tumor tissues from regular ones. To minimize positive and negative errors, it is also important to have techniques allowing simultaneous detection of several important biomarkers. For example, prostate-specific antigen and prostatic acid phosphatase are important markers for prostate cancer cells (*130*). In order to overcome the non-specific binding between aptamers and cell surfaces, a counter selection is usually performed to access oligonucleotides that can only interact with target cells but not with the control cells (*131–133*).

Table III lists the sequences of some aptamers that are used to prepare functional Au NPs for specific targeting of cancer cells (*134–141*). These selected sequences have high affinities towards specific cell lines with dissociation constant values ranging from 0.76 to 95 nM. The Apt–Au NPs are assembled on the surface of a specific type of cancer cell through the recognition of the aptamer to its target on the cell membrane, leading to a significant shift in the absorption band (*135*). Using 20-nm functional Au NPs, detection of human acute lymphoblastic leukemia CCRF-CEM cells and human burkitt's lymphoma Ramos cells down to about 90 cells is realized (*135*). SK-BR3 cells, a human breast cancer cell over expressing human epidermal growth factor receptor 2 (HER2) on their membrane, were detected using multifunctional oval-shaped Au NPs (*137*). An aptamer-conjugated nanoparticle strip biosensor was demonstrated for the specific and rapid sensing of Ramos cells (Figure 8), with an LOD of 4000 cells (*139*). A capture aptamer TE02 is bioconjugated onto a strip through streptavidin-biotin interaction. On the other hand, thiolated aptamer TD05 is modified on the surface of Au NPs to function as a detection probe. This approach provides a linear range of 8×10^3 to 4×10^5 for the detection of Ramos cells.

Table III. Sequence of cancer cell aptamers

Targets	Sequences	K_d (nM)	References
Human burkitt's lymphoma cell line (Ramos cells)	5′-AGG CAG TGG TTT GAC GTC CGC ATG TTG GGA ATA GCC ACG CCT-3′	0.76	(138)
	5′-AAC ACC GTG GAG GAT AGT TCG GTG GCT GTT CAG GGT CTC CTC CCG GTG-3′	75	(138)
Human acute lymphoblastic leukemia (CCRF-CEM)	5′-TTT AAA ATA CCA GCT TAT TCA- ATT AGT CAC ACT TAG AGT TCT AGC TGC TGC GCC GCC GGG AAA ATA CTG TAC GGA TAG ATA GTA AGT GCA ATC T-3′	n. s.[a]	(134)
Prostate specific membrane antigen on prostate cancer cells	5′-CGA CGA CGA CGA CGA CGA CGA-3′	n. s.[a]	(140)
Human breast cancer cell overexpressing HER2 (SK-BR3 cells)	5′- TGG ATG GGG AGA TCC GTT GAG TAA GCG GGC GTG TCT CTC TGC CGC CTT GCT ATG GGG-3′	95	(136)

[a] n. s., not specified.

DNA-Templated Silver Nanoclusters (DNA–Ag NCs)

Preparation

The optical properties of DNA–Ag NCs are highly dependent on their size and DNA sequences (17, 64). In other words, the optical properties of DNA–Ag NCs are tunable by controlling their size, which can be achieved by varying several important parameters such as DNA sequences, the molar ratio of bases/metal ions, reducing agents, buffer pH, and ionic strength (76–79, 142–145). In general, DNA that forms stable complexes with Ag^+ ions are used as templates for the preparation of stable and fluorescent DNA–Ag NCs, mainly because of the formation of cytosine–Ag^+–cytosine (C–Ag^+–C) coordination that stabilizes the DNA duplexes (76–79). The presence of both carbonyl oxygen and doubly-bonded ring nitrogen in cytosine plays an important role in determining their interaction strength with Ag^+ ions (146). Using single-stranded 5′-AGGTCGCCGCCC-3′ as a template and NaBH4 as a reducing agent, a mixture of several DNA–Ag NCs were prepared that exhibit emissions centered at 640 nm (76). The formation of DNA–Ag NCs changes the structure of DNA, as evident from the NMR and circular dichroism (CD) data (76). The Ag NCs are encapsulated in short single-stranded DNA, thus preventing their aggregation. A small number (≤ 4) of silver atoms bounded with the single-stranded DNA template was demonstrated by electrospray ionization-mass spectrometry (ESI-MS) measurements (*76*).

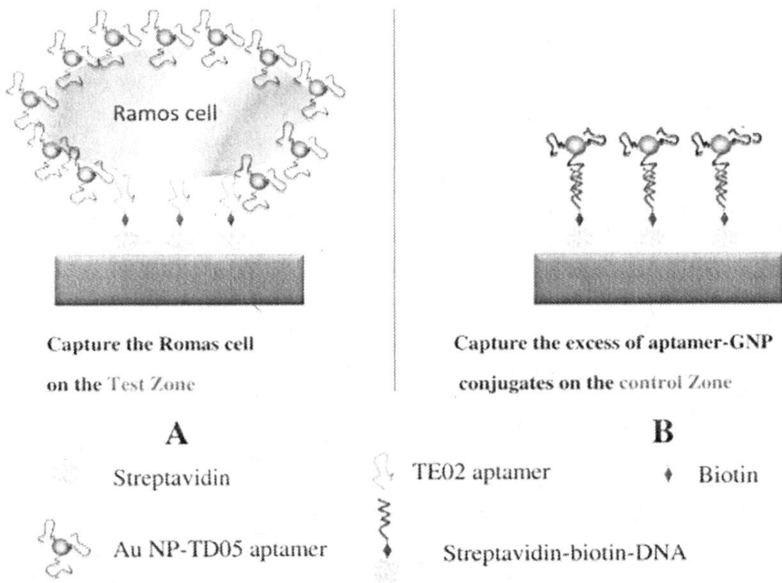

Figure 8. Schematic diagram of the detection of Ramos cells on aptamer–nanoparticle strip biosensor (ANSB). (A) capturing Au NP–aptamer–Ramos cells on the test zone of ANSB through specific aptamer–cell interactions and (B) capturing the excess of Au NP–aptamer on the control zone of ANSB through aptamer-DNA hybridization reaction. Reprinted with permission from reference (139). Copyright 2009 American Chemical Society.

Polycytosine (C_{12}) oligonucleotide is a useful template for the preparation of C_{12}–Ag NCs that are stable at neutral pH (*77*). The time-dependent density functional theory implies that Ag NCs (Ag_1 to Ag_6) prefer to bind with the double bonded ring nitrogen (*146*). The fluorescence intensity of C_{12}–Ag NCs increases upon increasing pH from 3.0 to 5.5. Notably, the midpoint at pH 4.0 overlaps with the pK_a of the N3 of cytosine, suggesting that deprotonation of N3 is associated with Ag NCs binding (*77*). Among the loop of hairpin sequences, 5′-TATCCGTX$_5$ACGGATA-3′ (X = C, G, A, or T for C-, G-, A-, or T-loop, respectively), C- and G-loops rich in cytosine or guanine generate similar fluorescence intensity of the DNA–Ag NCs; while the emission from A-loop-DNA–Ag NCs is 10-fold weaker and T-loops does not yield emissive Ag NCs, implying differential binding affinities of bases to Ag$^+$ ions (*78*). The C_{12}–Ag NCs contain 2–7 Ag atoms (multiple species), as evident from the number of transitions exhibited in the fluorescence (excitation and emission maxima of 650 and 750 nm, respectively) and absorption spectra (*77*).

DNA–Ag NCs with distinct emission properties have been synthesized by tuning the bases of corresponding five DNA sequences (*78, 142–144*). Five distinct and spectrally pure DNA–Ag NCs were synthesized in bulk solutions, which have fluorescence colors throughout the visible and near-IR regions (Figure 9) (*142*). Blue and green-emitting species with poor photostability are only formed in un-buffered solutions. While, three highly emitting species (yellow, red and near-IR) are formed in phosphate (20 mM, pH 6.5–8.0), citrate (20 mM, pH 5.0–7.0), and ammonium acetate (20 mM, pH 6.5–8.0) solutions, respectively. Relative to organic dyes, the DNA–Ag NCs provide stronger fluorescence intensity (*65, 147, 148*). In order to synthesize DNA–Ag NCs with high quantum yields, four long DNA strands (22, 29, 34 and 46 bases) with different sequences and lengths were selected to prepare DNA–Ag NCs (*143*), which emit in a range of wavelengths from green to near-infrared region. The highest quantum yield of the Ag NCs is about 50%.

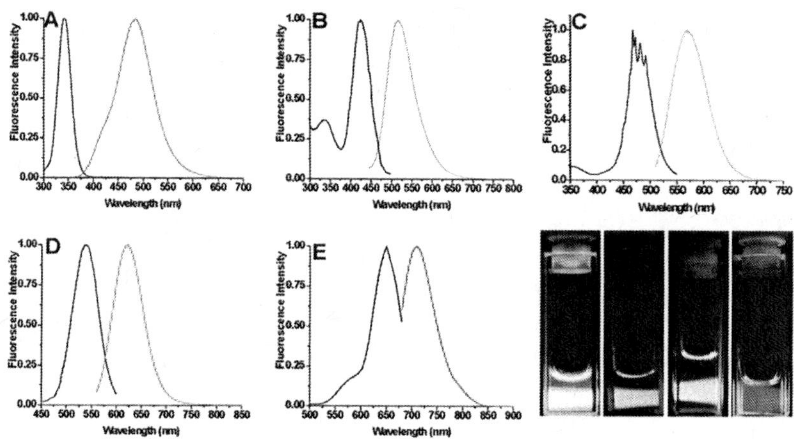

Figure 9. Schematic diagram of excitation and emission spectra for five distinct DNA–Ag NCs. (A) Blue emitters created in 5'-CCCTTTAACCCC-3', (B) green emitters created in 5'-CCCTCTTAACCC-3', (C) yellow emitters created in 5'-CCCTTAATCCCC-3', (D) red emitters created in 5'-CCTCCTTCCTCC-3', and (E) near-IR emitters created in 5'-CCCTAACTCCCC-3'. Photograph of emissive solutions in A-D. Reprinted with permission from reference (142). Copyright 2008 American Chemical Society.

Repeated cytosine units also play an important role in determining the quantum yields of DNA–Ag NCs (*144*). Among a set of tested DNA sequences, the oligonucleotide 5'-CCC(TTCC)$_2$TT(CCAA)2CCC-3' (DNA$_{TAr2}$) as the scaffold allows preparation of DNA$_{TAr2}$–Ag NCs that provide the highest quantum yield (61%) at 608 nm when excited at 540 nm (*144*). Moreover, the presence of cytosine 3-mers at the 5' and 3' ends of the DNA scaffold stabilize the DNA–Ag NCs as a result of entropic effects. CD spectra results reveal that DNA$_{TAr2}$–Ag NCs exhibit negative ellipticities because the hairpin-like structure formed through C–Ag$^+$–C coordination induces the formation of non-planar and tilted orientations of the bases relative to the helical axis. However, one disadvantage of the preparation of DNA–Ag NCs is the long preparation time (usually 5 days) (*149, 150*). To shorten reaction time, a one-pot approach was demonstrated separately for the preparation of fluorescent DNA-templated copper/silver and gold/silver NCs from Cu^{2+}/Ag^+, Au^{3+}/Ag^+, and DNA (5'-CCCTTAATCCCC-3') in the presence of $NaBH_4$ as a reducing agent at varied buffered systems (*149, 150*). Among these buffered systems, phosphate and citrate are found to be more suitable for preparing the two NCs with high fluorescence quantum yields. The DNA–Au/Ag NCs are even more stable in high-ionic-strength media. The preparation time is only about 90 min. Both C and G strands that are predominantly present as individual strands in the solution produce bright fluorescent DNA–Ag NC solutions (*151*), while the double-stranded DNA (dsDNA) formed by the C and G strands produces negligible visible fluorescence, mainly because the site specific binding of Ag atoms to ssDNA is rendered inaccessible by Watson-Crick base pairing. DNA scaffolds containing dsDNA with an extra six-base cytosine loop (C_6 loop) and a mismatched double-stranded DNA were used as templates for the preparation of DNA–Ag NCs, with results showing the important roles of the specific interaction between $C–Ag^+–C$ coordination and structural flexibility of the region near the mismatched site playing in determining their fluorescence properties (*152, 153*). The two as-prepared DNA–Ag NCs fluoresce at 572 and 570 nm upon excited at 520 nm. In addition to cytosine-rich oligonucleotides, thymine-rich oligonucleotides such as T_{12} and $T_4C_4T_4$ have the potential to form blue/green-emitting DNA–Ag NCs (*154*). Unfortunately, the as-prepared emitters are only stable for a few days.

A DNA combined C-rich sequence and G-rich sequence for i-motif and/or G-quadruplex structure was used for the preparation of DNA–Ag NCs, based on the fact that an Ag^+ ion can stabilize C–C pair, analogous to a proton in i-motif structure (*155, 156*). In a typical preparation procedure, $(TA_2C_4)_4$ and $(C_4A_2)_3C_4$ each with a common C_4 i-motif core were used as templates for the preparation of red- and green-emitting Ag NCs, respectively (*155*). Notably, the stability of red emitters of Ag NCs in these i-motif sequences is pH sensitive, with the best conditions being at pH 6.0. The resulting DNA–Ag NCs are possibly encapsulated within i-motif structures. Alternatively, G-quadruplex as an effective template was used to prepare G-quadruplex-encapsulated Ag NCs, possessing multi-emission wavelengths that are highly dependent on the excitation wavelengths (E_x/E_m 325/510, 325/420, and 510/680 nm) (*156*).

In addition to the one-pot synthesis of fluorescent DNA–Ag NCs, Dickson and co-workers demonstrated the synthesis of fluorescent DNA–Ag NCs by converting their templates poly(acrylic acid) (PA) with the DNA (*157*). The Ag NCs prepared in the presence of PA as a template are stable, but with a low quantum yield. Through a shuttle-based fluorescent NC transferring process, PA–Ag NCs acting as a fluorophore unit are transferred from a low molecular weight shuttle to a C-rich ssDNA tag on the cellular protein of interest, ascribed to the enhanced affinity and fluorescence quantum yield (*157*). Dark DNA–Ag NCs are transformed into bright red-emitting NCs when they are placed in proximity to G-rich DNA sequences (Figure 10) (*158*). Ag NCs with weak green emission were formed on an ssDNA strand that has a 12-base oligonucleotide sequence and a 30-base hybridization sequence. Upon addition and hybridization of a complement with the guanine, proximity-induced fluorescence enhancement (ca. 500-fold) along with a strong red fluorescence emission is observed after duplex formation, revealing that the proximity to guanine induce the transformation of Ag NCs from a dark species to a bright red-emitting species.

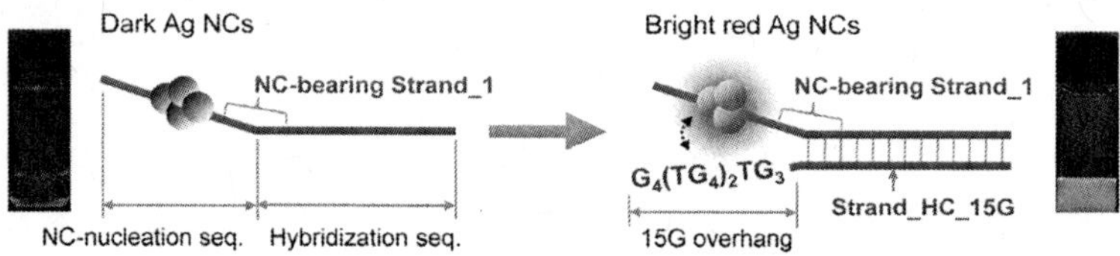

Figure 10. Schematic diagram of red fluorescence enhancement of DNA–Ag NCs through proximity with a G-rich overhang, 3'-G$_4$(TG$_4$)$_2$TG$_3$-5', caused by DNA hybridization. Photograph of the resulting emission under UV (366 nm) irradiation. Reprinted with permission from reference (158). Copyright 2010 American Chemical Society.

Optical Properties and Characterization

When the sizes of Au and Ag NCs are down to few atoms, the continuous density of states broke up into discrete energy levels (*17, 66*). Owing to the interaction with light through electronic transitions between energy levels, sharp absorption and emission bands are formed. The size-dependent electronic structures and relative electronic transitions of small NCs are described by the jellium approximation (*159, 160*). X-ray photoelectron spectroscopy (XPS), which allows determination of the oxidation state of Ag NC surfaces, reveals the existence of Ag^0 and Ag^+. The peak for Ag $3d_{5/2}$ in DNA–Au/Ag NCs occurs at 368.7 eV, revealing the existence of Ag^+ ions and atoms in the DNA scaffold (*150*). The molar absorptivity of DNA templates at 350 nm increases upon interaction with Ag^+ ions, which is attributed to the new and overlapping electronic bands of DNA–Ag NCs (*65*).

Relative to most of the Au NCs, DNA-templated Ag NCs have higher QYs. A microsecond fluorescence intermittency ("blinking") dynamics observed for DNA–Ag NCs at single molecular level corresponds to a spectrally observable transient dark state (*161*). In general, the transient dark state is formed when the photoexcited charge of Ag NCs transfers to the encapsulating single-stranded DNA that provides a stable, photoaccessible charge accepting site, enabling photoexcited back-electron transfer. The ellipticity of dsDNA at 265 nm is typically smaller than that of ssDNA (*76, 77*). A decrease in the ellipticity at 265 nm suggests weaker interactions of DNA strands with Ag^0 than with Ag^+ ions (Figure 11) (*76*). The CD data reveal that the interaction of DNA templates with Ag NCs is weaker than that with Cu/Ag NCs (*162, 163*). Ag K-edge x-ray absorption fine structure (EXAFS) analysis of DNA-templated Ag NCs provides insights into bonding, size, and structural correlations with the fluorescence of

Ag NCs (*149, 150, 164, 165*). The observed Ag–Ag bond distance is shorter than the value of 2.89 Å for that in Ag metal, due to the surface tension of particles and other effects (*166*). Thiol compounds such as 3-mercaptopropionic acid and cysteine can induce the fluorescence quenching of DNA–Ag NCs, with support from CD results (*167, 168*). Since the thiol compounds interact strongly with Ag NCs, they weaken the interactions between DNA templates and Ag NCs. In addition, they cause partial oxidation of Ag NCs.

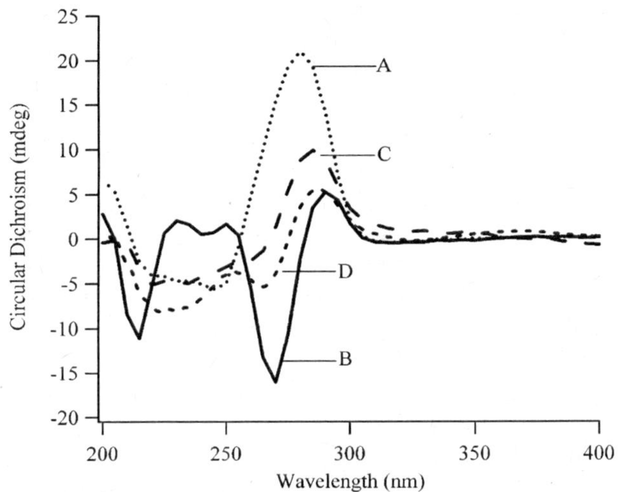

Figure 11. Response of the circular dichroism associated with electronic transition of the oligonucleotide bases in association with Ag$^+$ and Ag nanoclusters. Following are the conditions for the spectra: A (dotted line), 10 μM oligonucleotide solution; B (solid line), oligonucleotide with 60 μM Ag$^+$ (molar ratio of Ag$^+$/bases is 1/2); C (coarse dashed line), 120 min after adding BH$_4^-$ and Ag$^+$ (molar ratio of BH$_4^-$/Ag$^+$/ is 1/1) to the oligonucleotide/Ag$^+$ solution; D (fine dashed line), 4300 min after adding BH$_4^-$ to the oligonucleotide/Ag$^+$ solution. Reprinted with permission from reference (76). Copyright 2004 American Chemical Society.

The mass spectra of reduced silver and protection groups generally show several few-atoms containing species, such as DNA–Ag, DNA–Ag$_2$, DNA–Ag$_3$, DNA–Ag$_4$, and DNA–Ag$_5$ NCs (*76*). Upon increasing the number of repeated C–C pairs, more Ag atoms assemble into the DNA scaffolds, leading to the formation of DNA–Ag NCs of greater sizes (*144*). The ESI-MS data obtained for the repeated cytosine units involved in the preparation of the DNA–Ag NCs reveal that there are 2–6 Ag atoms in each DNA scaffold (*144*). The ESI-MS and fluorescence data confirm that the emission wavelength undergoes a red shift upon increasing the number of Ag atoms. More recent studies on different NCs emitters demonstrated that near IR emission events are usually observed from 9–11 atom cluster sizes (*169, 170*).

Analytical Application of DNA–Ag NCs

The highly fluorescent and water-soluble DNA–Ag NCs offer several features including long Stokes shift, long lifetime, with their sizes comparable to those of biopolymers, biocompatibility, and ease in bioconjugation (*41, 63–65*). Having such advantages, they have become attractive for the detection of ions, small organic molecules, DNA, and proteins and for cell staining. Recent advances in the application of DNA–Ag NCs for the detection of various analytes and cell imaging are discussed in this section.

Detection of Metal Ions and Anions

1. Mercury (Hg^{2+}) Ions

DNA–Ag NCs have been utilized for acute monitoring of Hg^{2+} ions based on the quenching of their fluorescence (*144, 171*). C_{12}-protected Ag NCs are capable of sensing Hg^{2+} ions based on the high-affinity metallophilic interactions between Hg^{2+} and Ag atoms. Recent studies suggest that the dispersion forces between closed-shell metal atoms are highly specific and strong between Hg^{2+} ions ($4f^{14}5d^{10}$) and silver atoms ($6d^{10}$) (*144*). The highly fluorescent DNA–Ag NCs are sensitive and selective for the detection of Hg^{2+} in real samples, with an LOD of 0.9 nM (0.18 ppb), much lower than its maximum level (2.0 ppb) permitted by EPA in drinking water (*144*). Alternatively, a novel strategy for the "turn-on" detection of Hg^{2+} using C_6-loop–Ag NCs was demonstrated. Through the formation of T–Hg^{2+}–T mismatches, hybridized DNA duplexes is formed, leading to increase in the fluorescence (*172*). This approach allows the detection of Hg^{2+} ions down to 10 nM.

2. Copper (Cu^{2+}) Ions

Two approaches using DNA–Ag NCs have been demonstrated for the sensitive and selective determination of the concentrations of Cu^{2+} ions in environmental samples (*149, 167*). The sensing mechanism of the first approach is based on Cu^{2+} induced fluorescence increases in the DNA–Ag NCs (*149*). The introduction of Cu^{2+} results in the formation of DNA–Cu/Ag NCs with more rigid structure from DNA templates and thus enhance the fluorescence, with strong support from the CD data (*162, 167*). The ESI-MS data also reveal that there are three Ag and one Cu ions/atoms per DNA template (*162*). The DNA–Ag NC probe provides high sensitivity (LOD = 8 nM) and selectivity towards Cu^{2+}. This approach allows the detection of Cu^{2+} ions in Montana soil (SRM 2710) and pond water (*149*). Another approach is based on the suppression of 3-mercaptopropionic acid (MPA) induced fluorescence quenching in the DNA–Cu/Ag NCs by Cu^{2+} ions. The MPA induced fluorescence quenching is due to its strong interaction with Cu/Ag ions/atoms, resulting in the change in DNA conformation. In the presence of Cu^{2+} ions, Cu–MPA complexes (intermediate) are formed, which are further oxidized to form disulfide compounds. Decreasing MPA concentration in the mixture suppresses its quenching to the emission of DNA–Cu/Ag NCs. Through the specific interaction of Cu^{2+} ions with MPA, the fluorescence is turn-on in the presence of Cu^{2+} ions (Figure 12) (*167*). This approach provides an LOD of 2.7 nM for Cu^{2+} ions, with greater selectivity towards Cu^{2+} ions relative to the other approach (*162*).

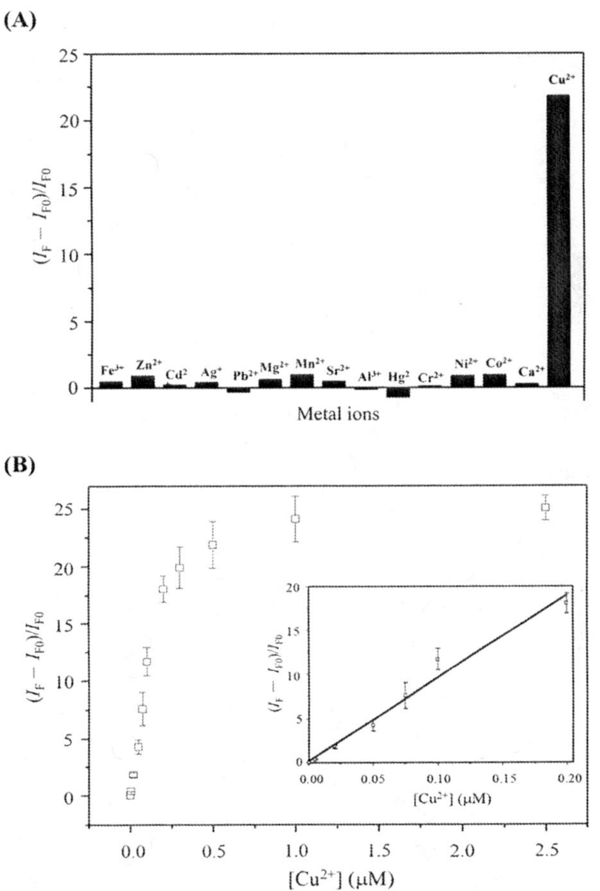

Figure 12. The DNA–Cu/Ag NC probe in the presence of thiols for the detection of Cu^{2+} ions. (A) Concentrations of 0.5 and 50 µM for Cu^{2+} ions and each of the other metal ions, respectively. (B) Error bars represent standard deviations from three repeated measurements. I$_{F0}$ and I$_F$ are the fluorescence intensities of the DNA–Cu/Ag NCs in the absence and presence of Cu^{2+} ions, respectively. Inset: Linear range of the plot of (I$_F$ – I$_{F0}$)/I$_{F0}$ against the Cu^{2+} ion concentration (0–0.2 µM). Reprinted with permission from reference (167). Copyright 2010 American Chemical Society.

3. Anions

Sulfide (S^{2-}) ions are widely distributed in natural water and wastewater, and thus the concentration of S^{2-} ions is an important environmental index (*173, 174*). With the aid of 12-base oligonucleotides (5'-CCCTTAATCCCC-3'), a one-pot approach was applied to prepare fluorescent DNA–Au/Ag NCs from Au^{3+} and Ag^+ using $NaBH_4$ as the reductant in citrate buffered system at pH 5.0 (*150*). The ESI-MS data reveal that the DNA–Au/Ag NCs contain 2 Au atoms and 1 Ag atom per DNA strand. Through the interaction of the Au/Ag atoms/ions with S^{2-} ions, the fluorescence of the DNA–Au/Ag NCs is quenched. The formation of Au_2S and Ag_2S causes the DNA scaffolds to revert to random coiled structures, leading to the exposure of the Au/Ag NCs to the bulk solution and thus decreases in the fluorescence. In order to reduce the interference from I^- ions, $S_2O_8^{2-}$ ions as a masking agent is essential. Having an LOD of 0.83 nM and selectivity, this approach allows the determination of the concentrations of S^{2-} ions in hot spring and seawater samples.

DNA, RNA, Protein, and Small Biomolecules

1. Small Organic Molecules

Biothiols such as cysteine (Cys), homocysteine (Hcy), and glutathione (GSH) play critical roles in reversible redox reactions and important cellular functions including detoxification and metabolism. DNA–Ag NCs are quenched by the formation of non-fluorescent complexes with biothiols, allowing the detection of Cys, GSH and Hcy with LOD values of 4.0, 4.0 and 200 nM, respectively (*168*). By virtue of the template-dependent fluorescence properties of DNA–Ag NCs and their specific response to thiol compounds, a fluorescent turn-on assay for biothiols with high specificity and sensitivity was developed (*175*). The emission of some C_{12}-DNA–Ag NCs can be enhanced in the presence of thiol compounds, mainly through the charge transfer from the ligands to the metal center due to the formation of the Ag–S bonds. This turn-on assay enables a sensitive determination of GSH with an LOD as low as 6.2 nM. This study also reveals that Ag NCs formed in different DNA templates have different fluorescence response patterns to analytes, suggesting the opportunity to modulate the response patterns of DNA–Ag NCs to a specific analyte simply by selecting appropriate DNA templates for the synthesis of NCs. Using aptamers (G-rich DNA), one approach was developed for the detection of small molecules such as cocaine and ATP (*176*). In the absence of the analytes, the two DNA strands are in free-state, and the Ag NCs formed through the reduction of Ag^+ by $NaBH_4$ show weak fluorescence emission. However, in the presence of the analytes, the two G-rich sequence fragments are brought to proximity, leading to significant enhancement in the fluorescent intensity of DNA–Ag NCs. This turn-on fluorescent method allows the detection of cocaine and ATP, with LODs of 0.1 and 0.2 μM, respectively.

2. DNA and RNA

Developing highly sensitive, selective, and rather inexpensive approaches for the detection of DNA is of paramount importance in basic research as well as in many applied areas including clinical diagnoses of infectious and genetic diseases, forensics, food analysis, and environmental monitoring (*177–179*). By taking the advantage of the emission properties of DNA–Ag NCs depending on the local environment and nucleotide sequences, several sensing systems for DNA have been developed (*180*). An Ag NC-based fluorescent assay capable of specifically identifying single nucleotide modifications in DNA was developed, based on the decreases in fluorescence as a result of the formation of duplex scaffolds through DNA hybridization (*152, 180*). Even a single-nucleotide mismatch located two bases away from the NCs formation site can prohibit the generation of fluorescent Ag NCs. The formation of fluorescent Ag NCs in hybridized DNA duplex scaffolds is highly sequence-dependent, which allows the single nucleotide polymorphism (SNP) analysis of hemoglobin beta chain (HBB) gene in sickle cell mutation (*152*). Based on the "light-up" of DNA–Ag NCs in proximity to G-rich DNA sequences, two short, linear DNA probes (a NC probe and a G-rich probe) that are brought into proximity through hybridization with target DNA were used for the preparation of DNA–Ag NCs. The as-prepared DNA–Ag NCs are sensitive for the detection of an influenza target (a sequence from influenza A virus (S-OIV) (H1N1)) (*158*). DNA comprising two functional regions (a specific DNA sequence for recognition of target DNA and a scaffold (C_{12}) for the preparation of DNA–Ag NCs) was used in the preparation of functional DNA–Ag NCs (Figure 13A) (*180*). The ESI-MS data reveal that the DNA–Ag NCs contain 3 Ag atoms per DNA strand. The fluorescence intensity of the DNA–Ag NCs is highly quenched in the solution containing 150 mM NaCl, especially in the absence of its perfect match DNA (DNA_{pmt}). The fluorescence quenching of DNA–Ag NCs by Cl– ions through the strong interaction of Ag^+/Cl^- is minimized in the presence of DNA_{pmt} (*180*). Based on the different stabilities (fluorescence intensities) of DNA–Ag NCs in solutions containing 150 mM NaCl, this probe allows the detection of the target gene of fumarylacetoacetate hydrolase (FAH) (Figure 13B), with a linear response ranging from 25 to 1000 nM and an LOD of 14 nM. The probe also allows differentiation of genes with single-base mismatches, showing potential for SNP analysis.

Detection of microRNAs (miRNAs) is also of importance because they play key roles in regulation of gene expression (*181, 182*). A DNA–Ag NC-based probe has been demonstrated for the detection of miRNAs (*183*). A DNA sequence having a specific complementary sequence against a target miRNA and a 12-nucleotide Ag NC-encoding sequence was used for producing bifunctional DNA–Ag NCs with red emission. When the target miRNA is present, the emission of the DNA–Ag NC probe is significantly quenched *via* the hybridization of miRNA with its complementary sequence of the DNA–Ag NCs. This turn-off DNA–Ag NC probe allows detection of the target miRNAs down to 20 nM (*183*). The DNA–Ag NC probe is practical for the analysis of targeted miRNA in whole plant (Arabidopsis thaliana) endogenous RNA.

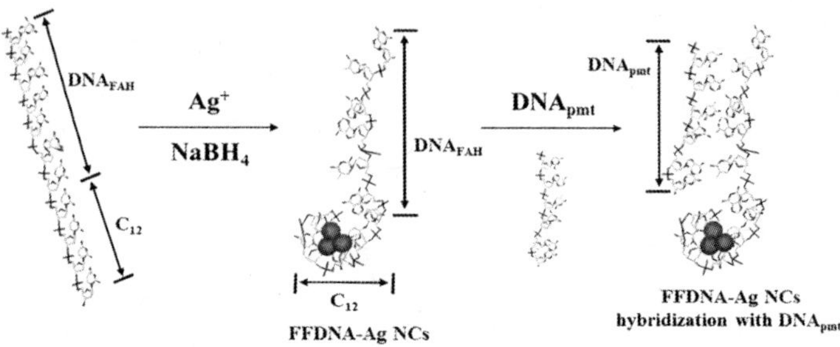

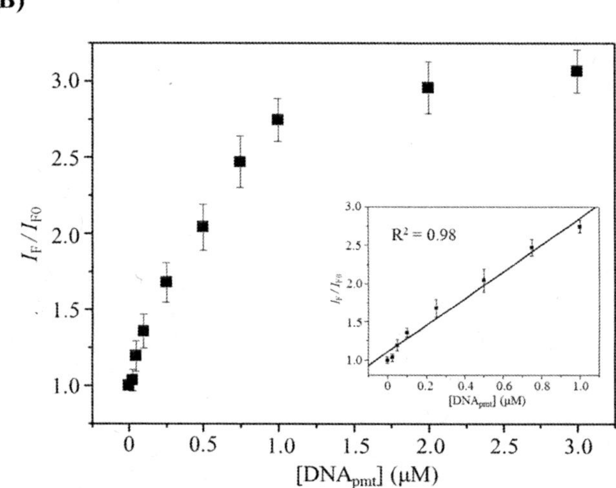

Figure 13. (A) Schematic representation of the preparation of FFDNA–Ag NC probe and the detection of target DNA. (B) Fluorescence ratios (I_F/I_{F0}) of the DNA–Ag NC solutions are plotted against the DNA$_{pmt}$ concentration. Inset: Linear plot of the fluorescence ratio against the DNA$_{pmt}$ concentration. Reprinted with permission from reference (180). Copyright 2011 Elsevier.

3. Proteins

DNA–Ag NCs are useful for the detection of proteins, mainly based on the specific interactions of the aptamers with their target proteins that induce the conformational changes in the aptamers and thus the fluorescence (*184*). The aptamer combines a C-rich DNA sequence (12-mer) for producing highly fluorescent Ag NCs with a thrombin-binding aptamer sequence (29-mer) for specifically recognizing thrombin. The fluorescence of DNA–Ag NCs is quenched upon binding thrombin through a static quenching mechanism. The

DNA–Ag NCs provides an LOD of 1 nM for thrombin. DNA–Cu/Ag NCs with a QY of 51.2% were employed for the detection of single-stranded DNA binding protein (SSB) (*162*). SSB has strong interactions with DNA and plays an essential role in DNA replication, recombination, and repair. When SSB interacts with DNA–Cu/Ag NCs, the DNA template changes their conformation, weakening its interaction with Cu/Ag NCs and thus leading to a corresponding decrease in the fluorescence. This sensing strategy is sensitive (LOD of 0.2 nM) for the detection of SSB.

4. Cell Imaging

Functional Ag NCs have been employed in cell imaging (*41, 65, 147, 148, 157, 185, 186*). For instance, DNA–Ag NCs possess high molar extinction coefficients leading to efficient photon absorption at either very low fluorophore concentration, or at a very low excitation light intensity. Thioflavin T conjugated Ag NCs were first applied to obtain images of amyloid fribrils (*120*). The Ag NCs show stronger staining of nucleoli relative to other organelles based on the preferential binding of Ag^+ ions with argyrophilic proteins, which are located in nucleoli (*157, 185*). Since the high density of DNA limits its intracellular availability, positively charged ligands such as lipofectamine and cell-penetrating peptide were used to conjugate DNA–Ag NCs for cytosolic delivery (*147, 148*). Peptide-protected Ag NCs are primarily localized in the cytosol for staining live and fixed NIH 3T3 cells, while the C_{20}–Ag NCs stain nucleoli strongly in fixed NIH 3T3 cells. In the NIH 3T3 and bovine pulmonary artery endothelial cells (BPAEC) cells, penetratin–C_{12} Ag NCs are initially bound with and eventually cross the cell membrane to locate inside their nuclei within few minutes (*148*). Lipofectamine as a transfection agent was employed to incorporate red emitting C_{24}–Ag NCs (excitation and emission wavelengths 650 and 715 nm, respectively) inside living HeLa cells (*147*). The red emitting C_{24}–Ag NC unit is also conjugated to antibody targeting cellular surfaces (*186*). Anti-heparin/heparan sulfate antibody (anti-HS) interacts with heparin/heparin sulfate on the NIH 3T3 cell surface. The uptake of antibody–DNA–Ag NC conjugates is through endocytosis, resulting in bright emission in the nucleus. DNA–Ag NCs (Sgc8c–Ag NCs) were applied for the specific targeting of the nucleus and surface of tumor cells (*187, 188*). The Sgc8c aptamer specifically targets the human protein tyrosine kinase 7 once it passes through the cell membrane, which is mediated by a receptor, and finally it is localized in the endosome (*189*). AS1411 (also named AGRO100) is an 26-base G-rich oligonucleotide that can bind to nucleolin overexpressed in cancer cells such as breast (MDA-MB-231 and MCF-7) and HeLa cancer cell lines (*190*). AS1411 with a C-rich T_5 loop scaffold was used to prepare AS1411-T_5-templated Ag NCs (*191*). The DNA–Ag NCs with a G-quadruplex structure relative to free AS1411 have enhanced inhibition efficiency of the growth of cancer cells. The DNA–Ag NCs are internalized into the cells and stain the nucleus *via* receptor-mediated endocytosis (Figure 14).

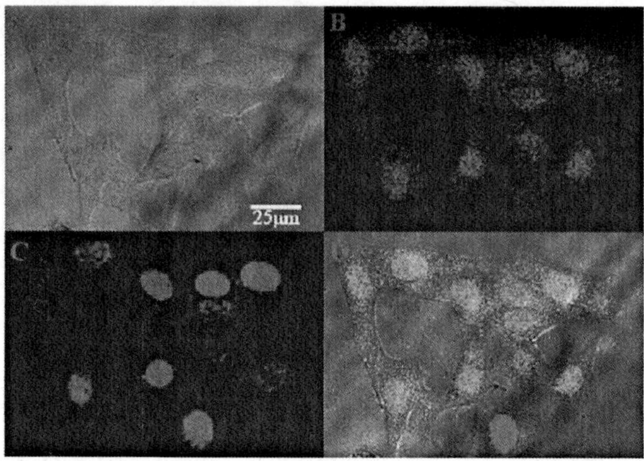

Figure 14. Intracellular distribution of internalized NC-AS1411-T₅-templated Ag NCs. MCF-7 cells were incubated with Ag NCs at 4 °C for 30 min. After being fixed with paraformaldehyde for 20 min, the cells were nuclear-staining with DAPI for 10 min. The fluorescence imaging was then recorded by confocal laser microscopy. (A) Brightfield images; (B) fluorescence images of Ag NCs (red); (C) fluorescence images with DAPI nuclear staining (blue); (D) overlap of corresponding fluorescence image and bright-field image. Scale bar, 25 μm. Reprinted with permission from reference (191). Copyright 2010 American Chemical Society.

Conclusion and Perspective

A wide range of DNA based recognition elements have been used for the synthesis of a variety of Au and Ag functional NMs that hold great potential for the detection of vital biomolecules such as nucleic acids, proteins, and biothiols. In particular, DNAzymes and aptamers that can act as specific recognition elements for the target molecules ranging from metal ions to proteins, and to cells are suitable for this purpose. DNAzymes and aptamers are commonly conjugated to Au NPs through Au–S or Au–N bonding. Alternatively, they can be used as templates for the preparation of DNA–Ag NCs and as recognition elements and sensitive probes for targeting analytes of interest. In the past decade we have witnessed many excited and successful examples of DNA functionalized Au NPs and DNA–Ag NCs for the detection of various analytes from small ions and molecules, to DNA and proteins, and to cells. These successful examples have shown great potential of the DNA functionalized Au NPs and DNA–Ag NCs for diagnosis of diseases, the analysis of biological and environmental samples, and cell imaging.

However, most of these DNA-based Au and Ag NMs can only be performed in controlled buffered systems, limiting their practicality in biological systems. They are unstable in high-ionic-strength media, leading to weak fluorescence and loss in their recognition capability. In order to stabilize the DNA functional Au

314

NPs, control of the DNA surface density and introduction of other hydrophilic compounds such as proteins, thiol compounds, and thiol-derivatized hydrophilic polymers are important. More stable DNA–protein complexes relative to DNA can be used as templates to prepare more stable DNA–Ag NCs in high-ionic-strength media. Metal nanocomposites such as fluorescent DNA–Au/Ag NCs relative to DNA–Ag NCs have shown great stability in high-ionic-strength media, while preparation of DNA functional bimetallic NCs that provide stronger fluorescence intensity and higher stability is another trend. Because the quantum yield depends on the structure of DNA templates, hybridized DNA with various conformations such as bulge loops may be useful for the preparation of functional Ag NMs.

Although DNA functional NMs have shown selectivity toward some analytes, available aptamers are still rare. As a result, only few DNA functional NMs have been shown for targets of interest. Relative to antibodies, the binding strength of the aptamers toward their targets is relatively weaker. Thus, it is still an urgent need to screen for more selective aptamers with higher binding strength. In order to further improve the affinity of DNA functional NMs, self-assembly of the aptamers on the DNA conjugated Au NPs through DNA hybridization is effective (*42, 141*). The flexibility of the aptamers on the Au NPs surface through self-assembly relative to those directly attached to the Au NP surfaces is higher, leading to greater affinity. In addition, use of more than two aptamers to prepare functional Au NPs is also possible for enhancing the selectivity. It is also possible to further improve the selectivity by using combination of aptamers with small molecule, polymers, or proteins to prepare functional Au and Ag NMs.

Multifunctional NPs (*e.g.*, targeting and inhibition) have become important materials in the targeting and inhibition of cancer cells (*192–197*). For example, DNA-functional Au NPs containing doxorubicin (Dox) have been used for targeting the cancer cells and then for the suppression of topoisomerase II in T-cell acute lymphoblastic leukemia CCRF-CEM cells (*198*). We can foresee more and more versatile multiple functional Au and Ag NMs will soon become a reality. For example, various DNAs and RNAs that recognize, and suppress/kill cancer cells or bacteria can be used to prepare functional Au NPs. Introduction of small molecules, polymers, and or proteins (drugs) into the functional Au NPs to provide stronger biological activity as a result of synergetic or multivalency is also becoming popular. Similar strategies can also be applied to prepare DNA–Ag NCs. For example, DNA having multiple sequences of aptamers (recognition), templates, and drug binding sites can be designed and used for the preparation of stable and multiple functional DNA–Ag NCs.

Acknowledgments

This study was supported by the National Science Council of Taiwan under Contracts NSC 98-2113-M-002-011-MY3, and by the National Health Research Institutes,Taiwan, under Contract NHRI-EX100-10047NI.

References

1. Talapin, D. V.; Lee, J.-S.; Kovalenko, M. V.; Shevchenko, E. V. *Chem. Rev.* **2010**, *110*, 389–458.
2. Ray, P. C. *Chem. Rev.* **2010**, *110*, 5332–5365.
3. Roduner, E. *Chem. Soc. Rev.* **2006**, *35*, 583–592.
4. Daniel, M.-C.; Astruc, D. *Chem. Rev.* **2004**, *104*, 293–346.
5. Saha, K.; Agasti, S. S.; Kim, C.; Li, X.; Rotello, V. M. *Chem. Rev.* **2012**, *112*, 2739–2779.
6. Lin, C.-A. J.; Lee, C.-H.; Hsieh, J.-T.; Wang, H.-H.; Li, J. K.; Shen, J.-L.; Chan, W.-H.; Yeh, H.-I.; Chang, W. H. *J. Med. Biol. Eng.* **2009**, *29*, 276–283.
7. Medintz, I. L.; Mattoussi, H. *Phys. Chem. Chem. Phys.* **2009**, *11*, 17–45.
8. Medintz, I. L.; Uyeda, H. T.; Goldman, E. R.; Mattoussi, H. *Nat. Mater.* **2005**, *4*, 435–446.
9. Lu, A.-H.; Salabas, E. L.; Schüth, F. *Angew. Chem., Int. Ed.* **2007**, *46*, 1222–1244.
10. Ho, D.; Sun, X.; Sun, S. *Acc. Chem. Res.* **2011**, *44*, 875–882.
11. Markowska-Szczupak, A.; Ulfig, K.; Morawski, A. W. *Catal. Today* **2011**, *169*, 249–257.
12. Zou, H.; Wu, S.; Shen, J. *Chem. Rev.* **2008**, *108*, 3893–3957.
13. Baker, S. N.; Baker, G. A. *Angew. Chem., Int. Ed.* **2010**, *49*, 6726–6744.
14. Hsu, P.-Z.; Shih, Z.-Y.; Lee, C.-H.; Chang, H.-T. *Green Chem.* **2012**, *14*, 917–920.
15. Castro Neto, A. H.; Guinea, F.; Peres, N. M. R.; Novoselov, K. S.; Geim, A. K. *Rev. Mod. Phys.* **2009**, *81*, 109–162.
16. Rao, C. N. R.; Sood, A. K.; Subrahmanyam, K. S.; Govindaraj, A. *Angew. Chem., Int. Ed.* **2009**, *48*, 7752–7777.
17. Díez, I.; Ras, R. H. A. *Nanoscale* **2011**, *3*, 1963–1970.
18. Lu, Y.; Liu, J. *Acc. Chem. Res.* **2007**, *40*, 315–323.
19. Baron, R.; Willner, B.; Willner, I. *Chem. Commun.* **2007**, 323–332.
20. Wang, Z.; Lu, Y. *J. Mater. Chem.* **2009**, *19*, 1788–1798.
21. About the Wiley-VCH – Books, URL http://www.wiley-vch.de/books/sample/3527311157_c01.pdf.
22. About the Biomedical Engineering - Frontiers and Challenges Books, URL http://cdn.intechopen.com/pdfs/17655/InTech-The_application_of_biomolecules_in_the_preparation_of_nanomaterials.pdf.
23. Podsiadlo, P.; Sinani, V. A.; Bahng, J. H.; Kam, N. W. S.; Lee, J.; Kotov, N. A. *Langmuir* **2008**, *24*, 568–574.
24. Mout, R.; Moyano, D. F.; Rana, S.; Rotello, V. M. *Chem. Soc. Rev.* **2012**, *41*, 2539–2544.
25. Yeh, Y.-C.; Creran, B.; Rotello, V. M. *Nanoscale* **2012**, *4*, 1871–1880.
26. Liu, C.-W.; Hsieh, Y.-T.; Huang, C.-C.; Lin, Z.-H.; Chang, H.-T. *Chem. Commun.* **2008**, 2242–2244.
27. Lin, Y.-W.; Liu, C.-W.; Chang, H.-T. *Anal. Methods* **2009**, *1*, 14–24.
28. Tuerk, C.; Gold, L. *Science* **1990**, *249*, 505–510.
29. Ellington, A. D.; Szostak, J. W. *Nature* **1990**, *346*, 818–822.
30. Gopinath, S. C. B. *Anal. Bioanal. Chem.* **2007**, *387*, 171–182.

31. Breaker, R. R.; Joyce, G. F. *Chem. Biol.* **1994**, *1*, 223–229.

32. Breaker, R. R.; Joyce, G. F. *Chem. Biol.* **1995**, *2*, 655–660.

33. Breaker, R. R. *Nat. Biotechnol.* **1997**, *15*, 427–431.

34. Breaker, R. R. *Chem. Rev.* **1997**, *97*, 371–390.

35. Mirkin, C. A.; Letsinger, R. L.; Mucic, R. C.; Storhoff, J. J. *Nature* **1996**, *382*, 607–609.

36. Balamurugan, S; Obubuafo, A; Soper, S. A.; Spivak, D. A. *Anal. Bioanal. Chem.* **2008**, *390*, 1009–1021.

37. Liu, J.; Cao, Z.; Lu, Y. *Chem. Rev.* **2009**, *109*, 1948–1998.

38. Wang, H.; Yang, R.; Yang, L.; Tan, W. *ACS Nano* **2009**, *3*, 2451–2460.

39. Hamula, C. L. A.; Guthrie, J. W.; Zhang, H.; Li, X.-F.; Le, X. C. *Trends Anal. Chem.* **2006**, *25*, 681–691.

40. Koo, H.; Huh, M. S.; Ryu, J. H.; Lee, D.-E.; Sun, I.-C.; Choi, K.; Kim, K.; Kwon, I. C. *Nano Today* **2011**, *6*, 204–220.

41. Choi, S.; Dickson, R. M.; Yu, J. *Chem. Soc. Rev.* **2012**, *41*, 1867–1891.

42. Shiang, Y.-C.; Hsu, C.-L.; Huang, C.-C.; Chang, H.-T. *Angew. Chem., Int. Ed.* **2011**, *50*, 7660–7665.

43. Brenan, J. L.; Hatzakis, N. S.; Tshikhudo, T. R.; Dirvianskyte, N.; Razumas, V.; Patkar, S.; Vind, J.; Svendsen, A.; Nolte, R. J. M.; Rowan, A. E.; Brust, M. *Bioconjugate Chem.* **2006**, *17*, 1373–1375.

44. Giljohann, D. A.; Seferos, D. S.; Prigodich, A. E.; Patel, P. C.; Mirkin, C. A. *J. Am. Chem. Soc.* **2009**, *131*, 2072–2073.

45. Jans, H.; Huo, Q. *Chem. Soc. Rev.* **2012**, *41*, 2849–2866.

46. Lee, J. H.; Yigit, M. V.; Mazumdar, D.; Lu, Y. *Adv. Drug. Delivery Rev.* **2010**, *62*, 592–605.

47. Du, Y.; Li, B.; Wang, E. *Bioanal. Rev.* **2010**, *1*, 187–208.

48. Nam, J.-M.; Thaxton, C. S.; Mirkin, C. A. *Science* **2003**, *301*, 1884–1886.

49. Elghanian, R.; Storhoff, J. J.; Mucic, R. C.; Letsinger, R. L.; Mirkin, C. A. *Science* **1997**, *277*, 1078–1081.

50. Sato, K.; Hosokawa, K.; Maeda, M. *J. Am. Chem. Soc.* **2003**, *125*, 8102–8103.

51. Wang, L.; Liu, X.; Hu, X.; Song, S.; Fan, C. *Chem. Commun.* **2006**, 3780–3782.

52. Wang, W.; Chen, C.; Qian, M.; Zhao, X. S. *Anal. Biochem.* **2008**, *373*, 213–219.

53. Wang, J.; Shan, Y.; Zhao, W.-W.; Xu, J.-J.; Chen, H.-Y. *Anal. Chem.* **2011**, *83*, 4004–4011.

54. Chen, J.-W.; Liu, X.-P.; Feng, K.-J.; Liang, Y.; Jiang, J.-H.; Shen, G.-L.; Yu, R.-Q. *Biosens. Bioelectron.* **2008**, *24*, 66–71.

55. Wang, Y.; Irudayaraj, J. *Chem. Commun.* **2011**, *47*, 4394–4396.

56. Sato, Y.; Sato, K.; Hosokawa, K.; Maeda, M. *Anal. Biochem.* **2006**, *355*, 125–131.

57. Song, J.; Li, Z.; Cheng, Y.; Liu, C. *Chem. Commun.* **2010**, *46*, 5548–5550.

58. Su, C.-L.; Tseng, W.-L. *Anal. Chem.* **2007**, *79*, 1626–1633.

59. Huang, Y.-F.; Chang, H.-T. *Anal. Chem.* **2007**, *79*, 4852–4859.

60. Kawasaki, H.; Sugitani, T.; Watanabe, T.; Yonezawa, T.; Moriwaki, H.; Arakawa, R. *Anal. Chem.* **2008**, *80*, 7524–7533.

61. Shang, L.; Dong, S.; Nienhaus, G. U. *Nano Today* **2011**, *6*, 401–418.

62. Guo, S.; Wang, E. *Nano Today* **2011**, *6*, 240–264.

63. Latorre, A.; Somoza, Á. *ChemBioChem* **2012**, *13*, 951–958.

64. Han, B.; Wang, E. *Anal. Bioanal. Chem.* **2012**, *402*, 129–138.

65. Shiang, Y.-C.; Huang, C.-C.; Chen, W.-Y.; Chen, P.-C.; Chang, H.-T. *J. Mater. Chem.* **2012**, *22*, 12972–12982.

66. Zheng, J.; Nicovich, P. R.; Dickson, R. M. *Annu. Rev. Phys. Chem.* **2007**, *58*, 409–431.

67. Lu, Y.; Chen, W. *Chem. Soc. Rev.* **2012**, *41*, 3594–3623.

68. Krückeberg, S.; Dietrich, G.; Lützenkirchen, K.; Schweikhard, L.; Walther, C.; Ziegler, J. *Int. J. Mass. Spectrom. Ion Process.* **1996**, *155*, 141–148.

69. Krückeberg, S.; Dietrich, G.; Lützenkirchen, K.; Schweikhard, L.; Walther, C.; Ziegler, J. *J. Chem. Phys.* **1999**, *110*, 7216–7227.

70. Schooss, D.; Gilb, S.; Kaller, J.; Kappes, M. M.; Furche, F.; Köhn, A.; May, K.; Ahlrichs, R. *J. Chem. Phys.* **2000**, *113*, 5361–5371.

71. Xie, J.; Zheng, Y.; Ying, J. Y. *J. Am. Chem. Soc.* **2009**, *31*, 888–889.

72. Lin, Y.-H.; Tseng, W.-L. *Anal. Chem.* **2010**, *82*, 9194–9200.

73. Wen, F.; Dong, Y.; Feng, L.; Wang, S.; Zhang, S.; Zhang, X. *Anal. Chem.* **2011**, *83*, 1193–1196.

74. Liu, C.-L.; Wu, H.-T.; Hsiao, Y.-H.; Lai, C.-W.; Shih, C.-W.; Peng, Y.-K.; Tang, K.-C.; Chang, H.-W.; Chien, Y.-C.; Hsiao, J.-K.; Cheng, J.-T.; Chou, P.-T. *Angew. Chem., Int. Ed.* **2011**, *50*, 7056–7060.

75. Wang, Y.; Chen, J.; Irudayraj, J. *ACS Nano* **2011**, *5*, 9718–9725.

76. Petty, J. T.; Zheng, J.; Hud, N. V.; Dickson, R. M. *J. Am. Chem. Soc.* **2004**, *126*, 5207–5212.

77. Ritchie, C. M.; Johnsen, K. R.; Kiser, J. R.; Antoku, Y.; Dickson, R. M.; Petty, J. T. *J. Phys. Chem. C* **2007**, *111*, 175–181.

78. Gwinn, E. G; O'Neill, P. R.; Guerrero, A. J.; Bouwmeester, D.; Fygenson, D. K. *Adv. Mater.* **2008**, *20*, 279–283.

79. O'Neill, P. R.; Velazquez, L. R.; Dunn, D. G.; Gwinn, E. G.; Fygenson, D. K. *J. Phys. Chem. C* **2009**, *113*, 4229–4233.

80. Enüstün, B. V.; Turkevich, J. *J. Am. Chem. Soc.* **1963**, *85*, 3317–3328.

81. Kimling, J.; Maier, M.; Okenve, B.; Kotaidis, V.; Ballot, H.; Plech, A. *J. Phys. Chem. B* **2006**, *110*, 15700–15707.

82. Liu, C.-W.; Huang, C.-C.; Chang, H.-T. *Langmuir* **2008**, *24*, 8346–8350.

83. Demers, L. M.; Mirkin, C. A.; Mucic, R. C.; Reynolds, R. A., III; Letsinger, R. L.; Elghanian, R.; Viswanadham, G. *Anal. Chem.* **2000**, *72*, 5535–5541.

84. Myroshnychenko, V.; Rodríguez-Fernándes, J.; Pastoriza-Santos, I.; Funston, A. M.; Novo, C.; Mulvaney, P.; Liz-Marzán, L. M.; García de Abajo, F. J. *Chem. Soc. Rev.* **2008**, *37*, 1792–1805.

85. Ghosh, S. K.; Pal, T. *Chem. Rev.* **2007**, *107*, 4797–4862.

86. Huang, C.-C.; Yang, Z.; Chang, H.-T. *Langmuir* **2004**, *20*, 6089–6092.

87. Huang, Y.-F.; Huang, K.-M.; Chang, H.-T. *J. Colloid Interface Sci.* **2006**, *301*, 145–154.

88. Huang, Y.-F.; Lin, Y.-W.; Chang, H.-T. *Nanotechnology* **2006**, *17*, 4885–4894.

89. Pyykkö, P. *Angew. Chem., Int. Ed.* **2004**, *43*, 4412–4456.

90. Jin, R. *Nanoscale* **2010**, *2*, 343–362.

91. Lukaski, H. C.; Klevay, L. M.; Milne, D. B. *Eur. J. Appl. Physiol.* **1988**, *58*, 74–80.

92. Georgopoulos, P. G.; Roy, A.; Yonone-Lioy, M. J.; Opiekun, R. E.; Lioy, P. J. *J. Toxicol. Environ. Health, Part B* **2001**, *4*, 341–394.

93. Tchounwou, P. B.; Ayensu, W. K.; Ninashvili, N.; Sutton, D. *Environ. Toxicol.* **2003**, *18*, 149–175.

94. Needleman, H. *Annu. Rev. Med.* **2004**, *55*, 209–222.

95. About the EPA's Safe Drinking Water, URL http://www.epa.gov/safewater/.

96. Lee, J.-S.; Han, M. S.; Mirkin, C. A. *Angew. Chem. Int. Ed.* **2007**, *46*, 4093–4096.

97. Xue, X.; Wang, F.; Liu, X. *J. Am. Chem. Soc.* **2008**, *130*, 3244–3245.

98. Li, L.; Li, B.; Qi, Y.; Jin, Y. *Anal. Bioanal. Chem.* **2009**, *393*, 2051–2057.

99. Liu, J.; Lu, Y. *J. Am. Chem. Soc.* **2003**, *125*, 6642–6643.

100. Mazumdar, D.; Liu, J.; Lu, G.; Zhou, J.; Lu, Y. *Chem. Commun.* **2010**, 1416–1418.

101. Brown, A. K.; Li, J.; Pavot, C. M.-B.; Lu, Y. *Biochemistry* **2003**, *42*, 7152–7161.

102. Wang, Z.; Lee, J. H.; Lu, Y. *Adv. Mater.* **2008**, *20*, 3263–3267.

103. Chen, Y.-Y.; Chang, H.-T.; Shiang, Y.-C.; Hung, Y.-L.; Chiang, C.-K.; Huang, C.-C. *Anal. Chem.* **2009**, *81*, 9433–9439.

104. Xu, X.; Daniel, W. L.; Wei, W.; Mirkin, C. A. *Small* **2010**, *6*, 623–626.

105. Huizenga, D. E.; Szostak, J. W. *Biochemistry* **1995**, *34*, 656–665.

106. Li, Y.; Geyer, C. R.; Sen, D. *Biochemistry* **1996**, *35*, 6911–6922.

107. Li, Y.; Sen, D. *Nat. Struct. Biol.* **1996**, *3*, 743–747.

108. Travascio, P.; Li, Y.; Sen, D. *Chem. Biol.* **1998**, *5*, 505–517.

109. Stojanovic, M. N.; de Prada, P.; Landry, D. W. *J. Am. Chem. Soc.* **2000**, *122*, 11547–11548.

110. Stojanovic, M. N.; de Prada, P.; Landry, D. W. *J. Am. Chem. Soc.* **2001**, *123*, 4928–4931.

111. Song, K.-M.; Cho, M.; Jo, H.; Min, K.; Jeon, S. H.; Kim, T.; Han, M. S.; Ku, J. K.; Ban, C. *Anal. Biochem.* **2011**, *415*, 175–181.

112. Liu, J.; Lu, Y. *Angew. Chem., Int. Ed.* **2006**, *45*, 90–94.

113. Famulok, M.; Mayer, G. *Acc. Chem. Res.* **2011**, *44*, 1349–1358.

114. Chen, S.-J.; Huang, Y.-F.; Huang, C.-C.; Lee, K.-H.; Lin, Z.-H.; Chang, H.-T. *Biosens. Bioelectron.* **2008**, *23*, 1749–1753.

115. Chen, S.-J.; Huang, C.-C.; Chang, H.-T. *Talanta* **2010**, *81*, 493–498.

116. Liu, J.; Lu, Y. *Anal. Chem.* **2004**, *76*, 1627–1632.

117. Andrae, J.; Gallini, R.; Betsholtz, C. *Genes Dev.* **2008**, *22*, 1276–1312.

118. Huang, C.-C.; Huang, Y.-F.; Cao, Z.; Tan, W.; Chang, H.-T. *Anal. Chem.* **2005**, *77*, 5735–5741.

119. Huang, C.-C.; Chiu, S.-H.; Huang, Y.-F.; Chang, H.-T. *Anal. Chem.* **2007**, *79*, 4798–4804.

120. Becker, H.-C.; Norden, B. *J. Am. Chem. Soc.* **1997**, *119*, 5798–5803.

121. Huang, C.-C.; Chiang, C.-K.; Lin, Z.-H.; Lee, K.-H.; Chang, H.-T. *Anal. Chem.* **2008**, *80*, 1497–1504.

122. Wiegand, T. W.; Williams, P. B.; Dreskin, S. C.; Jouvin, M. H.; Kinet, J. P.; Tasset, D. *J. Immunol.* **1996**, *157*, 221–230.

123. Wang, J.; Munir, A.; Li, Z.; Zhou, H. S. *Biosens. Bioelectron.* **2009**, *25*, 124–129.

124. Cox, J. C.; Ellington, A. D. *Bioorg. Med. Chem.* **2001**, *9*, 2525–2531.

125. Deng, C.; Chen, J.; Nie, L.; Nie, Z.; Yao, S. *Anal. Chem.* **2009**, *81*, 9972–9978.

126. Bock, L. C.; Griffin, L. C.; Latham, J. A.; Vermaas, E. H.; Toole, J. J. *Nature* **1992**, *355*, 564–566.

127. Tasset, D. M.; Kubik, M. F.; Steiner, W. *J. Mol. Biol.* **1997**, *272*, 688–698.

128. Pavlov, V.; Xiao, Y.; Shlyahovsky, B.; Willner, I. *J. Am. Chem. Soc.* **2004**, *126*, 11768–11769.

129. Chen, C.-K.; Shiang, Y.-C.; Huang, C.-C.; Chang, H.-T. *Biosens. Bioelectron.* **2011**, *26*, 3464–3468.

130. Stamey, T. A.; Yang, N.; Hay, A. R.; McNeal, J. E.; Freiha, F. S.; Redwine, E. *N. Engl. J. Med.* **1987**, *317*, 909–916.

131. Morris, K. N.; Jensen, K. B.; Julin, C. M.; Weil, M.; Gold, L. *Proc. Natl. Acad. Sci. U.S.A.* **1998**, *95*, 2902–2907.

132. Shangguan, D.; Cao, Z. C.; Li, Y.; Tan, W. *Clin. Chem.* **2007**, *53*, 1153–1158.

133. Lee, J. F.; Stovall, G. M.; Ellington, A. D. *Curr. Opin. Chem. Biol.* **2006**, *10*, 282–289.

134. Shangguan, D.; Li, Y.; Tang, Z.; Cao, Z. C.; Chen, H. W.; Mallikaratchy, P.; Sefah, K.; Yang, C. J.; Tan, W. *Proc. Natl. Acad. Sci. U.S.A.* **2006**, *103*, 11838–11843.

135. Medley, C. D.; Smith, J. E.; Tang, Z.; Wu, Y.; Bamrungsap, S.; Tan, W. *Anal. Chem.* **2008**, *80*, 1067–1072.

136. King, S. H.; Huh, Y. M.; Kim, S.; Lee, D.-K. *Bull. Korean Chem. Soc.* **2009**, *30*, 1827–1831.

137. Lu, W.; Arumugam, S. R.; Senapati, D.; Singh, A. K.; Arbneshi, T.; Khan, S. A.; Yu, H.; Ray, P. C. *ACS Nano* **2010**, *4*, 1739–1749.

138. Tang, Z.; Shangguan, D.; Wang, K.; Shi, H.; Sefah, K.; Mallikaratchy, P.; Chen, H. W.; Li, Y.; Tan, W. *Anal. Chem.* **2007**, *79*, 4900–4907.

139. Liu, G.; Mao, X.; Phillips, J. A.; Xu, H.; Tan, W.; Zeng, L. *Anal. Chem.* **2009**, *81*, 10013–10018.

140. Lupold, S. E.; Hicke, B. J.; Lin, Y.; Coffey, D. S. *Cancer Res.* **2002**, *62*, 4029–4033.

141. Kim, D.; Jeong, Y. Y.; Jon, S. *ACS Nano* **2010**, *4*, 3689–3696.

142. Richards, C. I.; Choi, S.; Hsiang, J.-C.; Antoku, Y.; Vosch, T.; Bongiorno, A.; Tzeng, Y.-L.; Dickson, R. M. *J. Am. Chem. Soc.* **2008**, *130*, 5038–5039.

143. Sharma, J.; Yeh, H.-C.; Yoo, H.; Werner, J. H.; Martinez, J. S. *Chem. Commun.* **2010**, *46*, 3280–3282.

144. Lan, G.-Y.; Chen, W.-Y.; Chang, H.-T. *RSC Adv.* **2011**, *1*, 802–807.

145. Pal, S.; Varghese, R.; Deng, Z.; Zhao, Z.; Kumar, A.; Yan, H.; Liu, Y. *Angew. Chem., Int. Ed.* **2011**, *50*, 4176–4179.

146. Soto-Verdugo, V.; Metiu, H.; Gwinn, E. *J. Chem. Phys.* **2010**, *132*, 195102–195110.

147. Antoku, Y.; Hotta, J.-i.; Mizuno, H.; Dickson, R. M.; Hofkens, J.; Vosch, T. *Photochem. Photobiol. Sci.* **2010**, *9*, 716–721.

148. Choi, S.; Yu, J.; Patel, S. A.; Tzeng, Y.-L.; Dickson, R. M. *Photochem. Photobiol. Sci.* **2011**, *10*, 109–115.

149. Lan, G.-Y.; Huang, C.-C.; Chang, H.-T. *Chem. Commun.* **2010**, *46*, 1257–1259.

150. Chen, W.-Y.; Lan, G.-Y.; Chang, H.-T. *Anal. Chem.* **2011**, *83*, 9450–9455.

151. Loo, K.; Degtyareva, N.; Park, J.; Sengupta, B.; Reddish, M.; Rogers, C. C.; Bryant, A.; Petty, J. T. *J. Phys. Chem. B* **2010**, *114*, 4320–4326.

152. Guo, W.; Yuan, J.; Dong, Q.; Wang, E. *J. Am. Chem. Soc.* **2010**, *132*, 932–934.

153. Huang, Z.; Pu, F.; Hu, D.; Wang, C.; Ren, J.; Qu, X. *Chem. Eur. J.* **2011**, *17*, 3774–3780.

154. Sengupta, B.; Ritchie, C. M.; Buckman, J. G.; Johnsen, K. R.; Goodwin, P. M.; Petty, J. T. *J. Phys. Chem. C* **2008**, *112*, 18776–18782.

155. Sengupta, B.; Springer, K.; Buckman, J. G.; Story, S. P.; Abe, O. H.; Hasan, Z. W.; Prudowsky, Z. D.; Rudisill, S. E.; Degtyareva, N. N.; Petty, J. T. *J. Phys. Chem. C* **2009**, *113*, 19518–19524.

156. Ai, J.; Guo, W.; Li, B.; Li, T.; Li, D.; Wang, E. *Talanta* **2012**, *88*, 450–455.

157. Yu, J.; Choi, S.; Dickson, R. M. *Angew. Chem., Int. Ed.* **2009**, *48*, 318–320.

158. Yeh, H.-C.; Sharma, J.; Han, J. J.; Martinez, J. S.; Werner, J. H. *Nano Lett.* **2010**, *10*, 3106–3110.

159. Zheng, J.; Zhang, C.; Dickson, R. M. *Phys. Rev. Lett.* **2004**, *93*, 077402–077404.

160. Narayanan, S. S.; Pal, S. K. *J. Phys. Chem. C* **2008**, *112*, 4874–4879.

161. Patel, S. A.; Cozzuol, M.; Hales, J. M.; Richards, C. I.; Sartin, M.; Hsiang, J.-C.; Vosch, T.; Perry, J. W.; Dickson, R. M. *J. Phys. Chem. C* **2009**, *113*, 20264–20270.

162. Lan, G.-Y.; Chen, W.-Y.; Chang, H.-T. *Analyst* **2011**, *136*, 3623–3628.

163. Shemer, G.; Krichevski, O.; Markovich, G.; Molotsky, T.; Lubitz, I.; Kotlyar, A. B. *J. Am. Chem. Soc.* **2006**, *128*, 11006–11007.

164. Jentys, A. *Phys. Chem. Chem. Phys.* **1999**, *1*, 4059–4063.

165. Kip, B. J.; Duivenvoorden, F. B. M.; Konigsberger, D. C.; Prins, R. *J. Catal.* **1987**, *105*, 26–38.

166. Neidig, M. L.; Sharma, J.; Yeh, H.-C.; Martinez, J. S.; Conradson, S. D.; Shreve, A. P. *J. Am. Chem. Soc.* **2011**, *133*, 11837–11839.

167. Su, Y.-T.; Lan, G.-Y.; Chen, W.-Y.; Chang, H.-T. *Anal. Chem.* **2010**, *82*, 8566–8572.

168. Han, B.; Wang, E. *Biosens. Bioelectron.* **2011**, *26*, 2585–2589.

169. Petty, J. T.; Fan, C.; Story, S. P.; Sengupta, B.; Iyer, A. St. J.; Prudowsky, Z.; Dickson, R. M. *J. Phys. Chem. Lett.* **2010**, *1*, 2524–2529.

170. Petty, J. T.; Fan, C.; Story, S. P.; Sengupta, B.; Sartin, M.; Hsiang, J.-C.; Perry, J. W.; Dickson, R. M. *J. Phys. Chem. B* **2011**, *115*, 7996–8003.

171. Guo, W.; Yuan, J.; Wang, E. *Chem. Commun.* **2009**, 3395–3397.

172. Deng, L.; Zhou, Z.; Li, J.; Li, T.; Dong, S. *Chem. Commun.* **2011**, *47*, 11065–11067.

173. Morse, J. W.; Millero, F. J.; Cornwell, J. C.; Rickard, D. *Earth Sci. Rev.* **1987**, *24*, 1–42.

174. Bagarinao, T. *Aquat. Toxicol.* **1992**, *24*, 21–62.

175. Huang, Z.; Pu, F.; Lin, Y.; Ren, J.; Qu, X. *Chem. Commun.* **2011**, *47*, 3487–3489.

176. Zhoua, Z.; Du, Y.; Dong, S. *Biosens. Bioelectron.* **2011**, *28*, 33–37.

177. Debouck, C.; Goodfellow, P. N. *Nat. Genet* **1999**, *21*, 48–50.

178. Staudt, L. M. *Trends. Immunol* **2001**, *22*, 35–40.

179. Kim, S.; Misra, A. *Annu. Rev. Biomed. Eng.* **2007**, *9*, 289–320.

180. Lan, G.-Y.; Chen, W.-Y.; Chang, H.-T. *Biosens. Bioelectron.* **2011**, *26*, 2431–2435.

181. Wark, A. W.; Lee, H. J.; Corn, R. M. *Angew. Chem., Int. Ed.* **2008**, *47*, 644–652.

182. Davies, B. P.; Arenz, C. *Angew. Chem., Int. Ed.* **2006**, *45*, 5550–5552.

183. Yang, S. W.; Vosch, T. *Anal. Chem.* **2011**, *83*, 6935–6939.

184. Sharma, J.; Yeh, H.-C.; Yoo, H.; Werner, J. H.; Martinez, J. S. *Chem. Commun.* **2011**, *47*, 2294–2296.

185. Makarava, N.; Parfenov, A.; Baskakov, I. V. *Biophys. J.* **2005**, *89*, 572–580.

186. Yu, J.; Choi, S.; Richards, C. I.; Antoku, Y.; Dickson, R. M. *Photochem. Photobiol.* **2008**, *84*, 1435–1439.

187. Sun, Z.; Wang, Y.; Wei, Y.; Liu, R.; Zhu, H.; Cui, Y.; Zhao, Y.; Gao, X. *Chem. Commun.* **2011**, *47*, 11960–11962.

188. Yin, J.; He, X.; Wang, K.; Qing, Z.; Wu, X.; Shi, H.; Yang, X. *Nanoscale* **2012**, *4*, 110–112.

189. Huang, Y.-F.; Shangguan, D.; Liu, H.; Phillips, J. A.; Zhang, X.; Chen, Y.; Tan, W. *ChemBioChem* **2009**, *10*, 862–868.

190. Ireson, C. R.; Kelland, L. R. *Mol. Cancer Ther.* **2006**, *5*, 2957–2962.

191. Li, J.; Zhong, X.; Cheng, F.; Zhang, J.-R.; Jiang, L.-P.; Zhu, J.-J. *Anal. Chem.* **2012**, *84*, 4140–4146.

192. Farokhzad, O. C.; Cheng, J.; Teply, B. A.; Sherifi, I.; Jon, S.; Kantoff, P. W.; Richie, J. P.; Langer, R. *Proc. Natl. Acad. Sci. U.S.A.* **2006**, *103*, 6315–6320.

193. Peer, D.; Karp, J. M.; Hong, S.; Farokhzad, O. C.; Margalit, R.; Langer, R. *Nat. Nanotechnol.* **2007**, *2*, 751–760.

194. Nie, S.; Xing, Y.; Kim, G. J.; Simons, J. W. *Annu. Rev. Biomed. Eng.* **2007**, *9*, 257–288.

195. Bhirde, A. A.; Patel, V.; Gavard, J.; Zhang, G.; Sousa, A. A.; Masedunskas, A.; Leapman, R. D.; Weigert, R.; Gutkind, J. S.; Rusling, J. F. *ACS Nano* **2009**, *3*, 307–316.

196. Wang, A. Z.; Langer, R.; Farokhzad, O. C. *Annu. Rev. Med.* **2012**, *63*, 185–198.

197. Xiao, Z.; Levy-Nissenbaum, E.; Alexis, F.; Lupták, A.; Teply, B. A.; Chan, J. M.; Shi, J.; Digga, E.; Cheng, J.; Langer, R.; Farokhzad, O. C. *ACS Nano* **2012**, *6*, 696–704.

198. Luo, Y.-L.; Shiao, Y.-S.; Huang, Y.-F. *ACS Nano* **2011**, *5*, 7796–7804.

Chapter 13

Colloidal Quantum Dots: The Opportunities and the Pitfalls for DNA Analysis Applications

Katrin Pechstedt and Tracy Melvin*

Optoelectronics Research Centre, University of Southampton,
Highfield, SO17 1BJ, United Kingdom
*E-mail: tm@orc.soton.ac.uk

Colloidal quantum dots (QDs) have received considerable attention as luminescent probes for DNA analysis applications. The underlying photophysical and photochemical properties of these probes need consideration when designing assays for DNA analysis. These properties include intermittent fluorescence often termed 'blinking', photobleaching, photoinduced fluorescence enhancement and, as a result of recent evidence, photo-induced generation of reactive oxygen species leading to DNA damage. Even though the design of assays for DNA analysis using QDs needs care, QDs do provide advantages over fluorophores for many emerging DNA analysis methods where low copy numbers of DNA are present. Many of the more traditional DNA assay methods using fluorophore labeled probes either cannot be translated, or show no benefit in using QDs as the lumophore.

Introduction

Colloidal quantum dots (QDs) were first discovered in the early 1980s, and by the 2000s these nanoparticles had attracted overwhelming attention from the bioanalytical community searching for alternative luminescent probes to organic fluorescent dyes. The high quantum yields, spectral absorption properties and the tunability of the narrow emission spectra of colloidal QDs provided opportunities for developing new technologies for bioanalysis *in vitro* and *in situ*.

© 2012 American Chemical Society

In some respects new bioanalytical formats have emerged at a rate faster than the knowledge of the photophysical and photochemical properties of these nanomaterials. It is now clear that bioanalytical assays which exploit colloidal QDs must be done with wisdom and regard for the photophysical, photochemical and physical properties of these nanostructures. The purpose of this review is to provide readers with an overview of current understanding of the photophysical and photochemical properties of colloidal QDs as compared to fluorophores, a discussion of the differences in these properties and where there are pitfalls for the application of QDs instead of fluorophores, especially where these are important to DNA bioanalytical analysis applications. We include a brief overview of some of the methodologies where colloidal QD probes provide advantages over fluorescent dyes and give a perspective for the future.

Photophysical Properties of Fluorescent Organic Dyes in Comparison to Colloidal QDs

There is a long-history of the use of fluorescent dyes for DNA analysis applications and as a consequence of this a myriad of different fluorescent probes is available. These can be classified into various classes (i) DNA intercalators (such as ethydium bromide, TOTO-1) (*1–7*), (ii) major or minor groove binding (i.e. Methyl Green, DAPI, SYBR green) (*8, 9*) and (iii) non-interacting dyes (i.e. fluorescein) (*10*). Intercalators and groove binding fluorophores have higher quantum yields of fluorescence upon binding to DNA (*11–13*). The optical properties of fluorescent dyes depend on their electronic structure, which can be manipulated by their chemical composition. The energy states of organic fluorophores consist of an electronic singlet ground state (S_0) and electronic singlet excited states (S_n), each with vibrational states (v=0,1,2,...) (Figure 1).

Absorption occurs from $S_0(v=0) \rightarrow S_1(v=n)$, while fluorescence occurs from $S_1(v=0) \rightarrow S_0(v=n)$ due to fast non-radiative vibrational relaxation to $S_1(v=0)$ following absorption. The absorption spectrum of most common fluorophores, such as cyanine dyes, fluorescein or Alexa dyes, is asymmetric, reflecting the transition probability between $S_0(v=0) \rightarrow S_1(v=n)$. The fluorescence spectrum is typically a mirror image of the absorption spectrum and crosses the absorption band at the wavelength with an energy which corresponds to the $S_0(v=0) \rightarrow S_1(v=0)$ transition (Figure 1). Consequently the separation in wavelength for the maxima for the absorption and emission spectra (Stokes shift) is limited. Alternative non-fluorescence pathways are non-radiative decay and intersystem crossing to a lower energy triplet state followed by phosphorescence (Figure 1).

There is a wealth of fluorescence probe technologies which allow excitation across the whole spectrum from the UV wavelengths right into the infrared. The full width half maximum (FWHM) for the absorption and emission spectra of fluorophores with shorter emission wavelengths is ~35nm (*i.e.* fluorescein) to ~50 nm FWHM for fluorophores with longer emission wavelengths (*i.e.* Cy7). The extinction coefficients for absorption maximum of fluorescent dyes are typically of the order of 10^4-10^5 M^{-1} cm^{-1} (*15*). The fluorescence quantum yield in the visible range varies considerably for organic fluorophores, ranging from 0.04 for Cy3 to 0.97 for fluorescein (*15*). The fluorescence lifetime of most fluorophores used for DNA bioanalysis applications is less than 10ns. It should be noted that there is a range of fluorophore-doped particles, most notable are those of nanometer dimensions termed transFluoroSpheres (*16–22*), these are outside the scope of this review.

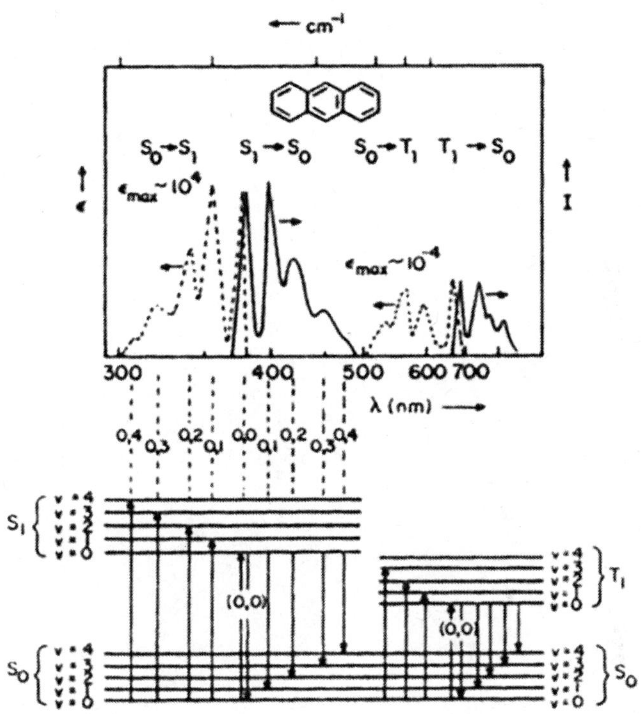

Figure 1. Anthracene: absorption (dotted lines) and emission (solid lines) spectrum (upper figure), and assigned vibrational modes (lower figure). Reproduced with permission from Reference (14). Copyright 1991 University Science Books.

A quantum dot (QD) is a semiconductor nanocrystal of a few nanometers in diameter and, for biological applications, commonly consist of a cadmium selenide (CdSe) core surrounded by a zinc sulfide (ZnS) shell. The optical properties of QDs are determined by the core material, the core size, the number of surface defects and the QD size distribution within a QD ensemble. The optical properties of QDs arise from quantum confinement, which is the confinement of excitons in all three spatial dimensions inside a nanocrystal of the order of the bulk exciton

Bohr radius (R_B). Quantum confinement results in the quantization of energy states into discrete, well-separated energy levels (Figure 2(a)) and an energy gap between the conduction and valence band that is dependent on crystal composition, as in the bulk material, as well as, due to quantum confinement, on the nanocrystal core size. For instance, in the strong confinement regime, where for a spherical QD the core radius $R < R_B$, the electron and hole can be considered as two independent particles, as the kinetic energy associated with confinement exceeds the Coulombic attraction, so that the energy gap can be approximated to equal (*23–25*)

$$E_g(QD) = E_g(bulk) + \frac{\hbar^2 \pi^2}{2R^2}\left(\frac{1}{m_e^*} + \frac{1}{m_h^*}\right) - 1.786\frac{e^2}{\varepsilon R} \qquad (1)$$

where the first term in Equation 1 is the bulk crystal band gap, the second term is the lowest kinetic energy associated with quantum confinement of the excited electron and excited hole, of effective mass m_e^* and m_h^*, respectively, and the third term is the Coulomb attraction, where ε is the dielectric constant (Eqn. 1).

As QD emission occurs by electron-hole recombination across the energy gap, quantum confinement allows for precise tuning of the emission peak simply by controlling the QD size, providing luminescent probes over the entire visible range. For instance, the emission peak of CdSe/ZnS core/shell QDs can be tuned from about 490 nm to 625 nm by a change in core diameter from 2.3 nm to 5.5 nm, respectively (Figure 2(c)) (*26*).

The molar extinction coefficient of QDs at the first exciton absorption peak exceeds that of fluorophores at the absorption maximum. However for QDs the absorption spectrum is broad, spanning a large spectral range (shown in Figure 2(b)) (*15, 27*). QDs have a broad absorption spectrum that increases with energy above the first absorption band (Figure 2(b)), and a narrow, symmetric emission peak (Figure 2(c)). The emission peak width is mainly determined by the QD size distribution within a QD ensemble, for instance, a 5% QD size variation yielding a 30 nm FWHM (*28*). Typically, the emission spectra of QDs have a Gaussian profile with FWHM of 30-90nm. The emission lifetime for QDs is typically 10-100ns (*29–32*). An additional difference between QDs as compared to fluorophores is that for QDs there is a large energy separation between the excitation wavelength and fluorescence peak (Stokes-shift) of up to 400 nanometers (*27*). This can simplify the optical set-ups required for assays.

The quantum yield of emission can be as low as 0.05 for CdSe core QDs (*26*). Trap states on the core surface, such as surface defects and dangling (unpassivated) atoms, provide non-radiative de-excitation pathways (*33, 34*). Coating the core QD with an inorganic, wide band gap semiconductor shell, such as ZnS (over the CdSe core) or/and an organic capping layer passivates the surface trap states and increases the quantum yield to 0.50 for CdSe/ZnS core/shell QDs (figure 2(c)) (*26*). Although coating with a passivation shell increases the quantum yield, there is also a slight red-shift of the emission peak due to partial extension of the electron wave function into the shell (Figure 2(c)) (*26*). Optimized surface passivation and growth parameters have been reported to increase the quantum yield to up to 0.85 (*35, 36*), although typically quantum yields of up to 0.50 are found for commercially available core/shell QDs (*26*).

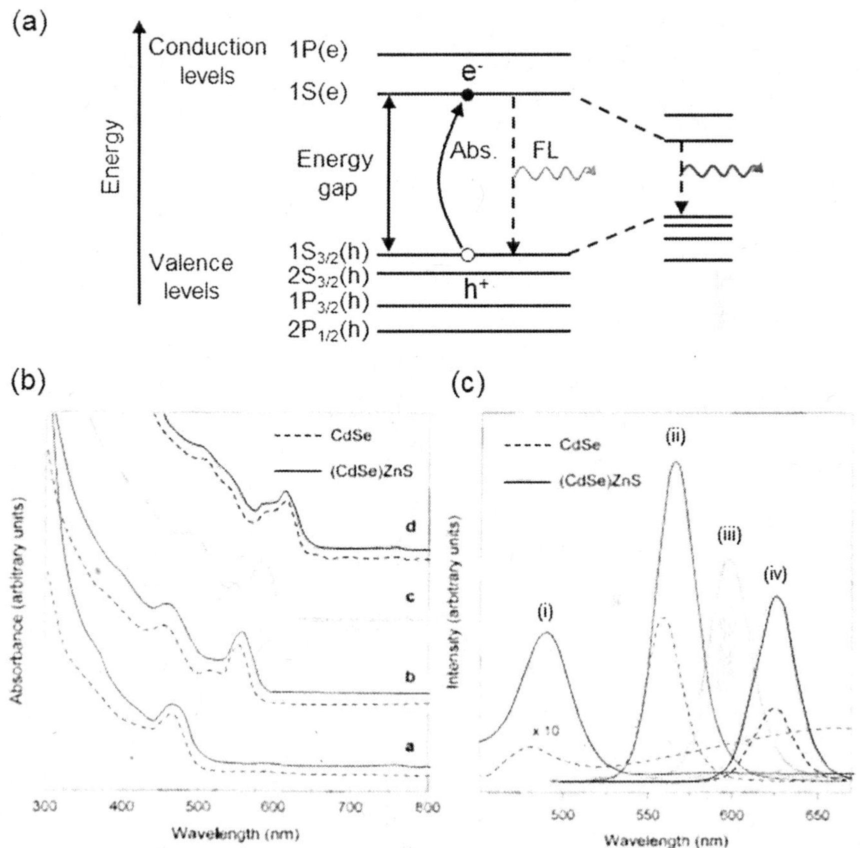

Figure 2. QD optical properties: (a) schematic diagram of core QD energy levels, illustrating discrete energy levels (nL_F), labeled with quantum numbers n, L and F, and lowest energy absorption (Abs.) and emission (FL) across the energy gap for a QD of diameter equal to 2.3 nm (left) and 5.5 nm (right;) (b) absorption spectra and (c) emission spectra of core CdSe QDs (dashed line) and core/shell CdSe/ZnS QDs (continuous line) of core diameter equal to (i) 2.3 nm (ii) 4.2 nm (iii) 4.8 nm (iv) 5.5 nm. Reproduced with permission from Reference (26). Copyright 1997 American Chemical Society.

Consideration of the Photophysical and Photochemical Properties of Fluorescent Organic Dyes in Comparison to Colloidal QDs with Respect to Bioanalytical Applications with DNA

Fluorophores have found broad application as labels for DNA analysis applications including polymerase chain reaction (PCR) probes (*37*), fluorescence *in situ* hybridization (FISH) (*38, 39*), DNA microarray technologies (*40*), assays involving fluorescence resonance energy transfer (FRET) and DNA sequencing, to name a few (*37, 41*). The important factor which limits the use of fluorophores for many applications is the photostability. Photo-excitation of fluorescent dyes in aqueous environments are known to yield reactive oxygen species (ROS) and subsequent photo-oxidation, photo-bleaching or photo-degradation of organic fluorescent dyes (*42–44*).

As with fluorophores, photo-induced oxidation of QDs results in photo-bleaching (*45–50*). A thick, dense epitaxial semiconductor shell, such as ZnS around a CdSe core, may hinder oxygen permeation through to the QD core and thus inhibit photo-oxidation (*26, 51, 52*). This and the addition of a polymer coating makes core/shell QDs more resistant to photo-bleaching than organic fluorophores (Figure 3) (*53*). The improved photostability of QDs is one of the key factors that has attracted the attention of many for the application of these luminescent probes in place of fluorophores for bioanalytical applications.

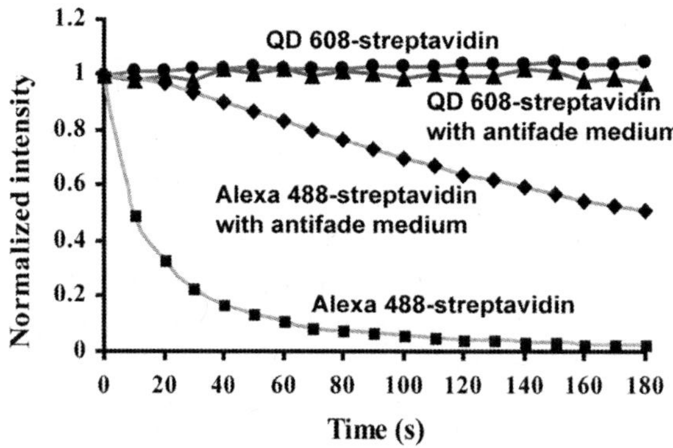

Figure 3. Photostability of CdSe/ZnS QDs (with an octylamine-modified polyacrylic acid (OPA) based polymer coating) compared to organic fluorophores: normalized fluorescence intensity of QD-608-streptavidin bound to microtubules (red) and Alexa 488-streptavidin bound to nuclear antigens (green) as a function of illumination time. Reproduced with permission from Reference (53). Copyright 2003 Nature Publishing Group.

The photophysical and photochemical properties of QDs and fluorophores should be considered when designing a bioanalytical assay for DNA analysis. The interpretation of single molecule data, from both organic fluorescent probes and QDs, is problematic as a result of random switching between an emitting ("bright") state and non-emitting ("dark") state on the time-scales of the kinetic processes studied with continuous excitation; this phenomenon is known as "blinking" (Figure 4) (*42, 54–64*).

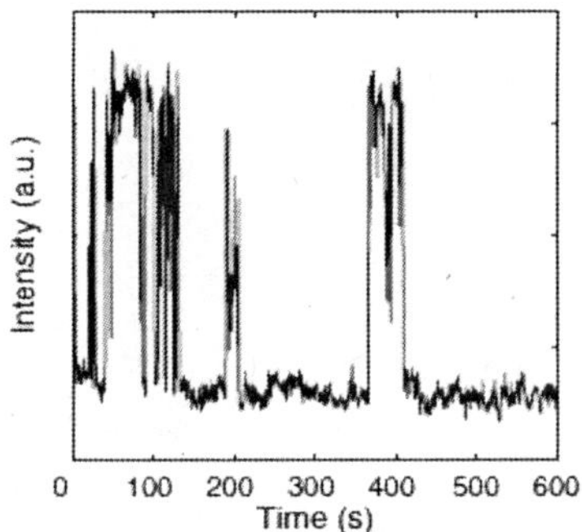

Figure 4. Fluorescence time trace of a continuously excited single CdSe QD. Reproduced with permission from Reference (64). Copyright 2003 American Physical Society.

An ideal luminescent probe for the majority of DNA analysis applications would be one where there are no (i) photoprocesses that lead to photobleaching or damage to the DNA molecules, and (ii) no 'blinking' pathways. There are exceptions to this, the most notable is where photo-switchable probes and thus the 'blinking' properties, are exploited for single molecule studies (*65*). Although fluorophores have found broad application for DNA assays where multiple copies of the DNA molecules and fluorophore probes are present, perhaps the most problematic assay format for the use of fluorophores is FISH or single molecule analysis methods. Bioanalytical methods for the analysis of single DNA molecules are emerging (and will be described later), though historically single molecule studies were mostly performed on protein systems. Typical studies have included protein folding (*55*), RNA folding (*54*) and enzyme kinetics (*66*).

As already mentioned, the interpretation of single molecule data, from both organic fluorescent probes and QDs, is problematic as a result of random switching between an emitting ("bright") state and non-emitting ("dark") state on the time-scales of the kinetic processes studied (*42, 54–64*) - the process known as 'blinking' (Figure 4). The dark states that have been identified for organic fluorophores, include (i) a metastable triplet state formed by inter-system-crossing from the singlet excited state (Figure 5(a)) (*42, 60, 66, 67*), (ii) a one electron oxidized fluorophore formed by electron transfer from the triplet state to an acceptor in the environment (Figure 5(b)) (*42, 59, 60, 68*), via electron-tunneling to surrounding trap states (*44*) or direct photoionization at high excitation intensities (*69*) or a one electron reduced fluorophore formed by electron transfer to the excited state fluorophore from a donor in the environment, (Figure 5(b)) (*42, 59, 60, 68*) and (iii) fluorophore isomerization, in which isomerization to an isomer with a different absorption spectrum is no longer excited under the illumination conditions, and resulting in a 'dark isomer state' (*60, 66*).

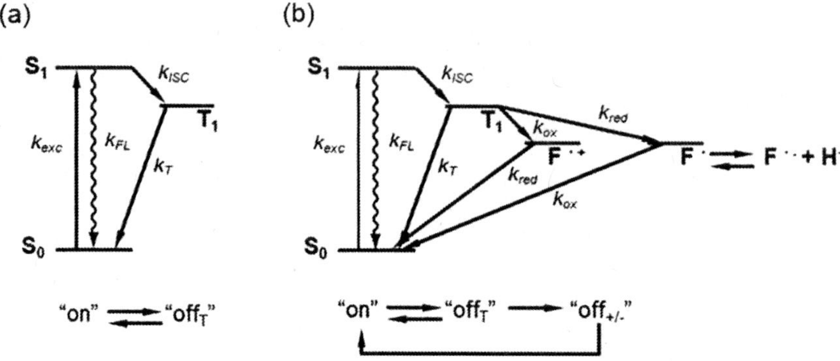

Figure 5. Two blinking mechanisms in organic fluorophores (a) a three-state model for triplet-state formation: excitation from the singlet ground state (S_0) to the first excited singlet state (S_1), is followed either by fluorescence, at rate k_{FL}, or inter-system-crossing (ISC) to the dark triplet state (T_1)("off$_T$"), at rate k_{ISC}, adapted from Weston et.al (67). (b) four-state-model for ionic fluorophore formation: oxidation or reduction of the fluorophore triplet state, at rate k_{ox} or k_{red}, respectively, forms a dark radical cation $F^{·+}$ or dark radical anion $F^{·-}$, respectively("off$_{+/-}$"), which returns to the ground state by reduction or oxidation, respectively. Adapted with permission from Reference (42). Copyright 2008 John Wiley & Sons.

Some of the pathways involved in blinking and photobleaching of fluorophores are closely related (*42, 44, 70*). The excited triplet states (T_1) of fluorophores are reactive, this is in part due to the longer lifetime (*42*) and because one electron oxidized or reduced fluorescent dyes tend to be very reactive and undergo irreversible chemical processes leading to photobleaching; the majority of these processes involve ROS (*42, 44, 69–71*). ROS comprise of oxygen radical anions, singlet oxygen and hydroxy radicals, and are generated by photo-excited species in oxygenated aqueous solutions.

Ground state molecular oxygen ($^3O_2$) is a well-known triplet state quencher (*72*); the quenching of the dark triplet state of the fluorophore results in suppressed fluorophore blinking (*67*). However, triplet state quenching is accompanied by the formation of singlet oxygen (Eqn. 2) (*72*),

$$T_1 + {}^3O_2 \rightarrow S_0 + {}^1O_2{}^* \qquad (2)$$

where T_1 and S_0 is the fluorophore dark triplet state and fluorophore singlet ground state, respectively and $^3O_2$ and $^1O_2{}^*$ is the oxygen triplet ground state and singlet excited state oxygen, respectively. (The two lowest excited states of oxygen

($^1O_2^*$) are the singlet states $^1\Delta_g$ and $^1\Sigma^+_g$, located 0.98 eV and 1.6 eV above the oxygen triplet ground state $^3\Sigma^-_g$, respectively (*73, 74*). The higher energy singlet state $^1\Sigma^+_g$ quickly decays to $^1\Delta_g$; $^1\Delta_g$ is a metastable state as the transition from the singlet excited state $^1\Delta_g$ to the triplet ground state is spin-forbidden (*74*)). Singlet oxygen formation can occur from excited singlet states of organic molecules, but this is generally a minor process (*72, 74, 75*).

Singlet oxygen is highly reactive and tends to oxidize fluorescent dyes, leaving them permanently non-fluorescent (*44, 76*). Chemical quenching of a triplet state fluorophore by ground state oxygen can also lead to the formation of a one electron oxidized fluorophore, $F^{+\cdot}$, (Eqn. 3) (*71*)

$$ T_1 + {}^1O_2 \rightarrow F^{+\cdot} + O_2^{-\cdot} \tag{3} $$

The one electron-oxidized fluorophore radical $F^{+\cdot}$ can in turn form a non-fluorescent dye molecule via further reaction (*71*). Enhanced singlet oxygen and oxygen radical anion generation by organic fluorophores has been demonstrated by close proximity to metal nanostructures, which is attributed to a plasmon-induced increase in fluorophore triplet yield (*77, 78*). Although oxygen has been shown to supress fluorophore blinking (*42, 67*), it simultaneously enhances photobleaching (*42–44, 67, 71*).

An approach to provide a fluorophore molecule that is photostable in the presence of oxygen is by using scavenging systems for reactive oxygen species (*76*). Although in this case, the removal of oxygen leads to an increased fluorophore triplet state lifetime, and thus blinking is not suppressed. An alternative triplet quenching method suitable for suppressing blinking requires the presence of both reducing and oxidizing agents. This occurs via two consecutive electron-transfer reactions, the first one quenching the triplet state by generating a non-fluorescent ionic fluorophore, and the second one returning the ionic fluorophore to the ground state (Figure 5(b)) (*42*). Fluorophore blinking has additionally been suppressed by direct linkage of the fluorophore to small triplet-state-quenching molecules, such as cyclooctatetraene or Trolox (*43*).

The dark state in QDs, responsible for the blinking effect, is a charged QD core containing only a single delocalized charge carrier (*81*). An exciton excited inside such a charged QD core quickly, within picoseconds, non-radiatively transfers its recombination energy to the extra charge carrier (radiationless Auger recombination), instead of slower, nanosecond timescale radiative recombination across the energy gap (Figure 6(a)) (*62, 81*).

A charged QD core results from the loss of a charge carrier from a neutral QD core to trap states located on the QD surface or in the QD environment (*62, 80, 82*). Charge carrier trapping can be induced by high excitation intensities that increase the probability of multi-exciton formation and subsequent radiationless Auger recombination followed by phonon-assisted carrier tunneling to a trap state (Figure 6(b)) (*80, 82*). A proposed QD charging mechanism for a QD core containing a single exciton only (instead of multiexcitons) is Auger-assisted hole trapping. Here the trapping of the excited hole to surface trap state, with energy levels within the QD energy gap, is assisted by the simultaneous excitation of the electron to a higher energy state (Auger process) (*62*).

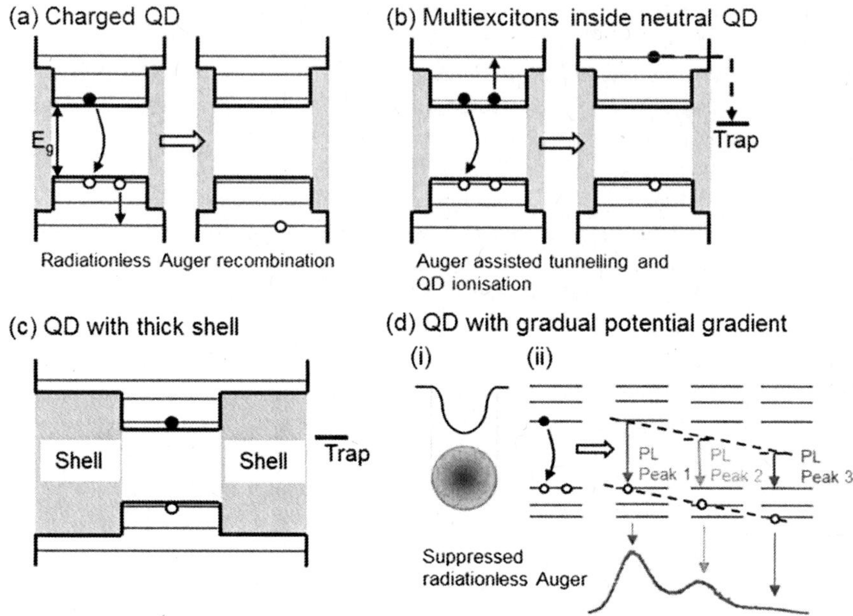

Figure 6. Blinking in QDs: (a) in a charged QD core radiationless Auger non-radiatively transfers the recombination energy of an exciton to the extra charge carrier inside the QD core. Adapted with permission from Reference (79). Copyright 2009 Nature Publishing Group. (b) Auger-assisted QD charging by excitation of a charge carrier to a higher energy state by radiationless Auger, followed by either non-radiative relaxation (not shown), or scattering to a trap state via phonon-assisted tunneling (dashed line), leaving behind a charged, non-emissive QD core; Adapted with permission from Reference (80). Copyright 2005 American Chemical Society; (c) a thick CdS shell around a CdSe QD core hinders trapping to a surface trap state, reducing blinking (d) blinking elimination by (i) a gradual potential energy gradient between a CdZnSe core and ZnSe shell, which (ii) suppresses radiationless Auger in positively charged QDs, and instead the ion recombination energy (dashed parallel lines) is shared between radiative recombination and the extra hole inside the QD core. Adapted with permission from Reference (79). Copyright 2009 Nature Publishing Group.

A passivating ZnS shell around the CdSe QD core is known to reduce blinking compared to a core CdSe QD, but a large lattice mismatch of 10% between CdSe and ZnS induces high pressures in the CdSe core and ZnS shell, which are released by the formation of defects in the ZnS shell, which can act as trap sites and induce blinking (*83, 84*). However, in the CdSe/CdS system the lattice mismatch is only 4% (*83*). As demonstrated by Mahler *et al.* (*83*) and Chen *et al.* (*52*) simultaneously, cadmium sulfide (CdS) shells as thick as 5 nm (*83*) or 19 monolayers (*52*) can be grown around a CdSe QD core with no detectable defects (*83*). The thick shell should also suppress charge carrier trapping to the surface (Figure 6(c)) (*81*). Thick shell CdSe/CdS core/shell QDs have shown reduced blinking (*51, 52, 83*), with over 60% of CdSe/CdS QDs not blinking when observed individually for 5 mins at 33kHz (*83*), as well as enhanced photostability, with no photobleaching being observed for illuminations of several hours (*52*). This can be compared to CdSe/ZnS QDs, where a large portion of which can spend a considerable amount of time in the dark state (*52, 83*).

Blinking kinetics are very sensitive to the emitter environment. QD blinking, for instance, has been suppressed by the elimination of surface trap states such as by the addition of surface ligands (*58, 85–88*). Specifically, electron-donating thiol groups added to a solution of bare CdSe QDs suppress QD blinking by filling vacant electron traps with electrons (*58*). However, Jeong *et al.* showed that the passivation of surface traps by thiol groups is dependent on the pH of the QD solution; A reduction in photoluminescence yield under basic conditions (pH=9.2) is attributed to the formation of thiolate providing hole traps. Consistent with this hypothesis, an increase and temporally stable photoluminescence in acidic conditions (pH=5.5) is observed (*89*). Other ligand molecules can also suppress QD blinking, such as the organic, electron-donating ligand oligo(phenelyne vinyline)(OPV) conjugated to bare CdSe QDs (*86, 87*), or propyl gallate added to a solution of CdSe/ZnS QDs (*85*). There has been a wide range of capping chemistries applied to QDs with varying degrees of value. For instance, some result in a reduction of photoluminescence yield, poor short-term stability and a tendency for the QDs to aggregate and precipitate, as recently reviewed (*88, 90, 91*). Polymer capping/coating and silica coating of QDs has also been applied (*92–95*), but it is clear that the addition of a capping/coating layer will result in QDs of a larger size and this size increase can be detrimental for some applications, as to be discussed later.

Surface ligands and thick shells do not stop QD blinking completely, since the cause of blinking, fast radiationless Auger in a charged QD, is not eliminated. Wang *et al.*, in contrast, have eliminated QD blinking for up to several hours by growing CdZnSe/ZnSe core/shell QDs which yields a gradual change in potential energy between QD core and shell (Figure 6 (d)) (*79*). The gradual potential energy gradient reduces radiationless Auger recombination in positive trions (an exciton plus an extra hole inside the QD core), yielding emission from both neutral and charged QDs (Figure 6(d)) (*79*). However, instead of a single fluorescence peak, CdZnSe/ZnSe core/shell QDs have multi-peaked emission as the recombination energy of a trion is redistributed between radiative recombination and the extra hole (Figure 6 (d)).

A blinking suppression method that is effective for both QDs and organic fluorophores is close proximity to nanostructured metal substrates (*60, 96–99*), which, in the case of organic fluorophores, has been shown to also increase photostability (*60*). QD blinking is also reduced by proximity to a flat gold surface (*97*). The underlying metal-induced blinking suppression mechanism is suggested to be the inhibition of electron transfer in fluorophores (*60*), and in QDs the neutralization of charged QDs by charge transfer between charged QD and metal surface (*97*). Alternatively, blinking suppression by metal proximity is suggested to be due to additional de-excitation pathways that compete with radiationless

Auger, such as (i) energy transfer to the metal, inhibiting the formation of a dark, charged QD state (*99*), and (ii) concurrent to energy transfer to the metal, more efficient radiative decay due to coupling with surface plasmon resonances supported by the rough metal surface, converting an otherwise dark charged QD into a emissive charged state (*96*).

Reports on the effect of oxygen on QD blinking are highly contradictory. On the one hand, dark state depopulation by electron transfer from a charged dark QD core to oxygen adsorbed on the QD surface has been reported, which is facilitated by the presence of water molecules (*100*). The oxygen induced depopulation of the dark state, either being a charged QD (*100*) or a triplet fluorophore state (*42, 67*), is common in both QDs and fluorophores. However, in contrast Koberling *et al.* reported a reduced on-time duration under a dry oxygen atmosphere, which is attributed to bright-state depopulation by Auger electron transfer to surface adsorbed oxygen (*101*).

Direct electron transfer from a QD core to an oxygen molecule adsorbed on the QD surface requires the $O_2 \rightarrow O_2^{\cdot -}$ redox potential of -0.15 eV to be greater than the conduction band potential of the semiconductor material of the QD (Figure 7(a)) (*100, 102*). Theoretically CdS, with a conduction band potential of -0.4 eV, is sufficient to reduce oxygen, while that of CdSe, with a conduction band potential of -0.1 eV, is not (*102*). The addition of a 2 monolayer or thicker ZnS shell around the QD should confine both the electron and hole to the QD core, further reducing the probability of $O_2^{\cdot -}$ generation due to the additional requirement of electron-tunneling through a 4 eV energy barrier (*102*). In agreement with theory, $O_2^{\cdot -}$ generation by CdS QDs but not CdSe or CdSe/ZnS QDs in aqueous solution and illuminated by UV light, has been experimentally verified (*102*). However, unfortunately contrasting experimental evidence exists with respect to $O_2^{\cdot -}$ generation by CdSe/ZnS QDs in aqueous solution. Products of ROS have been detected for CdSe/ZnS QDs that were illuminated with UV light or kept in the dark (*103*). Oxygen radical anions or peroxy radicals are suggested to be formed by CdSe/ZnS QDs under illumination in aqueous solution (*104*). In addition to these observations, Koberling *et al.* (*101*) and Müller *et al.* (*100*) have provided indirect evidence that the $O_2^{\cdot -}$ is formed by illumination of CdSe/ZnS QDs under oxygen-containing atmospheres from blinking studies. Müller *et al.* suggested that direct electron transfer from a CdSe QD core to surface adsorbed oxygen is facilitated by the presence of surrounding polar water molecules that broaden the O_2 lowest unoccupied molecular orbital (LUMO) levels, and that an increasing shell thickness, ranging from 1.2 to 4.2 monolayers, increasingly inhibits direct electron transfer (*100*). An alternative mechanism to direct electron transfer is suggested by Koberling *et al.*, where $O_2^{\cdot -}$ formation occurs as a result of ejection of an Auger electron from a neutral QD core of a CdSe/ZnS QD (*101*).

It is well-established that in the presence of water, $O_2^{\cdot -}$ can spontaneously dismutate to hydrogen peroxide (H_2O_2) (*105, 106*). H_2O_2, in turn, is known to oxidize ZnS particles to zinc oxide (ZnO) (*107*). Thus, X-ray photoelectron spectroscopy studies of CdSe/ZnS QDs illuminated in a humid oxygen environment, revealed oxidation of the epitaxial ZnS shell to ZnO (*50*), this suggests that photoillumination of CdSe/ZnS QDs indirectly generates H_2O_2, possibly via spontaneous dismutation of oxygen radical anions (*101*). Singlet oxygen generation is closely linked to oxygen radical anion formation, as singlet oxygen can be formed from an oxygen radical anion (*104*) (Eqns. 4):

$$O_2^{-} + H^+ \leftrightarrow HO_2^{\cdot}\,, pKa=4.8 \qquad 2HO_2^{\cdot} \left(or\ O_2^{-}\right) \rightarrow H_2O_2\ (or\ O_2\,) + {}^1O_2 \qquad (4)$$

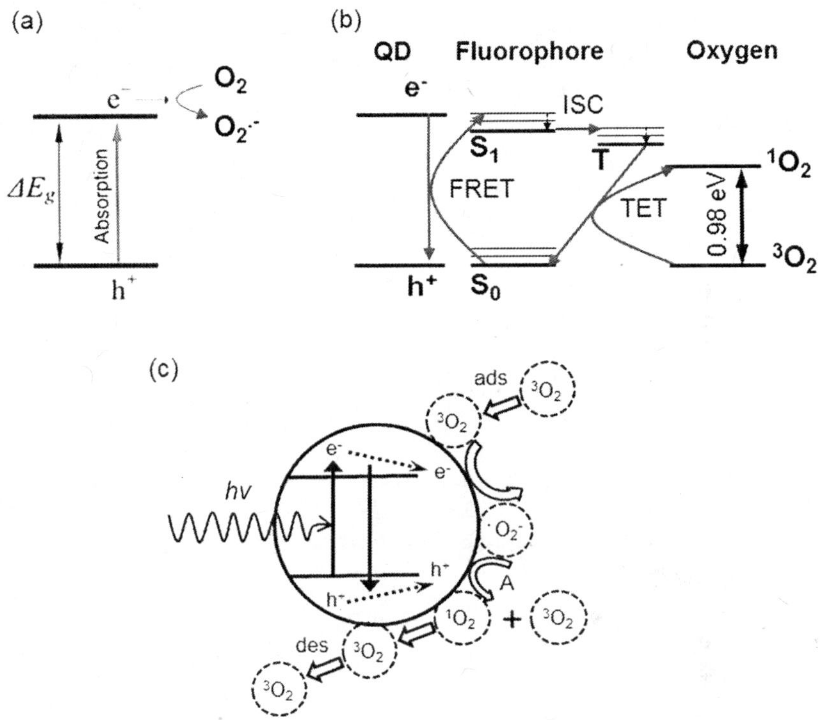

Figure 7. Schematic illustration of the generation of (a) radical oxygen anion
(O_2^{-}) by electron transfer from the QD core to molecular oxygen adsorbed
onto the QD surface, and of singlet oxygen ($^1O_2$) either (b) indirectly via
a photosensitizer molecule, involving FRET from an excited state QD to
the photosensitizer, yielding the first excited singlet state (S_1), followed by
inter-system-crossing (ISC) to the photosensitizer triplet state (T), from where
triplet-energy-transfer (TET) to ground state triplet oxygen ($^3O_2$) excites the
ground state oxygen molecule to the lowest excited singlet oxygen state ($^1O_2$), or
via (c) oxidation of an oxygen radical anion formed on the surface of a titanium
dioxide nanoparticle (TiO$_2$ NP): TiO$_2$ NP excitation forms a conduction band
electron (e^-) and a valence band hole (h^+), followed by either electron-hole
recombination or electron transfer to ground state oxygen ($^3O_2$) adsorbed
onto the NP surface, forming an oxygen radical anion (O_2^{-}), pathway A is the
oxidation of O_2^{-} to yield singlet oxygen ($^1O_2$). Adapted with permission from
Reference (108). Copyright 2005 American Chemical Society.

Core QDs, such as CdSe QDs (*109*) and cadmium telluride (CdTe) QDs (*48, 110*), under illumination in toluene and water/deuterated water, respectively, are capable of direct $^1O_2$ generation (*74, 111*). In general, direct singlet oxygen generation occurs by triplet energy transfer (TET) from a triplet excited state donor molecule, such as an organic fluorophore or QD (with an energy of at least 0.98 eV above the ground state) to an oxygen molecule in the ground state, which is a triplet state (Figure 7(b)) (*73, 74*). TET only occurs between a triplet donor and a triplet oxygen molecule in contact with each other, such as by collision (*73, 74*). In the case of QDs TET occurs from the lowest QD electronic excited state, which is a triplet state (*34*), to $^3O_2$ molecules intercalated into the tri-n-octylphosphine oxide (TOPO) or thiol-capping layer on the QD surface (*48, 109*). The $^1O_2$ yield by core QDs is low, and is typically between 1% to 5% (*109, 110*). Although under certain circumstances high yields of $^1O_2$ can be generated by excitation of QDs. This is achieved through energy transfer from an excited singlet state QD to a ground state photosensitizer and then the resulting fluorophore singlet excited state undergoes intersystem crossing (ISC) to yield the triplet excited photosensitiser which undergoes TET with oxygen to generate $^1O_2$ (Figure 7(b)) (*110, 111*).

Oxygen radical anions were reported to be formed by core/shell CdSe/ZnS QDs illuminated in aqueous solution, but surprisingly singlet oxygen was not detected in these studies (*104*). This apparent paradox is suggested to be due to quick recombination of the generated oxygen radical anions, which is suggested to be a result of interactions of the oxygen radical anions with the QD hole in a "shuttling process", inhibiting the conversion of oxygen radical anion to singlet oxygen (in contrast to process A, figure 7(c)) (*104*). This can be contrasted to titanium dioxide (TiO$_2$) nanoparticles which have been shown to generate $^1O_2$ in relatively high yields - typically 12% to 38%, due to oxygen radical anion conversion to singlet oxygen (*108*). The process of singlet oxygen formation in TiO$_2$ nanoparticles is illustrated in figure 7(c) and shows the formation of radical oxygen anion by electron transfer from an excited state TiO$_2$ nanoparticle to surface adsorbed oxygen ($^3O_2$), leaving behind a single hole inside the TiO$_2$ nanoparticle, which is followed by singlet oxygen generation by oxidation of the oxygen radical (see pathway A, Figure 7(c)) (*108*).

Unlike fluorophores, QDs exhibit photo-induced fluorescence enhancement (PFE), which is the gradual increase in the emission yield under continuous excitation (Figure 8) . PFE is due to a light-induced decrease in carrier loss to non-radiative de-excitation pathways; the reduced carrier loss has been hypothesized to be caused by mechanisms, such as (i) the passivation of surface trap states by surface-adsorbed water molecules (photo-activation) (*112*), (ii) the stabilization of surface trap states by water molecules and resulting repopulation of the lowest exciton state (*50*), (iii) the stabilization of surface trap states by photo-induced re-arrangement of capping agents on the QD surface (*113*), (iv) photo-neutralization of charged sites in and outside the QDs (*114, 115*) (v) photo-ionization of QDs resulting in Coulomb blockage by ejected electrons and reduced ionization probability of remaining neutral QDs (photo-electrification) (*116–120*) and (vi) the removal of surface-roughness trap states in the early stage of photo-oxidation (photo-corrosion) (*121*).

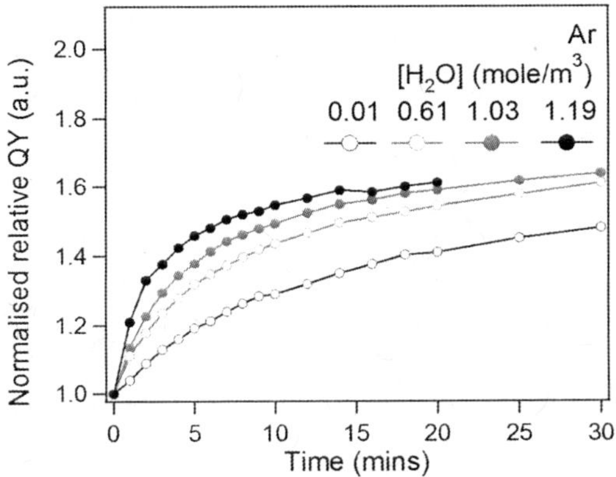

Figure 8. Photo-induced fluorescence enhancement of CdSe/ZnS QDs under an argon atmosphere. Plot of quantum yield versus time for continued illumination (532 nm), and PFE effect as a function of humidity in an argon atmosphere (normalized at time =0). Reproduced with permission from Reference (50). Copyright 2010 American Chemical Society.

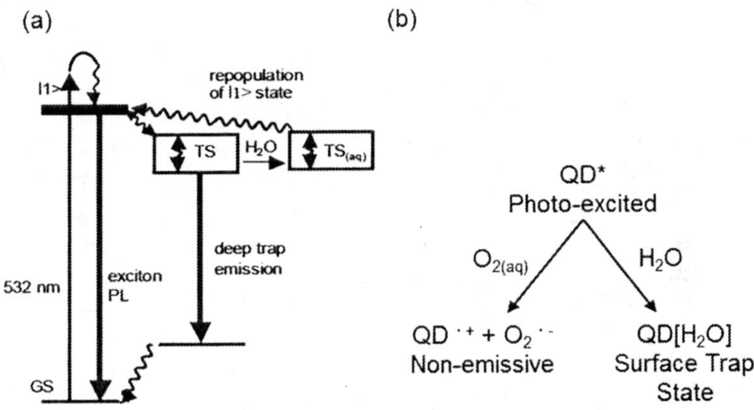

Figure 9. Schematic illustration of the PFE effect by water molecules and QD quenching by hydrated oxygen molecules: (a) a QD is excited from the ground state (GS) to the first singlet excited state (|1>) by light absorption, from where the QD can either return to the GS via fluorescence or transfer to a trap state (TS), from where deep trap emission or non-radiative decay occurs. Water-molecules present nearby a QD in the TS state can stabilize the TS state to form a longer-lived TS state (TS$_{aq}$), from where the QD can repopulate the singlet excited state with subsequent fluorescence decay, (b) a photo-excited CdSe/ZnS core/shell QD can interact with hydrated oxygen, forming a non-emissive, positively charged QD and an oxygen radical anion, or transfer to a water-stabilized surface trap state. Reproduced with permission from Reference (50). Copyright 2010 American Chemical Society.

An increased PFE has been observed with increasing water molecule concentration in the QD environment in the absence of oxygen (Figure 9) (*50*). In the presence of oxygen at low humidity levels an initial decrease in emission quantum yield is seen with continuous illumination, followed by an increase in yield. But at high humidity levels there is no decrease in yield, but a significant PFE is observed. Under prolonged illumination under oxygen/high humidity conditions the emission yield drops significantly; this is suggested to be due to degradation of the QDs (*50*). The water-induced PFE effect is suggested to be due to the stabilization of a charge-separated, surface trap state by the surrounding water molecules; this results in an increased repopulation of the first excited state from the stabilized trap state and facilitates fluorescence emission (Figure 9(a)).

In the presence of oxygen electron transfer from a neutral QD to nearby oxygen molecules results in a non-emissive charged QD and an oxygen radical anion in competition with water stabilization of a surface trap state (Figure 9(b)). The reaction with oxygen is dominant at low water humidity levels, resulting in an initial decrease in fluorescence yield upon illumination, while the water-induced PFE effect dominates at high water humidities. Furthermore, XPS studies have confirmed that the presence of oxygen molecules under high water humidity levels leads to the oxidation of the ZnS shell to ZnO, which is suggested to lead to the subsequent degradation of the QD core (*50*). Recently the PFE has been investigated for CdSe and CdSe/ZnS QDs with various different surface modifications and it was established that modification of the surface with ligands greatly affected the PFE process, with photooxidation being a major process implicated in the photoactivation (*122*). Surface modification of the QDs with mercaptoacetic acid or polymers was found to have a negative impact on the stability and\or photoluminescence properties of the QDs (*88, 90, 91, 123*).

The generation of ROS by bare, core CdSe QDs in aqueous solution and under UV illumination has been established (*48, 102, 109, 110, 124*), while experimental results on ROS formation by CdSe/ZnS core /shell QDs in aqueous solution and under UV illumination has been contrasting, with reports claiming both ROS formation (*103, 104*) and no detectable ROS formation (*102*).

In summary, highly reactive oxygen species can photo-bleach organic fluorophores and QDs. ROS, such as oxygen radical anions or singlet oxygen, can be generated by irradiated organic fluorophores and bare core QDs. However, there have been contrasting reports about the ability of core/shell QDs to generate ROS, which might be due to experimental conditions used for the different studies including (i) QD energy levels, such as due to variations in QD size, surface ligands, pH and the presence of polar water molecules and (ii) variations in electron/hole confinement inside the QD due to variations in shell thickness, as well as (iii) variations in the shell growth and hence shell quality/permeability.

What is clear is that the emission properties of QDs are more robust than fluorophores with respect to reaction with the ROS that are generated by photoexcitation. This could be because multiple ROS 'reactions' might be required leading to partial or complete degradation of the shell for photobleaching. What should not be overlooked is the fact that ROS not only damage the QDs and the coatings (*125*), but are also likely to damage the biomolecules in the surroundings. Notably, oxidation of amines within macromolecules such as nucleic acids and proteins by singlet oxygen (*74, 126*) which can eventually result in cell damage and cell apoptosis *in vivo* (*111, 124, 126–128*). Furthermore, ROS also have a high chemical reactivity with biomolecules, such as DNA yielding nucleic acid base damage as well as DNA strand breaks and abasic sites (*129*). Semiconductor QDs were proposed to induce DNA strand breaks via a free radical process (*130*). The damage was detected in samples exposed to UV light as well as in the dark and thus the damage is proposed to occur as a result of photo-generated and surface oxidized species from the QDs. Following this publication (*130*) a series of contradictory reports have been published and it has been suggested that shell coated CdS, CdSe and CdTe are less susceptible to the release of cadmium ions (*102, 131*). It should be noted that oxidative conditions can also give rise to the release of toxic cadmium ions from the QD core due to oxidative shell degradation by ROS (*49, 50*). Cadmium is known to induce oxidative damage to DNA (as well as lipids) (*132*). The formation of DNA strand breaks (*130, 131, 133*) which then lead to chromosomal aberrations is well documented (*134*).

Differences in the Interaction of Fluorescent Dyes and QDs with DNA

Fluorescent dyes used for DNA based bioanalytical methods are of the order 0.5nm in size and most are only slightly larger than a DNA base pair. QDs that are applied to DNA analysis applications have a size of 2-60nm and are thus large with respect to the size of a nucleic acid base pair within DNA. The relatively large size of QDs, as compared to the DNA double helix, is sometimes easy to forget - most QDs used for DNA analysis applications have a diameter which is at least twice that of two helical pitches of a DNA helix!

Because there is a long history of use of organic fluorophores for bioanalytical applications means that there are a plethora of fluorescent probes available which are suitable for both (i) covalent attachment to single DNA strands and (ii) direct labeling of DNA by binding or intercalation (*37, 135*). Covalent attachment can be achieved by bioconjugation methods to the 3' or 5' end of DNA oligonucleotides (*136*), enzymatic methods, for instance addition of fluorophore-labeled nucleotide triphosphates, as used for real-time DNA sequencing (*137*), or during DNA oligonucleotide synthesis using fluorescent analogues of nucleosides (*138*). On the whole the attached fluorophores provide labels for detection, however, it should be noted that fluorophores such as Cy3 have also been applied for time resolved approaches whereby photo-induced isomerization can be exploited for single-molecule methods (*139–142*). As already mentioned, direct labeling of DNA can also be achieved using fluorophores which interact with DNA by intercalation (or bis-intercalation) or by binding in the major or minor grooves (*6, 7*). As a result of these types of interactions the quantum yield and fluorescence lifetime of these fluorophores are greatly enhanced. An example is the dye TOTO-1 which binds to DNA by bis-intercalation and at higher concentrations additionally binds into the DNA grooves. Bis-intercalation into DNA enhances the fluorescence quantum yield of TOTO-1 by 1,100 fold compared to when free in solution (*13*). In summary, there are a large range of approaches available for labeling DNA with fluorophores; this, in addition to the comparatively smaller size (as compared to QDs) is advantageous for DNA analysis applications - to be discussed later.

QDs approach the size of histone proteins within chromosomal DNA, which are ~11nm diameter (*143*). Modification of the surface of CdSe/ZnS QDs with a coating containing positively charged amine groups results in electrostatic association of DNA (*144*) and thus provides a 'pseudo-QD-histone-mimic'. Even though QDs can be surface modified for DNA association, it has long been established that DNA binds in a non-specific manner to unfunctionalized nanometer-sized particles composed of cadmium sulfide (*145*). With the interaction of calf-thymus DNA, there is a decrease in the photoluminescence yield of the cationic CdS QDs and this interaction is considered to be entropically driven; the binding constants are of the order of 10^3-10^4 M^{-1} which are in the range of protein-DNA interactions (*146, 147*). Thus it is believed that the interaction occurs as a result of release of the counter ions and water from the DNA interface (*146*). Although colloidal CdS nanoparticles have a net negative charge (*148*), the activation with hydroxide ions prior to the addition to the DNA, is suggested to be what drives the charge induced association (*146*). Recently the interaction of QDs with DNA has been exploited for reducing non-specific amplification during PCR. The majority of the work in this area has been achieved with CdTe QDs, where the specificity of the PCR at different annealing temperatures and with different DNA templates could be enhanced (*149, 150*). Although one study has involved coated CdS QDs (*151*). It should be noted that nanoparticles made from other materials such as gold are well known to enhance the efficiency of PCR (*152*).

The photophysical properties of a CdSe/ZnS QD-DNA oligonucleotide are reported to be affected by hybridization with a complementary sequence of DNA. In this case both fluorescence quenching and a reduction of the QD fluorescence lifetime occurs (*153*). It is believed that the changes in QD photophysical properties upon target DNA hybridization are due to the electric field introduced by the negatively charged target DNA, which induces non-radiative Auger recombination in the QD (*153*).

The larger size of QDs cannot be overlooked, CdSe/ZnS QDs of 6nm diameter coated with a soluble polymer coating were found to have a diffusion coefficient of 29 $\pm$ 0.5 μm^{2}s^{-1} relating to a hydrodynamic radius of 7.4 $\pm$ 0.6 nm (*154*). This isn't always a disadvantage, for instance *Giraud et al.* have detected hybridization of unlabelled DNA to CdSe/ZnS QD - DNA conjugates by measuring the change in rotational correlation time upon hybridization (*155*). The QDs were polymer coated and streptavidin functionalized for water solubility and to allow conjugate formation with single-stranded biotinylated probe DNA. The rotational correlation time of the QD-ssDNA complex upon complementary target hybridization was experimentally shown to increase from 330 ns to 1,300 ns, corresponding to a theoretical construct diameter of 14.2 nm and 23 nm, respectively. A further excellent example is that of Zhang *et al.* where mapping the DNA quantity was achieved using QDs labeled with nanotethers for specific DNA sequences and according to the quantity of complementary sequences present, an electrophoretic mobility difference (mobility shift assay) was detected. The approach was applied to evaluate DNA copy number as well as DNA methylation levels (after bisulfite treatment) in specific sequences (*156*). Clearly for these examples, the large QD size and the potential for attachment of multiple probe DNA oligonucleotides on the surface is an advantage.

The attachment of DNA oligonucleotides to QDs is currently defined by the capping or coating layer used (*86, 87, 89*). There are a number of approaches for the attachment of DNA oligonucleotides, but direct covalent linkage generally involves bioconjugate chemistry methods or addition of a DNA oligonucleotide end-labeled with a biotin moiety to streptavidin coated QDs (*136, 157*). In any case, the surface of the QD, when modified with a capping or coating is accessible to attachment at multiple chemically equivalent sites. Also as Boeneman *et al.* have elegantly shown, the conjugation chemistry can determine the final structure of a QD-DNA conjugate (*158*). To prepare QDs functionalized with a single DNA oligonucleotide it is possible to mix equi-molar (or less than equi-molar) amounts of QDs (with the appropriately functionalized surface for chemical attachment) and DNA oligonucleotides (with end-labeled amino or thiol linkers) but there is always a probability of a fraction of the QD-DNA conjugates being formed with two or more DNA strands attached onto the same QD. The fraction of DNA labeled QDs to unlabelled QDs can be enhanced by exploiting the differences in solubility (*159*), however a more accurate approach to isolate QDs conjugated to a single DNA oligonucleotide involves gel electrophoresis and is based on a method developed originally for selection of single gold nanoparticle-DNA oligonucleotides (*160, 161*). The formation of a QD-DNA conjugate consisting of a single QD conjugated to a single long DNA strand has been demonstrated by He *et al* (*162*). This particular one-to-one QD-single long DNA conjugate was formed by PCR amplification of octylamine-modified polyacrylic acid (OPA) coated QDs covalently linked to 5' amine-modified forward primers. The emission spectrum of the OPA coated QDs, after conjugation to forward primers and after PCR amplification is well conserved. However there is a small decrease in the emission yield (*162*). It should be noted that this approach will provide a QD with multiple short DNA strands (primers) and one extended DNA strand per QD.

Assessment of the Use of Colloidal QDs As Compared to Fluorophores for DNA Analysis Applications

QDs are being increasingly used in biological applications instead of organic fluorescence dyes (*15, 53, 163–166*). For instance, Gerion *et al.* have developed a microarray and CdSe/ZnS QD based DNA detection system that allows for simultaneous detection of single nucleotide polymorphisms (SNP) in the human p53 tumor suppressor gene, hepatitis B virus (HBV) and hepatitis C virus (HCV) (Figure 10) (*167*).

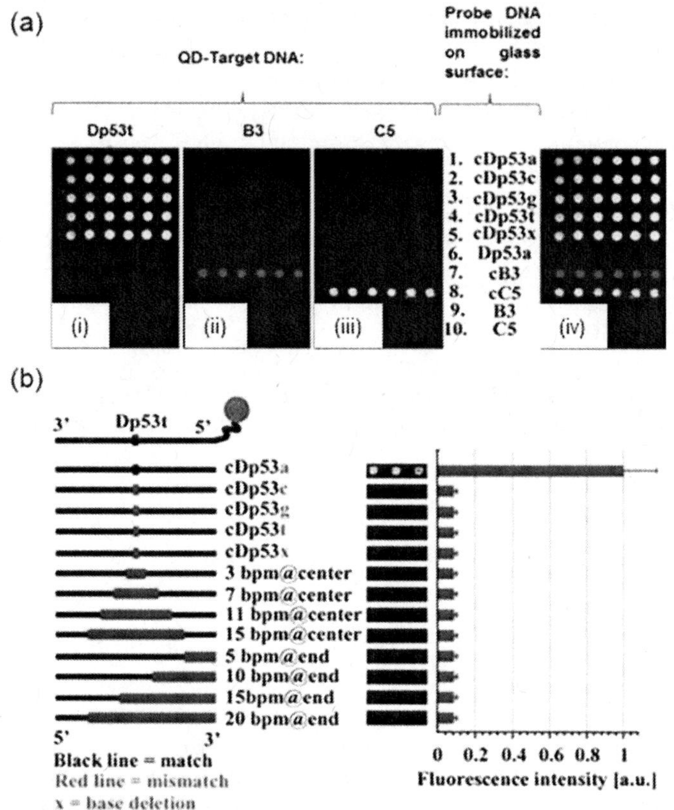

Figure 10. QD based DNA detection in a microarray format: (a) Microarray printed with the DNA sequences (row 1-5) 25-mer (related to the sequence on exon 7 of the human oncogene p53 called Dp53) mutated in the middle with ACGT or deleted(x), (row 6) Dp53g, the complementary sequence to cDp53c (row 7) cB3, the reverse complement to a portion of the 3'-end of HBV, (row 8) C5, the reverse complement to a portion of the 3'-end of HCV (row 9) cB3 the complement to the sequence cB3 (row 10) C3 the complement to the sequence cC3, C5; (i) D53t labeled with yellow QDs targeted to mutations, (ii) red QD-DNA sequences target to HBV (iii) yellow QD-DNA sequences target to HCV and (iv) overlay of (i), (ii) and (iii) (b) left: schematic of QD-target Dp53t sequence and the DNA probes used, including the complementary sequence cDp53a, and sequence with mismatches (marked in red). right: QD fluorescence intensity of the spots on the microarray after probe-target incubation. Adapted with permission from Reference (167). Copyright 2003 American Chemical Society.

The 2D probe DNA microarray contains 10 rows and 6 columns and is formed by printing single stranded, amino-modified probe DNA onto an aldehyde-activated glass substrate, forming a covalent link between the aldehyde group and amino end group. Each row of the microarray is functionalized with a specific probe DNA sequence. Single stranded target DNA is covalently linked with a silica shell that contains CdSe/ZnS core/shell QDs. Incubation of the QD-target complex onto the array results in probe-target hybridization but with low homologies. It should be noted that the Dp53t probes all show a false positive (Figure 10(a) lanes 2-5) and that non-standard hybridization conditions (with low salt and formamide) were required to provide selectivity for SNPs with these QD-DNA probes. (Figure 10(b))

It should be noted that the surface density of complementary strands of fluorescently labeled oligonucleotides on a DNA functionalized glass surface is 2.3 x 10^{12} oligonucleotides cm^{-2} (*168*) and on a silicon substrate 6.7 x 10^{12} oligonucleotides cm^{-2} (and this equates to one oligonucleotide per 15 square nanometers) (*169*). The surface density of complementary strands of QD labeled oligonucleotides to DNA oligonucleotide sequences attached to surfaces will be limited by the particle size and is expected to be far lower than can be achieved using fluorophore labeled DNA probes, but needs to be verified.

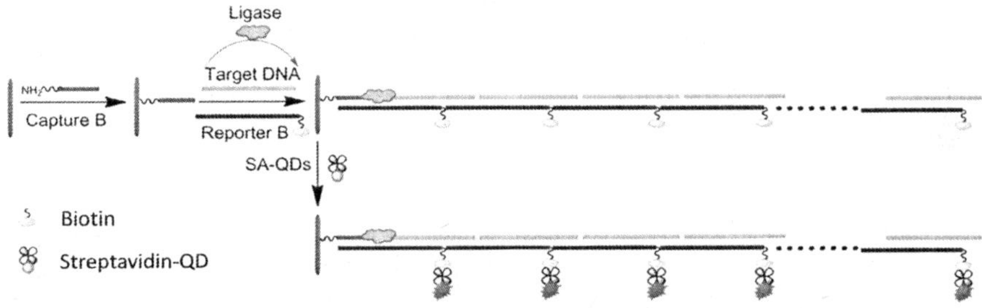

Figure 11. Schematic illustration of signal amplification by DNA chain elongation. Reproduced with permission from Reference (170). Copyright 2012 Royal Society of Chemistry.

Increased target detection sensitivities of up to 10 fM target DNA have been experimentally obtained with a signal amplification scheme via DNA chain elongation (Figure 11) (*170*). Specifically, the DNA assay consists of a surface tethered capture DNA and a biotin labeled reporter DNA (Figure 11). By choice of appropriate capture and reporter DNA sequence, a ternary hybridization complex forms between capture, reporter and target DNA, with the end of the target DNA left free to hybridize to another reporter DNA, which in turn has an end left free to hybridize to a further target DNA *etc*, setting off a cascade of hybridization events between reporter and target DNA (Figure 11). The addition of streptavidin coated QDs allows a biotin-streptavidin link between the QD and amplified reporter DNA strands to be formed, resulting in an increased fluorescence signal. The addition of T4 ligase allows a link between capture and target DNA to be formed (Figure 11), further augmenting the detection limit to 10 fM target DNA (*170*).

The broad absorption spectrum of QDs (see figure 2(b)) allows for excitation with a large range of possible excitation wavelengths. This provides the possibility for co-excitation of a range of differently sized QDs at a single excitation wavelength. By contrast, color multiplexing applications using fluorophores are more limited due to a narrow absorption spectrum, a narrow Stokes shift and a broader, asymmetric fluorescence peak (figure 1), with FWHMs ranging between 35-100 nm (*15*). Thus the generation of multiplex labeling systems using QDs has been possible (*171, 172*).

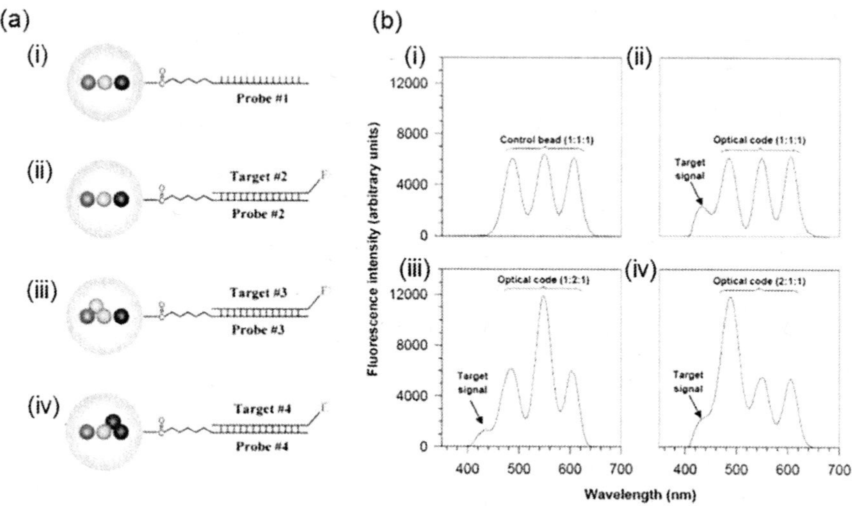

Figure 12. QD-based optical bar-coding DNA assay: (a) schematic representation of QDs of different colors at precisely controlled ratios embedded inside polymer beads, and conjugated with probe oligonucleotides, allowing for hybridization of sequence specific, fluorescently labeled target sequences (b) experimental single bead fluorescence spectroscopy results following probe-target hybridization for a single bead. Reproduced with permission from Reference (171). Copyright 2001 Nature Publishing Group.

There are a few fluorophore dye combinations which can be excited by a single wavelength, for instance the 6-carboxyfluorescein (FAM), 6-carboxy-4', 5'-dichloro-2', 7'- dimethoxyfluorescein) (JOE), 6 - (Tetramethylrhodamine - 5 - (and - 6) - carboxamido) hexanoic acid (TAMRA) and 6-carboxy-X-rhodamine (ROX), which can be excited at 488nm for DNA sequencing applications (*173*). However, the broad absorption spectrum and narrow fluorescence spectrum of QDs, compared to the narrow absorption spectrum and broader fluorescence spectrum of organic fluorophores, has allowed QDs to be employed more easily in multiplexing applications, in which specific DNA target sequences can be detected and identified (*171, 172*). An excellent example includes the use of QD embedded polymeric microbeads; these microbeads contain QDs of different sizes at varying, precisely controlled ratios thus providing a range of 'optical barcodes' where the emission varies in intensity and wavelength as a function of the QD content (Figure 12(a)) (*171, 172*). Probe-target hybridization between microbead-probe DNA and complementary, fluorophore-labeled target DNA (Figure 12(a)) results in the appearance of the dye fluorescence peak within the optical barcode specific to a certain probe and hence target sequence, allowing for the detection of the presence or absence of specific target sequences (Figure 12 (b)) (*171, 172*).

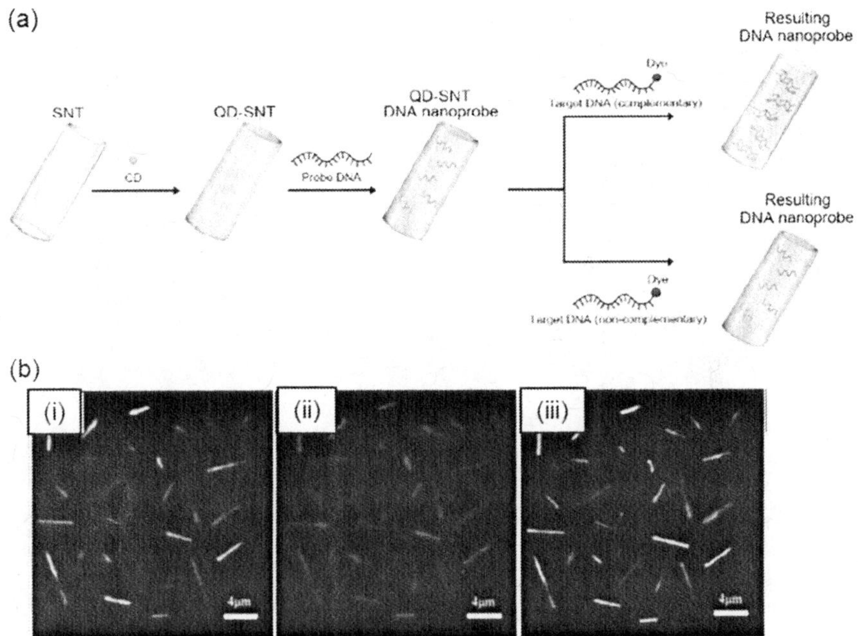

Figure 13. (a) Schematic illustration of QD-SNT-DNA probe based colorimetric target DNA detection: SNTs are embedded with QDs of a single color, and functionalized with single-stranded probe DNA. Hybridization with complementary target DNA results in a color change from green to yellow, while in the presence of non-complementary target DNA no color change of the green SNTs occurs (b) fluorescence images of (i) green-colored QD-SNT-DNA probes (ii) red-colored TAMRA labeled complementary target DNA and (iii) merged yellow color after probe-target DNA hybridization between QD-SNT-DNA probes in (i) and dye-target DNA in (ii). Adapted with permission from reference (174). Copyright 2011 IOPscience.

Liu *et al.* recently developed a colorimetric, high selectivity DNA detection system providing a detection sensitivity of up to ~100 attomole (*174*). The sensor probe consists of QD-embedded silica nanotubes (SNT) surface immobilized with single stranded probe DNA (Figure 13(a)). Mixing in solution with complementary target DNA labeled with a different fluorophore with a different emission wavelength (Figure 13 (b)) results in an easily detectable color change in a color image due to 'merging' of the QD and fluorophore emission upon probe-target DNA hybridization (Figure 13(b)).

QDs as probes for bioanalysis applications have emerged concurrently with optical approaches suitable for detecting and imaging fluorescently labeled biological structures at high spatial resolution. In many respects these techniques are blighted by the photostability of conventional fluorophores, so QDs have risen to the forefront for many applications and are sometimes advantageous as fluorescent probes. At this stage these methods will be considered in two categories (i) systems for the detection of co-localized probes and (ii) systems suitable for imaging separated probes at high spatial resolution.

Perhaps the most well known approach for the detection of co-localized probes (and for evaluation of their close spatial separation) is FRET and DNA hybridization with a specific complementary sequence provides the co-localization of the FRET pair (*175–177*). DNA analysis methods which are based upon FRET principles using fluorophore pairs are well established, most notably as molecular beacons and PCR probes such as Scorpion and Taqman probes (*178–182*). Ideal FRET pairs are those where (i) only the donor, but not the acceptor, is excited by the excitation wavelength, (ii) the donor emission has significant overlap with the acceptor absorption spectrum and where (iii) there is good separation of the donor\acceptor emission spectra, and (iv) where there is a close spatial separation between the donor and acceptor. (further discussion has been provided in the excellent review by Algar *et al* (*177*)).

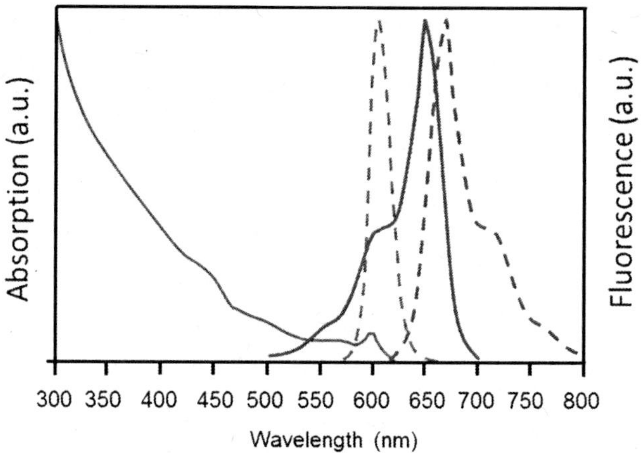

Figure 14. Schematic figure showing QD absorption spectrum (blue solid line), QD emission spectrum (blue dashed line), absorption spectrum of the fluorophore acceptor (red solid line) and emission spectrum of the acceptor (dashed red line).

Because QDs have narrow emission spectra and a broad absorption spectrum, this makes them ideal donors for FRET with fluorophores but not with QDs (*183–186*). A FRET-based DNA assay usually consists of a QD energy donor, which is conjugated to a single stranded probe DNA, and an energy acceptor, which is a fluorescent dye conjugated either directly to the target DNA strand

(Figure 15(a)) (*185*) or to a reporter probe DNA strand that is complementary with part of the target sequence to be detected (Figure 15(b)) (*183, 184*). The presence of a target DNA sequence complementary to the probe sequence leads to probe-target hybridization and FRET between donor and acceptor, yielding a change in donor and acceptor fluorescence intensities and donor excited-state lifetime (*183–185*). A detection sensitivity of up to 4.8 fM target DNA has been experimentally demonstrated by acceptor signal amplification, achieved by an increased FRET efficiency due to increasing the number of capture probe DNA strands that bind to a single CdSe/ZnS QD via the biotin-streptavidin interaction (*183*).

The extension of these FRET assays to detect two different target sequences simultaneously using FRET from a single CdSe/ZnS QD to multiple, target-specific fluorophores, and two-color coincidence detection, employing two different fluorophores of differing fluorescence peaks for each target DNA sequence has been experimentally demonstrated (*184*). However, multiple target detection via dye fluorescence multiplexing, employing a CdSe/ZnS QD donor and multiple, target-specific organic fluorophore acceptors, is limited by the broad fluorescence spectra of organic fluorophores.

Gill *et al.* have probed DNA hybridization and cleavage using FRET methods. CdSe/ZnS core/shell QDs were functionalized with thiol-end labeled oligonucleotide strands (Figure 15(a)) (*185*). The estimated QD surface coverage is 6 oligonucleotide strands per QD. Upon addition of a complementary sequence labeled with the FRET acceptor fluorophore Texas Red (Figure 15(a)), the QD emission was quenched by 60% and the Texas Red fluorescence intensity increased. The QD emission decrease and the Texas Red emission increase is considered to be due to efficient FRET from the QD donor to the fluorophore acceptor (Figure 15(a)). Addition of the hydrolytic enzyme DNase I, resulted in the partial recovery of QD emission yield and a reduction in the dye fluorescence. DNase I is believed to reduce the FRET efficiency by cleaving the double stranded DNA-QD-fluorophore complex. Only partial QD fluorescence recovery suggests non-specific adsorption of the Texas red dye onto the QD surface (*185*). An alternative format applied using FRET methods for DNA analysis is shown in Figure 15(b), here a sandwich hybrid is formed by hybridization of a reporter probe with a fluorophore and a biotin labeled capture probe. The sandwich hybrid is associated to the streptavidin labeled QD (*183*).

An alternative format for DNA analysis based upon FRET principles is where a QD donor is used with a quencher acceptor. The most well known configuration for this is the molecular beacon (*187*), where a fluorophore instead of a QD is used (*188*). The application of a silica coated QD is suggested to provide an enhanced photostability to the molecular biosensor for *in vivo* applications (*189*).

FRET based DNA assays generally require covalent attachment between the QD and the probe DNA, the need for which has been circumvented by the employment of an electrostatic linker. For this case, negatively charged CdTe QD donors were first immersed in a cationic polymer solution, then the negatively charged probe DNA conjugated to a fluorophore-acceptor added and associated on the surface (Figure 16) (*186*). By this type of approach complementary target hybridization to the fluorophore labeled probe DNA results in a decrease in fluorophore acceptor fluorescence intensity due to an increased donor-acceptor separation induced by the increased rigidity of the double stranded probe-target DNA complex, compared to single-stranded probe DNA. This electrostatic linker-based FRET approach allowed for target detection in the nMolar range (*186*).

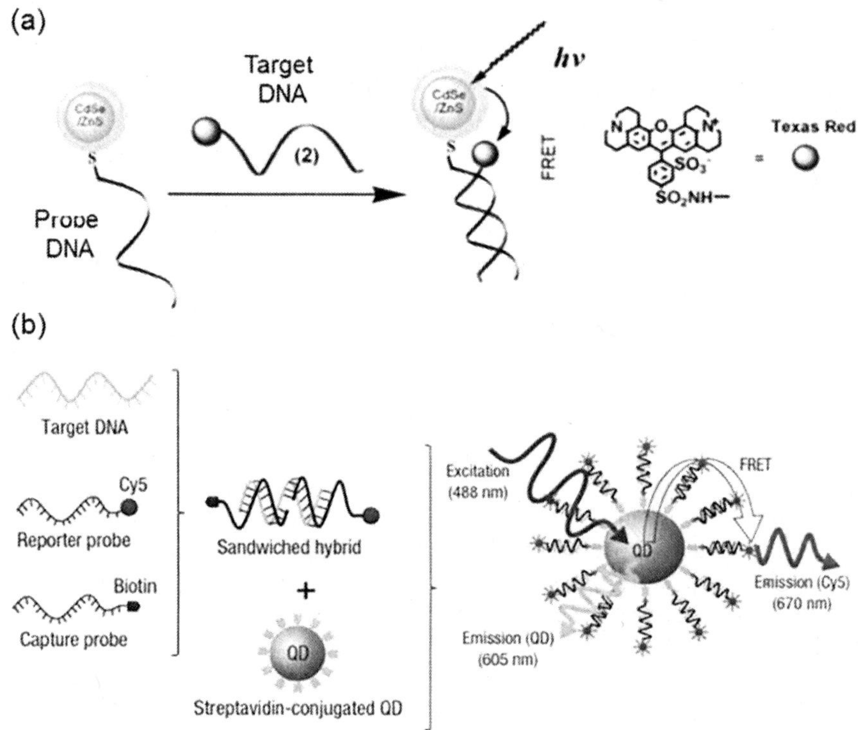

Figure 15. Schematic illustration of FRET-based DNA assays (a) hybridization between QD-labeled DNA probe and complementary, fluorophore-labeled target DNA leads to FRET between QD and fluorophore (185), (b) hybridization of capture probe and reporter probe to complementary target DNA forms a sandwiched hybrid, which attaches to a QD surface via biotin-streptavidin linkage, resulting in FRET between QD donor and fluorophore acceptor. Reproduced with permission from Reference (183). Copyright 2005 Nature Publishing Group.

An alternative, non-covalent binding mechanism which does not employ an electrostatic polymer linker has been described by Lee *et al.* (*144*). In their approach, the QDs themselves were given a positively charged surface coating by ligand exchange of TOPO-coated QDs with a dihydrolipoic acid ligand derivative which contains positively charged amine groups. Further addition of polyethylene glycol molecules to the QD surface enhances QD water solubility and inhibits non-specific DNA adsorption. Compact, electrostatic conjugate formation of the positively charged QD with negatively charged, fluorophore labeled probe DNA results in efficient FRET between QD donor and fluorophore acceptor. Target DNA hybridization with the probe DNA results in QD fluorescence recovery, allowing target DNA detection of to up to 200 nM.

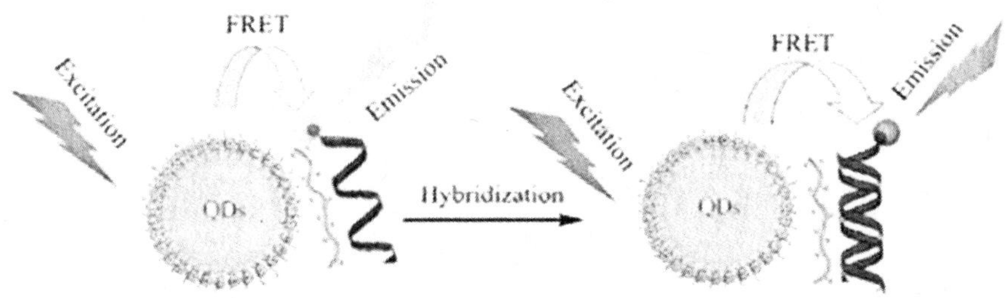

Figure 16. Schematic of a FRET based non-covalent DNA nanosensor employing a cationic polymer. Reproduced with permission from Reference (186). Copyright 2007 American Chemical Society.

The use of QDs as a FRET acceptor is less common and the reason for this is that QDs are more easily excited than most fluorophore donors and the lifetime of QD emission is much longer (10-100ns) than typical fluorophores (<5ns). However these problems have been overcome for a small number of studies, although not for true DNA analysis applications so far (examples include (*190*)) There are approaches emerging where QDs are applied as both the donor and acceptor (*i.e.* (*191*)); it will be interesting to see how these approaches evolve and whether there will be value for application for analysis of DNA in the future.

The use of fluorescence quenching gold nanoparticles in conjunction with QDs for the detection of specific DNA sequences has been demonstrated by Dyadyusha *et al* (*192*). A single CdSe/ZnS QD attached to the 3'-end of a single-stranded 16-mer oligonucleotide strand has a strong QD emission peak (Figure 17). The addition of complementary DNA labeled with a single gold nanoparticle (GNP) at the 3'-end results in quenching of the QD fluorescence by 85%. This fluorescence quenching was attributed to a fluorescence energy transfer type interaction between the QD and nearby GNP upon DNA hybridization and resulting proximity of QD and GNP. The addition of a 10x excess of unlabeled complementary oligonucleotides partially reverses the QD quenching due to hybridization of the unlabeled complementary oligonucleotides with the QD-oligonucleotides.(Figure 17(b))

Gueroui *et al.* first studied the distance dependence of quenching of CdSe QD-DNA conjugates by hybridization with GNP-DNA conjugates and suggested a Förster type process (*193*). Subsequent to these studies, Pons *et al.* (*194*) and Li *et al.* (*195*) have shown that the dynamics of CdSe/ZnS QD fluorescence quenching by GNP and gold nanorods, respectively, was more consistent with the nanometal surface energy transfer (NSET) model, displaying a $1/R^4$ separation dependence, rather than the FRET model between two dipoles, which exhibit a $1/R^6$ separation dependence. According to the NSET model, a point dipole interacting with an infinite metal surface induces currents in the metal surface, which eventually dissipate due to ohmic loss, resulting in fluorescence quenching (*196*). The application of QD-GNP pairs for DNA analysis applications has evolved significantly, with the demonstration of QD-GNP pairs for DNA analysis (*in vitro* and *in vivo*) for both emission quenching and enhancement (*197, 198*).

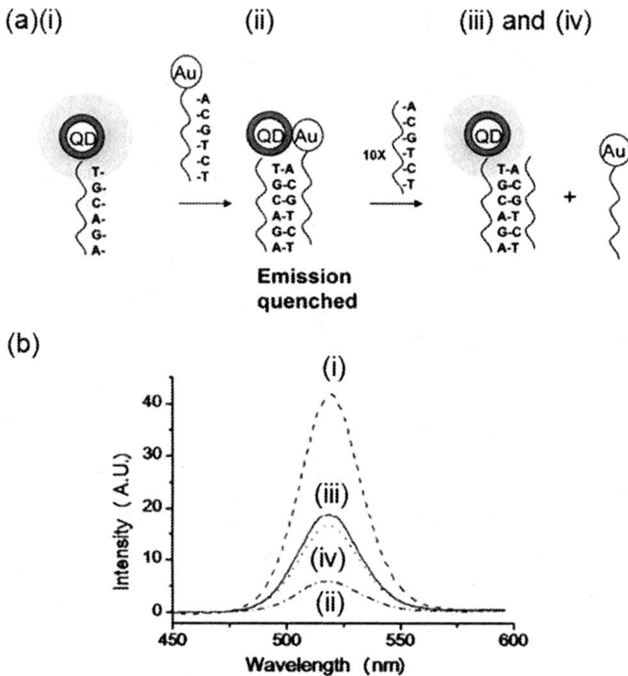

Figure 17. QD-gold nanoparticle based DNA detection scheme in solution: (a) schematic illustration and (b) corresponding QD fluorescence spectrum of (i) QD-ssDNA (ii) formation of a (QD-ssDNA)-(GNP-complementary ssDNA) hybridization complex with close proximity between the QD and gold nanoparticle after the addition of GNP-complementary ssDNA (iii) displacement of the GNP-ssDNA by unlabelled complementary ssDNA upon addition of 10x excess of unlabelled complementary ssDNA after (iii) 1 hour and (iv) 2.5 hours. Adapted with permission from Reference (192). Copyright 2005 Royal Society of Chemistry.

The disadvantage of FRET-based assays with QDs is that rather complicated labeling steps of the QD and the fluorophore with the DNA are required to yield a QD-fluorophore separation of less than the Förster radius (between 4-7 nm) (for efficient FRET to occur). This has been shown to limit the length of target DNA that can be detected in certain assay formats (*153, 188*). In addition for improved solubilization, the QDs are coated with an amphiphilic polymer or enclosed within a micelle structure - this then means the QD donor - fluorophore acceptor separation exceeds the Förster radius. To circumvent this problem high fluorescent probe-to-QD ratios are used to improve the FRET efficiencies. This of course means that high quantities of target DNA sequences are required (*199*). An approach to overcome this problem is the use of a compact QD-DNA oligonucleotide linker and this provides a reduction of the Förster radius of the QD-fluorophore to ~ 4.2 nm (*199*). The addition of either (i) fluorescently tagged target DNA or (ii) unlabeled target DNA and the dsDNA intercalator ethidium bromide to the QD-probe DNA conjugate allowed for FRET based detection of DNA sequences in the nM range (*199*).

Due to the challenges of highly efficient quenching of single QDs with single quenchers (as compare to fluorophore-quencher pairs), the creation of PCR probes that are homologues of Taqman or Scorpion probes (*178–181*) containing QDs instead of fluorophores appears to be currently impractical. However, a number of approaches using indirect PCR based assays have been developed using a QD as the probe. This includes the reported study of Bailey *et al.* where FRET using a QD-fluorophore pair was combined with methylation specific (MS) to enable the detection of DNA methylation (*200*). The method for FRET-based DNA methylation detection, termed methylation-specific QD fluorescence resonance energy transfer (MS-qFRET), has demonstrated detection sensitivities of up to 15 pM methylated DNA sequences among a 10,000 fold excess of unmethylated sequences (*200*). In addition, next generation sequencing approaches based upon FRET have emerged and involve the use of QD-labeled polymerase enzymes, for instance the approach of Life Technologies (*201, 202*). The DNA molecule to be sequenced is ligated to a surface attached oligonucleotide and then addition of a primer and subsequent extension results in incorporation of fluorescently labeled nucleotides one-by-one by the QD-labeled polymerase. This interaction is read optically by a reduction in the QD (donor) emission yield and the emission from the fluorophore (acceptor) attached to the nucleotide rises.

Perhaps the most promising application for QDs over fluorophores is for detection of DNA sequences within a single DNA molecule greater than 10nm apart, i.e. for establishing the location of fluorescent probes on larger DNA structures, *i.e.* chromosome spreads for FISH applications (*203–206*), and stretched long fragments of DNA (*207*). At this time QDs are showing future potential for analysis of DNA sequences in stretched long fragments of DNA by association at specific sites along the strand. To accurately determine the separation at nanometer resolution of two luminescent probes with the same emission wavelength there are a number of techniques that have been developed, as reviewed by Toprak and Selvin (*208*). However, there is a need for better accuracy for locating probes of different emission wavelengths, particularly for DNA analysis applications. As elegantly stated by Antleman *et al.* issues such as chromatic aberrations, alignment of excitation beams or image registration need to be considered for many conventional microscopy methods (*160*). By using (i) QDs or (ii) silica beads composed of FRET pairs of fluorophores (often called tranFluoroSpheres) (*16–18*), it is possible to apply a common excitation wavelength. By using this format it is possible to achieve a resolution of less than

10nm. Improvements in the passivation of QD 'blinking' have recently allowed for improvements in imaging as well as distance measurements between QDs attached to DNA sequences (*160*).

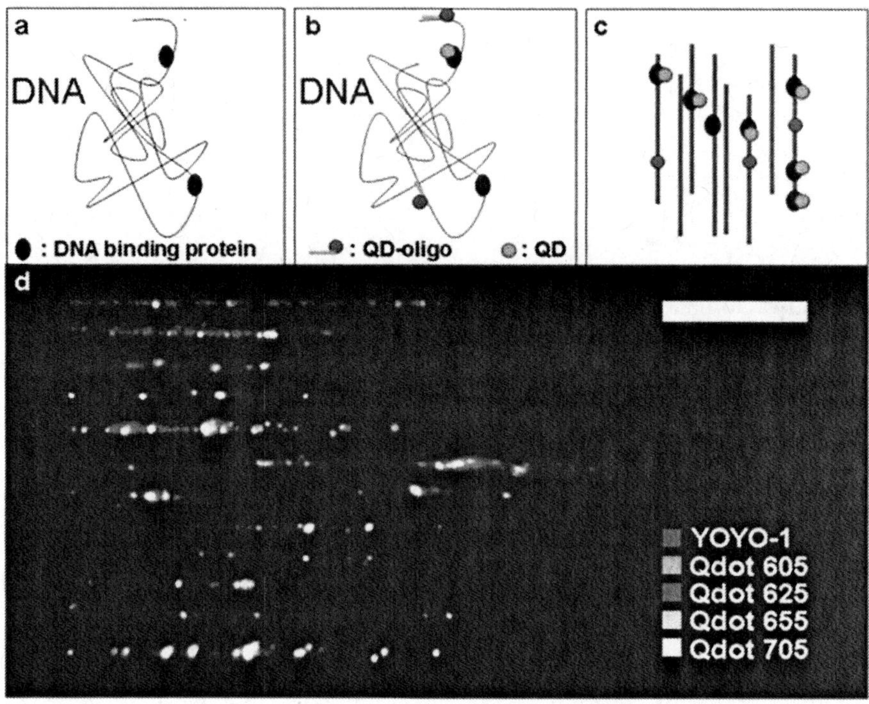

Figure 18. Localization of DNA binding proteins using QDs as fluorescent labels: schematic of (a) cross-linking of DNA binding proteins to DNA (b) labeling of bound proteins with QDs (green) and of specific reference sequences with QDs (red), staining of DNA backbone (blue) (c) stretching and alignment of labeled DNA-protein-QD complexes on a glass slide for fluorescence imaging (d) fluorescence image of linearized stained DNA (blue) cross-linked with the T7-RNA polymerase, scale bar=10 μm. Reproduced with permission from Reference (209). Copyright 2009 American Chemical Society.

Ebenstein *et al.* have employed QDs for high resolution, multicolor localization of DNA binding proteins bound to a DNA template (*209*) and enzymatically incorporated genomic tags for optical mapping of DNA-binding proteins (*210*). DNA binding proteins cross-linked to DNA (Figure 18(a)) were labeled with QDs via a biotin-strepatvidin link, while the DNA backbone was stained with a DNA staining fluorophore and specific reference sequences on the DNA were also labeled with QDs.(Figure 18(b)) Linearization of the stained DNA-protein-QD complex by molecular combing, which uses a receding meniscus to linearly extend DNA strands, allowed direct visualization of the location of the DNA binding proteins along the DNA strand using fluorescence microscopy (Figure 18(c) and (d)).

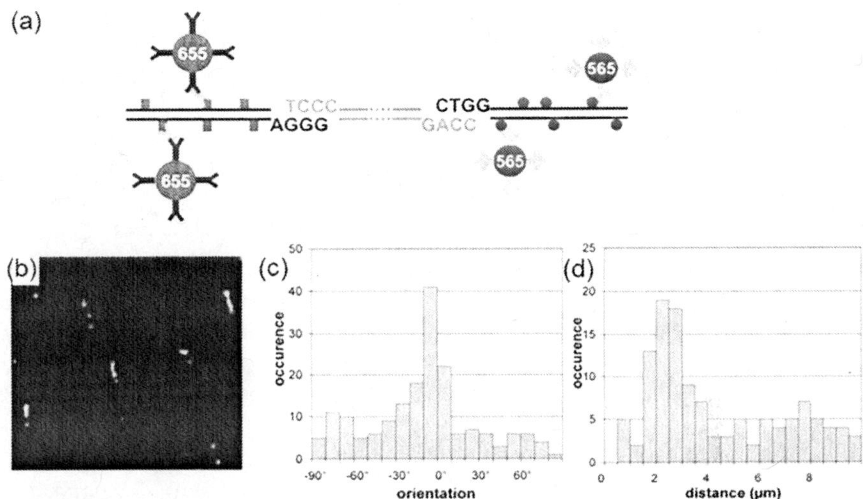

Figure 19. QD-method for the determination of the DNA sequence location and orientation:(a)scheme for dual-color labeling, showing the 6.3 kb long DNA restriction fragment (green) ligated at either end to short 0.5 kb long DNA fragments (black) modified either with biotin (red circles) to which streptavidin-coated QD565 can bind, or with digoxigenin (blue squares), to which a layer of mouse-anti-digoxigenin (light blue) can bind, followed by a layer of anti-mouse QD656 (b) fluorescence image of DNA linearized by combing over glass and labeled at one end with (biotin)-(streptavidin-QD565) (red), and at the other end with (digoxigenin)-(mouse-anti-digoxigenin)-(anti-mouse QD655) (blue), the DNA backbone was also stained in these images with the DNA staining dye YOYO-1 (green); analysis of the (c) angle and (d) distance between red-blue QD pairs for digoxigenin-biotin modified combed DNA. Reproduced with permission from Reference (211). Copyright 2005 Oxford University Press.

Fluorophores used for FISH and fiberFISH usually suffer from poor stability to photobleaching. DNA-staining-free detection of the position and orientation of elongated, linearized DNA molecules by multicolor QD labeling has been experimentally demonstrated by Crut *et al* (*211*). In this scheme, both ends of a DNA molecule were ligated with biotin and/or digoxigenin modified short DNA fragments that were produced by PCR, and linearized by combing over a glass substrate. The orientation and position of the elongated modified DNA strands were detected by fluorescence microscopy using streptavidin functionalized (CdSe/ZnS) QD565 on the biotin-labeled DNA and a two layer scheme, which consists of one layer of antidigoxigenin and a second layer of secondary-antibody coated QD655, on the other, digoxigenin-labeled DNA (Figure 19(a) and (b)). The distance between fluorescing 565 and 655 QD pairs and their and angle relative to the combing direction can then be analyzed using fluorescence microscopy (Figure 19 (c) and (d)).

Currently there are significant efforts to extend these approaches for mapping of DNA binding proteins in which QDs are applied as the photoluminescent label (*210*). Although in these cases the use of QDs for DNA analysis has been achieved on DNA strands which have been combed, there is a growing trend towards microfluidic and nanofluidic based systems for DNA analysis applications using proteins or DNA probes that are specific for DNA sequences (*212, 213*).

Conclusions

QDs provide better luminescent probes to organic fluorophores for DNA analysis methods where there is an advantage for single wavelength excitation of QDs with distinctly different emission spectra and where the analysis is based upon steady-state exposure conditions, *i.e.* for multiplexing applications and for emerging methods based upon DNA mapping (*160, 209, 210*). It is clear that the larger size of QDs, as compared to conventional fluorophores poses limitations for labeling a single DNA oligonucleotide with a single QD, as described by Algar *et al.* (*157*). However, the major problem of QDs, although partly solved, is the intermittence in fluorescence intensity, termed 'blinking'. Current approaches to improve both the photostability and blinking properties of these QDs requires the addition of a capping or coating layer which increases the size of the QD. Capping and\or coating of the QD increases the separation of donor-acceptor pairs applied to FRET and thus both limits and complicates DNA analysis assays as compared to fluorophore based FRET approaches. Currently there has been little research on the DNA damage caused by ROS that are generated in the presence of photo-illuminated QDs; it is well established that ROS yield DNA damage such as 8-oxoguanosine from guanosine residues and that this damaged residue pairs with adenosine in DNA (*214*). If approaches for single DNA molecule sequence analysis are to be advanced, then a better understanding of the photoprocesses that lead to ROS from QDs and approaches to suppress these are required. At this juncture there is still need of prudence in the use of QDs instead of fluorophores for DNA bioanalysis approaches, but it is clear that the necessary improvements will be driven by emerging genetic, epigenetic and diagnostic applications.

Acknowledgments

The Biotechnology and Biological Sciences Research Council (Funding reference BB/I023720/1 and BB/I022791/1) is acknowledged for funding the research in the field of this review. We are grateful to Alex Clark, Michael Carter and Tom Brown for helping to improve the clarity of the text in our manuscript.

References

1. Reinhardt, C. G.; Krugh, T. R. *Biochemistry* **1978**, *17* (23), 4845–4854.
2. Spielmann, H. P.; Wemmer, D. E.; Peter, J. *Biochemistry* **1995**, *34*, 8542–8553.
3. Larsson, A.; Carlsson, C.; Jonsson, M.; Albinsson, B. *J. Am. Chem. Soc.* **1994**, *116*, 8459–8465.
4. Jacobsen, J. P.; Pedersen, J. B.; Hansen, L. F.; Wemmer, D. E. *Nucleic Acids Res.* **1995**, *23* (5), 753–760.
5. Bunkenborg, J.; Stidsen, M. M.; Jacobsen, J. P. *Bioconjugate Chem.* **1999**, *10* (5), 824–831.
6. Kapuscinski, J. *Biotech. Histochem.* **1995**, *70* (5), 220–233.
7. Ploeger, L. S.; Dullens, H. F. J.; Huisman, A.; van Diest, P. J. *Biotech. Histochem.* **2008**, *83* (2), 63–69.
8. Tanious, F. A.; Veal, J. M.; Buczak, H.; Ratmeyer, L. S.; Wilson, W. D. *Biochemistry* **1992**, *31*, 3103–3112.
9. Pjura, P. E.; Grzeskowiak, K.; Dickerson, R. E. *J. Mol. Biol.* **1987**, *197* (2), 257–271.
10. Telser, J.; Cruickshank, K. A.; Morrison, L. E.; Netzel, T. L. *J. Am. Chem. Soc.* **1989**, *111* (18), 6966–6976.
11. Nygren, J.; Svanvik, N.; Kubista, M. *Biopolymers* **1998**, *46*, 39–51.
12. Kapuscinski, J.; Skoczylas, B. *Nucleic Acids Res.* **1978**, *5*, 3775–3799.
13. Rye, H. S.; Y., S.; Wemmer, D. E.; Quesada, M. A.; Haugland, R. P.; Mathies, R. A.; Glazer, A. N. *Nucleic Acids Res.* **1992**, *20* (11), 2803–2812.
14. Turro, N. J. *Modern Molecular Photochemistry*; University Science Books: Sausalito, California, 1991; p 628.
15. Resch-Genger, U.; Grabolle, M.; Cavaliere-Jaricot, S.; Nitschke, R.; Nann, T. *Nat. Methods* **2008**, *5* (9), 763–775.
16. Lacoste, T. D.; Michalet, X.; Pinaud, F.; Chemla, D. S.; Alivisatos, A. P.; Weiss, S. *Proc. Natl. Acad. Sci. U.S.A.* **2000**, *97* (17), 9461–9466.
17. Michalet, X.; Lacoste, T. D.; Pinaud, F.; Chemla, D. S.; Alivisatos, A. P.; Weiss, S., Ultrahigh resolution multicolor colocalization of single fluorescent nanocrystals. In *Nanoparticles and Nanostructured Surfaces: Novel Reporters with Biological Applications*; Murphy, C. J., Ed.; SPIE: 2001; Vol. 2, pp 8–15.
18. Michalet, X.; Lacoste, T. D.; Weiss, S. *Methods* **2001**, *25* (1), 87–102.
19. Yan, J.; Estévez, M. C.; Smith, J. E.; Wang, K.; He, X.; Wang, L.; Tan, W. *Nano Today* **2007**, *2*, 44–50.
20. Zhao, X.; Tapec-Dytioco, R.; Tan, W. *J. Am. Chem. Soc.* **2003**, *125*, 11474–11475.
21. Zhou, X.; Zhou, J. *Anal. Chem.* **2004**, *76*, 5302–5312.
22. Santra, S.; Zhang, P.; Wang, K.; Tapec, R.; Tan, W. *Anal. Chem.* **2001**, *73*, 4988–4993.
23. Einevoll, G. T. *Phys. Rev. B* **1992**, *45* (7), 3410–3417.
24. Brus, L. *J. Phys. Chem.* **1986**, *90* (12), 2555–2560.
25. Brus, L. E. *J. Chem. Phys.* **1983**, *79* (11), 5566–5571.

26. Dabbousi, B. O.; Rodriguez-Viejo, J.; Mikulec, F. V.; Heine, J. R.; Mattoussi, H.; Ober, R.; Jensen, K. F.; Bawendi, M. G. *J. Phys. Chem. B* **1997**, *101*, 9463–9475.

27. Xing, Y.; Rao, J. *Cancer Biomarkers* **2008**, *4*, 307–319.

28. Montón, H.; Nogués, C.; Rossinyol, E.; Castell, O.; Roldán, M. *J. Nanobiotechnol.* **2009**, *7*, 4.

29. Grecco, H. E.; Lidke, K. A.; Heintzmann, R.; Lidke, D. S.; Spagnuolo, C.; Martinez, O. E.; Jares-Erijman, E. A.; Jovin, T. M. *Microsc. Res. Tech.* **2004**, *65*, 169–179.

30. Schlegel, G.; Bohnenberger, J.; Potapova, I.; Mews, A. *Phys. Rev. Lett.* **2002**, *88* (13), 137401.

31. Zhang, K.; Chang, H.; Fu, A.; Alivisatos, A. P.; Yang, H. *Nano Lett.* **2006**, *6* (1), 843–847.

32. Brokmann, X.; Messin, G.; Desbiolles, P.; Giacobino, E.; Dahan, M.; Hermier, J. P. *New Journal of Physics* **2004**, *6* (99), 1–8.

33. Pokrant, S.; Whaley, K. B. *Eur. Phys. J. D* **1999**, *6*, 255–267.

34. Underwood, D. F.; Kippeny, T.; Rosenthal, S. J. *J. Phys. Chem. B* **2001**, *105*, 436–443.

35. Qu, L.; Peng, X. *J. Am. Chem. Soc.* **2002**, *124* (9), 2049–2055.

36. Talapin, D. V.; Mekis, I.; Goetzinger, S.; Kornowski, A.; Benson, O.; Weller, H. *J. Phys. Chem. B* **2004**, *108*, 18826–18831.

37. Ranasinghe, R. T.; Brown, T. *Chem. Commun.* **2011**, *47* (13), 3717–3735.

38. Raap, A. K. *Mutat. Res., Fundam. Mol. Mech. Mutagen.* **1998**, *400* (1-2), 287–298.

39. Raap, A. K.; Tanke, H. J. *Cytogenet. Genome Res.* **2006**, *114* (3-4), 222–226.

40. Stoughton, R. B. Applications of DNA microarrays in biology. *Annu. Rev. Biochem.* **2005**, 74, 53–82.

41. Ranasinghe, R. T.; Brown, T. *Chem. Commun.* **2005** (44), 5487–5502.

42. Vogelsang, J.; Kasper, R.; Steinhauer, C.; Person, B.; Heilemann, M.; Sauer, M.; Tinnefeld, P. *Angew. Chem. Int. Ed.* **2008**, *47*, 5465–5469.

43. Altman, R. B.; Terry, D. S.; Zhou, Z.; Zheng, Q.; Geggier, P.; Kolster, R. A.; Zhao, Y.; Javitch, J. A.; Warren, J. D.; Blanchard, S. C. *Nat. Methods* **2012**, *9* (1), 68–73.

44. Hoogenboom, J. P.; Dijk, E. M. H. P. v.; Hernando, J.; Hulst, N. F. v.; Garcia-Parajo, M. F. *Phys. Rev. Lett.* **2005**, *95*, 097401.

45. Sark, W. G. J. H. M. v.; Frederix, P. L. T. M.; Bol, A. A.; Gerritsen, H. C.; Meijerink, A. *ChemPhysChem* **2002**, *3*, 817–879.

46. Sark, W. G. J. H. M. v.; Frederix, P. L. T. M.; Heuvel, D. J. V. d.; Gerritsen, H. C. *J. Phys. Chem. B* **2001**, *105*, 8281–8284.

47. Katari, J. E. B.; Colvin, V. L.; Alivisatos, A. P. *J. Phys. Chem.* **1994**, *98*, 4109–4117.

48. Zhang, Y.; He, J.; Wang, P.-N.; Chen, J.-Y.; Lu, Z.-J.; Lu, D.-R.; Guo, J.; Wang, C.-C.; Yang, W.-L. *J. Am. Chem. Soc.* **2006**, *128*, 13396–13401.

49. Metz, K. M.; Mangahm, A. N.; Bierman, M. J.; Jin, S.; Hamers, R. J.; Pedersen, J. A. *Environ. Sci. Technol.* **2009**, *43*, 1598–1604.

50. Pechstedt, K.; Whittle, T.; Baumberg, J.; Melvin, T. *J. Phys. Chem. C* **2010**, *114*, 12069–12077.

51. Nirmal, M.; Dabbousi, B. O.; Bawendi, M. G.; Macklin, J. J.; Trautman, J. K.; Harris, T. D.; Brus, L. E. *Nature* **1996**, *383*, 802–804.

52. Chen, Y.; Vela, J.; Htoon, H.; Casson, J. L.; Werder, D. J.; Bussian, D. A.; Klimov, V. I.; Hollingsworth, J. A. *J. Am. Chem. Soc.* **2008**, *130*, 5026–5027.

53. Wu, X.; Liu, H.; Liu, J.; Haley, K. N.; A.Treadway, J.; Larson, J. P.; Ge, N.; Peale, F.; Bruchez, M. P. *Nat. Biotechnol.* **2003**, *21*, 41–46.

54. Lee, T.-H.; Lapidus, L. J.; Zhao, W.; Travers, K. J.; Herschlag, D.; Chu, S.; Zhang, Y. *Biophys. J.* **2007**, *92*, 3275–3283.

55. Briggs, M.; Roder, H. *Proc. Natl. Acad. Sci. U.S.A.* **1992**, *89*, 2017–2021.

56. Veitshans, T.; Klimov, D.; Thirumalai, D. *Folding Des.* **1997**, *2* (1), 1–22.

57. Sabanayagam, C. R.; Eid, J. S.; Meller, A. *J. Chem. Phys.* **2005**, *123*, 224708.

58. Hohng, S.; Ha, T. *J. Am. Chem. Soc.* **2004**, *126*, 1324–1325.

59. Yeow, E. K. L.; Melnikov, S. M.; Bell, T. D. M.; Schryver, F. C. D.; Hofkens, J. *J. Phys. Chem. A* **2006**, *110*, 1726–1734.

60. Fu, Y.; Zhang, J.; Lakowicz, J. R. *Langmuir* **2008**, *24* (7), 3429–3433.

61. Kuno, M.; Fromm, D. P.; Hamann, H. F.; Gallagher, A.; Nesbitt, D. J. *J. Chem. Phys.* **2000**, *112* (7), 3117–3120.

62. Frantsuzov, P. A.; Marcus, R. A. *Phys. Rev. B* **2005**, *72*, 155321.

63. Margolin, G.; Protasenko, V.; Kuno, M.; Barkai, E. *J. Phys. Chem. B* **2006**, *110*, 19053–19060.

64. Brokmann, X.; Hermier, J.-P.; Messin, G.; Desbiolles, P.; Bouchaud, J.-P.; Dahan, M. *Phys. Rev. Lett.* **2003**, *90* (12), 120601.

65. Ha, T.; Tinnefeld, P. Photophysics of Fluorescent Probes for Single-Molecule Biophysics and Super-Resolution Imaging. *Annu. Rev. Phys. Chem.* **2012**, *63*, 595–617.

66. Bagshaw, C. R.; Cherny, D. *Biochem. Soc. Trans.* **2006**, *34* (5), 979–982.

67. Weston, K. D.; Carson, P. J.; DeAro, J. A.; Buratto, S. K. *Chem. Phys. Lett.* **1999**, *308*, 58–64.

68. Zondervan, R.; Kulzer, F.; Orlinskii, S. B.; Orrit, M. *J. Phys. Chem. A* **2003**, *107*, 6770–6776.

69. Widengren, J.; Chmyrov, A.; Eggeling, C.; Lofdahl, P.-A.; Seidel, C. A. M. *J. Phys. Chem. A* **2007**, *111*, 429–440.

70. Schuster, J.; Cichos, F.; Borczyskowski, C. v. *Opt. Spectrosc.* **2005**, *98* (5), 712–171.

71. Song, L.; Hennink, E. J.; Young, T.; Tanke, H. J. *Biophys. J.* **1995**, *68*, 2588–2600.

72. Wilkinson, F.; McGarvey, D. J.; Olea, A. F. *J. Am. Chem. Soc.* **1993**, *115*, 12144–12151.

73. Juzenas, P.; Chen, W.; Sun, Y.-P.; Coelho, M. A. N.; Generalov, R.; Generalova, N.; Christensen, I. L. *Adv. Drug Delivery Rev.* **2008**, *60*, 1600–1614.

74. DeRosa, M. C.; Crutchley, R. J. *Coord. Chem. Rev.* **2002**, *233-234*, 351–371.

75. Olea, A. F.; Wilkinson, F. *J. Phys. Chem.* **1995**, *99*, 4518–4524.

76. Aitken, C. E.; Marshall, R. A.; Puglisiz, J. D. *Biophys. J.* **2008**, *94*, 1826–1835.

77. Zhang, Y.; Aslan, K.; Previte, M. J. R.; Geddes, C. D. *Appl. Phys. Lett.* **2007**, *91*, 023114.

78. Zhang, Y.; Aslan, K.; Previte, M. J. R.; Geddes, C. D. *Proc. Natl. Acad. Sci. U.S.A.* **2008**, *105* (6), 1798–1802.

79. Wang, X.; Ren, X.; Kahen, K.; Hahn, M. A.; Rajeswaran, M.; Maccagnano-Zacher, S.; Silcox, J.; Cragg, G. E.; Efros, A. L.; Krauss, T. D. *Nature* **2009**, *459*, 686–689.

80. Kraus, R. M.; Lagoudakis, P. G.; Muller, J.; Rogach, A. L.; Lupton, J. M.; Feldmann, J.; Talapin, D. V.; Weller, H. *J. Phys. Chem. B* **2005**, *109*, 18214–18217.

81. Efros, A. L. *Nat. Mater.* **2008**, *7*, 612–613.

82. Lounis, B.; Bechtel, H. A.; Gerion, D.; Alivisatos, P.; Moerner, W. E. *Chem. Phys. Lett.* **2000**, *329*, 399–404.

83. Mahler, B.; Spinicelli, P.; Buil, S.; Quelin, X.; Hermier, J.-P.; Dubertret, B. *Nat. Mater.* **2008**, *7*, 659–664.

84. Yu, Z.; Guo, L.; Du, H.; Krauss, T.; Silcox, J. *Nano Lett.* **2005**, *5* (4), 565–570.

85. Fomenko, V.; Nesbitt, D. J. *Nano Lett.* **2008**, *8* (1), 287–293.

86. Hammer, N. I.; Early, K. T.; Sill, K.; Odoi, M. Y.; Emrick, T.; Barnes, M. D. *J. Phys. Chem. B* **2006**, *110*, 14167–14171.

87. Early, K. T.; McCarthy, K. D.; Hammer, N. I.; Odoi, M. Y.; Tangirala, R.; Emrick, T.; Barnes, M. D. *Nanotechnology* **2007**, *18*, 424027.

88. Zhang, Y. J.; Clapp, A. *Sensors* **2011**, *11* (12), 11036–11055.

89. Jeong, S.; Achermann, M.; Nanda, J.; Ivanov, S.; Klimov, V. I.; Hollingsworth, J. A. *J. Am. Chem. Soc.* **2005**, *127*, 10126–10127.

90. Obonyo, O.; Fisher, E.; Edwards, M.; Douroumis, D. *Crit. Rev. Biotechnol.* **2010**, *30* (4), 283–301.

91. Kuang, H.; Zhao, Y.; Ma, W.; Xu, L. G.; Wang, L. B.; Xu, C. L. *TrAC, Trends Anal. Chem.* **2011**, *30* (10), 1620–1636.

92. Quarta, A.; Curcio, A.; Kakwere, H.; Pellegrino, T. *Nanoscale* **2012**, *4* (11), 3319–3334.

93. Bingbo, Z.; Da, X.; Chao, L.; Fangfang, G.; Peng, Z.; Xuejun, W.; Zhihao, B.; Donglu, S. *J. Nanopart. Res.* **2011**, *13* (6), 2407–2415.

94. Darbandi, M.; Urban, G.; Krueger, M. *J. Colloid Interface Sci.* **2012**, *365* (1), 41–45.

95. Koole, R.; van Schooneveld, M. M.; Hilhorst, J.; Donega, C. d. M.; t Hart, D. C.; van Blaaderen, A.; Vanmaekelbergh, D.; Meijerink, A. *Chem. Mater.* **2008**, *20* (7), 2503–2512.

96. Shimizu, K. T.; K.Woo, W.; Fisher, B. R.; Eisler, H. J.; Bawendi, M. G. *Phys. Rev. Lett.* **2002**, *89* (11), 117401.

97. Ito, Y.; Matsuda, K.; Kanemitsu, Y. *Phys. Rev. B* **2007**, *75*, 033309.

98. Fu, Y.; Zhang, J.; Lakowicz, J. R. *Chem. Phys. Lett.* **2007**, *44*, 96–100.

99. Yuan, C. T.; Yu, P.; Tang, J. *Appl. Phys. Lett.* **2009**, *94*, 243108.

100. Müller, J.; Lupton, J. M.; Rogach, A. L.; Feldmann, J.; Talapin, D. V.; Weller, H. *Appl. Phys. Lett.* **2004**, *85* (3), 381–383.

101. Koberling, F.; Mews, A.; Basch, T. *Adv. Mater.* **2001**, *13* (9), 672–676.

102. Ipe, B. I.; Lehnig, M.; Niemeyer, C. M. *Small* **2005**, *1* (7), 706–709.

103. Green, M.; Howman, E. *Chem. Commun.* **2005** (1), 121–123.

104. Cooper, D. R.; Dimitrijevic, N. M.; Nadeau, J. L. *Nanoscale* **2010**, *2*, 114–121.

105. Aust, A. E.; Eveleigh, J. F. *Proc. Soc. Exp. Biol. Med.* **1999**, *222*, 246–252.

106. Chen, J.-Y.; Lee, Y.-M.; Zhao, D.; Mak, N.-K.; Wong, R. N.-S.; Chan, W.-H.; Cheung, N.-H. *Photochem. Photobiol.* **2010**, *86*, 431–437.

107. Duran, J. D. G.; Guindo, M. C.; Delgado, A. V. *J. Colloid Interface Sci.* **1995**, *173*, 436–442.

108. Daimon, T.; Nosaka, Y. *J. Phys. Chem. C* **2007**, *111*, 4420–4424.

109. Samia, A. C. S.; Chen, X.; Burda, C. *J. Am. Chem. Soc.* **2003**, *125*, 15736–15737.

110. Ma, J.; Chen, J.-Y.; Idowu, M.; Nyokong, T. *J. Phys. Chem. B* **2008**, *112*, 4465–4469.

111. Tsay, J. M.; Trzoss, M.; Shi, L.; Kong, X.; Selke, M.; Jung, M. E.; Weiss, S. *J. Am. Chem. Soc.* **2007**, *129*, 6865–6871.

112. Cordero, S. R.; Carson, P. J.; Estabrook, R. A.; Strouse, G. F.; Buratto, S. K. *J. Phys. Chem. B* **2000**, *104*, 12137–12142.

113. Jones, M.; Nedeljkovic, J.; Ellingson, R. J.; Nozik, A. J.; Rumbles, G. *J. Phys. Chem. B* **2003**, *107*, 11345–11352.

114. Oda, M.; Hasegawa, A.; Iwami, N.; Nishiura, K.; Ando, N.; Nishiyama, A.; Horiuchi, H.; Tani, T. *Colloids Surf., B* **2007**, *56*, 241–245.

115. Oda, M.; Tsukamoto, J.; Hasegawa, A.; Iwami, N.; Nishiura, K.; Hagiwara, I.; Ando, N.; Horiuchi, H.; Tani, T. *J. Lumin.* **2007**, *122-123*, 762–765.

116. Uematsu, T.; Maenosono, S.; Yamaguchi, Y. *J. Phys. Chem.* **2005**, *109*, 8613–8618.

117. Maenosono, S.; Ozaki, E.; Yoshie, K.; Yamaguchi, Y. *Jpn. J. Appl. Phys.* **2001**, *40*, L638–L641.

118. Maenosono, s. *Chem. Phys. Lett.* **2005**, *405*, 182–186.

119. Kimura, J.; Uematsu, T.; Maenosono, S.; Yamaguchi, Y. *J. Phys. Chem. B* **2004**, *108*, 13258–13264.

120. Uematsu, T.; Kimura, J.; Yamaguchi, Y. *Nanotechnology* **2004**, *15*, 822–827.

121. Wang, Y.; Tang, Z.; Correa-Duarte, M. A.; Pastoriza-Santos, I.; Giersig, M.; Kotov, N. A.; Liz-Marzan, L. M. *J. Phys. Chem. B* **2004**, *108*, 15461–15469.

122. Valledor Llopis, M.; Campo Rodriguez, J. C.; Ferrero Martin, F. J.; Maria Coto, A.; Fernandez-Argueelles, M. T.; Costa-Fernandez, J. M.; Sanz-Medel, A. *Nanotechnology* **2011**, *22* (38), 385703.

123. Zhang, Y.; Clapp, A. *Sensors* **2011**, *11* (12), 11036–11055.

124. Chan, W.-H.; Shiao, N.-H.; Lu, P.-Z. *Toxicol. Lett.* **2006**, *167*, 191–200.

125. Li, Y.; Zhang, W.; Li, K.; Yao, Y.; Niu, J.; Chen, Y. *Environ. Pollut.* **2012**, *164*, 259–266.

126. Kim, S. Y.; Kim, E. J.; Woo, J. *J. Biochem. Mol. Biol.* **2002**, *35* (4), 353–357.

127. Triantaphylides, C.; Krischke, M.; Hoeberichts, F. A.; Ksas, B.; Gresser, G.; Havaux, M.; Breusegem, F. V.; Mueller, M. J. *Plant Physiol.* **2008**, *148*, 960–968.

128. Choi, A. O.; Cho, S. J.; Desbarats, J.; Lovrić, J.; Maysinger, D. *J. Nanobiotechnol.* **2007**, *5* (1).

129. Von Sonntag, C. *The Chemical Basis of Radiation Biology*; Taylor and Francis: London, 1987.

130. Green, M.; Howman, E. *Chem. Commun.* **2005** (1), 121–123.

131. Anas, A.; Akita, H.; Harashima, H.; Itoh, T.; Ishikawa, M.; Biju, V. *J. Phys. Chem. B* **2008**, *112* (32), 10005–10011.

132. Cuypers, A.; Plusquin, M.; Remans, T.; Jozefczak, M.; Keunen, E.; Gielen, H.; Opdenakker, K.; Nair, A. R.; Munters, E.; Artois, T. J.; Nawrot, T.; Vangronsveld, J.; Smeets, K. *Biometals* **2010**, *23* (5), 927–940.

133. Yamazaki, Y.; Zinchenko, A. A.; Murata, S. *Nanoscale* **2011**, *3* (7), 2909–2915.

134. Hartwig, A. *Environ. Health Perspect.* **1994**, *102*, 45–50.

135. Kricka, L. J.; Fortina, P. *Clin. Chem.* **2009**, *55* (4), 670–683.

136. Hermanson, G. T. *Bioconjugate Techniques*; Academic Press: New York, 1996.

137. Eid, J.; Fehr, A.; Gray, J.; Luong, K.; Lyle, J.; Otto, G.; Peluso, P.; Rank, D.; Baybayan, P.; Bettman, B.; Bibillo, A.; Bjornson, K.; Chaudhuri, B.; Christians, F.; Cicero, R.; Clark, S.; Dalal, R.; Dewinter, A.; Dixon, J.; Foquet, M.; Gaertner, A.; Hardenbol, P.; Heiner, C.; Hester, K.; Holden, D.; Kearns, G.; Kong, X. X.; Kuse, R.; Lacroix, Y.; Lin, S.; Lundquist, P.; Ma, C. C.; Marks, P.; Maxham, M.; Murphy, D.; Park, I.; Pham, T.; Phillips, M.; Roy, J.; Sebra, R.; Shen, G.; Sorenson, J.; Tomaney, A.; Travers, K.; Trulson, M.; Vieceli, J.; Wegener, J.; Wu, D.; Yang, A.; Zaccarin, D.; Zhao, P.; Zhong, F.; Korlach, J.; Turner, S. *Science* **2009**, *323* (5910), 133–138.

138. Sinkeldam, R. W.; Greco, N. J.; Tor, Y. *Chem. Rev.* **2010**, *110* (5), 2579–2619.

139. Murphy, M. C.; Rasnik, I.; Cheng, W.; Lohman, T. M.; Ha, T. *Biophys. J.* **2004**, *86*, 2530–2537.

140. Iqbal, A.; Arslan, S.; Okumus, B.; Wilson, T. J.; Giraud, G.; Norman, D. G.; Ha, T.; Lilley, D. M. J. *Proc. Natl. Acad. Sci. U.S.A.* **2008**, *105* (32), 11176–11181.

141. Sanborn, M. E.; Connolly, B. K.; Gurunathan, K.; Levitus, M. *J. Phys. Chem. B* **2007**, *111*, 11064–11074.

142. Bates, M.; Huang, B.; Dempsey, G. T.; Zhuang, X. *Science* **2007**, *317* (5845), 1749–1753.

143. Horn, P. J.; Peterson, C. L. *Science* **2002**, *297* (5588), 1824–1827.

144. Lee, J.; Choi, Y.; Kim, J.; Park, E.; Song, R. *ChemPhysChem.* **2009**, *10*, 806–811.

145. Mahtab, R.; Harden, H. H.; Murphy, C. J. *J. Am. Chem. Soc.* **2000**, *122* (1), 14–17.

146. Berg, M. A.; Coleman, R. S.; Murphy, C. J. *Phys. Chem. Chem. Phys.* **2008**, *10* (9), 1229–1242.

147. Kotov, N. A. *Science* **2010**, *330* (6001), 188–189.

148. Mahtab, R.; Sealey, S. M.; Hunyadi, S. E.; Kinard, B.; Ray, T.; Murphy, C. J. *J. Inorg. Biochem.* **2007**, *101* (4), 559–564.

149. Wang, L.; Zhu, Y.; Jiang, Y.; Qiao, R.; Zhu, S.; Chen, W.; Xu, C. *J. Phys. Chem. B* **2009**, *113* (21), 7637–7641.

150. Liang, G.; Ma, C.; Zhu, Y.; Li, S.; Shao, Y.; Wang, Y.; Xiao, Z. *Nanoscale Res. Lett.* **2011**, *6*.

151. Ma, L.; He, S.; Huang, J.; Cao, L.; Yang, F.; Li, L. *Biochimie* **2009**, *91* (8), 969–973.

152. Li, M.; Lin, Y. C.; Wu, C. C.; Liu, H. S. *Nucleic Acids Res.* **2005**, *33* (21), e184.

153. Lim, S. H.; Buchy, P.; Mardy, S.; Kang, M. S.; Yu, A. D. C. *Anal. Chem.* **2010**, *82*, 886–891.

154. Liedl, T.; Dietz, H.; Yurke, B.; Simmel, F. C. *Small* **2007**, *3* (10), 1688–1693.

155. Giraud, G.; Schulze, H.; Bachmann, T. T.; Campbell, C. J.; Mount, A. R.; Ghazal, P.; Khondoker, M. R.; Ember, S. W. J.; Ciani, I.; Tlili, C.; Walton, A. J.; Terry, J. G.; Crain, J. *Chem. Phys. Lett.* **2010**, *484*, 309–314.

156. Zhang, Y.; Liu, K. J.; Wang, T.-L.; Shih, I.-M.; Wang, T.-H. *ACS Nano* **2012**, *6* (1), 858–864.

157. Algar, W. R.; Susumu, K.; Delehanty, J. B.; Medintz, I. L. *Anal. Chem.* **2011**, *83* (23), 8826–8837.

158. Boeneman, K.; Deschamps, J. R.; Buckhout-White, S.; Prasuhn, D. E.; Blanco-Canosa, J. B.; Dawson, P. E.; Stewart, M. H.; Susumu, K.; Goldman, E. R.; Ancona, M.; Medintz, I. L. *ACS Nano* **2010**, *4*, 7253–7266.

159. Dyadyusha, L.; Yin, H.; Jaiswal, S.; Brown, T.; Baumberg, J. J.; Booy, F. P.; Melvin, T. *Chem. Commun.* **2005**, 3201–3203.

160. Antelman, J.; Ebenstein, Y.; Dertinger, T.; Michalet, X.; Weiss, S. *J. Phys. Chem. C* **2009**, *113* (27), 11541–11545.

161. Zanchet, D.; Micheel, C. M.; Parak, W. J.; Gerion, D.; Williams, S. C.; Alivisatos, A. P. *J. Phys. Chem. B* **2002**, *106* (45), 11758–11763.

162. He, S.; Huang, B.-H.; Tan, J.; Luo, Q.-Y.; Lin, Y.; Li, J.; Hu, Y.; Zhang, L.; Yan, S.; Zhang, Q.; Pang, D.-W.; Li, L. *Biomaterials* **2011**, *32*, 5471–5477.

163. Pinaud, F.; Michalet, X.; Bentolila, L. A.; Tsay, J. M.; Doose, S.; Li, J. J.; Iyer, G.; Weiss, S. *Biomaterials* **2006**, *27*, 1679–1687.

164. Pons, T.; Medintz, I. L.; Wang, X.; English, D. S.; Mattoussi, H. *J. Am. Chem. Soc.* **2006**, *128*, 15324–15331.

165. Sarkar, R.; Narayanan, S. S.; Palsson, L.-O.; Dias, F.; Monkman, A.; Pal, S. K. *J. Phys. Chem. B* **2007**, *111*, 12294–12298.

166. Chan, W. C. W.; Nie, S. *Science* **1998**, *281*, 2016–2018.

167. Gerion, D.; Chen, F.; Kannan, B.; Fu, A.; Parak, W. J.; Chen, D. J.; Majumdar, A.; Alivisatos, A. P. *Anal. Chem.* **2003**, *75*, 4766–4772.

168. Strother, T.; Hamers, R. J.; Smith, L. M. *Nucleic Acids Res.* **2000**, *28* (18), 3535–3541.

169. Yin, H. B.; Brown, T.; Wilkinson, J. S.; Eason, R. W.; Melvin, T. *Nucleic Acids Res.* **2004**, *32* (14), e118–7.

170. Song, W.; Lau, C.; Lu, J. *Analyst* **2012**, *137*, 1611–1617.

171. Han, M.; Gao, X.; Su, J. Z.; Nie, S. *Nat. Biotechnol.* **2001**, *19*, 631–635.

172. Xu, H.; Sha, M. Y.; Wong, E. Y.; Uphoff, J.; Xu, Y.; Treadway, J. A.; Truong, A.; O'Brien, E.; Asquith, S.; Stubbins, M.; Spurr, N. K.; Lai, E. H.; Mahoney, W. *Nucleic Acids Res.* **2003**, *31* (8), e43.

173. Carson, S.; Cohen, A. S.; Belenkii, A.; Ruizmartinez, M. C.; Berka, J.; Karger, B. L. *Anal. Chem.* **1993**, *65* (22), 3219–3226.

174. Liu, Y.-H.; Tsai, Y.-Y.; Chien, H.-J.; Chen, C.-Y.; Huang, Y.-F.; Chen, J.-S.; Wu, Y.-C.; Chen, C.-C. *Nanotechnology* **2011**, *22*, 155102.

175. Medintz, I. L.; Mattoussi, H. *Phys. Chem. Chem. Phys.* **2009**, *11* (1), 17–45.

176. Algar, W. R.; Tavares, A. J.; Krull, U. J. *Anal. Chim. Acta* **2010**, *673* (1), 1–25.

177. Algar, W. R.; Krull, U. J. *Anal. Bioanal. Chem.* **2008**, *391* (5), 1609–1618.

178. Solinas, A.; Brown, L. J.; McKeen, C.; Mellor, J. M.; Nicol, J. T. G.; Thelwell, N.; Brown, T. *Nucleic Acids Res.* **2001**, *29* (20), e96.

179. Thelwell, N.; Millington, S.; Solinas, A.; Booth, J.; Brown, T. *Nucleic Acids Res.* **2000**, *28* (19), 3752–3761.

180. Cardullo, R. A.; Agrawal, S.; Flores, C.; Zamecnik, P. C.; Wolf, D. E. *Proc. Natl. Acad. Sci. U.S.A.* **1988**, *85* (23), 8790–8794.

181. Holland, P. M.; Abramson, R. D.; Watson, R.; Gelfand, D. H. *Proc. Natl. Acad. Sci. U.S.A.* **1991**, *88* (16), 7276–7280.

182. Guo, J.; Ju, J. Y.; Turro, N. J. *Anal. Bioanal. Chem.* **2012**, *402* (10), 3115–3125.

183. Zhang, C.-Y.; Yeh, H.-C.; Kuroki, M. T.; Wang, T.-H. *Nat. Mater.* **2005**, *4*, 826–831.

184. Zhang, C.-Y.; Hu, J. *Anal. Chem.* **2010**, *82*, 1921–1927.

185. Gill, R.; Willner, I.; Shweky, I.; Banin, U. *J. Phys. Chem. B* **2005**, *109*, 23715–23719.

186. Peng, H.; Zhang, L.; Kjallman, T. H. M.; Soeller, C.; Travas-Sejdic, J. *J. Am. Chem. Soc.* **2006**, *129*, 3048–3049.

187. Tyagi, S.; Kramer, F. R. *Nat. Biotechnol.* **1996**, *14* (3), 303–308.

188. Kim, J. H.; Morikis, D.; Ozkan, M. *Sens. Actuators, B* **2004**, *102* (2), 315–319.

189. Wu, C. S.; Oo, M. K. K.; Cupps, J. M.; Fan, X. D. *Biosens. Bioelectron.* **2011**, *26* (9), 3870–3875.

190. Hildebrandt, N.; Charbonniere, L. J.; Loehmannsroeben, H.-G. *J. Biomed. Biotechnol.* **2007**, 79169.

191. Algar, W. R.; Wegner, D.; Huston, A. L.; Blanco-Canosa, J. B.; Stewart, M. H.; Armstrong, A.; Dawson, P. E.; Hildebrandt, N.; Medintz, I. L. *J. Am. Chem. Soc.* **2012**, *134* (3), 1876–1891.

192. Dyadyusha, L.; Yin, H.; Jaiswal, S.; Brown, T.; Baumberg, J. J.; Booy, F. P.; Melvin, T. *Chem. Commun.* **2005**, 3201–3203.

193. Gueroui, Z.; Libchaber, A. *Phys. Rev. Lett.* **2004**, *93* (16), 166108.

194. Pons, T.; Medintz, I. L.; Sapsford, K. E.; Higashiya, S.; Grimes, A. F.; English, D. S.; Mattoussi, H. *Nano Lett.* **2007**, *7* (10), 3157–3164.

195. Li, X.; Qian, J.; Jiang, L.; He, S. *Appl. Phys. Lett.* **2009**, *94* (6), 063111.

196. Jennings, T. L.; Schlatterer, J. C.; Singh, M. P.; Greenbaum, N. L.; Strouse, G. F. *Nano Lett.* **2006**, *6* (7), 1318–1324.

197. Yeh, H.-Y.; Yates, M. V.; Mulchandania, A.; Chen, W. *Chem. Commun.* **2010**, *46* (22), 3914–3916.

198. Li, Y.-Q.; Guan, L.-Y.; Zhang, H.-L.; Chen, J.; Lin, S.; Ma, Z.-Y.; Zhao, Y.-D. *Anal. Chem.* **2011**, *83* (11), 4103–4109.

199. Zhou, D.; Ying, L.; Hong, X.; Hall, E. A.; Abell, C.; Klenerman, D. *Langmuir* **2008**, *24*, 1659–1664.

200. Bailey, V. J.; Easwaran, H.; Zhang, Y. *Genome Res.* **2009**, *19*, 1455–1461.

201. Thompson, J. F.; Milos, P. M. *Genome Biol.* **2011**, *12* (2), 217.

202. Pennisi, E. *Science* **2010**, *327* (5970), 1190–1190.

203. Ioannou, D.; Tempest, H. G.; Skinner, B. M.; Thornhill, A. R.; Ellis, M.; Griffin, D. K. *Chromosome Res.* **2009**, *17* (4), 519–530.

204. Xing, M.; Shen, H. B.; Zhao, W.; Liu, Y. F.; Du, Y. D.; Yu, Z. X.; Chen, X. *Biotechniques* **2011**, *50* (4), 259–261.

205. Bentolila, L. A. Direct In Situ Hybridization with Oligonucleotide Functionalized Quantum Dot Probes. In *Fluorescence in Situ Hybridization*; Bridger, J. M., Volpi, E. V., Eds.; Humana Press Inc: Totowa, 2010; Vol. 659, pp 147–163

206. Choi, Y.; Kim, H. P.; Hong, S. M.; Ryu, J. Y.; Han, S. J.; Song, R. *Small* **2009**, *5* (18), 2085–2091.

207. Ioannou, A.; Eleftheriou, I.; Lubatti, A.; Charalambous, A.; Skourides, P. A. *J. Biomed. Biotechnol.* **2012**.

208. Toprak, E.; Selvin, P. R. *Annu. Rev. Biophys. Biomol. Struct.* **2007**, *36*, 349–369.

209. Ebenstein, Y.; Gassman, N.; Kim, S.; Antelman, J.; Kim, Y.; Ho, S.; Samuel, R.; Michalet, X.; Weiss, S. *Nano Lett.* **2009**, *9* (4), 1598–1603.

210. Kim, S.; Gottfried, A.; Lin, R. R.; Dertinger, T.; Kim, A. S.; Chung, S.; Colyer, R. A.; Weinhold, E.; Weiss, S.; Ebenstein, Y. *Angew. Chem., Int. Ed.* **2012**, *51* (15), 3578–3581.

211. Crut, A.; Geron-Landre, B.; Bonnet, I.; Bonneau, S.; Desbiolles, P.; Escude, C. *Nucl. Acids Res.* **2005**, *33* (11), e98.

212. Tza-Huei, W.; Bailey, V.; Liu, K. *Proc. SPIE - Int. Soc. Opt. Eng.* **2011**, *8031*.

213. Sano, T.; Okamoto, Y.; Kaji, N.; Tokeshi, M.; Baba, Y. *Micro Total Anal. Syst.* **2011**, 647–649.

214. Epe, B. *Chem.-Biol. Interact.* **1991**, *80* (3), 239–260.

Chapter 14

Complement Sensing of Nanoparticles and Nanomedicines

Peter P. Wibroe[1,2] and S. Moein Moghimi*[,1,2]

[1]Centre for Pharmaceutical Nanotechnology and Nanotoxicology, Faculty of Health and Medical Sciences, University of Copenhagen, DK-2100 Copenhagen Ø, Denmark
[2]NanoScience Centre, University of Copenhagen, DK-2100 Copenhagen Ø, Denmark
***E-mail: moien.moghimi@sund.ku.dk**

Nanoparticles are being increasingly implemented in drug delivery and diagnostic systems, as they show promising features in protected, targeted and sustained delivery of active pharmaceutical ingredients and contrast agents. Following intravenous injection, nanoparticles, depending on their physicochemical properties, may trigger the complement system, which is a major part of the innate immune system. Complement activation and fixation can prime the surface of nanoparticles for rapid recognition and clearance by complement receptor bearing cells such as blood monocytes and macrophages of the reticuloendothelial system. Uncontrolled complement activation can also induce adverse reactions, and these have been reported with regulatory approved nanomedicines, including stealth therapeutics. We discuss on the interaction between the complement system and synthetic surfaces, and comment on general physicochemical parameters that incite complement. State-of-the-art approaches for evaluation and determination of complement activation and sensing are also described.

© 2012 American Chemical Society

Introduction

With potential improvements of drug solubility, targeting and controlled release, nanoparticle-based drug delivery systems have been forecasted to make a considerable impact in the pharmaceutical market (*1, 2*). The advancement of nanomedicines into applicable pharmaceutical products, however, not only depends on detailed understanding of physicochemical, biological and interfacial factors that control nanoparticle pharmacokinetics after single and repeated dosing regimens at set intervals, but also on their safety (*3*). Man-made nanoparticles often share many features with pathogenic microorganisms such as size, shape and certain surface characteristics. Therefore, of prime importance in nanomedicine design and engineering is a clear understanding of factors that regulate nanoparticle sensing and triggering by the body's defenses (*1, 3, 4*). Certain pathogenic microorganisms have developed arrays of complex physicochemical and interfacial strategies that efficiently deter immune system recognition and triggering (*4*). Some of the microbial strategies have also been translated into nanoparticle engineering as in combating rapid clearance by the body's defense cells (e.g., macrophages), while maintaining target specificity (*3–5*). However, these fabrication procedures do not necessarily provide complete protection against different arms of the innate immunity, and consequently trigger adverse immune reactions (*3–9*).

The complement system is a key effector of both innate and cognate immunity, and is responsible for rapid detection and elimination of particulate intruders in nano- and micro-size ranges (*10*). Complement triggering primes the intruders' surface for rapid recognition and clearance by phagocytic cells. It also induces inflammatory responses, but such responses arising from uncontrolled complement activation could be life threatening. Complement recognizes danger signals primarily through pattern recognition. This is of prime concern in design and engineering of nanomedicines as these entities are often composed of polymeric components and other patterned nanostructures. Indeed, there is compelling evidence that complement activation may be a contributing factor in eliciting acute-like reactions to regulatory-approved nanomedicines, including stealth entities, in certain individuals (*11, 12*).

The interaction between nanoparticles and the complement proteins is complex and regulated by several interfacial dynamic forces and physicochemical factors (*13*). These include surface curvature and defects, and characteristics such as functional groups and their surface density as well as polymeric decoration and architectural displays. Some pathogens have evolved and deployed sophisticated surface and interfacial strategies to evade complement activation and destruction, and/or subvert complement-mediated opsonization processes (*4, 14*). Accordingly, understanding of these events is critical for design and engineering of immunologically safe nanomedicines for clinical practice. We discuss nanoscale interfacial factors and patterns that incite complement as well as state-of-the-art methodologies for monitoring these processes.

The Complement System

The complement system is a key part of the innate immune system, serving as the first line of defense against pathogens (*10*). It has three overall functions. The primary function is to coat pathogens with proteins known as opsonins (e.g., C1q, C3b, iC3b) that subsequently interact with their corresponding receptors expressed by the phagocytic cells, thereby enhancing their clearance through receptor-mediated endocytic and phagocytic processes. Secondly, it releases potent soluble mediators that recruit and activate immune cells that help to fight infection. Finally, following activation the terminal complement proteins polymerize to form pores in pathogenic membranes that can lyse the cells. The complement system consists of over 30 plasma and membrane proteins, many of them being proteases that are themselves activated by proteolytic cleavage, allowing rapid amplification of a response as each activated enzyme produces several activated products (Figure 1). There are three distinct pathways through which complement can be activated on a typical pathogen or a nanostructured surface. These pathways depend on different sensing molecules for their initiation, but they converge to generate the same set of effector molecules (*10, 13*).

Classical Pathway

The first component of the classical pathway of complement is C1, which is composed of a single C1q molecule bound to two molecules each of the zymogens C1r and C1s. C1q is composed of six identical subunits with cationic globular heads and long collagen-like tails. C1q can sense polyanionic surfaces as well as antigen-antibody (antibodies of IgM, IgG1, IgG2 and IgG3 classes) complexes. Each globular head of C1q can bind to the Fc region of immunoglobulin; the CH2 domain of IgG, or the CH3 domain of IgM, but multivalent attachment of C1q is required for C1 activation. A single pentameric IgM is sufficient to activate C1, which allows at least two C1q heads to bind to separate Fc pieces. Since IgG is monomeric at least two molecules are required to cross-link the globular head of C1q and activate C1. On a typical surface two IgG molecules must be within 10–40 nm of each other to form a stable binding site for C1q. Accordingly, on particulate systems IgG antibody is a far less efficient complement activator than IgM. Following binding to an appropriate surface C1q undergoes a conformational change, which results in a Ca^{2+}-dependent activation of C1r, which in turn cleaves its associated C1s to generate an active serine protease that acts on two natural substrates C4 and C2. The cleavage of C4 by C1s releases a 6 kDa fragment (C4a) from the amino-terminus of the α chain of C4, and an internal thioester bond in the α′ chain of the major product C4b is exposed. Nucleophilic attack of the thioester bond by exposed hydroxyl groups or amino groups on nearby surfaces results in the formation of ester or amide bonds, respectively. C4b, which does not form ester or amide bonds loses its binding site as the thioester reacts with water to become inactive fluid-phase C4b. Next, C2 binds to C4b in a Mg^{2+}-dependent manner forming a pro-convertase before being cleaved by C1s into C2b, which bears the original C4b binding site, and C2a which develops a binding site for C4b with serine protease activity. The C4b2a complex is the classical pathway C3 convertase, which cleaves the α chain of C3 to release anaphylatoxin C3a, from the amino-terminus and expose an internal thioester bond in the α chain of C3b. Similar to C4b, the exposed thioester becomes the subject of nucleophilic attack and forms covalent ester or amide bonds with surface hydroxyl or amino groups. Approximately, 10 percent of C3 binds covalently to a complement-activating surface; the thioester of the majority of the activated C3 reacts with water forming inactive fluid-phase C3b. Some C3b binds to C4b in C4b2a and acts as a receptor for C5. The new complex, C4b2a3b, is the classical pathway C5 convertase. This releases the highly potent anaphylatoxin C5a from the amino-terminus of C5 and the major cleavage product C5b. The latter fragment in turn initiates the assembly of the cytolytic membrane attack complex (MAC) by recruiting the terminal complement proteins (C6, C7, C8 and C9).

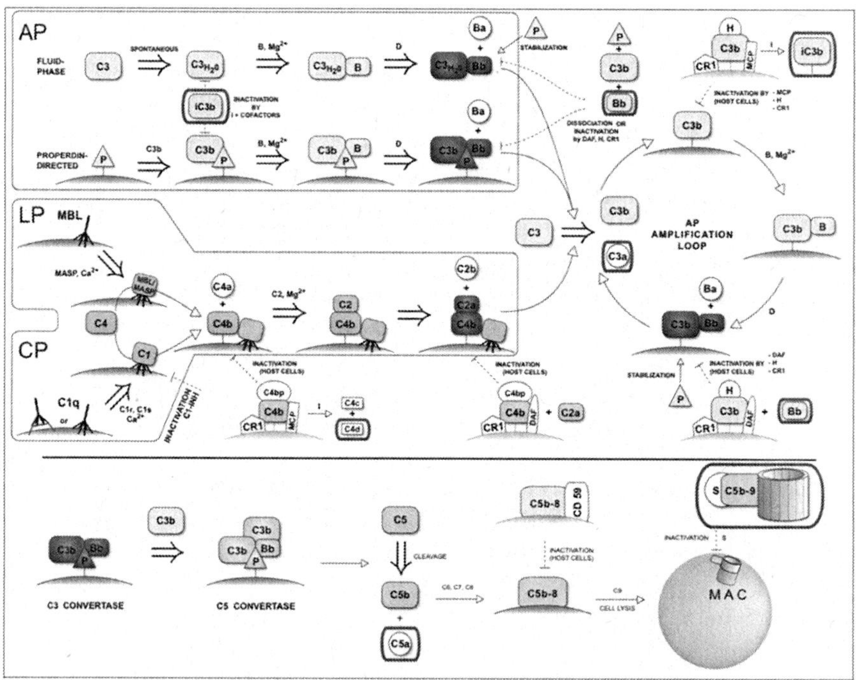

Figure 1. Complement activation pathways. The upper region shows alternative, classical and lectin pathways (AP, CP, LP), respectively. The lower region shows the stepS for the assembly of the membrane attack complex (MAC). Complement activation products C4d, Bb, iC3b, C3a, C5a, and SC5b-9 can be measured by ELISA shown in squares). Abbreviations used: CR1 = complement receptor 1; MCP = membrane cofactor of proteolysis; DAF = decay-accelerating factor.

Lectin Pathway

Activation of the lectin pathway proceeds following the binding of either mannose-binding lectin (MBL) to mannose expressing surfaces, or binding of ficolins (e.g., L-ficolin) to surfaces rich with *N*-acetylated sugars or other acetylated compounds (*10, 13*). The binding of MBL or ficolin to a surface results in the conversion of their associated serine proteases zymogens (mannose activating serine proteases (MASP) 1, 2 and 3) to active proteolytic enzymes. Following activation, MASP-2 cleaves C4, which in turn results in C2 cleavage and formation of a C3 convertase identical to that of the classical pathway.

Alternative Pathway

The alternative pathway is initiated through two known distinct modes. The first is based on spontaneous autoactivation (tick-over) of soluble C3, since the thioester of C3 undergoes slow spontaneous hydrolysis to form C3(H2O) that exposes a binding site for factor B. Bound factor B becomes susceptible to cleavage by factor D, forming an active fluid-phase C3 convertase, C3bBb (*10*). The second mode of triggering is based on the pattern-recognition molecule properdin (P) that binds to certain foreign or self-damaged surfaces (e.g., apoptotic cells). Soluble C3b then binds to the surface immobilized P causing recruitment of factor B, which in turn after cleavage forms the active, surface-bound C3 convertase C3bBbP (*15*). In addition to its initiating role, P can also stabilize already formed C3 convertases, extending their half-life from ~90 seconds to 7-15 minutes (*16*). As C3bBbP convertase also cleaves C3 into C3a and C3b, a positive amplification loop is established as one of the products (C3b) of the action of the enzyme on its substrate is a constituent of the enzyme itself (*10*). Turnover of this amplification loop is regulated by factors H and I (*10*). Factor H can bind to C3b in C3bBb and C3bBbP complexes to displace Bb from them, thus accelerating the decay of these two enzymes. In addition to this, factor H binds to particle-bound C3b and promotes inactivation of C3b by factor I. The inactivated product iC3b is unable to bind factor B so that convertase formation is prevented.

Physicochemical Parameters Affecting Complement Surface Sensing

Size versus Curvature

Sensing molecules such as C1q, antibodies, C-reactive protein, ficolins, MBL, C3b, and P vary in cross-sectional diameters from 7 nm as in the case C3b, and up to 37 nm for the pentameric IgM. With this cross-sectional size range, they are able to sense nanoparticles of different sizes and geometry, and as a result initiate complement responses that differ in terms of magnitude and activating pathway (*13*). There have been numerous studies relating particle size to complement activation, but the results have been somewhat confusing (*3, 17–19*). These discrepancies may be a reflection of surface curvature combined with the affinity, and imposed geometrical aspects/strain of the sensing molecules (*3, 17*). For instance, IgM antibodies to dextran potently activated complement on a dextran-coated nanoparticle of 250 nm in diameter, whereas for a 600 nm nanoparticle, which bound IgM as effectively as did the 250 nm particle, there was substantially less complement activation on the basis of surface area (*17*). In solution IgM is a planar structure composed of five structural units each containing four constant domains (Cμ1–4) and one variable domain in the heavy chain, and one constant and one variable domain in the light chain (*20*). For C1q binding, and hence complement activation, IgM must be in 'staple' conformation (*20, 21*). For IgM, the degree of strain (a measure of bending of the Cμ1 module of IgM and variable domains relative to Fc5 disk) and tendency towards 'staple' conformation grows rapidly as particle size decreases. For high-affinity interactions, it also seems possible to geometrically strain the IgM molecule on smaller particles (e.g., 40–60 nm) and activating C1q-dependent classical pathway (*17*). On the other hand, the extent of surface opsonization and clearance kinetics by the mononuclear phagocyte system tends to increase with particle size when normalizing surface area (*3, 4*). This may be a consequence of reduced local surface areas available for alternative pathway amplification loop on smaller particles. Indeed, in accordance with the crystal structure of C3b, each surface-bound C3b molecule may occupy an area of 40 nm² on a nanoparticle surface (*22*). With smaller particles, the bulk of activated C3 molecules will be released into the surrounding medium rather than deposited on the particle surface. These are examples of how nanostructures can manipulate protein structure and consequential immune cascades.

Surface Charge versus Surface Functional Groups

Particle charge has long been argued to be a contributing factor to complement activation (*13*). For instance, both negatively and positively charged liposomes have been shown to activate complement considerably more than zwitterionic vesicles (*23, 24*). Furthermore, the pathways of complement activation have also been shown to be different between liposomes of different lipid composition and electric charges (*25, 26*). Indeed, some anionic vesicles activate the classical pathway of complement purely following C1q binding. This may be attributed to the pattern recognition property of C1q arising from its cationic globular head, although hydrophobic interactions and hydrogen bonding have also been suggested to play some role (as in the case of C1q binding to liposomes enriched with cardiolipin) (*24, 27*). On the other hand, the extent and pathway of complement activation may be controlled by the nature and surface density of functional groups as they may display different affinities for both complement sensing molecules and naturally occurring complement inhibitors (e.g., C1-inhibitor). For example, anionic vesicles containing L-glutamic acid *N*-(3-carboxy-1-oxopropyl)-1,5-dihexadecyl ester in their lipid bilayer do not activate complement (*28*). Liposome-mediated complement activation may further be dependent on the binding of naturally occurring antibodies to phospholipid head-groups and cholesterol (*26*). Antibody binding to liposomes may further require additional epitopes arising from surface adsorption of other plasma proteins, such as apolipoprotein H. Conversely, surface deposition of apolipoproteins A-I and A-II may suppress complement activation and complement-mediated membrane damage, as these glycoproteins are known to inhibit C9 polymerization, thus preventing the assembly of MAC (*13*).

Recent studies also attest that cationic polymers such as polyethyleneimines and chitosans, which are used commonly for nucleic acid compaction activate complement through different pathways (*13, 29*). For instance chitosan, a biopolymer composed of repeating units of D-glucosamine, together with residual *N*-acetyl-D-glucosamine units triggers complement through lectin and alternative pathways.

Surface-Projected Polymers

Adsorption or surface grafting of poly(ethylene glycol)s, poly(ethylene oxide)-poly(propylene oxide) block copolymers, and related structures (e.g., Tweens) to nanoparticles is a common strategy that enhances nanoparticle stability in suspension and prolongs their circulation times in the blood (*3, 4*). The steric hindrance arising from surface projected polymers have long been thought to suppress opsonization processes as well as complement activation (*3, 4*). However, recent studies have challenged these concepts. One example is PEGylated liposomes. These entities are well capable of activating complement in human sera through both classical and alternative pathways (*30*). In some sera complement activation is due to the presence of antibodies that react with PEG. In the absence of such antibodies, complement activation still proceeds where a role for the anionic phosphate-oxygen moiety of the PEGylated phospholipid in complement triggering has been identified (*30*). Indeed, methylation of the phosphate-oxygen abolishes complement activation, which may sterically block the binding of sensing molecules such as antibodies, C1q and P to liposomes (*30*). On the other hand, poor complement activation by hyaluronic acid-coated liposomes has been reported (*31*). This observation further supports the notion that anionic charge per se is not the main contributor to complement activation, whereas the structural moieties and their surface density play a major role.

Complement activation is common with surfaces that display high density of repetitive epitopes (*13*). This is of concern when high concentrations of polymers are immobilized on a typical nanoparticle surface, which may resemble 'pathogen-mimicking' structures. Indeed, the role of polymer conformation in complement sensing and triggering was addressed recently (*6*). Generally, an increase in number of adsorbed polymers on a typical nanosphere may force a conformational change from 'mushroom' to the more dense 'brush' conformation. This architectural alteration, however, is accompanied by a significant change in both complement activation magnitude and sensing mechanism. For instance, poloxamine 908, a star-shaped poly(ethylene oxide)-poly(propylene oxide) block copolymer, has repetitive patterns of relative polarity and hydrophobicity, where the patterns can change with the density of poloxamine attached to a polystyrene nanoparticle surface (*32*). With surface projected poly(ethylene oxide) chains displaying transient 'mushroom-brush' and 'brush' configuration, nanoparticles exhibit prolonged circulation times in the blood, but the copolymer conformational changes affect the extent and pathways of complement activation (*6*). When surface poly(ethylene oxide) chains are spread laterally ('mushroom' pose) complement activation is C1q-dependent, which is also similar to complement activation by uncoated polystyrene nanospheres, whereas with chains displaying 'brush' architectures complement activation is predominantly through the lectin pathway. However, the extent of complement activation lessens with nanoparticles displaying poly(ethylene oxide) chains in the 'brush' conformation compared with entities exhibiting 'mushroom' pose. At the intermediate 'mushroom-brush' conformation, nanoparticles further incite complement through P binding. The reason for distinct shift from C1q-mediated complement triggering to the lectin pathway was suggested to arise from structural similarities between D-mannose and the terminal region of poly(ethylene oxide) moiety of poloxamine 908 (*6*). Accordingly, surface projected poly(ethylene oxide) chains in close proximity ('brush' conformation) may form dynamic 'pathogen-mimicking' clusters transiently resembling structural motifs of the D-mannose, which serves as a platform for MBL and/or ficolin docking (*6*). The role of lectin pathway in complement activation by PEGylated carbon nanotubes has also been demonstrated (*7, 9*).

Because of the dynamic nature of the polymers, it is rather difficult to deduce any meaningful relation between polymer design and complement activation (*33–36*). In addition, the extent of polymer hydration and possible formation of 'structured water' may aid complement triggering through the alternative pathway turnover. Accordingly, better strategies must be sought for future nanoparticle engineering to ensure immune safety. Some promising surface modification strategies in terms of minimizing complement activation may be based on the use of natural complement inhibitors (or their synthetic-recombinant derivatives), such as CD46 (membrane cofactor of proteolysis), decay-accelerating factor, CD59 (protectin), complement receptor 1, and factor H (*4, 37*).

Measuring Complement Activation

The interaction between the complement system and a foreign surface can be followed by different methodologies (*38–41*), and the most common approaches are described here.

Hemolytic Assay

A simple and commonly used method to evaluate complement activation by nanoparticles is the CH50 assay (*40*) (Figure 2). Nanoparticles are introduced to serum, and complement activation is allowed. As activation leads to irreversibly cleaved complement proteins, reactive particles will drain the serum for its later ability to perform complement activation. Sheep erythrocytes coated with IgG are therefore introduced to the complement-reacted sample, and the degree of MAC-induced hemolysis represents the unused complement capacity left in the serum. A high degree of hemolysis indicates an unreacted complement system, and thereby non-activating nanoparticles. By adding a fixed amount of coated erythrocytes to different aliquots of the reacted serum, the CH_{50} value given by the volume needed to lyse 50% of the erythrocytes can be found. It therefore provides a simple measure of evaluating complement consumption of nanoparticles relative to a standard activator like zymosan. The lysis of the IgG coated sheep erythrocytes is mediated by the classical pathway, and the CH_{50} is therefore a measure of this pathway only. To investigate the alternative pathway, rabbit erythrocytes are added to the serum-nanoparticle mixture diluted in presence of Mg^{2+} and EGTA, where the latter chelates Ca^{2+}, shutting down activation of both classical and lectin pathways of complement.

The hemolytic assay is a simple method that allows straightforward visualization and analysis, and is a reliable predictor of an overall complement response. Among the drawbacks is the limitation in monitoring specific complement components, and the non-standardized protocols, which do not allow comparisons of different experiments.

Binding Studies

Direct binding of the different pattern recognition molecules to various types of surfaces can be measured by techniques like surface plasmon resonance (*41*) and quartz crystal microbalance (*42, 43*). Detailed studies on binding kinetics and surface interaction can be resolved from these experiments, but information about the overall magnitude of activation is limited compared to the hemolytic assays. It may further impose challenges in simulating curvature-dependent effects, and may not detect activated compounds that are released into the fluid phase.

ELISA Measurements

Complement activation assessment based on ELISA techniques show great diversity with the possibility of quantifying several parameters simultaneously at high sensitivity. There are numerous commercial ELISA kits available for detecting plasma levels of complement proteins in human and animal models. Some kits contain microtitre strips coated separately with activators of the different pathways. By diluting unreacted serum in buffers selectively blocking the irrelevant pathways, the potential complement activation from each pathway can be detected. This approach is used to evaluate individuals overall ability to activate complement, and can reveal potential complement protein deficiencies (*44*).

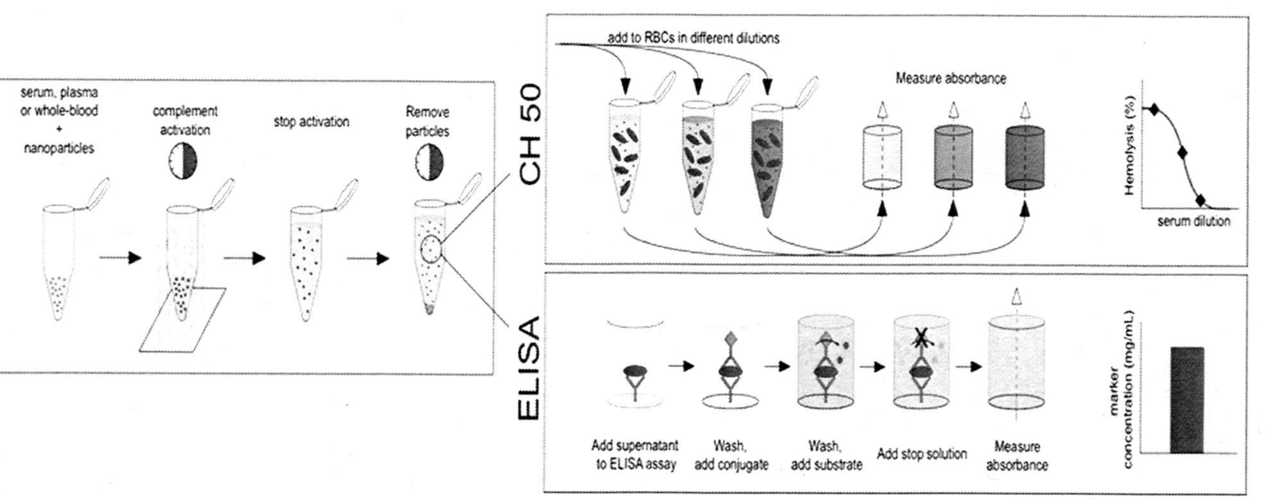

Figure 2. Basic principles behind two common in vitro complement detection methods. The CH_{50} assay (top) and the ELISA (bottom), measuring either the remaining functional complement activity (CH_{50}) or the concentration of specific activated complement markers.

Table 1. Complement activation products measurable by ELISA

Marker	Originated by	Pathway involved	Indication
C4d	cleavage of C4b by fI	CP[1]/LP[2]	activation through CP/LP
Bb	dissociation of C3 or C5 convertase	AP[3]	convertase formation
iC3b	cleavage of C3b by factor I[4]	AP	surface recognition & amplification loop
C3a	cleavage of C3 by C3 convertase[4]	all	surface recognition & amplification loop
C5a	cleavage of C5 by C5 convertase[4]	all	convertase stability
SC5b-9	inactivation by solubilizing MAC	all	convertase stability, C6-C9 stability, CD59 action

[1] CP represent the classical pathway, [2] LP represent the lectin pathway, and [3] AP represent the alternative pathway, respectively. [4] Extrinsic proteases such as kallikerin and thrombin can directly cleave C3 and C5.

Since complement proteins are activated through enzymatic cleavage, the activated substances expose epitopes that are not available on the native protein (termed neoepitopes). Assays have thereby been developed containing wells coated with capture antibodies specifically targeting neoepitopes of various complement proteins. Serum that has been allowed to react with the nanoparticle system of interest is added to the wells, followed by a detection antibody (Figure 2). The latter is conjugated to a detection system like horseradish peroxidase-mediated substrate conversion that generates a solution that absorbs light proportional to the amount of the captured complement marker of interest. Detection systems for numerous different markers (including end-point markers like C5a and SC5b-9) are available, allowing a complete profiling of potential particle-induced complement reactions. The activated complement complexes have relatively short half-lives, and therefore rapidly dissociate into the solution in fragments. Thus, most of the measurable markers are soluble byproducts of some of the above mentioned functional proteins. Table 1 outlines some markers that are measurable through commercially available ELISA kits and comments on their origin (Figure 1). To measure the overall magnitude of a response, the end-point markers C5a or SC5b-9 are usually quantified as they represent activation from all three pathways. If a particle system is activating complement, the responsible pathway can be elucidated by depletion of different pathway-dependent components. A simple approach is to allow activation exclusively through alternative pathway by chelating free Ca^{2+}, while retaining Mg^{2+} levels. Another overall approach to pathway investigation is to block specific complement proteins using monoclonal antibodies and/or employ sera depleted from certain complement proteins, and measure the following changes in complement development.

Monitoring Complement Responses in Vivo

Considering the complexity of the complement system and its interplay with other tissues and immune cells, it would be necessary to quantitatively and qualitatively assess nanomaterial-mediated complement activation in relevant animal models (38). However, the choice of animal model is important in relation to the pathological question and its interpretation to the human situation. Quantitative complement activation measures prior and after nanomaterial administration are essentially based on proper blood removal and treatment, and conduction of the abovementioned in vitro tests (38). A large body of clinical studies have strongly suggested that acute allergic reactions to infused nanomedicines comprise hemodynamic, respiratory, cutaneous and subjective manifestations, and may be related to the activation of the complement system among the sensitive individuals (11, 13, 45). These reactions are also common in pigs, sheep and dogs, and accordingly these species may serve as good models for measuring and evaluating such responses (45). For example, hemodynamic responses to liposome and polycation infusion in pigs include a rapid (within minutes of infusion) but massive rise in pulmonary arterial pressure, and a decline in systemic arterial pressure, cardiac output and left-ventricular end-diastolic pressure. Depending on material properties and reaction severity, ventricular fibrillation or cardiac arrest may even occur. In dogs, infusion reactions to nanomedicines are often associated with neuro-psychosomatic and vegetative responses (*45*).

New Methods in the Pipeline: Proteomics and Sequencing

A not yet fully developed but highly promising technique is the proteomic approach, where antibody-based microarrays allow high throughput protein expression profiling of plasma samples. The human plasma is, however, a complex protein source with over 1.000 proteins covering a dynamic concentration range over 10 orders of magnitude, where only a few proteins constitute 99% of the total protein mass (*46*). Despite these inherent challenges, quantitative proteomics technologies have been used for profiling of several human diseases, and could become the preferred method to discover biomarkers involved in pathological pathways like nanomaterial-mediated acute allergic reactions, as well as in selecting patients at risk prior to nanomedicine exposure (*12*).

Genetic sequencing of different patient reaction groups (e.g., responders/non-responders) could provide essential information about potential polymorphisms responsible for the large inter-individual differences in complement response (*12*). With advancing technology, fast and inexpensive sequencing is within reach, that potentially can make it a routine bench-top analysis (*47*).

Future Perspectives

The interplay between complement proteins and synthetic nanoparticulate surfaces dictates both circulating profile and adverse immune reactions for nanopharmaceuticals. With rapid developments in nanomedicine and nanopharmaceuticals, the incidence of acute allergic will increase and this may impose further health and financial burden. The best known approach to minimize unwanted interactions is the introduction of a polymeric coating layer, creating a dynamic, highly entropic interface. However, a recent study showed the role of polymer conformation in different triggering of the complement system (*6*). This observation has important bearings in design of safe stealth nanocarriers and for prevention of nanomedicine-mediated infusion reactions. Accordingly, these findings have raised further challenges for safe engineering of nanoparticle surfaces with polymers that was previously considered to be inert and safe. Future efforts may therefore take a lead from a wide spectrum of microbial approaches, which combat immune sensing and recognition, and translate them into safer nanomedicine design. Additionally, attention must also be paid to the role of other physical parameters, such as the shape and the extent of nanoparticle deformability to overcome problems with the immune system (*48*).

The techniques that are available to study complement activation and development has aided our understanding of the complex interplay between plasma proteins and synthetic surfaces, thereby allowing intelligent design of stealth nanoparticles with markedly improved immune-evading properties. Early-stage safety evaluations are, however, not fully predictive for clinical behaviour, suggesting the needs for more comprehensive immune assessment protocols. Considering the potential patient risk and financial burden associated with toxicity-related clinical failures, a reliable and validated immune assessment approach is important to properly select basic design of carriers and avoid late stage failures. Although complement directed, the adverse reaction magnitude may be controlled by downstream amplifying mediators like cytokine levels, and immune cell functions. High-throughput screening on the protein- or genetic level is an appealing method that can shed new light in our understanding of the complement reaction development introducing individual differences and potential underpinning polymorphisms. The fact that clinical studies have shown higher frequencies of elevated complement products in responsive patients suggest that complement is a contributing, but not necessarily a rate-limiting factor in nanomedicine infusion adverse reactions (*12*). This further supports a role for a spectrum of other mediators not yet listed in the conventional complement reaction development.

Acknowledgments

Financial support by the Danish Agency for Science, Technology and Innovation (Det Strategiske Forskningsrad), reference 2106-08-0081, is gratefully acknowledged.

References

1. Moghimi, S. M.; Peer, D.; Langer, R. Reshaping the future of nanopharmaceuticals: Ad iudicium. *ACS Nano* **2011**, *5*, 8454–8458.
2. Shi, J.; Xiao, Z.; Kamaly, N.; Farokhzad, O. C. Self-Assembled Targeted Nanoparticles: Evolution of Technologies and Bench to Bedside Translation. *Acc. Chem. Res.* **2011**, *44*, 1123–1134.
3. Moghimi, S. M.; Hunter, A. C.; Andresen, T. L. Factors controlling nanoparticle pharmacokinetics: an integrated analysis and perspective. *Annu. Rev. Pharmacol. Toxicol.* **2012**, *52*, 481–503.
4. Moghimi, S. M.; Hunter, A. C.; Murray, J. C. Long-circulating and target-specific nanoparticles: theory to practice. *Pharmacol. Rev.* **2001**, *53*, 283–318.
5. Moghimi, S. M.; Hunter, A. C.; Murray, J. C. Nanomedicine: current status and future prospects. *FASEB J.* **2005**, *19*, 311–330.
6. Hamad, I.; Al-Hanbali, O.; Hunter, A. C.; Rutt, K. J.; Andresen, T. L.; Moghimi, S. M. Distinct Polymer Architecture Mediates Switching of Complement Activation Pathways at the Nanosphere-Serum Interface:

Implications for Stealth Nanoparticle Engineering. *ACS Nano* **2010**, *4*, 6629–6638.

7. Hamad, I.; Hunter, A. C.; Rutt, K. J.; Liu, Z.; Dai, H.; Moghimi, S. M. Complement activation by PEGylated single-walled carbon nanotubes is independent of C1q and alternative pathway turnover. *Mol. Immunol.* **2008**, *45*, 3797–3803.

8. Moghimi, S. M.; Hunter, A. C. Complement monitoring of carbon nanotubes. *Nat. Nanotechnol.* **2010**, *5*, 382.

9. Andersen, A. J.; Wibroe, P. P.; Moghimi, S. M. Perspectives on carbon nanotube-mediated adverse immune effects. *Adv. Drug Delivery Rev.* **2012**, http://dx.doi.org/10.1016/j.addr.2012.05.005.

10. Ricklin, D.; Hajishengallis, G.; Yang, K.; Lambris, J. D. Complement: a key system for immune surveillance and homeostasis. *Nat. Immunol.* **2010**, *11*, 785–797.

11. Moghimi, S. M.; Andersen, A. J.; Hashemi, S. H.; Lettiero, B.; Ahmadvand, D.; Hunter, A. C.; Andresen, T. L.; Moghimi, S. M. Complement activation cascade triggered by PEG-PL engineered nanomedicines and carbon nanotubes: the challenges ahead. *J. Controlled Release* **2010**, *146*, 175–181.

12. Moghimi, S. M.; Wibroe, P. P.; Helvig, S. Y.; Farhangrazi, S.; Hunter, A. C. Genomic perspectives in inter-individual adverse responses following nanomedicine administration: the way forward. *Adv. Drug Delivery Rev.* **2012**, http://dx.doi.org/10.1016/j.addr.2012.05.010.

13. Moghimi, S. M.; Andersen, A. J.; Ahmadvand, D.; Wibroe, P. P.; Andresen, T. L.; Hunter, A. C. Material properties in complement activation. *Adv. Drug Delivery Rev.* **2011**, *63*, 1000–1007.

14. Lambris, J. D.; Ricklin, D.; Geisbrecht, B. V. Complement evasion by human pathogens. *Nat. Rev. Microbiol.* **2008**, *6*, 132–142.

15. Kemper, C.; Atkinson, J. P.; Hourcade, D. E. Properdin: emerging roles of a pattern-recognition molecule. *Annu. Rev. Immunol.* **2010**, *28*, 131–155.

16. Fearon, D. T.; Austen, K. F. Properdin: binding to C3b and stabilization of the C3b-dependent C3 convertase. *J. Exp. Med.* **1975**, *142*, 856–863.

17. Bradley, D. V.; Wong, K.; Serrano, K.; Chonn, A.; Devine, D. Liposome-complement interactions in rat serum: implications for liposome survival studies. *Biochim. Biophys. Acta* **1994**, *1191*, 43–51.

18. Pedersen, M. B.; Zhou, X.; Larsen, E. K. U.; Sorensen, U. S.; Kjems, J.; Nygaard, J. V.; Nyengaard, J. R.; Meyer, R. L.; Bosen, T.; Vorup-Jensen, T. Curvature of Synthetic and Natural Surfaces Is an Important Target Feature in Classical Pathway Complement Activation. *J. Immunol.* **2010**, *184*, 1931–1945.

19. Lunqvist, M.; Stigler, J.; Elia, G.; Lynch, I.; Cedervall, T.; Dawson, K. A. Nanoparticle size and surface properties determinethe protein corona with possible implications for biological impacts. *Proc. Natl. Acad. Sci. U.S.A.* **2008**, *105*, 14265–14270.

20. Perkins, S. J.; Nealis, A. S.; Sutton, B. J.; Feinstein, A. Solution structure of human and mouse immunoglobulin M by synchrotron X-ray scattering

and molecular graphics modeling. A possible mechanism for complement activation. *J. Mol. Biol.* **1991**, *221*, 1345–1366.

21. Roux, K. H. Immunoglobulin structure and function as revealed by electron microscopy. *Int. Arch. Allergy Immunol.* **1999**, *120*, 85–99.

22. Janssen, B. J.; Christodoulidou, A.; McCarthy, A.; Lambris, J. D.; Gros, P. Structure of C3b reveals conformational changes that underlie complement activity. *Nature* **2006**, *444*, 213–216.

23. Chonn, A.; Cullis, P.; Devine, D. The role of surface charge in the activation of the classical and alternative pathways of complement by liposomes. *J. Immunol.* **1991**, *146*, 4234–4241.

24. Bradley, A. J.; Brooks, D. E.; Norris-Jones, R.; Devine, D. V. C1q binding to liposomes is charge dependent and is inhibited by peptides consisting of residues 14–26 of the human C1Qa chain in a sequence independent manner. *Biochim. Biophys. Acta* **1999**, *1418*, 19–30.

25. Devine, D. V.; Marjan, J. M. J. The role of immunoproteins in the survival of liposomes in the circulation. *Crit. Rev. Ther. Drug Carrier Syst.* **1997**, *14*, 105–131.

26. Moghimi, S. M.; Hunter, A. C. Recognition of macrophages and liver cells of opsonized phospholipid vesicles and phospholipid headgroups. *Pharm. Res.* **2001**, *18*, 1–8.

27. Bradley, A. J.; Maurer-Spurej, E.; Brooks, D. E.; Devine, D. V. Unusual electrostatic effect on binding of C1q to anionic liposomes: role of anionic phospholipids domain and their line tension. *Biochemistry* **1999**, *38*, 8112–8123.

28. Sou, K.; Tsuchida, E. Electrostatic interactions and complement activation on the surface of phospholipid vesicle containing acidic lipids: Effect of the structure of acidic groups. *Biochim. Biophys. Acta* **2008**, *1778*, 1035–1041.

29. Merkel, O. M.; Urbanics, R.; Bedocs, P.; Rozsnyay, Z.; Rosivall, L.; Toth, M.; Kissel, T.; Szebeni, J. In vitro and in vivo complement activation and related anaphylactic effects associated with polyethylenimine and polyethylenimine-grafted-poly(ethylene glycol) block copolymers. *Biomaterials* **2011**, *32*, 4936–4942.

30. Moghimi, S. M.; Hamad, I.; Andresen, T. L.; Jorgensen, K.; Szebeni, J. Methylation of the phosphate oxygen moiety of phospholipid-methoxy(polyethylene glycol) conjugate prevents PEGylated liposome-mediated complement activation and anaphylatoxin production. *FASEB J.* **2006**, *20*, 2591–2593.

31. Mizrahy, S.; Raz, S. R.; Hasgaard, M.; Liu, H.; Soffer-Tsur, N.; Cohen, K.; Dvash, R.; Landsman-Milo, D.; Bremer, M. G. E. G.; Moghimi, S. M.; Peer, D. Hyaluronan-coated nanoparticles: the influence of the molecular weight on CD44-hyaluronan interactions and on the immune response. *J. Controlled Release.* **2011**, *156*, 231–238.

32. Al-Hanbali, O.; Rutt, K. J.; Sarker, D. K.; Hunter, A. C.; Moghimi, S. M. Concentration dependent structural ordering of poloxamine 908 on polystyrene nanoparticles and their modulatory role on complement consumption. *J. Nanosci. Nanotechnol.* **2006**, *6*, 3126–3133.

33. Hamad, I.; Hunter, A. C.; Szebeni, J.; Moghimi, S. M. Poly(ethylene glycol)s generate complement activation products in human serum through increased alternative pathway turnover and a MASP-2-dependent process. *Mol. Immunol.* **2008**, *46*, 225–232.

34. Moghimi, S. M.; Hunter, A. C.; Dadswell, C. M.; Savay, S.; Alving, C. R.; Szebeni, J. Causative factors behinf poloxamer 188 (Pluronic F68, Flocor™)-induced complement activation in human sera. A protective role against poloxamer-mediated complement activation by elevated serum lipoprotein levels. *Biochim. Biophys. Acta-Mol. Basis Dis.* **2004**, *1689*, 103–113.

35. Weiszhar, Z.; Czucz, J.; Revesz, C.; Rosivall, L.; Szebeni, J; Rozsnyay, Z. Complement activation by polyethoxylated pharmaceutical surfactants: Cremophor-EL, Tween-80 and Tween-20. *Eur. J. Pharm. Sci.* **2012**, *45*, 492–498.

36. Lettiero, B.; Andersen, A. J.; Hunter, A. C.; Moghimi, S. M. Complement system and the brain: selected pathologies and avenues toward engineering of neurological nanomedicines. *J. Control. Rel.* **2012**, *161*, 283–289.

37. Wu, Y-Q.; Qu, H. C.; Sfyroera, G.; Tzekou, A.; Kay, B. K.; Nilsson, B.; Ekdahl, K. N.; Ricklin, D.; Lambris, J. D. Protection of nonself surfaces from complement attack by factor H-binding peptides: implications for therapeutic medicine. *J. Immunol.* **2011**, *186*, 4269–4277.

38. Harboe, M.; Thorgersen, E. B.; Mollnes, T. E. Advances in assay of complement function and activation. *Adv. Drug Delivery Rev.* **2011**, *63*, 976–987.

39. Szebeni, J.; Baranyi, L.; Savay, S.; Milosevits, J.; Bodo, M.; Bunger, R.; Alving, C. R. The interaction of liposomes with the complement system: in vitro and in vivo assays. *Method Enzymol.* **2003**, *373*, 136–154.

40. Meerasa, A.; Huang, J. G.; Gu, F. X. CH50: A revisited hemolytic complement consumption assay for evaluation of nanoparticles and blood plasma protein interaction. *Curr. Drug Delivery* **2011**, *8*, 290–298.

41. Arima, Y.; Toda, M.; Iwata, H. Surface Plasmon resonance in monitoring of complement activation on biomaterials. *Adv. Drug Delivery Rev.* **2011**, *63*, 988–999.

42. Sellborn, A.; Andersson, M.; Hedlund, J.; Andersson, J.; Berglin, M.; Elwing, H. Immune complement activation on polystyrene and silicon dioxide surfaces Impact of reversible IgG adsorption. *Mol. Immunol.* **2005**, *42*, 569–574.

43. Andersson, M.; Andersson, J.; Sellborn, A.; Berglin, M.; Nilsson, B.; Elwing, H. Quartz crystal microbalance-with dissipation monitoring (QCM-D) for real-time measurements of blood coagulation density and immune complement activation on artificial surfaces. *Biosens. Bioelectron.* **2005**, *21*, 79–86.

44. Moghimi, S. M.; Hamad, I. Liposome-mediated triggering of complement cascade. *J. Liposome Res.* **2008**, *18*, 195–209.

45. Szebeni, J.; Muggia, F.; Gabizon, A.; Barenholz, Y. Activation of complement by therapeutic liposomes and other lipid excipient-based therapeutic products: prediction and prevention. *Adv. Drug Delivery Rev.* **2011**, *63*, 1020–1030.

46. Anderson, N.; Anderson, N. The human plasma proteome - History, character, and diagnostic prospects. *Mol. Cell Proteomic.* **2002**, *1*, 845–867.
47. Pareek, C.; Smoczynski, R.; Tretyn, A. Sequencing technologies and genome sequencing. *J. Appl. Genet.* **2011**, *52*, 413–435.
48. Mitragotri, S.; Lahann, J. Physical approaches to biomaterial design. *Nat. Mat.* **2009**, *8*, 15–23.

Chapter 15

Theranostic Applications of Plasmonic Nanosystems

Amit Singh,[1] Tatyana Chernenko,[1] and Mansoor Amiji*,[1]

[1]Department of Pharmaceutical Sciences School of Pharmacy, Northeastern University, Boston, Massachusetts 02115
*E-mail: m.amiji@neu.edu. Tel. (617) 373-3137. Fax (617) 373-8886

Metal nanoparticles have fascinated scientists since the middle ages due to their vibrant colors and have been used as colorant in glass windows and pottery. The colloidal chemistry resurged to popularity in the 1850s when Faraday first synthesized pure gold sol and called it "*activated gold*". Since then, scientists across the world have tried to understand the origin of color in the colloidal suspension and novel material synthesis from bottom up approach gained tremendous popularity for a variety of application. Gold and silver nanoparticles have been in the focus specifically due to their biocompatible nature and their potential biomedical applications have been envisioned. The last two decades have specially seen the synthesis of anisotropic noble nanostructures that show exotic optical properties with a strong absorption in the NIR region. The presence of the "*optically transparent window*" of biological materials between 800-1300 nm where absorption of water and oxygenated/deoxygenated hemoglobin is minimal, has led to exploitation of the NIR characteristic of noble metal nanoparticles for various imaging, diagnostic and therapeutic applications. This chapter details the fundamentals of plasmonic properties of noble metal nanoparticles, evolution of their surface chemistry, their promise as potential theranostic agents and safety considerations associated with their application.

© 2012 American Chemical Society

1. Introduction

The visionary speech by Richard Feynman titled *"There's plenty of room at the bottom"* laid the foundation of the emerging field of nanotechnology where the materials are manipulated at a miniaturized scale. As the materials are scaled down to the nanometer length, the properties are governed by quantum effects and are unique compared to the corresponding micro- and atomic properties. These properties strongly depend on the size, shape, inter-particle interaction and surface properties. The optical properties of noble metals of nanometer size have specifically attracted tremendous interest as decorative colorant since the ancient time (*1*). One of the most interesting examples is the Lycurgus Cup that appears ruby red in color in transmitted light but green in reflected light due to the presence of colloidal gold. The first rationalized synthesis of nanoparticles was however achieved by Faraday when he synthesized deep red color solution of gold nanoparticles. The optical properties of noble metal nanoparticles have been extensively studied since then to understand the origin of different colors in the colloidal solution of the gold nanoparticles. It was in 1908 that Gustav Mie described the phenomenon popularly known as *surface plasmon resonance* (SPR) to explain the optical properties of the noble metal nanoparticles (*2*).

The growing knowledge of synthesis methods of nanoparticles with precise control over shape and size as well as better understanding of the physical and chemical properties has led to their potential applications in different fields. Noble metal nanoparticles in particular show exotic shape and size dependent optical properties that have been exploited for various diagnostic and drug delivery applications. Anisotropic metallic nanostructures synthesized by physical, chemical or biological routes show strong optical absorption in the NIR region. Human cells have an *"optically transparent window"* where they do not show any absorption but the size and shape of the noble nanoparticles can be tuned to absorb in that region, which opens tremendous possibility of diagnostic, imaging and therapeutic possibility. The molar absorption and scattering coefficient of these nanoparticles have also been found to be several magnitudes higher than the conventional dyes that are routinely used for imaging application. This chapter will briefly discuss the origin of the shape and size dependent optical properties of noble metal nanostructures, the advancement in their surface functionalization and exploitation of these nanoparticles for theranostic application. The safety of these nanostructures in general and specifically for biomedical application has been a point of debate in the past two decades and this aspect has been discussed with contemporary knowledge and literature review.

2. Plasmonic Property of Metallic Nanosystems

2.1. Surface Plasmon Resonance

The interaction of an oscillating electromagnetic wave with the conduction band free-electrons of the metal nanoparticles results in collective coherent oscillation of these electrons with respect to the nanoparticle lattice. The displacement of the electron cloud due to the induced dipole results into a restoring attractive Coulomb force from the particle's nuclei core. The two opposing forces induce a collective coherent oscillation of the free-electrons. For spherical particles with size much smaller than the wavelength of the light, the surface plasmon oscillation is dominated by the dipolar mode given by the equation,

$$\alpha = 3\varepsilon_0 V \left(\frac{\varepsilon - \varepsilon_m}{\varepsilon + 2\varepsilon_m} \right)$$

where α gives the polarizability, ε_0 is the vacuum permitivity, V is the particle volume, ε_m is the dielectric constant of the medium and ε is the dielectric function of the metal [$\varepsilon(\omega)=\varepsilon_r(\omega) + i\varepsilon_i(\omega)$]. The resonance condition is met at a frequency ω, the SPR frequency where the $\varepsilon_r = -2\varepsilon_m$. The resonance frequency of the noble metals i.e., gold (Au), silver (Ag) and copper (Cu) meet this condition in the visible region of the spectrum imparting their nanoparticle colloidal solution a distinct color. The large optical polarization associated with the resonance results into an enhanced local electric field across the surface of the nanoparticle accompanied with a strong absorption and scattering of light at the SPR frequency (2, 3). The frequency of this oscillation depends on the density of the electron cloud, the effective electron mass, the size and shape of the charge distribution, dielectric constant of the medium and inter-particle plasmon coupling interactions (3). These factors can therefore be exploited to tailor the optical properties of the metallic system for specific applications. Figure 1 shows the surface plasmon resonance condition in spherical (top), hollow (middle) and rod (bottom) like nanoparticles under the influence of polarized light (adapted from Ref (4)). The bottom scheme shows only the longitudinal mode of resonance while the transverse mode is similar to that of spherical nanoparticles (top).

The plasmonic excitation of the electrons in a metal nanostructure decays into electron-hole excitation of either intra-band transition (within the conduction band) or inter-band transition including d-shells or conduction bands (5). This plasmonic decay could be non-radiative and comprises the absorption of light by the metallic nanoparticle. Instead, the oscillating electric field generated by polarized plasmon can radiate energy at a frequency same as the oscillating plasmon that constitutes the elastic/Rayleigh scattering by the nanoparticle. The absorption as well as the scattering by the particle together constitutes the optical extinction and according to Mie theory, for a metal nanostructure of size 2R (R being radius) much smaller than the wavelength of the light (λ), the particle's extinction coefficient is given by

$$C_{ext} = \frac{24\pi^2 R^3 \varepsilon_m^{3/2}}{\lambda} \frac{\varepsilon_i}{(\varepsilon_r + 2\varepsilon_m)^2 + \varepsilon_i^2}$$

where ε_m represents the dielectric function of the medium while ε_r and ε_i are the real and imaginary part of the dielectric function of the metal. The band maxima of the optical extinction is therefore achieved at the resonance condition where $\varepsilon_r=-2\varepsilon_m$ when the imaginary part of the dielectric function is small or weakly dependent on the frequency (*5*). A full solution of the Mie theory gives the absorption (C_{abs}) and scattering (C_{sca}) components of spherical gold nanoparticles including a dependence on the nanoparticle size (*3*). This equation can reasonably predict the position of the absorption band of spherical nanoparticles. The extinction band of the smaller nanoparticles is largely due to the absorption component but with the increasing size, the scattering components contribute towards the extinction coefficient as well.

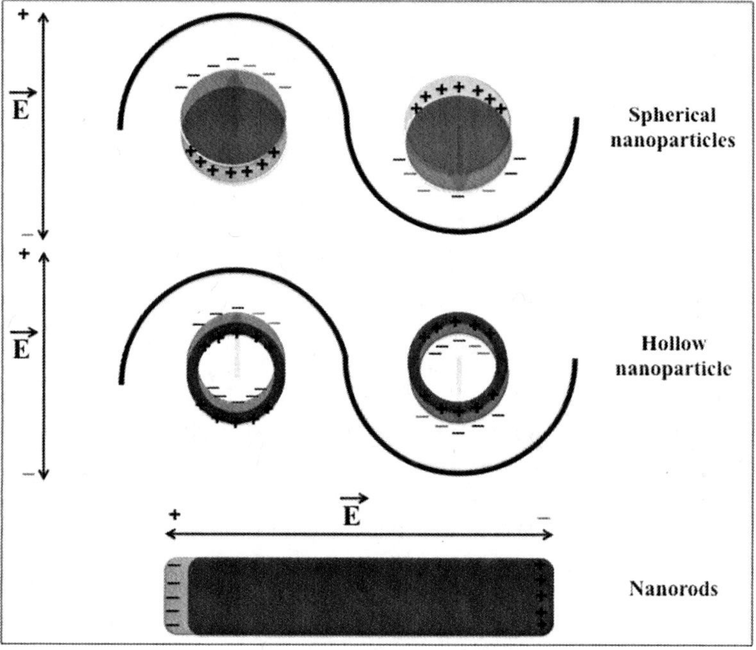

Figure 1. Schematic depicting the plasmon resonance frequency of metal spherical (top), hollow (middle) and rod (bottom) shaped nanoparticles upon incidence of the polarized light. In case of nanorods, the scheme shows only the longitudinal mode of resonance while the transverse mode is similar to the resonance of a spherical nanoparticle (top).

The absolute solutions of the Maxwell equations for non-spherical particles however are complicated and prediction of optical absorption characteristic of only certain non-spherical nanostructures such as spheroids (*6*) and infinite cylinders (*7*) has been realized. Therefore, the optical properties of the non-spherical nanoparticles could only be predicted with approximation and has been successfully done using discrete dipole approximation (DDA) (*3, 8,*

9) and finite-difference time-domain method (*10, 11*). The spherical particles are symmetric; show isotropic polarization with degenerate dipolar modes and therefore a single plasmon resonance frequency. The nanoparticles with arbitrary shapes alternatively show several non-degenerate dipolar modes with a complex optical absorption at several frequencies across a broad spectral region. A nanorod for example should be considered as a spherical particle that has been elongated in one dimension and therefore another low-energy dipolar resonance band arises in longitudinal direction while the original transverse dipolar resonance mode is retained. Therefore, a nanorod shows two absorption bands; transverse absorption at a lower wavelength and a longitudinal absorption at a higher wavelength. The miniaturization of the particles not only changes the optical but also other properties such as electrical, mechanical, catalytic etc. Owing to the change in these properties, nanoparticles have been exploited for several applications such as catalysis (*12*), fuel cells (*13*), heavy metal detection (*14*), photonic band-gap materials (*15*), single electron transistors (*16*), non-linear optical devices (*17*) and biomedical functions (*18*).

2.2. Noble Metal Nanoparticles

The research on optical properties of metal nanoparticles has largely focused on gold and silver nanoparticles due to the fact that their surface plasmon resonance frequency corresponds to the wavelength in the visible region of light. The spherical gold nanoparticles give plasmon absorption band around 520 nm with bright ruby-red color colloidal solution while spherical silver nanoparticles absorb at wavelength centered around 400 nm and their solution is typically yellow colored. Such interesting and exotic optical properties have inspired synthesis of noble metal nanoparticles and specifically gold nanoparticles of various shapes and sizes and theoretical calculations have been attempted to predict their absorption characteristics (*1*). Rods (*19*), triangles (*20*), stars (*21*), shells (*22*), hollow nanoparticles (*23, 24*), cubes (*25*), disks (*26*) and multipods (*27*) are some of the shapes that have been synthesized and their optical properties have been studied in great details. These structures have been prepared using variety of techniques including physical methods such as vapor deposition (*28*), thermal decomposition (*29*), spray pyrolysis (*30*), photo-irradiation (*20*), laser ablation (*31*), ultra-sonication (*32*), radiolysis (*33*) and solvated metal atom dispersion (*34*); chemical methods that include wet chemical synthesis (*35*), sol-gel method (*36*), solvo-thermal synthesis (*37*), micelles based synthesis (*38*) and galvanic replacement reaction (*24*); and biological routes using bacterial (*39, 40*), fungi (*41*) and plant extracts (*42*). Gold and silver nanoparticles are usually preferred as plasmonic materials over copper nanoparticles due to their better stability in colloidal solution. The advantages of noble metal nanoparticles for various applications are strongly derived from the size and shape and size dependent optical properties, ease of synthesis and well-researched surface chemistry. It is pertinent that these advantages are discussed in greater details in the further sections.

2.3. Tunable Optical Properties of Plasmonic Nanoparticles

One of the interesting properties of the gold and silver nanoparticles is the shift in the plasmon absorption band depending on their size. Spherical gold nanoparticles can be easily synthesized from 2-100 nm size range and their plasmon absorption band shifts to a higher wavelength with increasing size (*43*). Similarly, gold nanorods show a tunable longitudinal plasmon band that shows a red shift in the absorption profile with increasing aspect ratio of the particles (*2*). More recently, gold nanotriangles have proved to absorb strongly in the NIR-region of light and their longitudinal absorption band can be tuned by controlling the rate of reduction of gold salt (*44*) and temperature of reaction (*45*), which influences the aspect ratio of the nanotriangles. Inter-particle plasmon interaction is another property that has been extensively exploited to tailor the optical absorption characteristics of gold and silver nanoparticles (*46*). The application of nanoparticles as contrast agents largely depends on their optical properties and such freedom in tunability of the absorption characteristics make these nanoparticles a lucrative choice for such applications.

Besides, calculations from Mie theory illustrate some remarkable advantages of using nanoparticles as contrast agents over conventional fluorescence dyes, which are routinely employed as diagnostic tools in biomedical applications. According to the estimation from Mie theory, the optical cross section of a gold nanoparticle is four to five orders of magnitude higher than the dyes (*47*). A typical spherical gold nanoparticle of 40 nm size exhibit a calculated molar absorption coefficient (ε) of ~7.7 X 10^9 M^{-1} cm^{-1} with absorption maxima centered at 530 nm which is four orders of magnitude higher than that of rhodamine 6G ($\varepsilon = 1.2$ X 10^{-5} M^{-1} cm^{-1} at 530 nm) and malachite green ($\varepsilon = 1.5$ X 10^5 M^{-1} cm^{-1} at 617 nm) (*48*). Jain *et al* (*47*) have further calculated from Mie theory that when excited with same light intensity, the number of photons emitted from a fluorescein dye (emission coefficient ~ 9.2 X 10^5 M^{-1} cm^{-1} at 617 nm) (*48*) is five orders of magnitude smaller than the scattering from 80 nm spherical gold nanoparticles. They have further demonstrated that at same wavelength, light scattering from the 80 nm gold nanospheres are comparable to the scattering from 300 nm large polystyrene particles. It is also noteworthy that gold nanospheres below 20 nm show optical properties primarily dominated by SPR absorbance with insignificant contribution from scattering. However, the scattering contribution increases as the nanoparticle size is increases and therefore can be used as a tool to control scattering properties to design nanosystems suitable for scattering based imaging. These optical properties are extremely critical for biomedical imaging applications of noble metal nanoparticles and have been extensively leveraged for improved diagnostics. Figure 2a shows the dark-field image of light scattered by gold nanoparticles along with the SEM image of the corresponding nanoparticles. Figure 2b shows the scattering spectra from gold nanorods when the incident light is polarized along the length. The lines show the experimental data obtained from different aspect ratio nanorods while the circles depict the cross section calculated by an empirical formula (Figure duplicated from the Ref (*49*) with permission from the publisher)

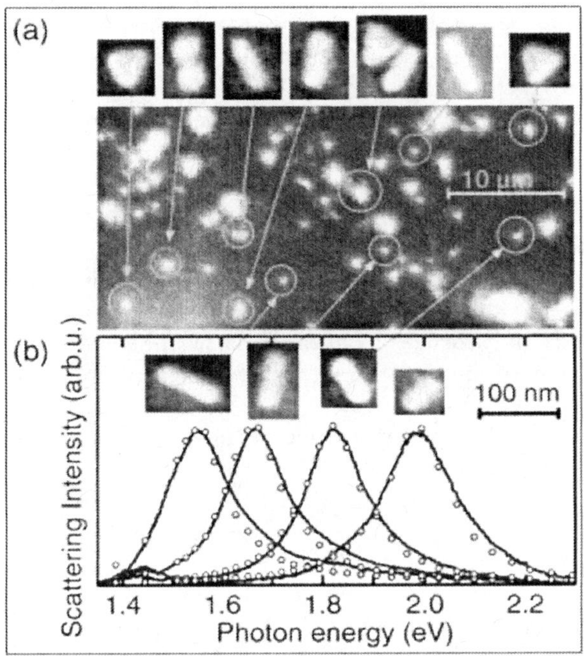

Figure 2. (a) An optical dark-field image of scattered light from Au nanoparticles, together with SEM images of the corresponding particles. (b) Scattering spectra from Au nanorods when the incident light is polarized parallel to the long axis; lines are experimental data, and circles are cross sections calculated by our empirical formula. Each spectrum is appropriately scaled. A higher aspect ratio gives lower resonant energy. (Adapted with permission from Reference (49). Copyright 2003, The American Institute of Physics.)

3. Surface Modifications for Biomedical Applications

Surface modification of noble metal nanoparticles is an important parameter in designing nanostructure for diagnostic and therapeutic application. They stabilize these particles in culture media and inside the cellular compartment as well as improve their uptake in the cells, show low cytotoxicity and improved biocompatibility. One of the popular surface modification strategies involves the use of polymers and surfactants that stabilize the nanoparticle surface by stearic stabilization and induce repulsion among the particles preventing aggregation. Such modifications also provide the functional groups of the polymers/surfactants that could be further leveraged for immobilization of cell targeting moieties for specific applications. The surface of a nanoparticle can be usually modified by a non-covalent binding exploiting physical interaction or by covalent attachments using chemical interactions. Non-covalent binding such as hydrophobic, ionic and π-π interactions has been exploited for surface modification of nanoparticles. Hydrophobic interactions are usually popular since they facilitate the binding of not only hydrophobic chemical (and polymers) but also biological entities that have hydrophobic domains in their structure. These physical interactions are however weak in nature and therefore the functionalized surfaces suffer from poor stability of capping layer. The physical interactions for surface modification have been exploited usually for polymeric or silica nanoparticles as well as carbon nanotubes and have not been particularly used for gold or silver nanoparticles (*50*). The common methods for non-covalent attachment that have been explored include plasma spraying (*51*) and ozone ablation (*52*).

Covalent attachment remains the most popular method of surface modification of nanoparticles since it relies on a much stable chemical interaction that provides a consistent and robust surface chemistry. The well-known thiol chemistry has been extensively used for gold and silver nanoparticles for covalent attachment of surface ligands. Thiols form a stable self-assembled monolayer (SAM) on the surface of these nanoparticle that are highly reproducible and well characterized. A typical SAM formation is achieved at room temperature by exposing the nanostructures to the solution of the desired thiol and the excess ligands are removed by a washing step in the same solvent. SAM formation is usually achieved by ligand-displacement reaction where the ligand with more affinity to the surface partially or completely replaces the pre-existing capping ligand (*53*). The work from Brust *et al* (*54*) pioneered the field and was followed up with tremendous work on use of monolayer protected gold nanoparticles for several applications. Amine based functionalization of gold nanoparticles (*55*) has also been a popular approach that has been realized by some seminal contributions from Sastry *et al.* Selvakannan *et al*, have shown that lysine capped gold nanoparticles render them highly water-dispersible and pH responsive. These nanoparticles tend to aggregate at alkaline pH but can be reversibly redispersed at acidic pH (*56*). Sastry *et al*, have further used green synthesis of amino acid capped gold (*57*) and silver (*58*) nanoparticles where the amino acids act as the reducing as well as the capping agent.

Poly(ethylene glycol) (PEG) modification of nanoparticles are certainly most popular for diagnostic and therapeutic application due to its ability to stabilize the nanoparticles against agglomeration. Thiol modified PEG is most commonly used to immobilize the protective coating layer on noble metal nanoparticles. Qian *et al*, have demonstrated that thiol-PEG modified gold nanoparticles were found to be stable in strong acids, strong bases, concentrated salts and organic solvents (*59*). It has also been accepted universally that PEG coating of nanoparticles prevents the opsonization in physiological conditions providing "stealth" effect and dramatically improves the circulation time of nanoparticles in blood. Shenoy *et al*, have used a hetero-bifunctional PEG derivative to immobilize coumarin (a fluorescent dye) modified PEG-thiol on to gold nanoparticles where PEG performs the dual role of nanoparticle surface stabilizer as well as spacer to prevent fluorescence quenching effect of gold nanoparticle (*60*). Gold nanorods with strong NIR absorption spectra have been explored for several diagnostic and therapeutic applications in cancer therapeutics. The popular seed-mediated synthesis of these nanorods however leaves hexadecyltrimethylammonium bromide (CTAB) coating, which renders them cytotoxic to cells. Niidome *et al*, have reported that surface modification of these nanorods by PEG not only considerably reduces the cytoxicity affects but also provides the "stealth" effect to the nanoparticles (*61*). Therefore, the affinity of the noble metal nanoparticle surface to thiols, amines, di-sulfides and dithio-carbamates opens up avenues for surface stabilization and bio-conjugation with proteins, peptides, and nucleic acid constructs such as DNA, RNA, peptide nucleic acids, and aptamers.

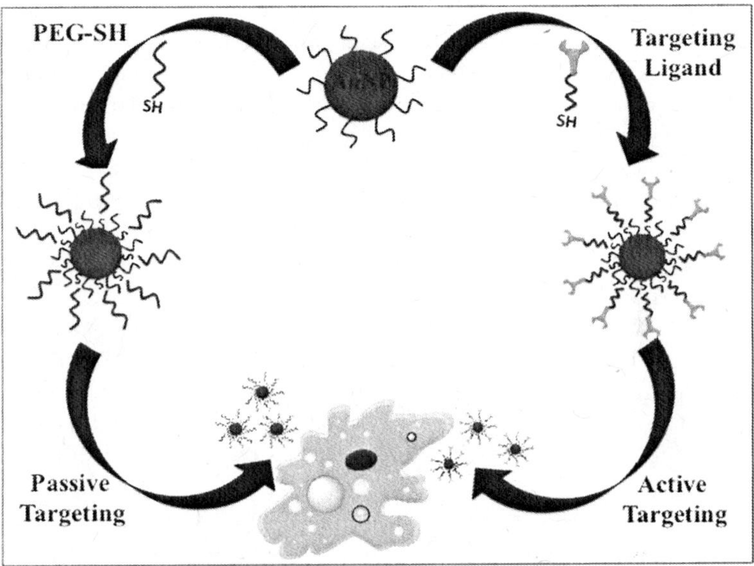

Figure 3. Schematic illustrating the passive (left arrow) and active targeting strategies adopted for delivery of the nanoparticle by EPR effect and ligand-receptor interaction respectively. These strategies enable enhanced delivery and accumulation of the particles at the desired site of their action.

4. Targeting of Plasmonic Nanoparticles

One of the greatest challenges in biomedical application of nanoparticles is a directed delivery of therapeutically relevant material to the desired site of action. Nanoparticles administered into the *in vivo* systems generally tend to distribute widely in different organs and are seldom localized at the target site. Therefore, several strategies have been developed to aid nanoparticle localization either by passive targeting where inert nanoparticles are concentrated at the target site or by active targeting where nanoparticles are modified by specific ligands to enhance uptake at specific sites. Figure 3 illustrates the essence of the targeting strategies to enhance the residence time of the nanoparticles inside the body as well as higher uptake due to ligand based targeting of cells. The details of the two different targeting strategies have been discussed in the next section for better understanding.

4.1. Passive Targeting

Passive targeting is often referred to the preferential concentration of the nanoparticles at the site of action in the absence of any targeting ligand on the surface. Nanoparticles size and surface properties are extremely critical for such non-targeted localization. Passive targeting is achieved by extravasation of nanoparticles across the leaky vasculature of blood capillaries in the vicinity of the unhealthy tissues such as tumors. Metal nanoparticles are designed in the size ranges such that they can effectively evade the retention into liver or spleen but are concentrated into sites of unhealthy tissues due to *enhanced permeability and retention* (EPR) effect, first described by Maeda's group (*62*). PEG modified nanoparticles have been shown to preferentially concentrate in the tumor tissues for a much longer time due to lack of lymphatic system while they are actively cleared from the normal tissues (*61*). O'Neal *et. al,* have demonstrated that gold nanoshells of 130 nm diameter can actively pass through the blood vessels and concentrate around the tumorous tissues (*63*).

4.2. Active Targeting

Passive concentration of nanoparticles in the site of action has drawbacks that can be circumvented by surface functionalization of actively targeting ligands. This approach allows a directed delivery of the nanoparticles at the desired site of action and is referred to as active targeting. Several novel ligands targeting specific surface receptors on the cells have been explored for such applications and the choice of ligand thus largely depends on the nature of the target cells. Cancerous cells over-express different receptors on their surface that have been lucrative choices for active targeting of tumors. Some commonly used protein based receptor targeting ligands are anti-EGFR antibody (*64*), herceptin (*65*) and transferrin (*66*). Folate receptor expressing cells have been targeted using folic acid (FA), their derivatives and methotrexate (MTX). Dixit et al. have demonstrated the successful delivery of FA conjugated to gold nanoparticles through a PEG spacer, to FR-positive KB cells while the FR-negative WI cells show minimal uptake (*67*). Chen et al. have similarly shown that MTX conjugated to gold nanoparticles effectively concentrates in several FR-positive cancer cell lines and exhibit tumor suppression in mouse ascites model of Lewis lung carcinoma (LL2) (*68*).

Besides such cell specific targeting, gold nanoparticles have also been modified by peptides ligands to target specific cell organelle based localization. Nuclear targeting has specifically been studied with great interest due to the possibility of direct delivery of therapeutic payload to the nucleus. Use of peptides has been the most popular approach for such application and tat-peptide has specifically been exploited for such applications. Fuente and Berry have demonstrated a successful localization of tat-peptide modified 2.8 nanometer size gold nanoparticles in to the human fibroblast cells (hTERT-BJ1) (*69*). In yet another approach, 30 nm size gold nanoparticles were co-functionalized with arginine-glycine-aspartic acid (RGD) peptide and a nuclear localizing signal (NLS) peptide to facilitate nuclear targeting in human oral squamous cell carcinoma (HSC) (*70*). Their finding however suggest that nuclear targeting of gold nanoparticles results in to cytokinesis arrest in cancer cells leading to incomplete cell division and eventual apoptosis. Tkachenko et al. designed bio-inspired peptides mimicking adenovirus proteins to target the nucleus of human hepatocarcinoma (HepG2) cells (*71*).

5. Biocompatibility of Plasmonic Nanosystems

5.1. Gold Nanoparticles

The first known attempt to use gold nanoparticles for clinical studies dates back to early 1920s when colloidal gold were used for treatment of tuberculosis and many rheumatic diseases (*72*). Since then, plasmonic nanoparticles have been used for a variety of biomedical application such as transfecting vectors (*73*), cell tracers (*74*), diagnostics (*75*), drug delivery (*76*), biosensing (*77*) and therapeutics (*78*). However, it is of paramount importance to ascertain the biocompatibility of gold nanoparticles before envisioning any potential biological applications. Several *in vitro* studies have been undertaken to understand the size, shape and surface modification dependent biocompatibility and cytotoxicity of gold nanoparticles in different types of cancer cell lines. Even though *in vitro* studies may not accurately predict the *in vivo* toxicity affects nanoparticles, they enable us to understand the mechanism of cytotoxicity and design routes to circumvent them.

The cytotoxicity effect of spherical gold nanoparticles has been extensively studied to observe the effect of size and surface modification on the biocompatibility. Results indicate that extremely small size gold nanoclusters (1.4 nm) show acute toxicity to human cells but larger sized nanoparticles have been found to be non-toxic. Tsoli *et. al* (*79*), studied the effect of 1.4 nm Au (*55*) nanoclusters on different animal cell lines to show a 100% cell death on exposure to 0.4 µM concentration of the nanoclusters. On the contrary, exposure of the same cells to 50 µM concentration of cisplatin, a potent anti-cancerous drug, showed 60% cell viability. This acute cell toxicity has been attributed to similar size of the gold nanoclusters and the major grooves of the DNA resulting into strong intercalation of these nanoclusters in the DNA of the cells. It has also been elucidated that the gold nanoclusters show a size dependent cellular cytotoxicity where 1.4 nm particles kill cells by necrosis; similar sized 1.2 nm clusters show cell death by apoptosis while 15 nm gold particles do not show cytotoxicity even up to 100 times higher concentration (*80*). Gu *et al,* have shown that 3.7 nm gold nanoparticles modified with PEG as stabilizing layer do not show toxicity in the human cervical cancer HeLa cells and are able to penetrate into the nucleus of the cells (*81*).

Tkachenko *et. al,* studied different protein-BSA-gold nanoparticle conjugates in 3 different human cell lines (Hela, NIH-3T3 and HepG2) to show that the cellular uptake, intra-cellular localization and cytotoxicity of the gold nanoparticles are strongly dependent on the surface modification (*82*). Goodman *et al,* performed a similar study using cationic and anionic gold nanoparticles to show that the cationic gold nanoparticles are moderately cytotoxic compared to their anionic counterparts (*83*). In yet another study, Connor et al. used biotin, cysteine, glucose, citrate and cetyltriethylammonium bromide (CTAB) as surface modifiers to different sized gold nanoparticles to show that biotin and citrate modified particles show minimal cytotoxicity; glucose and cysteine modified particles were moderately toxic while CTAB modified particles were found to be acutely toxic (*83*). Shukla *et al,* have demonstrated that besides cytotoxicity effects, exposure to gold nanoparticles can also affect the immunological response of the cells (*84*).

Shape is yet another parameter that has been a focus of cytotoxic studies owing to the fact that anisotropy in structure imparts unique plasmonic properties to gold nanoparticles with several potential applications. Loo *et al,* have shown that anti-HER2 modified gold nanoshells do not show any cytotoxicity effects in the SKBR2 breast cancer cells and are one of the promising candidates for hyperthermia-induced treatment of cancer tumors (*85*). Gold nanorods synthesized by seed-mediated chemistry show acute toxicity due to the presence of CTAB surface coating and it has been shown that further stabilization of their surface by PEG considerably reduces the surfactant induced cytotoxicity (*61*). The same group further shows that gold nanorods extracted from CTAB using phosphatidylcholine-containing chloroform show little cytotoxicity up to 0.72 mM concentration while CTAB removal by centrifugation and washing show considerable toxicity at lower concentrations (*86*). Leonov *et al,* further demonstrated that removal of CTAB from the gold nanorod surface using treatment with poly(styrene sulfonate) significantly reduces their cytotoxicity effects (*87*).

5.2. Silver Nanoparticles

Silver nanoparticles have gained popularity in biomedical applications due to their antimicrobial activity (*88*) and are actively used in impregnated catheters (*89*), contraceptive devices and wound dressings (*90*). Despite these potential applications of silver nanoparticles, very few endeavors have been made to study the cytotoxic effect of these nanoparticles. The available literature searches reveal contrasting observations of the probable toxic and non-toxic behavior of silver nanoparticles *in vitro*. Yen *et. al,* have performed a comparative study of cytotoxicity effect of similar sized Au and Ag nanoparticles in J774.A1 macrophage cell lines to show that Ag nanoparticles are less toxic compared to Au nanoparticles and do not show any significant immunological response (*91*). However, in a contrasting report, Hussain *et al,* show that silver nanoparticles initiate severe cytotoxicity accompanied with impaired mitochondrial function, leakage of lactose dehydrogenase (LDH) and abnormal cell morphology on exposure to BRL 3A rat liver cells (*92*). *In vivo* study also reveal that 7 days after inhalation, the ultrafine silver nanoparticles tend to localize in the liver of the rats apart from the lungs while most other organs and tissues show non-detectable limits (*93*). Detailed studies have revealed that the possible mechanism of cytotoxicity induced by silver nanoparticle could be mediated by dose dependent oxidative stress in the cells (*94, 95*). These studies indicate that the biological application of Ag nanoparticles could be limited due to their cytotoxic effects towards certain cell types in the body.

6. Applications of Plasmonic Nanosystems in Diagnosis and Imaging

6.1. Early Disease Diagnosis

Early disease detection - such as biological changes that are characteristics of cancer onset - is a major paradigm in the clinical establishment of an effective treatment plan. However, an impeding factor remains in detection, visualization and identification of such pathological tissue sites.

Invasive surgical procedures, such as attainment of a biopsy, remain highly implemented and thoroughly ingrained practices in the clinical setting for verification of a cancerous site. Subsequent manipulation of the extracted sample must be performed in the form of histochemical, immunohistochemical, or various other staining methods which aid in sample visualization and morphological contrast. These extensive procedures are required in order to gather accurate functional and biochemical information regarding a tumor site, categorizing and consequently staging it. Biopsy specimen, however, do not exhibit an overwhelming patient compliance; hence other avenues of cancer diagnosis are being explored.

One path offers itself in form of monitoring blood plasma, or *"liquid biopsy"*, samples of which require less invasive procedures. Extraction of relevant information from the blood regarding initial detection of cancer, however, has proven to be a daunting task. Peripheral blood samples from patients may harbor information on circulating tumor cells (CTCs), which are shed or seeded into the blood stream from a growing tumor.

There are a number of barriers to this method of diagnosis, however, even with the annotated existence of CTCs dating back to the late 19th century. One of the obstacles is the minute frequency of occurrence of CTCs, on the order one per $1 \times 10^{5-7}$ peripheral blood mononuclear cells (*96, 97*). Another obstacle lies in the extensive handling procedures from immunohistochemistry to PCR for validation. Such extensive and intensive methods of validation, however, require large sample quantities for an accurate result, which a patient blood sample intrinsically does not offer.

6.2. Imaging Beacons

Imaging has been termed a means of qualitative and quantitative analysis of the function of biological processes. A subset of such a method to assess the fundamental cellular and sub-cellular physiological processes has been emerging in a form of molecular imaging. This is a multidisciplinary field, which encompasses the need of in vivo analysis of a biological specimen along with chemical, physical, anatomic, and functional assessment of the system. With the convergence of biology and clinical imaging, there have been a number of improvements to existing systems as well as emergence of novel techniques for prognosis and treatment of disease. Greatest strides in improvements of imaging modalities are seen in magnetic resonance spectroscopy (MRI), computed tomography (CT), positron emission tomography (PET), optical coherence tomography (OCT), photo-acoustic (PA) therapy, and Raman spectroscopy, among many others.

The techniques enumerated above gain recognition and potential patient compliance due to ease of clinical implementation. For most of these imaging modalities, minimal preparation of the biological specimen is required. MRI and CT scans benefit the clinical field in their ability to generate functional information with relatively high spatial resolution. However, sensitivity is inadequate and image acquisition lasts from minutes to hours, which may cause discomfort to patients. PET imaging suffers from a low spatial resolution, but compensates with higher sensitivity then MRI and CT.

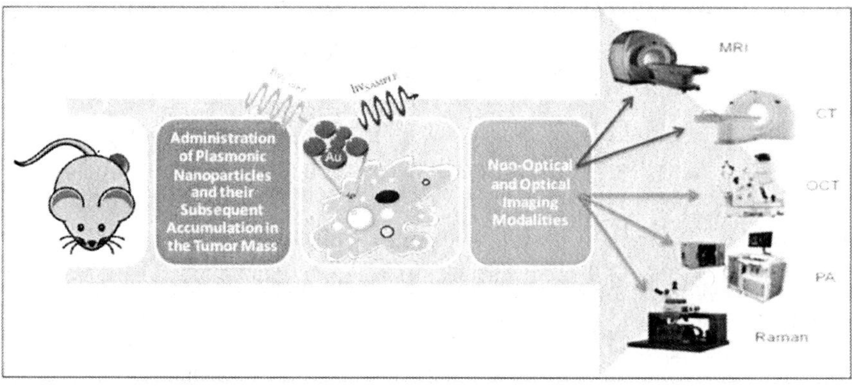

Figure 4. Schematic representation of administration and analysis of plasmonic nanoparticles in vivo, along with various imaging modalities implemented in the clinical setting. MRI and CT are nonoptical imaging modalities, while OCT, PA, and Raman spectroscopy are optical imaging techniques.

With all the technical improvements, however, imaging contrast agents, or beacons, are utilized throughout the modalities in vivo, where the metal nanoparticles play a fundamental role. Without contrast agents the various widely implemented methods of imaging suffer from poor spatial and temporal resolution, low signal-to-noise ratio and low sensitivity. These pose barriers to efficient detection and visualization of a diseased site, such as a growing tumor. Hence, there has been tremendous interest in implementation of nanotechnology in the clinical realm as chemotherapeutic delivery vehicle towards a targeted pathological tissue site as well as potential imaging contrast agent. A schematic of such an approach of tumor visualization and characterization is shown in Figure 4.

Gold and silver nanoparticles have demonstrated unsurpassed capability to be tailored for a multitude of applications. The fundamental and essential requirement for nanotechnology within the biological discipline is the need for an effective contrast agent in molecular imaging, in order to delineate healthy and diseased states accurately. Nanostructures also provide opportunity to encapsulate and distribute various chemotherapeutics to the tumor sites, bypassing accumulation within healthy tissue due to dysfunctional vasculature within the tumor interstitium.

6.2.1. Nonoptical Imaging Modalities

6.2.1.1. *Magnetic Resonance Imaging*

The primary anatomical imaging modalities implemented within the clinical application are MRI, CT and PET imaging. MRI offers a superior contrast to soft tissue, granting information in regard to the tumor homeostasis. Tumor perfusion, cellularity, and even grading may be feasible with MRI. This is usually established via following the water diffusion coefficients, which are low in highly cellular tumors. Hence, patient response to chemotherapy is usually followed via this method, since the gradation of the tumor cellularity is proportional to susceptibility of tumor to a therapeutic agent. Enhanced contrast to tissue may be achieved with preceding implementation of appropriate contrast agents. Such modalities have been aiding in determination of tumor volume and size, blood flow and perfusion, as well as integrity of blood vessels. The contrast agents of choice are iron oxide nanoparticles or paramagnetic gadolinium ions. MRI contrast and sensitivity, however, may be even further enhanced with the utility of gold nanospheres (*98*).

The major disadvantages to MRI as the anatomical imaging system stem from the high concentration of a contrast agent needed to be retained within the tumor site in order to yield relevant information. Thus sensitivity may be compromised if the reporter probes are in lower concentrations than the detection threshold, which is 10^{-3} - 10^{-5} mol/L of labeled contrast agent, namely gadolinium (*99*). The soft tissue contrast can be enhanced, however, with implementation of gold or silver nanoparticles as a means of delivering higher doses of the gadolinium ions to pathological areas (*100, 101*). Gold nanoshells themselves possess inherent magnetic behavior, which is of interest in the pharmaceutical community searching for optimal contrast-enhancing imaging agents. Morrigi *et. al.,* have shown thiolated gold nanoparticles and subsequently coated with Gd^{3+} chelates enhance MRI contrast without undesirable side effects.

The minimal requirements for optimal imaging are vast when compared with CT as morphological imaging modality. The detection for CT imaging is not as well characterized as for MRI; however modifications to the system are rivaling established anatomical imaging methods. CT images are acquired with a higher resolution then the MRI scans as well as much shorter dwell time (minutes compared to hours for some MRI scans). However, the major concern with patient compliance is the relatively high radiation dose (*99*). This may be ameliorated with the administration of gold nanostructures as a contrast agent, which compensates for the low soft tissue contrast of biological tissue of CT imaging.

Volumetric data of biological samples may be acquired with the CT scanner and significantly enhanced in an application designed by Bhatia *et. al,* who have utilized gold nanorods coated with PEG. Since CT imaging requires accumulation of large quantities of iodinated CT-probes for visualization of various biological processes (or tagged with another large atomic-number atom), PEG modified gold nanorods were seen to substitute for the dense X-ray absorbing agents. This alteration for the gold nanostructures would ultimately allow for a lower CT-probe dose administered to patients due to the achieved ~2-fold increase in soft tissue contrast (*102*).

6.2.2. *Optical Imaging Modalities*

Conventional optical imaging, unlike widely implemented anatomical and functional imaging techniques, is not currently used in many clinical applications, since it suffers mainly from the penetration depth due to scattering and absorption contaminations within the biological tissue (*103*). However, there are a number of emerging or already ingrained into the medical field optical imaging modalities, which rival conventional functional imaging of biological tissue, if hybridized with contrast agents, such as gold nanoparticles. Gold nanoparticles, due to their high absorption and scattering capabilities within the body, have been already discussed as ideal contrast agents for their utility in vivo. Gold nanoparticles possess another advantage, since they may be tuned to the NIR region for absorption, ideal for generation of high-resolution functional information via optical imaging techniques.

Optical imaging is a few orders of magnitude more sensitive than the conventional MRI imaging modality as well as much less expensive. Hence the potential benefits of optical imaging methods are very appealing to the broad field of medicinal sciences (*104*). Modalities which have high clinical translation potential (or have already been established in some form within the clinical setting) include OCT, PA, and Raman spectroscopy, among others.

Optical coherence tomography (OCT) is one such technique under the umbrella of optical imaging. It has been integrated into the clinic, but its utility is restricted mostly to ocular imaging. This is due to the intrinsic property of the technique, which measures interference signal of the reference beam and light backscattered from a biological sample. This yields low inherent contrast in soft tissue and a low sensitivity to minute morphological changes associated with disease. An improvement to the modality can be made with introduction of high contrast agents, such as gold nanostructures (*105, 106*).

A number of groups have been utilizing gold nanocages for enhancing tissue contrast of biological samples. A number of groups have unequivocally demonstrated the vast potential of these nanostructures as both diagnostic and therapeutic agents for OCT (*105, 107*). A proof-of-concept experiment was conducted using gold nanocages (35 nm) with a phantom tissue, which revealed large spectral modulations due to inherently strong absorption properties of the gold assemblies. Subsequent experimentation determined 5 orders of magnitude signal enhancement when compared with conventional fluorophores. Alternate experiments, with unique microneedle delivery of nanospheres to an oral tumor model, revealed a >150% contrast enhancement (*106*). This emphatically demonstrates the potential of gold nanostructures to be implemented as beacons or probes into optical imaging for image-guided diagnostics.

6.2.2.2. Photoacoustic Tomography

Another variant to optical imaging stems from the beneficial merging of optical with ultrasonic imaging techniques, yielding a system of photoacoustic imaging, or photoacoustic tomography. The modality allows much deeper penetration depths to be achieved (up to 50 mm) with relatively high contrast functional imaging (*108*). Photoacoustic (PA) imaging reaps the benefits of both the optical modalities, due to high optical contrast, and ultrasonic techniques, with high spatial resolution.

PA has been annotated in the late 19th century, however its implementation into the clinical field has not materialized until the last decade. Its utility emerges from ease of implementation of vast variations of contrast agents, including gold nanoparticles. A number of studies demonstrated nearly an 80% increase in signal contrast when PEG-ylated gold nanoparticles were administered systemically in order to observe cerebral cortex of rats (*109*). The hybrid technique has subsequently been recorded beneficial in analysis of cerebral cortex of small animals, tracing sentinel lymph nodes, as well as distinguishing various melanomas (*108, 110*).

Another optical imaging modality relies on the collection of backscattered radiation from a particular focal volume within the biological tissue sample. The technique - Raman micro-spectroscopy - is based on vibrational spectroscopy combined with optical microscopy. Raman imaging has been in utility as an analytical tool for assessing chemical components within a system. In the last few decades, since the discovery of resonance-based enhancement, such as surface enhanced Raman scattering (SERS) (*111, 112*), the technique has been implemented also in interrogation of biological environments with metallic tumor-targeting contrast agents. SERS is a plasmon resonance effect in which molecular species experience an enhancement in incident electromagnetic field due to them being adsorbed onto the surface of a noble metal. The enhancement in signal has been recorded to be several orders of magnitude higher than a regular Raman effect. Hence SERS has become an attractive technique to verify and monitor the uptake and kinetics of gold and silver nanoparticles, which would act as contrast agents.

Raman spectroscopy is an appealing approach to image-guided therapy since it offers rich chemical information regarding the contrast agent as well as the pathology of the surrounding tissue. There have been studies implementing Raman spectroscopy for identification of skin cancer, namely in basal cell carcinoma (*113*). In vivo studies have demonstrated the sensitivity of Raman imaging, as picomolar concentrations of administered nanospheres were detected (*114*).

As nanotechnology progresses, multimodal nano-contrast agents are being developed for preclinical applications in disease diagnosis and therapy. Noble metal nanoparticles offer an accommodating system of optical labels for molecular imaging of biological processes and architecture. The chemical environment in the immediate vicinity of the gold and silver nanoparticles bears information regarding uptake and sorting mechanisms. SERS nanoparticles have been recently embellished with active tumor-targeting moieties, such as to epidermal growth factor receptor (EGFR)-specific antibodies (*115*). This bimodal nanoparticle has the potential to guide functional imaging to the tumor site, accumulate in the tumor interstitium and delineate its boundaries.

6.3. Combination of Diagnostic, Imaging, and Therapeutic Applications

Early cancer screening suffers from lack of molecular profiling of the tumor. One example is with colon cancer, where only the anatomy of the colon is visualized. While the adenomatous polyps are usually benign and effortlessly excised, the prognosis of Stage IV cancer is poor due to the detection limit of lesions and metastatic foci. Hence, there is a growing need for accurate and effective methods of cancer detection and categorization. Various diagnostic imaging modalities, in form of contrast agents, are being constantly polished and perfected in order to aid the medicinal field in characterization and staging methods of cancer therapy. Detection and imaging of disease ideally should offer non-invasive methods of monitoring disease progression and treatment effects in real time, which would enhance patient consent.

Implementation of the unique properties of the gold nanoparticles in cancer therapy has only recently seen an explosion of interest due to increase in government funding and research interest from a range of sciences, all of which ensure progress in synthesis of various nano-therapeutic agents (*116*). Since mid 1990s, there has been a great increase in the development of novel nanoparticles as contrast and therapeutic agents in vivo, with first clinical trials in thermal therapy (*117*). Over the following years strides in photo-ablation therapy and photo-acoustic therapy have been gaining recognition, paving the way for utility of gold nanoparticles in cancer therapy. The combination of the diagnostic properties of an imaging modality of the nanoparticle with the therapeutic aspect results in an emergence of theranostics (*118*).

Versatility of gold nanoparticles stems from tunable properties of *Localized Surface Plasmon Resonance (LSPR)* peak into an excitation range acceptable for in vivo applications, where the penetration depth in soft tissue is at a maximum for optical imaging. Variation of shape and size of the gold nanostructures has allowed for a fine control of tuning LSPR properties to advance in vivo applications, since there are a number of requirements which the imaging contrast or a therapeutic agent must satisfy. One such requirement is the laser frequency employed for patient therapy, which would be in the NIR range, due to the optical window for biological applications (~800 - 1300 nm), where absorption of water and oxygenated/deoxygenated hemoglobin are minimal. Conventional gold nano-colloids do not satisfy easily this requirement, while nanorods, nanoshells, and nanocages have been able to step up to the task.

Gold nanoshells and nanocages have been analyzed for their potential in image-guided therapy, rendering contrast to intravascular systems of cerebral cortex (*119*) as well as sentinel lymph nodes (SLN) (*120*). Blood-brain-barrier is one of main fortifications of nanoparticle accumulation within the healthy brain tissue. However, tumor vasculature is known for its disfunction and permeability to nanostructures up <200 nm (*121*). This eases extravasation of various nanostructures into the surrounding tumor interstitium due to enhanced permeability and retention (EPR) effect, working as an advantage of aggregating nanoshells within diseased tissue. This yields a more accurate contrast of the brain vasculature, delineating with precision the boundaries of the tumor. A study performed by Wang's group demonstrated an increase of image contrast of >85% with implementation of gold nanoshells in blood, across a range of laser wavelength employed (~750 - 850 nm) (*109*). Thus, gold nanoshells have also been undergoing clinical trials for their utility in head and neck cancers, visualized via MRI and optical imaging (NanoSpectra NCT00848042).

Application of various gold nanostructures has been implemented in other aspects of cancer therapy, such as in identification of sentinel lymph nodes (SLNs). SLNs are needed to be excised because they receive first metastatic cancer cells drained from a tumor. This step also includes staging of the auxiliary lymph nodes. However, the difficulty lies in the identification of the primary SLN. Current methods of their identification use either dyes (isosulfan blue or methylene blue) or radioactive isotopes (technetium-99) (*122, 123*). However, these are highly invasive, hazardous, and expensive procedures. Hence, nanostructures have been recognized in this application as well, in their utility to non-invasively grant contrast to SLNs.

7. Therapeutic Applications of Plasmonic Nanosystems

7.1. Angiogenic Therapy

Angiogenesis is the process of formation of new blood vessel from the existing ones and plays as important role in the growth of tumor. Tumor growth is involved with abnormal angiogenesis to cope with the demand of the necessary nutrient and oxygen supplies for the rapidly dividing cells. Gold nanoparticles bind to the vascular permeability factor/vascular endothelial growth factor (VPF/VEGF)-165 and basic fibroblast growth factor resulting into inhibition of endothelial/fibroblast cell proliferation and angiogenesis (*124*). The authors show a dose-dependent inhibition in the phosphorylation of the growth factors with 40% inhibition at 67 nM and complete inhibition at 335 and 670 nM of gold nanoparticles. The proposed mechanism of this property has been attributed to the binding of the nanogold to the cysteine residues of heparin-binding growth factor by the gold-thiol interaction. This anti-angiogenic property of gold nanoparticles has high therapeutic potential due to the underlying importance of angiogenesis in cancer growth and progression and has been detailed in a recent review publication by the authors (*125*).

7.2. Photothermal Therapy

Gold nanoparticles have most widely been used for treatment of cancer. Conventional cancer therapy focuses on surgical excision of tumor that is largely applicable to large-sized confined tumors, chemotherapy that has broader impact but suffers from severe side effects and radiotherapy, which is detrimental to healthy tissue as well. The exploitation of the plasmonic property of noble metal nanoparticles has alternatively proved to be devoid of these drawbacks and photothermal therapy therefore is explored as stand-alone as well as in combination with existing conventional therapy. Besides being benign to the cells, gold nanoparticles provide better control on the optical property and a higher cross-section for radiation absorption. Irradiation of the gold nanoparticles at their plasmonic frequency results into increased absorption and conversion of light energy into heat that is dissipated into the surrounding (*63, 85*). Shape selective plasmon absorbance of gold nanoparticles further aids into photothermal therapy by tailoring the absorption maxima in the NIR range between 700-1000 nm that can be used for hyperthermic treatment of cancer cells (*126*). Gold nanoshells (AuNSs), nanorods (AuNRs) and nanocages (AuNCs) have therefore been synthesized with desired absorption maxima within the optical window, discussed previously (~800 - 1300 nm), and have been applied for photothermal therapy of cancer. Examples of therapeutic efficacy of plasmonic nanoparticles are provided in the form of angiogenesis and radiofrequency therapies. Figure 5 shows the scheme of events that occur in the photothermal ablation of the tumor by application of plasmonic nanoparticles. The laser of optimum wavelength matching to the absorption maximum of the particles is used to induce hyperthermia and subsequent cell death.

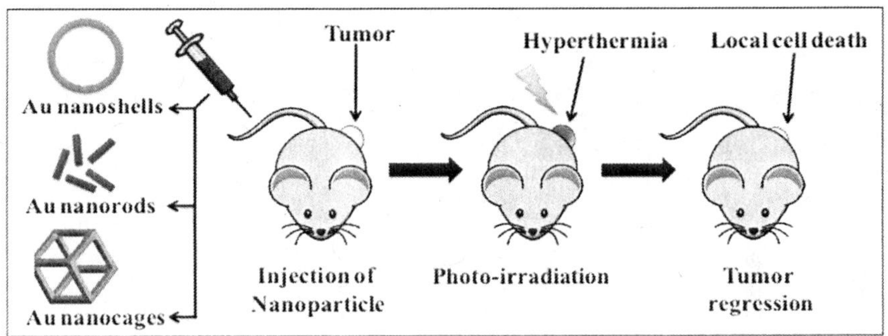

Figure 5. Schematic illustration of the events that occur during photothermal therapy of cancer. The desired nanoparticle suspension is delivered systemically followed by irradiation of the tumor site to enable hyperthermia-mediated cell death and subsequent tumor regression. AuNSs, AuNRs and AuNCs have commonly been employed for such therapy.

7.3. Radiofrequency Ablation Therapy

Although radiofrequency ablation is used for many therapeutic applications, they have been popularly implemented for the treatment of hepatic tumors as an adjunct therapy. A radiofrequency in the range of 10 kHz to 900 mHz has been applied for such applications. Initial studies involved the use of invasive RF needles into the core of the tumor but recent reports have explored non-invasive methods for selective heating of the tumor by the application of gold nanoparticles. Citrate-reduced gold nanoparticles injected into the tumor in rats followed by exposure to variable power (0-2 kW) RF radiation (13.56 mHz) lead to significant temperature increase and thermal injury at the site of injection *in vitro* as well as *in vivo* (Power ~ 35W) (*127*). In yet another report, treatment of Hep3B and Panc 1 cells with 67 µM/L concentration of Au nanoparticles lead to better than 98% cell cytotoxicity within 1 min exposure compared to the untreated cells that show significantly lower cytotoxicity (*128*).

8. Illustrative Examples of Theranostic Applications

Owing to their unique optical characteristics, a variety of metal nanoparticles of different shapes have been extensively used for therapeutic applications. Gold nanoshells (AuNSs), nanorods (AuNRs) and nanocubes (AuNCs) have particularly been explored for their potential therapeutic value in cancer treatment. The common underlying factors that make these materials an excellent choice for hyperthermic tumor ablation is their low cytotoxicity and tunable exotic optical properties in the NIR regime. Some illustrative examples of visualization of three classes of gold nanostructures are shown in Figure 6.

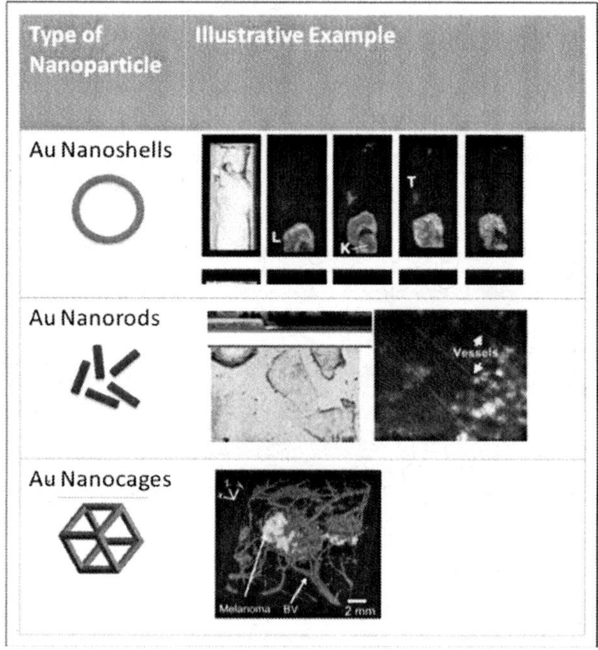

Figure 6. Illustrative examples of the three types of nanoparticles implemented in vitro and in vivo in the clinical setting. Gold nanoshells are currently the subject of research in their use in tumor accumulation subsequent to a tail-vain injection. Gold nanorods have been utilized in vitro cellular imaging with OCT as well as in vivo application to monitor perfusion within the mouse ear blood vessels. Gold nanocages have been implemented in delineating the architecture of melanomas with use of PA imaging.

8.1. Gold Nanoshells (AuNSs)

AuNSs are a novel class of composite material that consists of a core-shell type structure with a dielectric core and a metallic shell with thickness in nanometer range. These nanostructures offer an excellent control over the tunability of the plamonic absorption property by controlling the ratio of the core radius to the total radius of the particles. These plasmonic properties have been studied in details for silica core-Au shell (*22*, *129*) and Au$_2$S core-Au shell (*130*) and the peak positions and shifts with varying physical parameters could be accurately predicted by Mie theory. From the biomedicine perspective, the NIR absorption of the AuNSs has propelled them as a lucrative candidate for photothermal ablation of cancer cells. Exposure of human breast carcinoma cells to AuNSs *in vitro* followed by irradiation with NIR wavelength (820 nm, 35 W/cm^2) led to a significant increase in temperature and subsequent cell death while the untreated control cells do not show any irradiation induced cell death. Similar *in vivo* study in sub-cutaneous tumor bearing mice monitored by magnetic resonance thermal imaging (MRTI) indicated increase in local temperature to as high as 60 °C along with heat associated local cell death in tumors (*126*). The accurate measurement of local temperature is a key parameter and usefulness of MRTI for such application has been well characterized and standardized (*131*). The application of these AuNSs in imaging and diagnostic has been emphasized to obtain a multi-functional system for cancer theranostics. The scattering property of the AuNSs has been leveraged to develop optical coherence tomography (OCT) based imaging system that could be potentially applied towards early detection of cancer. OCT imaging of the tumor after systemic injection of 143 nm sized AuNSs showed a 56% enhancement in the contrast compared to the untreated tumor and their photo-irradiation resulted into complete regression of tumors in all but two mice (*132*, *133*). Photoacoustic tomography (*133*), two-photon induced photoluminescence (*134*) and dark field imaging (*85*) are other diagnostic modalities that have been explored to supplement the therapeutic potential of AuNSs. Yet another approach to realize the theranostic application of AuNSs has been by the synthesis of gold shell on iron oxide coated silica particles (*135*). The optical spectra of these nanoshells show absorption maxima at 700 nm and targeted delivery has been facilitated by functionalization with anti-HER2/neu (Ab$_{HER2/neu}$) antibody. These targeted magnetic AuNSs have been used for MRI based imaging and photothermal ablation based therapeutic treatment of tumors.

8.2. Gold Nanorods (AuNRs)

AuNRs are anisotropic nanostructures that show longitudinal plasmon absorption maxima in the NIR region and tunability of the peak could be achieved by changing the aspect ratio (*19*). The absorption capability of the AuNRs in the NIR wavelengths where the biological tissues are optically transparent makes them suitable candidate for application in hyperthermic treatment of the tumors. First *in vitro* application of AuNRs in imaging and photothermal therapy of cancer cells was illustrated by designing an anti-EGFR targeted system (*136*). One non-malignant epithelial cell line HaCaT (human keratinocytes) and two malignant epithelial cell lines, HOC313 clone 8 and HSC3 (human oral squamous cell carcinoma) were exposed to the anti-EGFR functionalized AuNRs with absorption maxima around 800 nm. The imaging modality could be exploited due to the strong red light scattering efficiency of the NRs under dark field microscope, which enabled visualization of EGFR expressing malignant cells lines. The exposure of the AuNRs treated cells to red laser at 800 nm shows that the malignant cells could be effectively killed at half the laser power (80 mW or 10 W/cm^2) compared to non-malignant cells (160 mW or 20 W/cm^2) due to very high accumulation of AuNRs.

PEG modified AuNRs have shown higher circulation time ($t_{1/2}$ = 17h compared to 4h) as well as better absorption and photothermal ablation ability (~ 6 times better than AuNSs) (*137*). While photothermal ablation remains the preferred therapeutic method for AuNRs, several other diagnostic modalities such as X-ray computed tomography (*137*), two-photon luminescence (*138*), SERS (*139*) and photoacoustic imaging (*140*) have been successfully demonstrated. Additionally, decorating the AuNRs with Fe$_3$O$_4$ nanoparticles to form the "nano-pearl-necklaces" (NPNs) render them suitable to MR imaging of the cancer cells (*141*). These NPNs show excellent stability in different solvents (PBS pH 7.4, chloroform, ethanol and dimethyl sulfoxide) as well as high salt concentrations (1M). *In vitro* MR analysis of Herceptin modified NPNs in HER2 over-expressing SK-BR-3 breast cancer cells showed improved r$_2$ value (248.1 mM^{-1}s^{-1}) compared to free iron oxide, suggesting that NPNs could be applied for *in vivo* MR imaging and tumor ablation by photothermal therapy.

8.3. Gold Nanocages (AuNCs)

AuNCs are another class of metallic nanostructures that show exciting optical properties and have therefore been looked upon as potential material for biomedical applications. They are most routinely synthesized by galvanic replacement reaction on the parent template of Ag nanocubes synthesized by polyol reduction (*25*). The reduction potential of AuCl$_4^-$/Au (0.99V vs SHE) is more positive than AgCl/Ag (0.22V vs SHE) and therefore Ag metal acts as a sacrificial template to reduce chloroaurate ions by the following reaction:

$$3Ag(s) + HAuCl_4 \longrightarrow Au(s) + 3AgCl(s) + HCl(aq)$$

The synthesized hollow AuNCs show a strong longitudinal plasmon absorption peak in the NIR region that strongly depends on the edge length. The bioinertness of gold makes them ideal for biomedical applications and their longer circulation and targeting capability can be easily achieved by surface modification with PEG and cell targeting ligand by using Au-thiol chemistry (*142*). Higher localization of AuNCs has been achieved by their functionalization with anti-HER2 antibody followed by HER2 receptor overexpressed by SK-BR-3 cells (*143*).

The large absorption and scattering cross-section makes AuNCs suitable for scattering based imaging modalities. OCT and SOCT based *in vitro* studies have revealed the applicability of these nanoparticles as targeted contrast enhancement agents (*143, 144*). *In vivo* injection of PEG modified AuNCs has shown 81% enhancement of contrast in cerebral cortex of rat brain by PAT imaging (*145*). Studies have also suggested that AuNCs are a much better candidate for PAT based imaging compared to AuNSs due to their smaller compact morphology (~50 nm compared to ~100 nm) and a larger optical absorption cross-section.

In vitro photothermal therapy has also enjoyed better success by the application of AuNCs. Treatment of SK-BR-3 cells with AuNCs of 45 nm edge length and LSPR wavelength of 810 nm following the irradiation with 810 nm laser at a low power density of 1.5 W/cm^2 for 5 min shows effective cell kill compared to 35 and 10 W/cm^2 reported for AuNSs (*126*) and AuNRs (*136*) respectively. AuNCs show tremendous promise and could therefore be an excellent choice to develop multi-modal theranostic agent for cancer treatment.

9. Conclusions

The unique optical property of metal nanoparticles coupled with their larger absorption and scattering cross-section render them as excellent candidate for imaging, diagnostic and therapeutic applications. Mathematical modeling and calculations have led to a much better understanding and prediction of the plasmonic characteristics of these nanomaterials. Evolution of the suitable surface chemistry of noble metals further enables successful surface functionalization to design passive and active targeting strategies and improve the residence time of the metallic nanostructure *in vivo*. Understanding the chemical and physical properties of these nanosystems and their interaction with the biological system has been of paramount importance to realize their potential as efficient theranostic agents. Along these lines, metal nanostructures have been engineered to impart attributes such as suitable choice of material, shape, size and charge, tunable plasmonic properties, choice of right targeting ligand and biocompatibility.

The recent advances of the diagnostic and therapeutic applications of noble metal nanoparticles is promising *in vitro* as well as *in vivo* but their success in the clinical setup is not yet realized. The major therapeutic route that has been extensively studied for plasmonic nanoparticles has been induction of hyperthermia in cancer tumors by photoirradiation, which is not capable of controlling the disease as a stand-alone therapy. There are several variables in this therapeutic approach such as penetration depth of the laser to reach deeply seated tumor, diffusion of heat to neighboring cells, ultimate route of clearance of the particles from the body etc. that will be of key importance in the clinical level application. Most importantly, safety considerations associated with the long-term theranostic application of these nanoparticles is still unclear. The answers to these predicaments have to be completely addressed to achieve desired success of the metal nanoparticles as theranostics agents.

References

1. Daniel, M. C.; Astruc, D. *Chem. Rev.* **2004**, *104*, 293–346.
2. Jain, P. K.; Huang, X.; El-Sayed, I. H.; El-Sayed, M. A. *Plasmonics* **2007**, *2*, 107–118.
3. Kelly, K. L.; Coronado, E.; Zhao, L. L.; Schatz, G. C. *J. Phys. Chem. B* **2003**, *107*, 668–677.
4. Zhang, J. Z.; Noguez, C. *Plasmonics* **2008**, *3*, 127–150.
5. Link, S.; El-Sayed, M. A. *Annu. Rev. Phys. Chem.* **2003**, *54*, 331–366.

6.   Asano, S.; Yamamoto, G. *Appl. Opt.* **1975**, *14*, 29–49.

7.   Botten, L. C.; Nicorovici, N. A.; Asatryan, A. A.; McPhedran, R. C.; de Sterke, C. M.; Robinson, P. A. *J. Opt. Soc. Am. A* **2000**, *17*, 2165–2176.

8.   Sosa, I. O.; Noguez, C.; Barrera, R. G. *J. Phys. Chem. B* **2003**, *107*, 6269–6275.

9.   Noguez, C. *J. Phys. Chem. C* **2007**, *111*, 3806–3819.

10.   Hao, E.; Li, S. Y.; Bailey, R. C.; Zou, S. L.; Schatz, G. C.; Hupp, J. T. *J. Phys. Chem. B* **2004**, *108*, 1224–1229.

11.   Payne, E. K.; Shuford, K. L.; Park, S.; Schatz, G. C.; Mirkin, C. A. *J. Phys. Chem. B* **2006**, *110*, 2150–2154.

12.   Lewis, L. N. *Chem. Rev.* **1993**, *93*, 2693–2730.

13.   Fichtner, M. *Adv. Eng. Mater.* **2005**, *7*, 443–455.

14.   Lu, Y.; Liu, J. *Acc. Chem. Res.* **2007**, *40*, 315–323.

15.   Moran, C. E.; Steele, J. M.; Halas, N. *Nano Lett.* **2004**, *4*, 1497–1500.

16.   Schmid, G. *Nanoparticles: from theory to application*; Wiley VCH: New York, 2006.

17.   Maier, S. A.; Brongersma, M. L.; Kik, P. G.; Meltzer, S.; Requicha, A. A. G.; Atwater, H. A. *Adv. Mater* **2001**, *13*, 1501–1505.

18.   Xie, J.; Lee, S.; Chen, X. *Adv. Drug Delivery Rev.* **2010**, *62*, 1064–1079.

19.   Jana, N. R.; Gearheart, L.; Murphy, C. J. *J. Phys. Chem. B* **2001**, *105*, 4065–4067.

20.   Jin, R.; Cao, Y. C.; Hao, E.; MÈtraux, G. S.; Schatz, G. C.; Mirkin, C. A. *Nature* **2003**, *425*, 487–490.

21.   Nehl, C. L.; Liao, H.; Hafner, J. H. *Nano Lett.* **2006**, *6*, 683–688.

22.   Oldenburg, S. J.; Averitt, R. D.; Westcott, S. L.; Halas, N. J. *Chem. Phys. Lett.* **1998**, *288*, 243–247.

23.   Liang, H. P.; Wan, L. J.; Bai, C. L.; Jiang, L. *J. Phys. Chem B* **2005**, *109*, 7795–7800.

24.   Sun, Y.; Mayers, B.; Xia, Y. *Adv. Mater.* **2003**, *15*, 641–646.

25.   Sun, Y.; Xia, Y. *Science* **2002**, *298*, 2176–2179.

26.   Chen, S.; Fan, Z.; Carroll, D. L. *J. Phys. Chem B* **2002**, *106*, 10777–10781.

27.   Hao, E.; Bailey, R. C.; Schatz, G. C.; Hupp, J. T.; Li, S. *Nano Lett.* **2004**, *4*, 327–330.

28.   Chaney, S. B.; Shanmukh, S.; Dluhy, R. A.; Zhao, Y. P. *Appl. Phys. Lett.* **2005**, *87*, 031908.

29.   Teng, X.; Black, D.; Watkins, N. J.; Gao, Y.; Yang, H. *Nano Lett.* **2003**, *3*, 261–264.

30.   Kim, J. H.; Germer, T. A.; Mulholland, G. W.; Ehrman, S. H. *Adv. Mater.* **2002**, *14*, 518–521.

31.   Amendola, V.; Meneghetti, M. *Phys. Chem. Chem. Phys.* **2009**, *11*, 3805–3821.

32.   Dhas, N. A.; Suslick, K. S. *J. Am. Chem. Soc.* **2005**, *127*, 2368–2369.

33.   Kurihara, K.; Kizling, J.; Stenius, P.; Fendler, J. H. *J. Am. Chem. Soc.* **1983**, *105*, 2574–2579.

34.   Davis, S. C.; Klabunde, K. J. *Chem. Rev.* **1982**, *82*, 153–208.

35.   Sau, T. K.; Murphy, C. J. *J. Am. Chem. Soc.* **2004**, *126*, 8648–8649.

36. Armelao, L.; Bertoncello, R.; De Dominicis, M. *Adv. Mater.* **1997**, *9*, 736–741.

37. Rosemary, M.; Pradeep, T. *J. Colloid Interface Sci.* **2003**, *268*, 81–84.

38. Bronstein, L. M.; Sidorov, S. N.; Valetsky, P. M.; Hartmann, J.; C^lfen, H.; Antonietti, M. *Langmuir* **1999**, *15*, 6256–6262.

39. Sarikaya, M. *Proc. Natl. Acad. Sci. U.S.A.* **1999**, *96*, 14183.

40. Sleytr, U. B.; Messner, P. *Annu. Rev. Microbiol.* **1983**, *37*, 311–339.

41. Mukherjee, P.; Ahmad, A.; Mandal, D.; Senapati, S.; Sainkar, S. R.; Khan, M. I.; Parishcha, R.; Ajaykumar, P.; Alam, M.; Kumar, R. *Nano Lett.* **2001**, *1*, 515–519.

42. Shankar, S. S.; Rai, A.; Ankamwar, B.; Singh, A.; Ahmad, A.; Sastry, M. *Nat. Mater.* **2004**, *3*, 482–488.

43. Link, S.; El-Sayed, M. A. *J. Phys. Chem B* **1999**, *103*, 4212–4217.

44. Shankar, S. S.; Rai, A.; Ahmad, A.; Sastry, M. *Chem. Mater.* **2005**, *17*, 566–572.

45. Rai, A.; Singh, A.; Ahmad, A.; Sastry, M. *Langmuir* **2006**, *22*, 736–741.

46. Sonnichsen, C.; Reinhard, B. M.; Liphardt, J.; Alivisatos, A. P. *Nat. Biotechnol.* **2005**, *23*, 741–745.

47. Jain, P. K.; Lee, K. S.; El-Sayed, I. H.; El-Sayed, M. A. *J. Phys. Chem B* **2006**, *110*, 7238–7248.

48. Smithpeter, C.; Dunn, A.; Drezek, R.; Collier, T.; Richards-Kortum, R. *J. Biomed. Opt.* **1998**, *3*, 429–436.

49. Kuwata, H.; Tamaru, H.; Esumi, K.; Miyano, K. *Appl. Phys. Lett.* **2003**, *83*, 4625.

50. Xu, T.; Zhang, N.; Nichols, H. L.; Shi, D.; Wen, X. *Mater. Sci. Eng., C* **2007**, *27*, 579–594.

51. Han, Y.; Xu, K.; Montay, G.; Fu, T.; Lu, J. *J. Biomed. Mater. Res.* **2002**, *60*, 511–516.

52. Chin, V.; Collins, B. E.; Sailor, M. J.; Bhatia, S. N. *Adv. Mater.* **2001**, *13*, 1877–1880.

53. Dahl, J. A.; Maddux, B. L. S.; Hutchison, J. E. *Chem. Rev.* **2007**, *107*, 2228–2269.

54. Brust, M.; Fink, J.; Bethell, D.; Schiffrin, D.; Kiely, C. *J. Chem. Soc., Chem. Commun.* **1995**, 1655–1656.

55. Joshi, H.; Shirude, P. S.; Bansal, V.; Ganesh, K.; Sastry, M. *J. Phys. Chem B* **2004**, *108*, 11535–11540.

56. Selvakannan, P.; Mandal, S.; Phadtare, S.; Pasricha, R.; Sastry, M. *Langmuir* **2003**, *19*, 3545–3549.

57. Selvakannan, P.; Mandal, S.; Phadtare, S.; Gole, A.; Pasricha, R.; Adyanthaya, S.; Sastry, M. *J. Colloid Interface Sci.* **2004**, *269*, 97–102.

58. Selvakannan, P.; Swami, A.; Srisathiyanarayanan, D.; Shirude, P. S.; Pasricha, R.; Mandale, A. B.; Sastry, M. *Langmuir* **2004**, *20*, 7825–7836.

59. Qian, X.; Peng, X. H.; Ansari, D. O.; Yin-Goen, Q.; Chen, G. Z.; Shin, D. M.; Yang, L.; Young, A. N.; Wang, M. D.; Nie, S. *Nat. Biotechnol.* **2007**, *26*, 83–90.

60. Shenoy, D.; Fu, W.; Li, J.; Crasto, C.; Jones, G.; DiMarzio, C.; Sridhar, S.; Amiji, M. *Int. J. Nanomed.* **2006**, *1*, 51–57.

61. Niidome, T.; Yamagata, M.; Okamoto, Y.; Akiyama, Y.; Takahashi, H.; Kawano, T.; Katayama, Y.; Niidome, Y. *J. Controlled Release* **2006**, *114*, 343–347.

62. Baban, D. F.; Seymour, L. W. *Adv. Drug Delivery Rev.* **1998**, *34*, 109–119.

63. O'Neal, D. P.; Hirsch, L. R.; Halas, N. J.; Payne, J. D.; West, J. L. *Cancer Lett.* **2004**, *209*, 171–176.

64. El-Sayed, I. H.; Huang, X.; El-Sayed, M. A. *Nano Lett.* **2005**, *5*, 829–834.

65. Jiang, W.; Kim, B. Y.; Rutka, J. T.; Chan, W. C. *Nat. Nanotechnol.* **2008**, *3*, 145–150.

66. Choi, C. H.; Alabi, C. A.; Webster, P.; Davis, M. E. *Proc. Natl. Acad. Sci. U.S.A.* **2010**, *107*, 1235–1240.

67. Dixit, V.; Van den Bossche, J.; Sherman, D. M.; Thompson, D. H.; Andres, R. P. *Bioconjugate Chem.* **2006**, *17*, 603–609.

68. Chen, Y. H.; Tsai, C. Y.; Huang, P. Y.; Chang, M. Y.; Cheng, P. C.; Chou, C. H.; Chen, D. H.; Wang, C. R.; Shiau, A. L.; Wu, C. L. *Mol. Pharm.* **2007**, *4*, 713–722.

69. de la Fuente, J. M.; Berry, C. C. *Bioconjugate Chem.* **2005**, *16*, 1176–1180.

70. Kang, B.; Mackey, M. A.; El-Sayed, M. A. *J. Am. Chem. Soc.* **2010**, *132*, 1517–1519.

71. Tkachenko, A. G.; Xie, H.; Coleman, D.; Glomm, W.; Ryan, J.; Anderson, M. F.; Franzen, S.; Feldheim, D. L. *J. Am. Chem. Soc.* **2003**, *125*, 4700–4701.

72. Bhattacharya, R.; Mukherjee, P. *Adv. Drug Delivery Rev.* **2008**, *60*, 1289–1306.

73. Sandhu, K. K.; McIntosh, C. M.; Simard, J. M.; Smith, S. W.; Rotello, V. M. *Bioconjugate Chem.* **2002**, *13*, 3–6.

74. Yi, H.; Leunissen, J. L. M.; Shi, G. M.; Gutekunst, C. A.; Hersch, S. M. *J. Histochem. Cytochem.* **2001**, *49*, 279–283.

75. Storhoff, J. J.; Mirkin, C. A. *Chem. Rev.* **1999**, *99*, 1849–1862.

76. Han, G.; Ghosh, P.; Rotello, V. M. *Nanomedicine* **2007**, *2*, 113–123.

77. Olofsson, L.; Rindzevicius, T.; Pfeiffer, I.; Käll, M.; Höök, F. *Langmuir* **2003**, *19*, 10414–10419.

78. West, J. L.; Halas, N. J. *Annu. Rev. Biomed. Eng.* **2003**, *5*, 285–292.

79. Tsoli, M.; Kuhn, H.; Brandau, W.; Esche, H.; Schmid, G. *Small* **2005**, *1*, 841–844.

80. Pan, Y.; Neuss, S.; Leifert, A.; Fischler, M.; Wen, F.; Simon, U.; Schmid, G.; Brandau, W.; Jahnen-Dechent, W. *Small* **2007**, *3*, 1941–1949.

81. Gu, Y. J.; Cheng, J.; Lin, C. C.; Lam, Y. W.; Cheng, S. H.; Wong, W. T. *Toxicol. Appl. Pharmacol.* **2009**, *237*, 196–204.

82. Tkachenko, A. G.; Xie, H.; Liu, Y.; Coleman, D.; Ryan, J.; Glomm, W. R.; Shipton, M. K.; Franzen, S.; Feldheim, D. L. *Bioconjugate Chem.* **2004**, *15*, 482–490.

83. Goodman, C. M.; McCusker, C. D.; Yilmaz, T.; Rotello, V. M. *Bioconjugate Chem.* **2004**, *15*, 897–900.

84. Shukla, R.; Bansal, V.; Chaudhary, M.; Basu, A.; Bhonde, R. R.; Sastry, M. *Langmuir* **2005**, *21*, 10644–10654.

85. Loo, C.; Lowery, A.; Halas, N.; West, J.; Drezek, R. *Nano Lett.* **2005**, *5*, 709–711.

86. Takahashi, H.; Niidome, Y.; Niidome, T.; Kaneko, K.; Kawasaki, H.; Yamada, S. *Langmuir* **2006**, *22*, 2–5.

87. Leonov, A. P.; Zheng, J.; Clogston, J. D.; Stern, S. T.; Patri, A. K.; Wei, A. *ACS Nano* **2008**, *2*, 2481–2488.

88. Kim, J. S.; Kuk, E.; Yu, K. N.; Kim, J. H.; Park, S. J.; Lee, H. J.; Kim, S. H.; Park, Y. K.; Park, Y. H.; Hwang, C. Y. *Nanomed.: Nanotechnol., Biol. Med.* **2007**, *3*, 95–101.

89. Samuel, U.; Guggenbichler, J. *Int. J. Antimicrobial Agents* **2004**, *23*, 75–78.

90. Tian, J.; Wong, K. K. Y.; Ho, C. M.; Lok, C. N.; Yu, W. Y.; Che, C. M.; Chiu, J. F.; Tam, P. K. H. *ChemMedChem* **2007**, *2*, 129–136.

91. Yen, H. J.; Hsu, S.; Tsai, C. L. *Small* **2009**, *5*, 1553–1561.

92. Hussain, S.; Hess, K.; Gearhart, J.; Geiss, K.; Schlager, J. *Toxicol. in Vitro* **2005**, *19*, 975–983.

93. Takenaka, S.; Karg, E.; Roth, C.; Schulz, H.; Ziesenis, A.; Heinzmann, U.; Schramel, P.; Heyder, J. *Environ. Health Perspect.* **2001**, *109*, 547–551.

94. Carlson, C.; Hussain, S.; Schrand, A.; Braydich-Stolle, L. K.; Hess, K.; Jones, R.; Schlager, J. *J. Phys. Chem B* **2008**, *112*, 13608–13619.

95. AshaRani, P.; Low Kah Mun, G.; Hande, M. P.; Valiyaveettil, S. *ACS Nano* **2008**, *3*, 279–290.

96. Sleijfer, S.; Gratama, J. W.; Seiuwerts, A. M.; Kraan, J.; Martens, J. W. M.; Foekens, J. A. *Eur. J. Cancer* **2007**, *43*, 2645–2650.

97. Alunni-Fabbronia, M.; Sandri, M. T. *Methods* **2010**, *50*, 289–297.

98. Garitaonandia, J. S.; Insausti, M.; Goikolea, E.; Suzuki, M.; Cashion, J. D.; Kawamura, N.; Ohsawa, H.; De Muro, I. G.; Suzuki, K.; Plazaola, F.; Rojo, T. *Nano Lett.* **2008**, *8*, 661–667.

99. Massoud, T. F.; Gambhir, S. S. *Genes Dev.* **2003**, *17*, 545–580.

100. Alric, C.; Taleb, J.; Le Duc, G.; Mandon, C.; Billotey, C.; Le Meur-Herland, A.; Brochard, T.; Vocanson, F.; Janier, M.; Perriat, P.; Roux, S.; Tillement, O. *J. Am. Chem. Soc.* **2008**, *130*, 5908–5915.

101. Moriggi, L.; Cannizzo, C.; Dumas, E.; Mayer, C. R.; Ulianov, A.; Helm, L. *J. Am. Chem. Soc.* **2009**, *131*, 10828–10829.

102. von Maltzahn, G.; Park, J.-H.; Agrawal, A.; Bandaru, N. K.; Das, S. K.; Sailor, M. J.; Bhatia, S. N. *Cancer Res.* **2009**, *69*, 3892–3900.

103. Zysk, A. M.; Boppart, S. A. *J. Biomed. Opt.* **2006**, *11*, 054015.

104. Nakajima, M.; Takeda, M.; Kobayashi, M.; Suzuki, S.; Ohuchi, N. *Cancer Sci.* **2005**, *96*, 353–356.

105. Hu, M.; Chen, J.; Li, Z.-Y.; Au, L.; Hartland, G. V.; Li, X.; Marquez, M.; Xia, Y. *Chem. Soc. Rev.* **2006**, *35*, 1084–1094.

106. Kim, C. S.; Ahn, Y.-C.; Wilder-Smith, P. ; Oh, S.; Chen, Z.; Kwon, Y. J. *J. Biomed. Opt.* **2009**, *14*, 034008.

107. Gobin, A. M.; Lee, M. H.; Halas, N. J.; James, W. D.; Drezek, R. A.; West, J. L. *Nano Lett.* **2007**, *7*, 1929–1934.

108. Li, W.; Brown, P. K.; Wang, L. V.; Xia, Y. *Contrast Media Mol. Imaging* **2011**, *6*, 370–377.

109. Yang, X.; Skrabalak, S. E.; Li, Z.-Y.; Xia, Y.; Wang, L. V. *Nano Lett.* **2007**, *7*, 3798–3802.

110. Olafsson, R.; Bauer, D. R.; Montilla, L. G.; Witte, R. S. *Opt. Express* **2010**, *18*, 18625–18632.

111. Matschulat, A.; Drescher, D.; Kneipp, J. *ACS Nano* **2010**, *4*, 3259–3269.

112. Kneipp, J.; Kneipp, H.; Wittig, B.; Kneipp, K. *Nanomed.: Nanotechnol., Biol. Med.* **2010**, *6*, 214–226.

113. Nijssen, A.; Bakker Schut, T. C.; Heule, F.; Caspers, P. J; Hayes, D. P; Neumann, M. H. A.; Puppels, G. J. *J. Invest. Dermatol.* **2002**, *119*, 64–69.

114. Keren, S.; Zavaleta, C.; Cheng, Z.; de la Zerda, A.; Gheysens, O.; Gambhir, S. S. *Proc. Natl. Acad. Sci. U.S.A.* **2008**, *105*, 5844–5849.

115. Qian, X.; Peng, X. H.; Ansari, D. O.; Yin-Goen, Q.; Chen, G. Z; Shin, D. M; Yang, L.; Young, A. N.; Wang, M. D; Nie, S. *Nat. Biotechnol.* **2008**, *26*, 83–90.

116. Chen, H.; Roco, M. C.; Li, X.; Lin, Y. Trends in nanotechnology patents. *Nat. Nanotechnol.* **2008**, *3*, 123–125.

117. Vernon, C. C.; Hand, J. W.; Field, S. B.; Machin, D.; Whaley, J. B.; van der Zee, J.; van Putten, W. L.; van Rhoon, G. C.; van Dijk, J. D.; Gonzalez Gonzalez, D.; Liu, F. F.; Goodman, P.; Sherar, M. *Int. J. Radiat. Oncol. Biol. Phys.* **1996**, *35*, 731–744.

118. Sumer, B.; Gao, J. M. *Nanomedicine (London)* **2008**, *3*, 137–140.

119. Yang, X.; Skrabalak, S. E.; Li, Z-Y.; Xia, Y.; Wang, L. V. *Nano Lett.* **2007**, *7*, 3798–3802.

120. Song, K. H.; Kim, C.; Cobley, C. M.; Xia, Y.; Wang, L. V. *Nano Lett.* **2009**, *9*, 183–188.

121. Torchilin, V. *Adv. Drug Delivery Rev.* **2010**, *63*, 131–135.

122. Morton, D. L.; Wen, D.-R.; Wong, J. H.; Economou, J. S.; Cagle, L. A.; Storm, F. K.; Foshag, L. J.; Cochran, A. J. *Arch Surg.* **1992**, *127*, 392–399.

123. Krag, D. N.; Weaver, D. L.; Alex, J. C.; Fairbank, J. T. *Surg. Oncol.* **1993**, *2*, 335–339.

124. Mukherjee, P.; Bhattacharya, R.; Wang, P.; Wang, L.; Basu, S.; Nagy, J. A.; Atala, A.; Mukhopadhyay, D.; Soker, S. *Clin. Cancer Res.* **2005**, *11*, 3530–3534.

125. Bhattacharya, R.; Mukherjee, P. *Adv. Drug Delivery Rev.* **2008**, *60*, 1289–1306.

126. Hirsch, L. R.; Stafford, R. J.; Bankson, J. A.; Sershen, S. R.; Rivera, B.; Price, R. E.; Hazle, J. D.; Halas, N. J.; West, J. L. *Proc. Natl. Acad. Sci. U.S.A.* **2003**, *100*, 13549–13554.

127. Cardinal, J.; Klune, J. R.; Chory, E.; Jeyabalan, G.; Kanzius, J. S.; Nalesnik, M.; Geller, D. A. *Surgery* **2008**, *144*, 125–132.

128. Gannon, C. J.; Patra, C. R.; Bhattacharya, R.; Mukherjee, P.; Curley, S. A. *J Nanobiotechnol.* **2008**, *6*, 2.

129. Oldenburg, S. J.; Jackson, J. B.; Westcott, S. L.; Halas, N. J. *Appl. Phys. Lett.* **1999**, *75*, 2897–2899.

130. Averitt, R. D.; Westcott, S. L.; Halas, N. J. *J. Opt. Soc. Am. B* **1999**, *16*, 1824–1832.

131. Elliott, A. M.; Stafford, R. J.; Schwartz, J.; Wang, J.; Shetty, A. M.; Bourgoyne, C.; O'Neal, P.; Hazle, J. D. *Med Phys* **2007**, *34*, 3102–3108.

132. Gobin, A. M.; Lee, M. H.; Halas, N. J.; James, W. D.; Drezek, R. A.; West, J. L. *Nano Lett.* **2007**, *7*, 1929–1934.

133. Wang, Y.; Xie, X.; Wang, X.; Ku, G.; Gill, K. L.; O'Neal, D. P.; Stoica, G.; Wang, L. V. *Nano Lett.* **2004**, *4*, 1689–1692.

134. Park, J.; Estrada, A.; Sharp, K.; Sang, K.; Schwartz, J. A.; Smith, D. K.; Coleman, C.; Payne, J. D.; Korgel, B. A.; Dunn, A. K.; Tunnell, J. W. *Opt. Express* **2008**, *16*, 1590–1599.

135. Kim, J.; Park, S.; Lee, J. E.; Jin, S. M.; Lee, J. H.; Lee, I. S.; Yang, I.; Kim, J. S.; Kim, S. K.; Cho, M. H.; Hyeon, T. *Angew. Chem., Int. Ed.* **2006**, *45*, 7754–7758.

136. Huang, X.; El-Sayed, I. H.; Qian, W.; El-Sayed, M. A. *J. Am. Chem. Soc.* **2006**, *128*, 2115–2120.

137. von Maltzahn, G.; Park, J. H.; Agrawal, A.; Bandaru, N. K.; Das, S. K.; Sailor, M. J.; Bhatia, S. N. *Cancer Res.* **2009**, *69*, 3892–3900.

138. Durr, N. J.; Larson, T.; Smith, D. K.; Korgel, B. A.; Sokolov, K.; Ben-Yakar, A. *Nano Lett.* **2007**, *7*, 941–945.

139. Huang, X.; El-Sayed, I. H.; Qian, W.; El-Sayed, M. A. *Nano Lett.* **2007**, *7*, 1591–1597.

140. Agarwal, A.; Huang, S.; O'Donnell, M.; Day, K.; Day, M.; Kotov, N.; Ashkenazi, S. *J. Appl. Phys.* **2007**, *102*, 064701–064704.

141. Wang, C.; Chen, J.; Talavage, T.; Irudayaraj, J. *Angew. Chem., Int. Ed.* **2009**, *48*, 2759–2763.

142. Xia, Y.; Rogers, J. A.; Paul, K. E.; Whitesides, G. M. *Chem. Rev.* **1999**, *99*, 1823–1848.

143. Chen, J.; Saeki, F.; Wiley, B. J.; Cang, H.; Cobb, M. J.; Li, Z. Y.; Au, L.; Zhang, H.; Kimmey, M. B.; Li, X. D.; Xia, Y. N. *Nano Lett.* **2005**, *5*, 473–477.

144. Cang, H.; Sun, T.; Li, Z. Y.; Chen, J.; Wiley, B. J.; Xia, Y.; Li, X. *Opt. Lett.* **2005**, *30*, 3048–3050.

145. Yang, X. M.; Skrabalak, S. E.; Li, Z. Y.; Xia, Y. N.; Wang, L. H. V. *Nano Lett.* **2007**, *7*, 3798–3802.

Editors' Biographies

Maria Hepel

Maria Hepel received the M.S. and Ph.D. degrees in chemistry from Jagellonian University in Krakow, Poland. She was a postdoctoral fellow at SUNY Buffalo. From 1985, she worked as the Faculty at the State University of New York at Potsdam where she is now a Distinguished Professor and Chair of the Department of Chemistry. She published over 144 papers, 25 chapters in books, and has made over 400 presentations at the international, national and regional symposia. She organized several symposia and has been the program chair of the 2010 North-East Regional Meeting of ACS. Her current research interests include DNA intercalation sensors, piezoimmunosensors, sensors for biomarkers of oxidative stress, fluorescence energy transfer (FRET and NSET), DNA-hybridization biosensors, photovoltaics, and electrochromic devices. She won the SUNY Potsdam President's Award for Excellence in Research and Creative Endeavor in 1995 and 2001, the SUNY Chancellor's Award for Excellence in Teaching in 1998 and SUNY Chancellor's Award for Research in 2003. She received the Northeast Region Award for Achievements in the Chemical Sciences at the Rochester Meeting NERM 2012 of the American Chemical Society.

Chuan-Jian Zhong

Chuan-Jian Zhong is Professor of Chemistry at State University of New York at Binghamton. He was Max-Planck-Society postdoctoral fellow at Fritz Haber Institute after receiving his Ph.D. at Xiamen University in 1989, and continued his postdoctoral work at University of Minnesota in 1991. He was associate scientist at Iowa State University/DOE-Ames Laboratory before joining the SUNY Binghamton faculty in 1998. His research interests include analytical and materials chemistry, catalysis, electrochemistry, and emerging fields of nanotechnology, focusing on challenging issues in chemical sensing and biomolecular recognition, and in green energy production, conversion and storage. He received NSF Career Award, SUNY Chancellor's Award for Excellence in Scholarship and Creative Activities, SUNY Innovation, Creation and Discovery Award, and 3M Faculty Research Award. He is the author of over 170 peer-reviewed research articles, the inventor of 11 U.S. patents, and has given over 100 invited talks at national/international conferences and university/industry/national lab seminars.

© 2012 American Chemical Society

Indexes

Author Index

Subject Index